# ENVIRONMENTAL RISKS AND HAZARDS

# ENVIRONMENTAL RISKS
# AND HAZARDS

Edited by

## Susan L. Cutter

*Rutgers University*

PRENTICE HALL, Upper Saddle River, NJ 07458

**Library of Congress Cataloging-in-Publication Data**
Environmental risks and hazards / edited by Susan L. Cutter.
        p.        cm.
     Includes bibliographical references.
     ISBN 0-13-753856-1 :
     1. Environmental risk assessment.     I. Cutter, Susan L.
   GE145.E58   1994
363.7--dc20                                          93-6925
                                                        CIP


Editorial/production supervision
   and interior design: *Joanne Jimenez*
Acquisitions editor: *Ray Henderson*
Cover design: *Jayne Conte*
Prepress buyer: *Paula Massenaro*
Manufacturing buyer: *Lori Bulwin*

Printed in the United States of America

10  9  8  7  6  5  4  3  2  1

ISBN   0-13-753856-1


Prentice-Hall International (UK) Limited,London
Prentice-Hall of Australia Pty. Limited, Sydney
Prentice-Hall Canada Inc., Toronto
Prentice-Hall Hispanoamericana, S.A., Mexico
Prentice-Hall of India Private Limited, New Delhi
Prentice-Hall of Japan, Inc., Tokyo
Pearson Education Asia Pte. Ltd., Singapore
Editora Prentice-Hall do Brasil, Ltda., Rio de Janeiro

# CONTENTS

# PREFACE

A year or so ago, a graduate student came into my office wanting to take his Ph.D. qualifying exams. Since he wished to specialize in hazards rather than broader environmental issues, he asked if I would provide an appropriate reading list. While I could immediately rattle off a number of important books, I had to think a little harder about articles. In true academic form, I said that I could provide the list, but it would take a little while. In compiling the list I wanted to make sure that all social science perspectives—traditions established by researchers in hazards (geography), disasters (sociology), and risk (economics, psychology)—were represented, not just geography. After rummaging through file drawers, bookshelves, and course syllabi, this seemingly little request was much harder and took more time than I had anticipated. Oh, how I wished that someone else had already gone through this exercise, so all I had to do was delete or add as I saw fit. This little episode provided the rationale for this collection of readings. A very selfish motive, to be sure, but one that I hope helps both my colleagues and their students.

The choice of articles was solely my responsibility. I did have conversations with colleagues in many far-flung venues—beers in San Diego, coffee in Washington, D.C., lunch in Boulder, and a bus ride in Czechoslovakia. As you can see, hazard researchers travel and eat together a lot. I always asked colleagues the same question: What are the "must read" articles that students of hazards should be familiar with? As you might expect, I got some pretty interesting answers.

This project would not be possible without the support and encouragement of Ray Henderson and his staff at Prentice Hall. Also crucial to the production of the book was the able assistance of Julie Tuason, who first and foremost organized me and then did all of the legwork in gathering the permissions to reprint the articles in this collection. This book is as much a product of her efforts as mine. Finally, I must acknowledge my gratitude to colleagues whose work is represented here. Just as space and cost pose constraints on hazards management, they provide equally important barriers in the publishing world. I apologize to those in advance whose articles I wasn't able to include. This simply means that there is so much good stuff out there, that we'll probably need another volume or two in the next few years!

*Susan L. Cutter*

# ISN'T ANY PLACE SAFE ANYMORE?

The 1990s are certainly competing for the title "Decade of Environmental Disasters." The tally of cataclysms includes the northern Iran earthquake; volcanic eruptions in the Philippines (Pinatubo) and Alaska (Redoubt Volcano); flooding along the Trinity, Red, and Arkansas rivers in Texas, and the perennial flood in Bangladesh. These were soon followed by the Persian Gulf War, the Oakland-Berkeley Hills fire in California, the cholera epidemics in Peru, the continuing droughts in California and in Eastern Africa. We also saw the first glimpses of the widespread environmental degradation in the former Soviet Union and Eastern and Central Europe, not to mention the depletion of the ozone layer and toxic contamination of military sites. As 1992 came to a close, Hurricane Andrew struck south Florida leaving 125,000 people homeless and billions of dollars worth of damage, making it the largest disaster in United States' history. Within weeks, another storm, Hurricane Iniki, devastated the Hawaiian island of Kauai.

Are these extreme events becoming more frequent? More severe? Or is the frequency of events remaining the same, but is society becoming more vulnerable to them? These are just a few of the questions hazard researchers attempt to answer.

## DISASTER TRENDS

### Natural Events

In a review of natural disaster trends from 1945 through 1986, Glickman, Golding, and Silverman (1992) make a number of very interesting observations. Worldwide there was an increase in the number of small (those with 25 or fewer fatalities) natural disaster events during this forty-year period. At the same time, the number of large events (more than 1,000 fatalities) has remained relatively steady. What has prompted the concern of hazard researchers is the uneven distribution of these disasters. Not only is the developing world bearing the brunt of fatalities from natural disasters, it is also having more disaster events. Most of the natural disasters were meteorological in origin—riverine flooding, coastal inundation from hurricanes and typhoons—accounting for more than a third of all disasters during this time period. Earthquakes, although far fewer in number, are more fatal. More than half of all deaths attributed to natural disasters result from earthquakes.

**TABLE 1**  Top Ten Natural Disasters, 1945–1990[a]

| Year | Location | Type | # Deaths |
|------|----------|------|----------|
| 1976 | China | Earthquake | 700,000 |
| 1970 | Bangladesh | Tropical cyclone | 500,000 |
| 1948 | Soviet Union | Earthquake | 110,000 |
| 1970 | Peru | Earthquake | 66,794 |
| 1949 | China | Flood | 57,000 |
| 1990 | Iran | Earthquake | 50,000 |
| 1965 | Bangladesh | Tropical cyclone | 36,000 |
| 1954 | China | Flood | 30,000 |
| 1965 | Bangladesh | Tropical cyclone | 30,000 |
| 1968 | Iran | Earthquake | 30,000 |
| 1971 | India | Tropical cyclone | 30,000 |

[a]Based on number of estimated fatalities
Source: Glickman, Golding, and Silverman 1992; UNEP 1991.

Recent years (1987–1989) show similar patterns (UNEP 1991). More than 90 percent of the natural disasters occurred outside North America, with slightly more than half occurring in developing countries. While the number of events may be holding steady, unlike the earlier time period, clearly the magnitude of their impacts *is* increasing. The "hit rate" (the tendency for natural events to adversely affect society) of disasters is greater now than in the past (UNEP 1991). The hit rate can be explained by the sheer increase in the world's population, thus increasing the likelihood that human beings will be affected by these natural events. It is also a function of population movements into more concentrated and higher density settlements, including cities, as well as movement into marginal lands with known seismic, flood, or coastal storm risks. Because of extraordinary population pressures, many parts of the world are extremely vulnerable to natural disasters such as earthquakes and tropical storms, even though the frequency and magnitude of these events have not changed radically in the last forty years. Population growth and distribution do not diminish the impact of these large-magnitude disasters (Table 1); they simply mean that we might anticipate greater fatalities in the future from both earthquakes and tropical storms.

North America suffers much less severely than the rest of the world from fatalities caused by natural disasters. Storms, including floods, are the most frequent event, but hurricanes are more deadly. Property damages, not fatalities, are more significant in North America, as the costs associated with Hurricane Hugo, the Loma Prieta earthquake, and the Oakland-Berkeley Hills fire attest.

## Industrial and Chemical Accidents

Two recent reviews of the historical trends in industrial accidents and chemical accidents reveal some interesting global patterns. Since 1945, the frequency of industrial accidents has increased steadily, peaking in 1979 (a result of better reporting) and then declining thereafter (Glickman, Golding, and Silverman 1992). Fatalities from industrial accidents are far fewer when compared to natural disasters (excluding the methyl isocyanate gas leak in Bhopal, India). Industrial accidents

**TABLE 2** Top Ten Industrial Disasters, 1945–1990[a]

| Year | Location | Type/Agent | # Deaths[b] |
|------|----------|------------|-------------|
| 1984 | Bhopal, India | Toxic vapor/methyl isocyanate | 2,750–3,849 |
| 1982 | Salang Pass, Afghanistan | Toxic vapor/carbon monoxide | 1,550–2,700 |
| 1956 | Cali, Colombia | Explosion/ammunitions | 1,200 |
| 1958 | Kyshtym, USSR | Radioactive leak | 1,118[c] |
| 1947 | Texas City, TX | Explosion/ammonium nitrate | 576 |
| 1989 | Acha Ufa, USSR | Explosion/natural gas | 500–574 |
| 1984 | Cubatao, Brazil | Explosion/gasoline | 508 |
| 1984 | St. Juan Ixhautepec, Mexico | Explosion/natural gas | 478–503 |
| 1983 | Nile River, Egypt | Explosion/natural gas | 317 |
| 1986 | Chernobyl, USSR | Explosion/radioactivity | 31–300[c] |

[a]Based on fatality estimates
[b]Estimates vary depending on the source(s) used, therefore ranges are provided where there are differences in the total.
[c]Total number of deaths are hard to gauge since the reported fatality figures only reflect immediate deaths, not the longer term deaths associated with radioactive exposures.
Sources: Glickman, Golding, and Silverman 1992; Cutter 1991, 1993.

have a smaller catastrophic potential. The most fatal (Bhopal) resulted in 4,000 deaths compared to the fatalities associated with the worst natural disasters, such as the 1970 tropical cyclone that killed 500,000 people in Bangladesh or the 1976 Chinese earthquake that claimed an estimated 700,000 lives (Table 2). Regionally, the United States and Canada have the greatest frequency of industrial accidents.

Equally interesting patterns emerge with regard to toxic clouds from industrial accidents. Using a longer time frame (1900–1990), Cutter (1991) found a dramatic increase in accidents in the 1970s and 1980s as the use of chemicals intensified in all areas of society. While most of the accidents occurred in the industrialized world, nearly 36 percent of all incidents during the twentieth century occurred in developing countries, especially those in Latin America and Asia. While the frequency of accidents increased, their severity in terms of immediate fatalities has decreased. However, the hazard potential of chemical accidents has actually increased as more incidents involving acutely toxic materials are recorded. Worldwide, about a third of the incidents involve vapor releases of acutely toxic substances—those that cause immediate physical harm and possibly death to exposed populations within minutes to hours (Cutter 1993). In the last twenty years, around 40 percent of all chemical accidents involved acutely toxic releases.

In the United States, the pattern is quite different. During the last 90 years, slightly less than half (48%) of the releases involved acute toxics. During the last twenty years, more than half (53%) of the incidents involved these deadly substances.

While the United States may be less at risk than many parts of the world from natural disasters such as earthquakes, floods, and hurricanes, the country does have more than its share of industrial disasters. As with natural disasters, the human impact (measured by deaths) of these industrial disasters is often much less than in other world regions. Why is the United States seemingly less vulnerable to these extreme events? What about chronic risks and hazards affecting modern living that have delayed or cumulative effects, such as drought or pollution? Are we becoming more or less vulnerable to these as well?

## THE CHANGING NATURE OF RISK AND HAZARDS

Traditionally, hazards research has been based on questions relevant to decision makers. What can be done to reduce the impacts of flooding? What types of policies are needed for longer term recovery from natural disasters? How effective is emergency preparedness, including warning systems in protecting the public from nuclear power plant accidents? Today these questions are still as relevant as they were fifty years ago. What has changed is the way we think about hazards.

During the last few years, two significant changes have occurred in our thinking about hazards (Mitchell 1989). The first is in the very nature of hazards. New risks and hazards are constantly identified and old ones rediscovered. The simple exercise of categorizing hazards as the consequence of extreme natural events (acts of God) or technological failures (acts of people) no longer works well. For example, is global warming a natural hazard, a technological hazard, or some combination of the two? Also, hazards are no longer viewed as singular events, but as complex interactions between natural, social, and technological systems. Even the basic textbooks in hazards now acknowledge this interrelatedness and increased complexity of the phenomena we study (Smith 1992, Burton, Kates, and White 1993).

A second major change is in how we respond to hazardous events. Researchers are no longer simply interested in the event itself, but are dwelling more and more on the context within which the event occurs. Hazards are imbedded in larger political, social, economic, and technological structures, and it is often impossible to separate these influences from the impacts of the event. In assessing societal vulnerability to hazards, it is impossible to divorce the context from the event (Cuny 1983, Cutter 1993). Similarly, it is virtually impossible to separate hazards from broader based environmental or social problems. As a consequence, the management of risks and hazards has also become quite complex, entailing new management systems at the local, regional, and global scale. More often than not, management alternatives no longer rely solely on technical solutions, as risks and risk reduction becomes increasingly politicized, and management options are debated in the court of public opinion.

The increasingly complex nature of hazards means that geography matters now more than ever. Scale is a crucial element in understanding the distribution, impact, and reduction of risks and hazards. While there is still a concern for hazards with localized impacts, more and more attention is directed at those risks and hazards that alter planetary systems and produce global environmental changes (Turner et al 1990a). The effects of global hazards are more dispersed and cumulative and require the globalization of management systems.

The changes in the hazards themselves have forced us to alter and adjust local, national, and international management systems (Mitchell 1990). For example, better scientific understanding of and data gathering techniques on the causes and consequences of hazards are needed. We also need more informed theoretical explanations on the processes by which hazards are produced, and the social options for recovery from them.

How do local risks and hazards become the driving forces for global environmental change (Turner et al 1990b)? How does sustainable development increase or decrease vulnerability to natural hazards? These are two questions that will occupy hazard researchers throughout the 1990s. Given these changes and trends in disasters, it is not coincidental that the United Nations has named the decade of 1990–2000 as the International Decade for Natural Disaster Reduction. This multinational effort acknowledges the importance of natural disasters in the decline of the human condition and is designed to reduce losses from sudden-impact events. Its intention is to reduce our vulnerability to natural disasters.

## LEARNING FROM THE PAST

While we are constantly reminded of the precarious nature of environmental risks and hazards and how they often dominate our everyday lives in the 1990s, formalized social science research in this area did not begin until the mid-twentieth century. Hazards coevolved along parallel paths in the social sciences under such diverse names as natural hazards, technological hazards, disaster research, and risk assessment and management.

Since our collective and institutional view of hazards is rapidly changing, it is important that we reflect on previous work in hazards research. We learn from past experiences and past mistakes. Hazards are complex phenomena, and it is important to understand all social science perspectives in order to bridge artificial intellectual boundaries between hazards, disasters, and risk communities. The purpose of this volume is to acquaint you with some seminal works in the fields of hazards, disasters, and risk management. The book is divided into five main sections. "Beginnings" examines the historical antecedents of social research into risks and hazards. "Theoretical Innovations" reveals the myriad perspectives social scientists use to understand human responses to disasters. "Responding to Threats" describes how people perceive threats and respond to hazards. Risk assessment methodology and its role in managing risks provide the focus for Part IV, "Improving Management." "Emerging and Recurring Issues" illustrates non-traditional hazards that are increasing in importance: chemical contamination, global environmental change, warfare. The Epilogue speaks for itself.

This is my interpretation of "hazards greatest hits." Colleagues will agree with some selections and argue with others. In fact, many of my own graduate students did not agree with my choices for the final selections! This intellectual pluralism provides one of the great strengths in the social science hazards research community. To facilitate your reading, I have included an introductory essay to each major section, complete with a selected list of the classic books in each area. I have also posed some questions for you to consider as you read through the articles.

Everything we do involves some risk. Since we all have different perceptions of risk, our responses to it are equally varied. Some risks and hazards are so highly politicized that it impedes their management. Others are unevenly distributed, and their effects are more severe in some places and on some people than others. Still others are completely ignored. The acceptability of risks and hazards and the social choices made as to their management informed my selection of articles for this volume. So you see, the question of whether any place is truly safe depends on your own point of view and your willingness and ability to live with risk and hazards.

# REFERENCES

Burton, I., R. W. Kates, and G. F. White, 1993 (forthcoming). *The Environment As Hazard*. New York: Guilford.

Cuny, F. C., 1983. *Disasters and Development*. New York: Oxford University Press.

Cutter, S. L., 1991. "Fleeing from harm: International trends in evacuations from chemical accidents," *International Journal of Mass Emergencies and Disasters* 9(2): 267–85.

———, 1993. *Living with Risk: The Geography of Technological Hazards*. London: Edward Arnold.

Glickman, T. S., D. Golding, and E. D. Silverman, 1992. "Acts of God and acts of man: Recent trends in natural disasters and major industrial accidents," Center for Risk Management Discussion Paper CRM 92-02, Washington D.C.: Resources for the Future.

Mitchell, J. K., 1989. "Hazards research," in Gary L. Gaile and Cort J. Willmott (Eds.), *Geography in America*. Columbus: Merrill, pp. 410–24.

———, 1990. "Human dimensions of environmental hazards: Complexity, disparity, and the search for guidance," in Andrew Kirby (Ed.), *Nothing to Fear: Risks and Hazards in American Society*. Tucson: University of Arizona Press, pp. 131–75.

Smith, K., 1992. *Environmental Hazards: Assessing Risk and Reducing Disaster*. London: Routledge.

Turner, B. L., R. E. Kasperson, W. B. Meyer, K. M. Dow, D. Golding, J. X. Kasperson, R. C. Mitchell, and S. J. Ratick, 1990a. "Two types of global environmental change: Definitional and spatial-scale issues in their human dimensions," *Global Environmental Change* 1(1): 14–22.

Turner, B. L., W. C. Clark, R. W. Kates, J. F. Richard, J. T. Mathews, and W. B. Meyer, 1990b. *The Earth As Transformed by Human Action*. Cambridge: Cambridge University Press.

UNEP (United Nations Environment Programme), 1991. *Environmental Data Report*. Oxford: Basil Blackwell.

# ENVIRONMENTAL RISKS AND HAZARDS

# PART I

# BEGINNINGS

$S$tarting in the 1930s, geographers such as Harlan Barrows were extremely interested in the interactions between human systems and the physical environment and tried to focus geographical inquiry along the lines of geography as human ecology. Gilbert White, a student of Barrows, became curious about how people cope with risk and uncertainty in the occurrence of natural events, especially floods. In his review of natural hazards (Reading 1), White illustrates the model of decision making and public policy that initiated what is now called natural hazards research. These elements include (1) the identification and distribution of the hazards; (2) the range of adjustments (under broad headings including those that modify the cause, modify the loss, or distribute the loss) that are available to individuals and society, and (3) understanding the differences in the choice of adjustments that are taken.

The advent of disaster studies in sociology paralleled hazards research in geography. Largely deriving from efforts to support civil defense during the Cold War, disaster studies were practical projects funded by the U.S. government, which had an immediate need to solve the problems of flooding, for example, or prepare for bombing attacks within the United States. Quarantelli (Reading 2) traces the evolution of disaster research within the United States. In his article, he candidly assesses how the applied focus had a major impact on sociological research on disasters. In addition to influencing methodology, the applied focus determined the very basic definition of what constitutes a disaster. The primary model was a bomb attack or other event with similar characteristics: sudden, unwarned events with a potential for extensive impact. Droughts, riverine flooding, or epidemics were not considered appropriate subject matter. Earthquakes, on the other hand, were. Other failings in the early development of the field included a preoccupation with the concentration of events in time and space, with very little concern on post-disaster recovery, a preoccupation with planning for disasters rather than managing them; and finally a strong bias toward American disasters to the almost total exclusion of disasters anywhere else in the world.

It is clear why there was little convergence between the two major hazard disciplines during their formative years. Historically, disaster research has always examined both natural and technological events. Geographers discovered technological hazards much later. Finally, during the 1980s, there was substantially more cross-fertilization between disaster and hazards researchers as hazards geographers expanded their interests to include industrial crises and disaster researchers looked at the rest of the world.

A third parallel development is the field of risk analysis and management, which had very early beginnings, as Covello and Mumpower illustrate (Reading 3). However, risk analysis did not become formalized as a legitimate reseach endeavor until the 1970s. The shifting nature of risk to more technological sources such as industrial failures (nuclear power plant accidents), complex technologies, and carcinogenic substances helped to establish the field. The development of analytical techniques (probabilistic risk assessments and quantitative risk assessments) enabled scientists to better identify and measure risks, which in turn led to more governmental involvement in regulating them. As new risks were "discovered," the regulatory environment shifted; as new analytical techniques were used, policy refinements were made. As the public became more concerned about risks, regulators had to defend programs and were often asked, to determine how safe is safe enough?

Chauncey Starr's 1969 seminal article (Reading 4) was the first attempt to address this question by using quantitative risk analysis measures—a type of risk-benefit analysis. Many important conclusions can be drawn from his work. The most important is that the public consistently accepts voluntary risks (such as smoking cigarettes) much more readily than risks involuntarily imposed on them (such as radon in the home) in spite of known consequences and ultimate outcomes such as death or serious injury. Risk analysis draws from a broad range of scientific disciplines, not just the social sciences. Besides disciplinary barriers, there is no clear reason why there has been little comingling of hazards and risk researchers or disaster researchers. In a speech before the Society for Risk Analysis, upon receiving its award for distinguished contributions, Gilbert White (Reading 5) illustrates how these barriers can be broken. Using his own experience, he illustrates that the parallel streams are not so different after all.

## DISCUSSION POINTS

1. What is the difference between a risk, a hazard, and a disaster?

2. Why did early work in hazards focus mostly on extreme natural events rather than on more chronic natural events like drought and soil erosion?

3. Gilbert F. White, E. L. Quarantelli, and Chauncey Starr are the grandfathers of risk and hazards research. How would each of them answer the question, How safe is safe enough?

4. What are some parameters that distinguish between different types of environmental risks, hazards, and disasters?

5. Is global climate change a technological or a natural risk/hazard? Why? How would you classify pollution? Are these distinctions important?

# ADDITIONAL READING

Burton, I., R. W. Kates, and G. F. White, 1978. *The Environment as Hazard.* New York: Oxford University Press.

Drabek, T. E., 1986. *Human System Responses to Disaster: An Inventory of Sociological Findings.* New York: Springer-Verlag.

Erikson, K. T., 1976. *Everything in Its Path: Destruction of Community in the Buffalo Creek Flood.* New York: Simon and Schuster.

Rowe, W. D., 1977. *An Anatomy of Risk.* New York: Wiley.

Schwing, R. C., and W. A. Albers, Jr. (Eds.), 1980. *Societal Risk Assessment: How Safe is Safe Enough?.* New York: Plenum.

White, G. F. (Ed.), 1974. *Natural Hazards: Local, National, Global.* New York: Oxford University Press.

White, G. F., and J. E. Haas, 1975. *Assessment of Research on Natural Hazards.* Cambridge: MIT Press.

Wright, J. D., and P. H. Rossi, 1981. *Social Sciences and Natural Hazards.* Cambridge: Abt Books.

# NATURAL HAZARDS RESEARCH

*Gilbert F. White*

To a remarkable degree during the 1960's, geographers turned away from certain environment problems at the same time that colleagues in neighboring fields discovered those issues. This cluster of problems relates to the relationship between man and his natural environment, with particular reference to the kinds of transactions into which man enters with biological and physical systems, and to the capacity of the earth to support him in the face of growing population and of expanding technological alteration of landscape. In their self-conscious concern for developing the theoretical lineaments of a discipline, geographers tended to overlook those problems with which they, by tradition, had been concerned and which do not fall readily into allotted provinces of other scientific enterprises.

By neglecting the theory of man–environment relationships and its applications to public policy, the geographer loses an opportunity to apply his knowledge, skills, and insights to fundamental questions of the survival and quality of human life. He also fails to sharpen and advance theoretical thinking by testing it

in a challenging arena of action. Any critical examination of man's activities as a dominant species in an ecosystem draws upon and invites refreshing appraisal by workers in other fields.

This argument is demonstrated by the line of natural hazard research as it has taken shape over the past fifteen years. It is presented here as an instance in which pursuit of a public policy issue led to a simple research paradigm and a model of decision-making dealing with how man copes with risk and uncertainty in the occurrence of natural events. The approach was refined and extended in a variety of situations, served to stimulate new methods of analyzing other geographical problems, and fostered a few changes in methods of environmental management by national and international agencies.

The study and policy activities related in this direction of research represent an attempt to deepen understanding of the decision-making process accounting for particular human activities at particular places and times. The research seeks application of new techniques to one of the old and recurring traditions of geographical enterprise—the ecology of human choice. The results are slim yet promising. The experience may point more to er-

Reprinted from *Directions in Geography*, R. J. Chorley (Ed.), (1973). London: Methuen, pp. 193–216. Used with permission of Methuen & Co. Ltd.

rors to be avoided than to procedures to be emulated. However, the approach deserves appraisal as a possibly fruitful way of orienting new research and teaching of an old problem.

Application of this model and paradigm does not require any drastic changes in institutionalized teaching and research. Nor does it claim to establish a new sector of geographical inquiry. Rather, it offers one device for bridging some of the divergent lines of current investigation.

## THE PROBLEM

How does man adjust to risk and uncertainty in natural systems, and what does understanding of that process imply for public policy? This problem, raised initially with respect to one uncertain and hazardous parameter of a geophysical system—floods—in one country—the United States—provides a central theme for investigation on a global scale the whole range of uncertain and risky events in nature.

## GENESIS OF THE RESEARCH

Definition of the problem had its genesis in observation of the results of a massive national effort in the United States to deal with the rising toll of flood losses. In 1927 the Corps of Engineers was authorized to conduct a series of comprehensive investigations to find means of managing the river basins of the United States for purposes of irrigation, navigation, flood control, and hydroelectric power. The legislative authorization called for the presentation to the Congress (the final decision-making body with respect to new construction projects on inter-state streams and tributaries thereof) of plans specifying the needs of each area, the types of engineering construction work which could be undertaken, and projects proposed for Federal or State investment, giving the estimated cost and ben-

efit. In the years following 1933 the so-called "308 reports" submitted to the Congress contained explicit benefit-cost analysis of possible construction projects.

In theory, to present a benefit-cost appraisal of a proposed project for a river basin required an analysis of the possible actions which man would take in managing the water and associated land resources of the area, and it also called for a systematic canvass of what, from the standpoint of society, would be the flows of social gains and losses to whomsoever they might accrue arising from any one of those interventions in the ecosystem. This was a monumental and presumptuous task.

Even in his most naive periods of technological mastery, man could not expect to understand the full set of consequences of any major interventions such as the channelization of the lower Mississippi River or the construction of a dam on the Upper Ohio or the building of a system of levees along the Sacramento. The investigator could make educated and hopefully intelligent guesses as to certian outcomes, e.g., alteration of stream regimen. He could not hope to identify all possible consequences. Measuring them would be still more difficult. Moreover, to complete a genuinely competent appraisal of possible lines of action would require canvass of the full range of possible activities which might be undertaken. A proposed dam then could be compared with other steps such as a levee, upstream management of vegetation, or downstream management of the flood plain. Yet, the practical engineering and administrative imperative was to go ahead with such investigations, using the best knowledge then available and applying an elementary kind of economic analysis in order to show for those items which could be readily quantified an estimate of prospective benefits and costs.

Thus, a program of planning took shape which was to have major consequences for resources planning and scientific work in other parts of the world as well as in the United States. Benefit-cost analysis of water projects in the United States became the most sophisticated piece of social impact investigation for several decades. There were more careful and detailed methodologies for computing water benefits and costs than for any other type of public investment. The procedures as first developed by the Corps of Engineers were later revised and embodied in rules and regulations issued by the Bureau of the Budget and approved by the Congress in two separate stages, and were the basis for extensive literature of economic analysis. The analysis was of a normative sort: it was designed to suggest ways by which estimates could be made as to the most effective investments to achieve specified public aims. Almost no time was given to finding out what in fact resulted from such investment. It was assumed that what was proposed—as, for example, the reduction of flood losses or the increase in waterway traffic—would in fact be realized if only the proper combination of technical means, discount rates, and time horizons could be found.

The 308 reports found their way into concrete action in a remarkably short period of time because they first appeared in the midst of the great economic depression of the 1930's and provided individual projects which could be used in mounting public works programs intended to relieve unemployment and stir economic recovery in the nation. The Tennessee Valley Authority was established with the intent, soon discarded, of using part of the Corps of Engineers 308 plan for that area. Large projects such as Grand Coulee Dam and the reservoirs in the Upper Ohio were authorized in the interest of revising a depressed economy.

Geographers early took an interest in this new line of planning but their more lively efforts either proved abortive or dwindled over a long period of time. They were active in the National Resources Planning Board—the first Federal agency in the United States to attempt to draw together the plans of independent state and national agencies into single, comprehensive river basin plans—and they joined in analysis of area economic and employment problems.

This interest stirred an investigation of the range of alternatives with respect to flood loss reduction (White 1942). It also stimulated the first comprehensive attempt to anticipate the full social impacts of a large impoundment. The impact study was carried out by the Bureau of Reclamation and associated agenices on the effects of the Grand Coulee Dam on the Columbia Basin (U.S. 1941). The latter work under the leadership of Harlan H. Barrows was not only a pioneer piece of interdisciplinary research, but defined in broad outlines and with notable gaps the problem of ecological impact which, while studied with considerable care for Grand Coulee, was not to be investigated again with similar energy or breadth until the late 1960's.

In 1936, following a series of disastrous floods affecting urban areas in the Mississippi system, the Congress authorized a national flood control policy which declared it to be the intent of the Federal government to contribute to the cost of flood control works wherever the anticipated benefits from such works would exceed the anticipated cost. In 1938 a supplemental act provided that where reservoirs were selected as a means of flood control no local contribution should be required to the cost of projects inasmuch as the allocation of benefits among the several state beneficiaries was so complicated that it seemed best to change it all on the Federal account.

Twenty years passed, more than five billion dollars were expended on new Federal flood control works, and in 1956 a geographic investigation was begun on what had happened in the urban flood plains of the nation as a result of the investments during the two intervening decades. That investigation was to be followed by more thoughtful and searching studies through which ran the common thread of a relatively consistent research paradigm.

# RESEARCH PARADIGM

In carrying out the 1956 appraisal of changes in land use in selected flood plains following the Flood Control Act of 1936, the geographic research group asked the following questions:

1. What is the nature of the physical hazard involved in extreme fluctuations in stream flow?
2. What types of adjustments has man made to those fluctuations?
3. What is the total range of possible adjustments which man theoretically could make to those fluctuations?
4. What accounts for the differences in adoption of adjustments from place to place and time to time?
5. What would be the effect of changing the public policy insofar as it constitutes a social guide to the conditions in which individuals or groups choose among the possible adjustments?

These questions were addressed to seven sites, chosen to give a diversity of conditions of floods, urban land use, and flood loss abatement measure (White, et al. 1958). A review also was made of the record for flood control expenditures and flood damages for the nation as a whole.

Adjustments were classified in three groups as shown in Table 9.1. From that view any human response to an extreme event in a natural system had the effect of (a) modifying the cause, (b) modifying the losses, or (c) distributing the losses.

A number of conclusions emerging from the field studies had an unsettling effect upon those who were responsible for Federal flood control programs, and triggered new investigations to probe unresolved questions. In brief, it was found that while flood-control expenditures had multiplied, the level of flood damages had risen, and that the national purpose of reducing the toll of flood losses by building flood-control projects had not been realized. Parts of valley bottoms were protected from floods, but increasing encroachment on the flood plain increased the damage potential from a smaller flood. One part of a city was protected by a levee, but new urban growth took place outside the levee. Works which controlled flood with a recurrence interval of 500 years were certain to fail with catastrophic consequences when the 1,000 year flood took place. The findings also indicated that because of the Federal government's concentration upon flood-control works and upstream water-management activities to the exclusion of other obvious but relatively unpracticed types of adjustments, the situation was becoming progressively worse and showed no promise of being improved by a continuation of the prevailing policies. It was recognized that a rising flood toll might be beneficial if accompanied by larger benefits from

**TABLE 9.1.** Types of Adjustments to Floods

| Modifying the Cause | Modifying the Loss | Distributing the Loss |
|---|---|---|
| Upstream land treatment | Flood protection works | Bearing the loss |
| | Dams | Public relief |
| | Levees | |
| | Channelization | Insurance |
| | Emergency measures | |
| | Flood warning | |
| Flood | Evacuation | |
| Proofing | Structural changes in buildings | |
| | Land elevation | |

flood plain use. However, the increased losses were contrary to the public expectation.

At that stage the study had (1) demonstrated that geographic research could have a direct bearing upon the formation of public policy in one country; and (2) posed a set of problems requiring further investigation if satisfactory policy readjustment was to be obtained. These problems centered on how to account for the differential behavior of individuals and groups in dealing with flood problems from one place to another. It had been shown that people did not behave as it had been expected that they would when the benefit-cost ratios for several thousand flood control projects had been drawn up. It was not equally clear why people had chosen the particular solutions they did and, therefore, what sorts of changes in public action would lead to genuine improvement in the character of their choices over a period of time. The effort to deal with this problem sastisfactorily demanded further inquiry.

## MODELS OF DECISION-MAKING

In all of the benefit-cost analysis and in the earlier work on changes in flood plain use, it was postulated that the choice made by people living on flood plains was essentially economic optimization. This was in the tradition of economic analysis, and conformed to the normative judgments on which the projects had been initiated. In essence, it assumed that individuals living in places of hazard would have relatively complete knowledge of the hazard and its occurrence, would be aware in some degree of the consequences, and would seek to make those adjustments which would represent an optimal resolution of the costs and benefits from each of the adjustments open to them. The ideal of the completely optimizing man was viewed as one rarely achieved in action, but as the framework within which a modified model, namely a model of subjec-

tive expected utility, might be explanatory. The subjective utility model held that man would seek to optimize but his judgement would be based on incomplete knowledge and upon his subjective view of the possible consequences. It would be expected that if the view people had of the expected effects of using a particular piece of flood plain could be ascertained, it would be possible to judge their probable response by selecting those solutions which would give them the maximum net utility.

Neither the optimizing model nor the subjective utility model seemed to explain much of the behavior observed in the study areas. For example, it was found that although people seemed to recognize distinct differences in hazard from one part of the flood plain to another, they did not readily translate that recognition into differentials in assigned valuation of the property. People oftentimes returned to the use of land which had been severely damaged by floods being aware of the consequences of a recurrence and facing probable disaster of either a personal or financial character from such recurrence. Adequately to describe behavior for purposes of predicting responses to changes in public policy required the use of some other kind of model. Experimentation began with other possibilities. The obvious direction in which to move was the model of bounded rationality as described by Simon and others (Simon 1959). It was proposed in a general sense for a variety of resource management decisions, and was developed in a more rigorous fashion by Kates in his study of Lafollette, Tennessee (Kates 1962). In examining the behavior and expressed perceptions of residents of a flood plain in the Tennessee Valley, Kates attempted to find out how people perceive the hazard, how they perceive the range of adjustments open to them, and what factors accounted for differences in their perceptions. This required measurement of clearly economic gains and losses as perceived by them, but also consideration of a number of other factors such as the information available to the individual, his personal experience, and the physical nature of the event.

In the following years additional efforts were made to refine a model of bounded rationality, the most recent being that developed by Kates in connection with the collaborative research on natural hazards (1970). It will be noted from a simplified version as presented in Fig. 9.1 that a resource-management decision may be hypothesized to involve the interaction of human systems and physical systems in terms of adjustment to a particular hazard. The interaction is represented as a choice-searching process as affected by personality, information, decision situation and managerial role. The result is a much more complicated model of how people make their choices in dealing with uncertainty and risk in the environment, one that did not lend itself as readily to careful field investigation, but that promised more revealing explanations of individual and group behavior.

## THE ROLE OF PERCEPTION

As it was recognized that the judgment of the resource manager could be more important than the judgment of the scientific observer,

other types of investigations were stimulated. It was clear in the most elementary way that the definition of a flood hydrograph as developed by the meteorologists and hydrologists would be important in the decisions by individuals only to the extent that the described flood parameters were meaningful to the user. This resulted in suggestions of different modes of describing floods and different modes of presenting the results of scientific analysis to flood plain occupants. For example, the question was raised as to what sort of graphic display of a flood hazard would be meaningful to a person considering building a house on an area in which he had a choice between land above and below the maximum flood. To investigate this subproblem required knowledge of the occupant's perception—that process by which individuals organize exterior stimuli in order to form some concept of an event or situation— and as to the relationship between such perceptions, verbalized expressions of attitudes, and behavior. Here was encountered one of the truly difficult problems of social science: the relationship, if any, between verbalized attitudes and actual behavior.

In dealing with perception, it was recognized that psychological studies had been

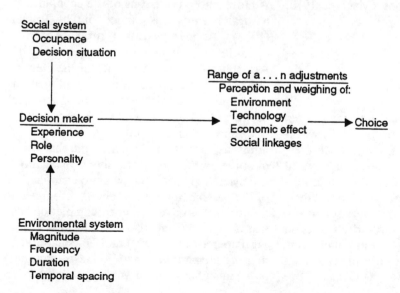

**FIGURE 9.1.** A Rough Model of Decision

largely within the laboratory and had dealt with very limited physical phenomena. There was doubt as to their applicability to observations of more complex and gross phenomena. Hewitt investigated the theoretical ground for expressing extreme events in probabilistic terms (1969). At this point the interest of geographers in problems of perception and attitude formation converged with those of psychologists, sociologists, city planners and architects who also were trying to specify perception and its implications (Burton and Kates 1964). Out of the concern for perception of floods came the first AAG symposium on problems of perception (Lowenthal 1967), a series of investigations dealing with perception of differing facets of the environment such as drought, recreational water, reservoirs, water supply alternatives, water recycling, and the like (Saarinen 1966; Baumann 1969), and Saarinen's geographic review of the perception literature (Saarinen 1969).

Out of the concern for attitudes developed a joint seminar between sociologists and geographers on problems of attitude formation, a joint investigation of attitudes toward water (White 1966), and a number of investigations bearing upon decision-making and public participation in such decisions (MacIver 1970; Johnson 1971).

## INTERDISCIPLINARY COOPERATION

Much of the research could not have taken place without strong cooperation with workers in other disciplines. Engineers were essential to appraisal of the effects of physical structures, and they often were the key professional group in applying geographic findings. Part of the field investigations were supported by the Tennessee Valley Authority's division of Local Flood Relations, and its engineering personnel were a chronic source of critical encouragement. Members of the Corps of Engineers engineering staff participated in a few studies (Cook and White 1962), and the agency later invited and used the results of appraisals of operating experience in flood plain management. Hydrologists from the U.S. Geological Survey shared in the design and assessment of flood plain mapping and its presentation to public agencies.

Wherever an urban area was studied there usually was collaboration with responsible city planning officials. Their critiques were illuminating, and in some cases produced interesting new ventures such as the combination of geographic planning and engineering skills in devising an urban redevelopment scheme for Waterloo, Iowa. However, it was necessary for geographic investigators to resist the temptation to become heavily involved in consulting activities. The pressures were to give time to applying the meager research findings rather than to expanding them.

Economists were drawn into the investigation, and contributed fresh insights into the process of optimal decisions. Unconventional views of flood losses as nature's rental for flood plains were developed by two of them (Renshaw 1961; Krutilla 1966). A refined method of assessing losses and benefits from land-use regulation was devised by Lind (1966). A more rigorous analysis of the economics of natural hazards was carried out by Russell (1970). An investigation of rural use of flood plains in the early 1960's was supported by the Agricultural Research Service of the Department of Agriculture, and brought geographers into working relations with agricultural economists (Burton 1962). However, it did not yield the anticipated refinements in economic aspects of flood hazard, and that sector of study awaits more intensive investigation.

Psychologists were drawn into examination of the personality traits affecting resource decisions. A simple sentence-completion test was devised by Sims (Natural Hazards Working Paper No. 16, 1970) and he collaborated in using thematic apperception tests (Saarinen and Sims 1969). A psychologist joined the research staff at the Department of Geography of the University of Toronto (Schiff 1971), and Kates collaborated with a psychology colleague in editing a review of hazard experience (Kates and Wohlwill 1966).

Interdisciplinary research in which workers in several fields genuinely interact is far more difficult to carry out than is research which draws from other fields at the pleasure of the investigator. In the latter case it is a readily manageable, sometimes gayly ostentatious, sometimes humbling exercise but always in the command of the investigator. When it is required by common commitment to solution of a problem, collaboration is not easily abandoned without personal hurt as well as cost to the whole enterprise.

## APPLICATIONS TO PUBLIC POLICY

It may help to briefly note a number of applications of this research paradigm and these models of decision-making to specific public policy issues. The interest in each case is twofold: (1) what was its use in forming public policy and (2) what feedbacks, if any, did it have upon geographic theory?

At the outset of the flood plain occupance studies in 1957 an attempt was made to enlist the collaboration of people who were directly concerned in drawing up and carrying out such plans, and it was recognized that a principal alternative to the construction of flood control works was the regulation of land use. A representative of the Corps of Engineers took a year for a study leave to work with the Chicago geographic group and to produce a critical appraisal of experience with flood plain regulations (Murphy 1958). This led to tentative hypotheses as to community response to regulations and it also stimulated a legal investigation of the constitutional and statutory grounds for such regulation (Dunham 1959). The latter became the standard legal work on the subject.

Discussions of how people perceive the range of adjustments contributed to the establishment by the Corps of Engineers of a system of "flood plain information reports" which since have become operating practice. Early appraisals of providing flood plain information to residents of such areas showed the importance of individual perception in contrast to that of the scientific observer (Roder 1961), and fostered detailed experiments with modes of mapping sponsored by the U.S. Geological Survey (Shaeffer 1964). The Chicago metropolitan area became the first metropolitan area in the world to be completely mapped in terms of flood hazard. In the course of promoting and carrying out flood plain mapping through the Northeastern Illinois Metropolitan Planning Commission, further inquiries were made into the decision process. It became apparent that merely publishing the maps would be unlikely to have any significant effect upon decisions made by individuals or public agencies whereas if specific and favorable situations could be found in which the maps would be made available, the decision making might be changed. Thus, Sheaffer arranged for the organized group of land appraisers to make systematic use of the maps so that they in turn could attach a judgment about flood hazard to each land value assessment submitted to financial and mortgage agencies in connection with the purchase of buildings or property.

In addition, Sheaffer carried the first academic study of the possibilities of flood proofing, working jointly with personnel of the Tennessee Valley Authority (Sheaffer 1960). In time, that experience was the groundwork for preparing for the Corps of Engineers the preliminary manual of procedures for flood proofing for use of engineers, architects, and other technicians concerned with those alternative adjustments to floods (Sheaffer 1967). Geographers joined in the studies of receptivity that lay the foundation for the first fully operative national program of flood insurance in the United States (Czamanske 1967). They

**TABLE 9.2.** Action Recommended by The Task Force on Federal Flood Control Policy

*To improve basic knowledge about flood hazard*

The immediate listing of all urban areas with flood problems to alert the responsible agencies.
Preparation on maps and aerial photographs by the U.S. Geological Survey of reconnaissance delimitation of hazard areas.
More floodplain information reports from the Corps of Engineers and Tenessee Valley Authority.
Agreement by federal agencies on a set of techniques to be used in determining flood frequencies.
A national program by the Corps of Engineers and Department of Agriculture for collecting more useful data on flood damages, using decennial appraisals, continuing records on sample reaches and special surveys after unusual floods.
Research by Department of Housing and Urban Development and USDA to gain greater knowledge on problems of floodplain occupance and on urban hydrology under the U.S. Geological Survey and HUD.

*To coordinate and plan new developments on the floodplain*

Specification by the Water Resources Council of criteria for regulation of floodplains and for treatment of floodplain problems.
Steps to assure that state and local planning would take proper and consistent account of flood hazard in:
    Federal mortgage insurance (Federal Housing Authority and Veterans Administration)
    Comprehensive local planning (HUD)
    Urban transport planning (Bureau of Public Roads)
    Recreational open space and development planning (Bureau of Outdoor Recreation)
    Urban open space acquisition (HUD)
    Urban renewal (Urban Renewal Administration and Corps of Engineers)
    Sewer and water facilities (HUD, USDA, Department of Health, Education and Welfare, and Economic Development Administration)
Consideration by Office of Emergency Planning, Small Business Administration, and Treasury Department of relocation and flood-proofing in rebuilding flooded areas.
A directive to all federal agencies to consider flood hazard in locating new facilities.

*To provide technical services to managers of floodplain property*

Collection and dissemination of information by Corps of Engineers in collaboration with USDA and HUD on alternative methods of reducing flood losses.
An improved system of flood forecasting under Environmental Sciences Service Administration.

*To move toward a practical national program for flood insurance*

A brief study by HUD on the feasibility of insurance.

*To adjust federal flood control policy to sound criteria and changing needs*

Broadened survey authorizations for Corps of Engineers and USDA.
Provision by the Congress for more suitable cost sharing by state and local groups.
Reporting of flood control benefits to distinguish protection of existing improvements from development of new property.
Authorization by the Congress to include land acquisition as part of flood control plans.
Authorization by the Congress of broadened authority to make loans to local interests for their contributions.

---

also helped sharpen the method of estimating flood losses (Kates 1965).

The eventual upshot of these investigations and their application in sample areas was the formulation of a new Federal flood policy under a task force in which geographers had a hand, established by the Executive Office of the President (U.S., 1966). The new policy involved basic changes in approaches and collaboration among nineteen different agencies, outlined a comprehensive effort by all interested agencies to deal with flood loss management, and inspired new lines of research and of data collection on their part (see Table

9.2). An Executive Order (Number 11296) at the same time required all government agencies responsible, directly or indirectly, for locating new buildings on flood plains to take account of flood hazard in the location decision.

It would be a mistake to suggest that the resulting policy has been fully or effectively translated into action in all responsible agencies. Any basic change in bureaucratic outlook is slow at best. Yet, part of the geographic view of flood plain adjustments had been adopted within four years. A geographer had been appointed to head the flood plain study section of the Corps of Engineers. While reasonable progress was made by most units of government, several agencies dragged their heels against revisions in their procedures, as when the Department of Agriculture committed itself to building flow regulation and land treatment structures to the virtual exclusion of other types of adjustments (White 1970a).

In a number of states such as Iowa, Nebraska, and Ohio, geographers played a part in instigating and carrying out state efforts to apply the same comprehensive approach to flood problems in their respective areas. Under the leadership of geographers the Center for Urban Studies at the University of Chicago initiated several appraisals of experience with flood plain management which assisted in revision of Federal operating policies.

Certain of the activities recommended by the Task Force found interested response in other countries. Thus, the preparation of maps of flood prone areas was undertaken in France under the sponsorship of the Ministere de l'Equipement et du Logement (France, n.d.; 1968). Studies of flood problems were sponsored by government agencies in Canada (Sewell 1965).

At the international level, the Department of Economic and Social Affairs of the United Nations joined with the Ministry of Reclamation and Water Management and other agencies in the USSR in sponsoring in 1969 a Seminar on Methods of Flood Loss Management. The Seminar brought together specialists, primarily engineers, from 28 developing countries and a number of consultants, including one geographer each from Canada, Japan, the United Kingdom, and the United States (White 1970b). They gave careful thought to the approaches initiated in the United States, and in some countries the effects are now observable in national study activity. The United Nations report on the Seminar gives the Seminar findings, points out the implications of geographic research, and suggests new flood loss reduction policies for informal guidance of officials coping with flood losses in developing countries.

The approach which seemed to be yielding results in the realm of flood losses was given application in several other sectors. The Office of Science and Technology established a special Task Force on Earthquake Problems which was patterned after the experience of the Task Force on Federal Flood Control Policy and which benefitted from the geographic contribution to the National Academy of Sciences report on the Alaska earthquake. Russell, Kates, and Arey pursued the problem of optimization in dealing with drought hazard as related to municipal water supply following the New England drought of 1965 (1970).

## COLLABORATIVE STUDIES ON NATURAL HAZARD

The various threads of inquiry were drawn together again in 1968 by a collaborative investigation of natural hazards supported by the National Science Foundation. Burton of Toronto and Kates of Clark joined with the author in examining the experience with a large array of hazards—drought, earthquake, flood, frost, landslide, hurricane, snow, tornado, and volcano—in a variety of settings (Burton et al. 1968).

Scientific as well as public policy response to these activities was sufficiently promising so that the International Geographical Union Commission on Man and Environment decided in 1969 to adopt as one of its two principal thrusts in the succeeding three-year period a program for international collaboration in the study of problems of environmental hazards. These joint investigations now comprise a number of comparative field observations and national studies as outlined in Table 9.3.

The selection of study areas and collaborators was in many instances fortuitous: areas were chosen in terms of the inherent interest of the occupance and environmental problems but with a practical eye to the availability of

**TABLE 9.3.** Field Studies as Part of Collaborative Research on Natural Hazards, 1971

*Comparative Observations of Small Areas*

| | |
|---|---|
| Coastal Erosion | Scotland |
| | United States |
| | Wales |
| Drought | Australia |
| | Brazil |
| | Mexico |
| | Tanzania |
| | United States |
| Earthquake | Peru |
| | Sicily |
| | United States |
| Flood | Canada |
| | Ceylon |
| | France |
| | India |
| | Japan |
| | United Kingdom |
| | United States |
| Frost | United States |
| Hurricane | Pakistan |
| | United States |
| Landslide | Japan |
| | United States |
| Snow | United States |

*National Studies of One Hazard*

| | |
|---|---|
| Drought | Australia |
| | Tanzania |
| Hurricane | East Pakistan |
| | United States |
| Flood | Ceylon |
| Air Pollution | United Kingdom |

competent personnel to carry out the work. It is hoped that from them will come a more rigorous and searching testing of a number of the hypotheses that slowly have emerged over the years since the first office analysis was made of the range of adjustments to floods. Some of the early findings no doubt will be reversed or discarded. The principal hypotheses are now under examination by the collaborators at Clark, Colorado and Toronto.

Perhaps the most fundamental of those hypotheses is that rational explanations can be found for the persistence of human occupance in areas of high hazard by examining the perception of the occupants of such areas and searching out their views of the alternative adjustments and the likely consequences of adopting any one of those opportunities.

In general, we suspect that there are three major types of response to natural hazards. Tentatively, we characterize these as follows: (1) Folk, or pre-industrial response, involving a wide range of adjustments requiring more modifications in behavior and harmony with nature than control of nature and being essentially flexible, easily abandoned, and low in capital requirements. (2) Modern technological, or industrial response involving a much more limited range of technological actions which tend to be inflexible, difficult to change, high in capital requirement, and to require interdependent social organization. (3) Comprehensive, or post-industrial, response combining features of both of the other types, and involving a larger range of adjustments, greater flexibility, and greater variety of capital and organizational requirements. We hypothesize that the United States currently is passing the peak of the modern technological type and is beginning to catch glimpses of the comprehensive type as it emerges here and elsewhere, but we do not suggest that there is a necessary sequence in the types of response.

It is also hypothesized that variations from place to place in hazard perception and estimation can be accounted for in considerable measure by a combination of factors embracing (1) certain physical characteristics of the hazard, (2) the recency and severity of personal experience with the hazard, (3) the situational characteristics of decisions regarding adjustments to the hazard, and (4) personality traits.

We have been inclined to try to describe choice of adjustment in terms of a perception model dealing with the individual manager's subjective recognition of the hazard, of the range of choice open to him, the availability of technology, the relative economic efficiency of the alternatives, and the likely linkages of his action with other people.

We further hypothesize that there are significant differences in the way in which these factors interact in relation to community action in contrast to individual action.

## THE ALTERNATIVES APPROACH

Another spinoff from the early flood plain investigations was application of the idea of range of alternative adjustment to other aspects of natural resources management. In elementary terms the alternatives approach in flood losses could be adopted to any other purposeful intervention in the environment. For example, in combatting stream pollution, the building of waste treatment works or of storage for diluting stream flow are only two of a much larger range of adjustments (Davis 1968). Alternatives would include such measures as controlling waste at the source, use of waste in agriculture, oxygenating streams, constructing special channels for waste transport, and the like.

This view was expressed in two reports from the National Academy of Sciences Committee on Water, with geographic participation (NAS 1966, 1968); that had a significant effect on national water policy in the late sixties. Attention to the full range of practicable adjustments converged with concern for systems analysis to produce water planning methods found in the North Atlantic Regional Water Study. In the North Atlantic study all interested Federal agencies and twelve states join in preparing river basin reports showing for each of three alternative aims (national economic efficiency, regional development, and environmental preservation) the range of possible activities, including non-structural devices, which might be undertaken to meet perceived needs.

The same approach was embodied in part of the High School Geography Project. Its unit on environmental study introduces the student to analysis of alternative ways of dealing with flood losses in an industrial area. The treatment there coincides with increasing emphasis in the social scene on consideration of the range of possible social action in contrast to dependence upon simple technological solutions.

## APPRAISAL

It is too early to venture an appraisal of how influential this direction of natural hazards research has been upon either public policy or geographic thinking; nor are we well equipped to try. In the short run it clearly has been linked to changes in methods of managing water and associated land resources in one nation, and to a smaller degree in several others. What effect those changes will have in the long run is impossible to predict. As of 1971 they pointed to more searching examination of the range of choice available to man in coming to terms with his environment. Although the research has made only modest contributions to a theory of man-environment relations, it supported new efforts to specify the nature of environmental perception, to recognize the process of decision making for resource management, and to identify the landscape consequences of alternative public policies.

In essence, the activity was problem oriented and interdisciplinary. Such work is often tiresome and sometimes exhilarating. It requires research findings in a form highly intelligible to workers in other fields. It ignores conventional divisions of an academic field.

One lesson emerges from this history of investigation of a single environmental problem, using a rather unsophisticated research paradigm. It is that if environmental problems are pursued rigorously enough and with sufficient attention to likely contributions from other disciplines they may foster constructive alternations in public policy but at the same time may stimulate new research and refinement of research methodology to the benefit of geographic discipline. Both may serve to advance man's painful, faltering, and crucial struggle to find his harmonious place in the global systems of which he is a part.

*Acknowledgments* The author is indebted to Ian Burton and Robert W. Kates for comments on an earlier draft.

## REFERENCES

Baumann, D. D. (1969) The recreational use of domestic water supply reservoirs: Perception and choice; *University of Chicago, Department of Geography Research Paper No. 121.*

Burton, I. (1962) Types of agricultural occupance of flood plains in the United States; *University of Chicago, Department of Geography Research Paper No. 75.*

Burton, I. (1965) Flood damage reduction in Canada; *Geophysical Bulletin 7*, 161–85.

Burton, I. and Kates, R. W. (1964) Perception of natural hazards in resources management; *Natural Resources Journal 3*, 412–41.

Burton, I., Kates, R. W. and Snead, R. E. (1969) The human ecology of coastal flood hazard in megalopolis; *University of Chicago, Department of Geography Research Paper No. 115.*

Burton, I., Kates, R. W. and White, G. F. (1968) The human ecology of extreme geophysical events; *Natural Hazard Research, Working Paper No. 1, Department of Geography, University of Toronto.*

Cook, H. L. and White, G. F. (1962) Making wise use of flood plains; In *United Nations Conference on Applications of Science and Technology*, (Government Printing Office, Washington), Vol. 1, 343–59.

Czamanske, D. V. (1967) *Receptivity to Flood Insurance*; (Master's Dissertation, University of Chicago).

Davis, R. K. (1968) *The Range of Choice in Water Management: A Study of Dissolved Oxygen in the Potomac Basin*; (Johns Hopkins Press, Baltimore).

Dunham, A. (1959) Flood control via the police power; *University of Pennsylvania Law Review 107*, 1098–132.

France, Ministère de l'Equipment et du Logement, *Inventaire des Zones Inondables*; (Paris: BCEOM, n.d.).

France, Ministère de l'Equipment et du Logement (1968) *Etats-Unis: Recherches Methodologiques sur la Rentabilite Economique des Mesures de la Controle des Crues a L'Etranger*; (Paris, BCEOM).

Goddard, J. E. (1971) Flood plain management must be ecologically and economically sound; *Civil Engineering* September, 81–5.

Hewitt, K. (1969) Probabilistic approaches to discrete natural events: A review and theoretical discussion; *Natural Hazard Research Working Paper No. 8, Department of Geography, University of Toronto.*

Johnson, J. F. (1971) Renovated waste water; *University of Chicago, Department of Geography Research Paper No. 135.*

Kates, R. W. (1964) Variation in flood hazard perception: Implications for rational flood plain use; In *Spatial Organization of Land Uses: The Willamette Valley*, (Oregon State University, Corvallis).

Kates, R. W. (1962) Hazard and choice perception in flood plain management; *University of Chicago, Department of Geography Research Paper No. 78.*

Kates, R. W. (1965) Industrial flood losses: Damage estimation in the Lehigh Valley; *University of Chicago, Department of Geography Research Paper No. 98.*

Kates, R. W. (1971) Natural hazard in human ecological perspective: Hypotheses and models; *Natural Hazard Research Working Paper No. 14, Department of Geography, University of Toronto.*

Kates, R. W. and Wohlwill, J. F. (Eds.) (1966) Man's response to the physical environment; *Journal of Social Issues 22*, 1–140.

Krutilla, J. V. (1966) An economic approach to coping with flood damage; *Water Resources Research 2*, 183–90.

Lind, R. C. (1966) *The Nature of Flood Control Benefits and the Economics of Flood Protection*; (Stanford University Institute for Mathematical Studies in the Social Sciences).

Lowenthal, D. (Ed.) (1967) Environmental perception and behavior; *University of Chicago, Department of Geography Research Paper No. 109.*

MacIver, I. (1970) Urban water supply alternatives: Perception and choice in the Grand Basin, Ontario; *University of Chicago, Department of Geography Research Paper No. 126.*

Miller, D. H. (1966) Cultural hydrology: A review; *Economic Geography 42*, 85–9.

Murphy, F. C. (1958) Regulating flood plain development; *University of Chicago, Department of Geography Research Paper No. 56.*

National Academy of Sciences Committee on Water (1966)

*Alternatives in Water Management*; Publication 1408, (Washington, D.C.).

National Academy of Sciences Committee on Water (1968) *Water and Choice in the Colorado Basin: An Example of Alternatives in Water Management*; Publication 1689, (Washington, D.C.).

National Hazard Research (1970) Suggestions for comparative field observations of natural hazards; *Natural Hazard Research Working Paper No. 16, Department of Geography, University of Toronto.*

Renshaw, E. F. (1961) The relationship between flood losses and flood control benefits; In 'Papers on Flood Problems', *University of Chicago, Department of Geography Research Paper No. 70.*

Roder, W. (1961) Attitudes and knowledge in the Topeka flood plain; In 'Papers on Flood Problems', *University of Chicago, Department of Geography Research Paper No. 70.*

Russell, C. S. (1970) Losses from natural hazards; *Journal of Land Economics*, Vol. 46, p. 38–00.

Saarinen, T. F. (1969) Perception of environment; *Association of American Geographers, Commission on College Geography*, (Washington, D.C.).

Saarinen, T. F. (1966) Perception of the drought hazard; *American Geographers, Commission on College Geography*, (Washington, D.C.).

Schiff, M. R. (1971) Psychological factors relating to the adoption of adjustments for natural hazards in London, Ontario; *Paper presented to Association of American Geographers, Boston, April 1971.*

Sewell, W. R. D. (1965) Water management and floods in the Fraser River Basin; *University of Chicago, Department of Geography Research Paper No. 100.*

Sheaffer, J. R. (1960) Flood proofing: An element in a flood damage reduction program; *University of Chicago, Department of Geography Research Paper No. 65.*

Sheaffer, J. R. (1964) Economic feasibility and use of flood maps; *Highway Research Record* 58, 44–6.

Sheaffer, J. R. (1967) *Introduction to Flood Proofing: An outline of principles and methods*; (University of Chicago, Center for Urban Studies).

Sims, J. and Saarinen, T. F. (1969) Coping with environmental threat: Great Plains farmers and the sudden storm; *Annals of the Association of American Geographers* 59, 677–86.

United States Bureau of Reclamation (1941) *Columbia Basin Joint Investigations: Character and Scope*; (Government Printing Office, Washington, D.C.).

United States, Eighty-ninth Congress, Second Session (1966) *A Unified National Program for Managing Flood Losses*; House Document 465, (Washington, D.C.).

White, G. F. (1942) Human adjustment to floods; *University of Chicago, Department of Geography Research Paper No. 29.*

White, G. F. (1964) Choice of adjustment to floods; *University of Chicago, Department of Geography Research Paper No. 93.*

White, G. F. (1966) Formation and role of public attitudes; In Jarrett (Ed.), *Environment Quality in a Growing Economy*, (Johns Hopkins Press, Baltimore).

White, G. F. (1966) Optimal flood damage management: Retrospect and prospect; In Kneese and Smith (Eds.), *Water Research*, (Johns Hopkins Press, Baltimore).

White, G. F., Calef, W. C., Hudson, J. W., Mayer, H. M., Scheaffer, J. R. and Volk, D. J. (1958) Changes in urban occupance of flood plains in the United States; *University of Chicago, Department of Geography Research Paper No. 57.*

White, G. F. (1969) *Strategies of American Water Management*; (University of Michigan Press, Ann Arbor).

White, G. F. (1970a) Flood loss reduction: The integrated approach; *Journal of Soil and Water Conservation* 25, 172–6.

White, G. F. (1970b) Recent developments in flood plain research; *Geographical Review* 60, 440–3.

# DISASTER STUDIES: AN ANALYSIS OF THE SOCIAL HISTORICAL FACTORS AFFECTING THE DEVELOPMENT OF RESEARCH IN THE AREA

*E. L. Quarantelli*

Very little has been written about the history of social science disaster research, the factors which have influenced the emergence of this field of study, and the ensuing theoretical and methodological consequences for scientific work on the human and group aspects of disasters (for passing observations, see, Fritz 1968; Quarantelli 1972; Quarantelli and Dynes 1977; Quarantelli and Wenger 1985; and Drabek 1986). In fact, apart from some of my earlier writings (Quarantelli, 1981) in response to about the only systematic effort ever made to examine some of the conditions involved in the development of the area (Kreps 1981), almost no one else has written at length or in depth on the topic. The field is only a little more than three decades old, which is not much but long enough both to allow and warrant an examination of the problem.

In a meeting in 1986 that focused on the relationships between basic and applied sociological research and disaster studies, we made four major points. First, we noted that disaster studies on behavioral and group aspects had their initial roots, almost exclusively, in rather narrowly focused applied questions or practical concerns. Second, we pointed out this led to certain kinds of selective emphases in terms of what and how the research was undertaken, with substantive consequences which we still see operative today. Third, we observed that nonetheless a basic sociological orientation and sociological ideas implicitly permeated much of the early research work and many of the answers that were offered. Fourth and last, we argued that the research approach initiated with a mixture of applied concerns and basic sociological questions, and continued now for about 35 years, has had primarily positive functional consequences on the development of the field of study of disasters.

In this article we elaborate only on the first two major points; points three and four are discussed in a related article (Quarantelli forthcoming). We essentially take a sociology of science approach to the problem, especially as has been developed in an offshoot of that orientation, namely the sociology of scientific knowledge (for the difference between the two see Tibbetts 1986). This kind of approach to the production of knowledge assumes that the social context of research activities is equally as important if not more

Reprinted from *International Journal of Mass Emergencies and Disasters* Volume 5 (3) (1988), pp. 285–310. Used with permission of the Research Committee on Disasters.

so than empirical data in influencing the growth of a field of study (see e.g., O'Neill 1981). This is at variance with the ideal but non-realistic notion that research findings or empirical observations are the prime movers in theory, model building or other scientific development (see e.g., Mannheim 1936; Kaplan 1964; Kuhn 1970; Johnson 1975). As such we try to emphasize the social factors or conditions operative in the early days of disaster research. Another consequence of this view is a downplaying of individual researchers. Thus, while an historical time frame is used to organize our remarks, this article is not meant to be a social history of the pioneering disaster researchers. Particular persons are named only if necessary for clarification of the exposition of the social factors affecting the development of the field of disaster research. The research, not the researchers, is our concern.

Our focus here is almost exclusively on the emergence of social science studies undertaken in the United States on natural and technological disasters. Thus we do not examine the initiation of work in the natural hazards area particularly the research on risk perceptions of floodplains (e.g., White 1964), a line of study out of this subfield of geography which partly converged with disaster studies in the early 1970s. Neither do we deal with accident research which later became partly embodied in risk analysis studies which in turn also came in part to converge with disaster research in the early 1980s. Nor do we look at the parallel pioneering effort in Canada in the very early 1950s (see Tyhurst 1950) and the independently initiated work in the very early 1960s in France (e.g., Chandessais 1966) and in Japan (see Okabe and Hirose 1985 for a short history of research in Japan since the 1960s). Without in any way denying the importance of these activities which we shall not discuss, we focus exclusively on the origins of what clearly is the historical core of what in the last three decades has developed and is known as the social science field of disaster

research today. In fact, one of our major purposes is to indicate the historical links between certain early studies we shall discuss and contemporary social science studies of disasters. The other intellectual stirrings we have just mentioned either are not in our view as directly important on the mainstream work or had their influence later than the early development we shall examine.

Many of the statements we make such as about the intellectual orientations or positions taken by many of the early researchers have been derived from personal involvement and observations, informal conversations, and a series of interviews for an oral history record we have initiated with the pioneers of disaster studies. As such they are impossible to reference directly although in time the oral history interviews being archived at the Disaster Research Center (DRC) library will become available for scholarly use. Similarly, many of the never publicly circulated documents which we cite, such as research proposals, organizational memos, field questionnaires, etc. are very fugitive with many of the only known copies in existence being in the personal possession of the author. These typed and written historical records are being slowly deposited in the archival collection of DRC and will also become accessible to interested scholars. It should be assumed that a non-referenced material (quotations, minutes of meetings, etc.) in the article is drawn either from these kinds of personal sources and/or non-printed records.

## THE APPLIED ORIENTATION OF THE EARLIEST STUDIES

The earliest disaster research in the social science area was almost exclusively supported by U.S.A. military organizations with very practical concerns about wartime situations. Who were the initial research sponsors and what were their interests? For our purposes, we can look at this from the perspective of the three roughly sequential sets of organized research activities from about 1950 to 1965.

## The Pioneering Field Teams

Unknown to many current disaster researchers, there were three different pioneering field team operations. The one that became famous in disaster circles was at the National Opinion Research Center (NORC) at the University of Chicago between 1950–1954. Its research was commissioned and supported by the Chemical Corps Medical Laboratories of the Army Chemical Center in Maryland.

Military personnel from this chemical center had looked at Donora, Pennsylvania, where in October, 1948, a combination of chemical fumes and a temperature inversion created a concentration of sulfur dioxide. Approximately 43 percent of the local population became ill and 25 persons died over the duration of several days. It was observed that some inhabitants of the area who had not been directly exposed apparently showed the same kind of symptoms as had victims who had been directly exposed. Seeking an explanation of this observation, the chemical center in 1949 approached NORC to do a retrospective study of the Donora episode. In joint discussions, this was eventually rejected as not worthwhile since any field work would have been done too far after the occurrence of the episode.

However, further contact between NORC and the Army Chemical Center led the latter to support a project by NORC on the study of natural and industrial disasters. As stated in the research proposal, "it is felt that empirical study of peacetime disasters will yield knowledge applicable to the understanding and control, not only of peacetime disasters, but also of those which may be anticipated in the event of another war." Elsewhere in the proposal, it is said that "careful selection of the natural or industrial disasters to be studied can furnish an approximation of the conditions to be expected in a war disaster." It was acknowledged that there are certain differences between war disaster and peacetime disasters, especially that in the latter, unlike the former, people's adherence to the cause for which the war is being fought will make them willing to make sacrifices on its behalf. Nevertheless, the proposal comes back a number of times to the idea that one could learn about the probable wartime behavior of a population from studying how they responded to natural and industrial disasters.

The Army Chemical Corps never had an opportunity to use its chemical weapons during World War II. Thus, its interest in the disaster area could be interpreted as an attempt by the organization to carve out a new future role for itself. Possibly more important was simply the widespread impression in the American military that the civilian population of the U.S.A. had never experienced a major external bombing raid and, therefore, there was consequent concern that civilians would react badly to future wartime attack that might involve the dropping of atomic bombs. That the U.S. Strategic Bombing Surveys (1947) done for the Air Force showed that civilian populations in Germany and Japan held up remarkably well under sustained bombing attacks was either unknown or ignored.

That primary interest was in the wartime implications can also be seen in two other aspects of the proposal. One is the emphasis on social control. The other is the implicit notion that the basic problems in disasters are to be found in the reactions of people to danger, loss and deprivation. Thus, it is observed that there is a need for "the reduction and control of panic reactions," that minimum elements in effective disaster control include "the securing of conformity to emergency regulations," that morale is "the key to disaster control; without it the cooperation and conformity needed from the public will not be forthcoming," and so on. Likewise, the research design focused on individual victims and the field instruments to be developed were aimed at answering five general questions:

1. Which elements in a disaster are most frightening or disrupting to people and how can these threats be met?
2. What techniques are effective in reducing or controlling fear?
3. What types of people are susceptible to panic and what types can be counted on for leadership in an emergency?
4. What aggressions and resentments are likely to emerge among victims of a disaster and how can these be prevented from disrupting the work of disaster control?
5. What types of organized effort work effectively and which do not?

The last question was conceived primarily in terms of "good disaster leadership" and not in organizational terms. Some informal interviewing of community leaders was projected, but this was to be done for the purpose of uncovering "more expert and informal accounts of the disaster, and description and analysis of public reactions to it, and of the adequacy of control measures, all of which information will be of great value in interpreting and evaluating the popular reactions uncovered by the systematic interviewing."

As one who was involved in the NORC project almost from its inception, we can attest that the actual field work generally proceeded more or less as indicated in the proposal. The effort made was to study peacetime disasters which appeared to have the closest parallel to a wartime situation (that is, a population subjected to some kind of sudden and widespread attack). The intent of the work was to find out how social control could be exercised by the authorities, and the assumption was made that disaster problems were primarily social psychological in nature, e.g., resulted from the internal states of the victim. However, as we shall note later, the sociological orientation of most of the researchers at NORC employed on the disaster project led in the course of the work to certain subtle changes in emphases and observations and perhaps even findings.

The NORC team undertook eight field studies of disasters ranging from an earth-quake in Bakersfield, California, to three consecutive airplane crashes in Elizabeth, New Jersey. The major work, however, was a very systematic population survey of 342 respondents (out of a strict probability sample of 362) in several towns and villages in northeast Arkansas hit by tornadoes in March 1952. Publications by project members from this study continued for some time after the end of the research (e.g., Bucher 1957; Fritz and Marks 1954; Quarantelli 1960; Schatzman 1960) although the final report itself was never put out in any regular published form (Marks et al. 1954).

An intended counterpart to the NORC work was that done at the University of Maryland in 1950–1954. This, too, was supported by the Army Chemical Corps and was aimed at studying in depth the psychiatric aspects of disasters as was partly indicated by the fact that the project was administrated through the Psychiatric Institute at the University of Maryland. The stated purpose of the work, as described in the contract was:

To study the psychological reactions and behavior of individuals and local population in disaster, for the purposes of developing methods for the prevention of panic, and for minimizing emotional and psychological failures.

In an Appendix to the research proposal under a heading of "Suggested Areas of Psychological Investigation" were listed:

A. Mass Population Behavior of Those Involved
   1. Herd Reaction
   2. Panic
   3. Emergence of Leaders
   4. Recommendations for Guidance and Control of Masses

Thus, even more so than in the NORC study, the University of Maryland work had a psychological emphasis and focused exclusively on individual victims. It is clear the findings were to be applied to a wartime civilian context. But like in the NORC work, and also partly perhaps because the projected multidisciplinary staff was never assembled, a somewhat different and more social science oriented end project was undertaken than probably had been originally intended by the research sponsor.

The field workers with, or supervised by, the University of Maryland study, undertook field studies of eleven different episodes. Major disasters studied were tornadoes in Arkansas; Worcester, Massachusetts; and Waco, Texas, but other emergencies researched included a chlorine gas episode, a hospital fire, a methyl alcohol poisoning episode and one of the Elizabeth plane crashes. University of Maryland field workers overlapped with NORC teams in the Arkansas tornadoes and the Elizabeth plane crash. The final report on the project, produced in mimeographed form, was about the only publication to result from the Maryland work (see Powell 1954a).

Finally, the third field team operation was at the University of Oklahoma. This was undertaken in 1950–1952 under a subcontract from the Operations Research Office at Johns Hopkins University which was conducting a much larger study of the effects of atomic weapons on troops in the field. As part of that effort by the military to understand the psychological aspects of exposure of soldiers to such weapons, researchers in the Department of Sociology at The University of Oklahoma were asked to do several things: to analyze afteraction reports, to observe troops in the field exposed to an atom bomb test explosion in a Nevada exercise, and also to study civilian behavior in extreme situations such as natural and industrial disasters.

All reports from this work were initially classified and not available to the general public for some years. Declassification of most of the written material (e.g., see, Logan et al. 1952) and discussions with the key researcher involved (the sociologist Lewis Killian) indicates that the findings of the research were intended almost exclusively for use by the Army with respect to the training of soldiers that might have to operate in a wartime setting

where atom bombs had been used. In fact, in the final report on the work, it is said that "this is a study of the effects of catastrophe among civilian groups, with the ultimate aim of extrapolation to military situations." Focus of the field work, both among the military and civilians, was on social psychological and psychological aspects of behavior under extreme stress. However, as we will again note later, this exclusively sociologically manned field work produced more theoretical results not part of the original research design with its very specific applied focus.

Civilian disaster situations systematically studied in the field included four tornadoes and a major fire in a college dormitory. By far the major study was a historical reconstruction done five years after the event of the Texas City ship explosion of 1947. The Oklahoma team overlapped in its field work with a NORC and a University of Maryland team in the third Elizabeth, New Jersey plane crash disaster.

## The Work at the National Academy of Sciences and the Diffusion of the Research Focus

The pioneering field team operations were followed by the work done at the National Academy of Sciences, first under the label of the Committee on Disaster Studies (1951– 1957), and later under the name of the Disaster Research Group (1957–1962). This work involved a variety of different activities ranging from a clearing house operation, to producing a publication series, and to supporting field studies by others outside of the Academy. A reading of the titles from the Disaster Study Series Publications gives a flavor of the multifaceted activities of this Committee and Group.

1. *Human Behavior in Extreme Situations: Survey of the Literature and Suggestions for Further Research*
2. *The Houston Fireworks Explosion*
3. *Tornado in Worcester: An Exploratory Study*

*of Individual and Community Behavior in an Extreme Situation*

4. *Social Aspects of Wartime Evacuation of American Cities*

5. *The Child and His Family in Disaster: A Study of the 1953 Vicksburg Tornado*

6. *Emergency Medical Care in Disasters, A Summary of Recorded Experience*

7. *The Rio Grande Flood: A Comparative Study of Border Communities in Disaster*

8. *An Introduction of Methodological Problems of Field Studies in Disasters*

9. *Convergence Behavior in Disasters: A Problem in Social Control*

10. *The Effects of a Threatening Rumor on a Disaster-Stricken Community*

11. *The Schoolhouse Disasters: Family and Community as Determinants of a Child's Response to Disaster*

12. *Human Problems in the Utilization of Fallout Shelters*

13. *Individual and Group Behavior in a Coal Mine Disaster*

14. *The Occasion Instant: The Structure of Social Responses to Field Studies of Disaster Behavior: An Inventory*

15. *Unanticipated Air Raid Warnings*

16. *Behavioral Science and Civil Defense*

17. *Social Organization Under Stress: A Sociological Review of Disaster Studies*

18. *The Social and Psychological Consequences of a Natural Disaster: A Longitudinal Study of Hurricane Audrey*

19. *Before the Wind: A Study of the Response to Hurricane Carla*

In a sense we see here the beginnings of a diffusion of the social science research focus in the disaster area as various tasks relevant to the development of an area of study were initiated.

Funding for the work at the Academy came from several sources, but the Committee work was initially supported until 1955 by the Surgeon General Office of the Army, Navy, and Air Force, and in 1955–1957 by the National Institute of Mental Health and the Ford Foundation. The later Disaster Research Group work was exclusively financed by the Federal Civil Defense Administration and the Office of Civil and Defense Mobilization. It should be remembered that in the years involved here, prior to 1962, civil defense in this country was basically wartime oriented.

It seems fair to say that insofar as the research supporters were concerned, the major interest was of an applied and wartime nature. In fact the Offices of the Surgeon Generals in its statement to the National Academy of Sciences had requested a program be initiated to conduct research and monitor scientific developments related to "problems that might result from disasters caused by enemy action." There was eventually a shift away from a direct military interest per se with the involvement of the federal civil defense organizations in supporting the work of the Disaster Research Group, but the basic thrust remained the same insofar as research sponsorship was concerned. The leadership in the Committee and the Group during most of its existence at the Academy was social science oriented and this had important consequences both inside and outside the Academy as we will discuss later. Even after the key leaders (Harry Williams and Charles Fritz) had left, the first annual meeting of the Group's OCDM-NRC Advisory Committee on Behavioral Research had as its objective "to stimulate both within and outside of the Office of Civil and Defense Mobilization behavioral research that will contribute to the Nation's civil defense." Given the kind of leadership left in the last two years and this kind of goal, it is perhaps not by chance that disaster work in the National Academy of Sciences had stopped within two years.

## The Establishment of the Disaster Research Center and Its Deepening of Work in the Disaster Area

The Disaster Research Center was established at Ohio State University in the fall of 1963 (DRC moved to the University of Delaware in 1985). That year, the Office of Civil Defense (OCD) gave the Center a rather large contract ($200,000) to initiate field studies of organizational functioning in disasters. It was explicitly stated that the field work was to be on civilian or peacetime disasters. But OCD's interest, and this was informally communicated to DRC, was in extrapolations from peacetime emergencies to wartime crises. In the research proposal from DRC to OCD (which had been indirectly discussed before formal submission) the wartime interest was only specifically alluded to in objective E of the proposed work (the only objective added at the explicit request of OCD). The introductory statement about objectives read:

### The General Proposal

It is proposed that there be established at The Ohio State University, a Disaster Research Center. The Center would focus on the study of organizations experiencing stress, particularly crisis situations. Generally speaking, the Center would have five major objectives:

A. To collate and synthesize findings obtained in prior studies of organizational behavior under stress.
B. To examine, both by field work and other means, pre-crisis organizational structures and procedures for meeting stress.
C. To establish a field research team to engage in immediate and follow-up studies of the operation of organizations in disaster settings, both domestic and foreign.
D. To develop, in coordination with a concurrent project, a program for field experiments and laboratory simulation studies or organizational behavior under stress.
E. To produce a series of publications on the basis of these four objectives, with special emphasis on recommendations concerning the effective emergency operations of organizations and other matters pertinent to civil defense planners.

It is not an accident that the fifth objective was only stated in this part of the proposal and, unlike the other four objectives which were discussed in great detail later, was not even alluded to anywhere else in the proposal. The wording essentially reflected the real in-

terests of the sociologists who wrote the proposal.

Irrespective of how the proposal may have read, there was no question the study was being supported only because of what it might say about a wartime situation. In actual fact it could not have been otherwise. At that time, OCD as a federal agency, was actually prohibited from direct participation in planning and/or response to civilian emergencies; the civilian area was the province at the national level of the Office of Emergency Planning (OEP), which significantly enough was not supporting any studies of peacetime disasters.

A few months after obtaining the contract from OCD, DRC received a grant from the Air Force Office of Scientific Research (AFOSR) to undertake laboratory or experimental studies of organizations under stress (what is alluded to in objective D of the OCD contract). This research was primarily and clearly seen as having possible consequences for military organizations. The Air Force never expressed any interest in results that might be applicable to civilian agencies or peacetime disasters. How closely it was viewed as related to Air Force interests is perhaps indicated by the fact that the grant was terminated in about five years, not because the research results (see Quarantelli 1967; Quarantelli and Roth 1969; Drabek and Haas 1969; Drabek 1970) were judged as invalid or uninteresting, but because the research as a whole was evaluated as not enough "mission oriented," that is, of very direct relevance for the operation of the Air Force.

DRC did continue to do research along the lines which had been initiated by the earlier pioneering field teams. The Center did build upon some, although not all, of the various disaster-related tasks originated in the research diffusion undertaken by the National Academy of Sciences. Namely, DRC initiated its own publication series and used the archives of the Academy Group to start creating a specialized social science disaster research library. It also, for the first time, deepened research in the disaster area by its continuous and concentrated studies on the planning and response, especially of emergency organizations at the local community level. It should be noted that most of these activities, for example, the publication series and the specialized library, were initiated by DRC. Directly, neither was supported by either funding or any material support from OCD or the AFOSR. Even the deepening of a research focus on organizations was also a DRC initiative, for along certain lines OCD seemed more interested in social psychological rather than social organizational problems. Put another way, many of the Center's activities were the result of the actions and decisions of the sociologists who directed DRC. The funding agencies at that time were almost exclusively concerned with the wartime or military organization extrapolations that could be made from peacetime or civilian groups. That overtly was their rationale for providing funding for disaster studies and they had no interest in directly supporting the Center in doing anything else. (It was about a decade before OCD began to exhibit a direct interest in peacetime disasters.)

The wartime orientation of OCD is illustrated in a statement covering the 1962 fiscal year (the year before DRC was established). It was written that insofar as OCD was concerned:

> The Social Sciences research program is responsible for (1) developing knowledge of the effects of war and tension upon society and its institutions; (2) determining the reactions of people to conditions before, during and after attack; (3) providing data for developing measures such as shelter, evacuation, and dispersion, for protecting the population; (4) developing data for planning relief and rehabilitation programs, embracing essential community and government functions; (5) determining effective means of securing active cooperation of people in promoting civil emergency planning measures throughout the nation.

There is no mention of civilian disasters any-
where in this 25-page summary of past and present social sciences research conducted by the then Office of Civil and Defense Mobilization and the Office of Civil Defense, Department of the Army.

Thus, in the first decade or so of disaster studies in the United States, the federal agencies supporting the research were primarily interested in wartime and/or military applications. There was no noticeable interest in civilian disasters per se; their study was undertaken to see what would be learned that could be extrapolated to a wartime or military setting. Such explicit statements as were made about the extrapolation almost always stressed that concern was with how the American population could be better prepared to withstand attacks from enemy sources. This position is well stated in remarks by the first head of the National Academy of Sciences group:

> Social science has been presented with several great challenges since World War II. Understanding the problems of technologic assistance to underdeveloped countries is one of these. Understanding psycho-cultural warfare and the true nature of subversion is another. A third great challenge is to develop a scientific understanding of the human effects and problems of disasters, both present and potential. One reason why this should be so is clear: American cities can now be attacked with the weapons which have led to dubbing our time the "age of mega-deaths." Such a prospect presents staggering problems—ranging from how to foster the most adaptive possible responses by threatened or stricken populations and how to care for millions of casualties and homeless persons, to the prospect of large-scale social, economic, and demographic reorganizations, if our urban complexes are gutted. Fundamentally, it has become necessary to know how Americans react to disaster and how they deal with it (Williams 1954; p. 5).

To the extent that the sponsoring agencies had any implicit disciplinary leanings, they were psychiatric, psychological or at best social psychological, rather than sociological. As for the implicit model of behavior under stress they operated with, it appeared to be one of personal breakdowns in disasters. The agencies also assumed that the purported problems which emerged in disasters were to be found in individuals, and the solution to such problems rested mostly in the imposition of directive social control (the command and control model which still prevails in certain disaster oriented circles today—for a discussion of this perspective see Dynes 1983)

It is possible to find some occasional references among funding agencies to an "offensive" rather than "defensive" use of extrapolations from peacetime to wartime situations. Thus, in one agency memo it is said:

> Not only do we need to know how to protect our soldiers and populace against the psychological ravages of an attack using chemical agents; in addition, we must know how to exploit to the utmost the psychological effects of toxic agents when used against an enemy.

Nonetheless, it is very important to stress that we are unaware of any instance in the past up to the present of where funding agencies have attempted to spell out the "offensive" possibilities. We have never encountered even an indirect reference to such possibilities in the disaster research literature per se. In fact, such use of research would be radically at variance with the ideological liberal or left tendencies of the large majority of American social scientists, especially sociologists. Nevertheless, all scientific knowledge can be put to "good" and "bad" purposes and it would be foolish to deny that disaster research could not also be used both ways. While this possibility does not seem to have affected researchers involved in studies of natural and technological disasters, the possibility has discouraged some students of collective stress situations from studying "terrorism." Although it is not our position, it is possible that some researchers may also be reluctant to expanding the disaster area to include "war" phenomena for the same reason.

## SOME IMPORTANT CONSEQUENCES OF THE APPLIED FOCUS

There were major consequences in the work done in the disaster area which resulted from the applied orientation of the sponsoring agencies. It is important to note that as a whole whatever influences there were from the research sponsor, they were indirect, not direct. This is true despite the fact that most of the funding for the research was of a contract, rather than grant nature, which might imply much directional and substantive control and supervision by the sponsoring groups and their officials. However, our conclusion from all the data we have examined is that there was very little effort made to direct what should be studied and/or how it should be studied.

The DRC's initial contract with OCD, for example, was the identical substantive proposal the Center had first submitted as a grant application to the National Science Foundation, except for the addition of objective #5. Informally, it was also understood that DRC should add a concluding chapter on possible extrapolations of its findings to wartime situations in the reports the Center would write about the behavior and problems of different kinds of emergency organizations in natural and technological disasters. The only administrative change in the shift from a grant to a contract proposal was that, at the suggestion of OCD, a substantial increase in both funds and duration of the project was requested and allowed.

At no time in the early days of the work did OCD attempt to dictate anything of a substantive nature. The only major problem that arose was OCD's refusal to allow the use of OCD funds for a DRC publication on the operations of the American Red Cross in disasters. The disallowance of publication stemmed from National Red Cross objections to publishing the Center's observations that

Red Cross disaster operations were negatively viewed by other organizations and the public at large. For political reasons OCD did not want such a finding, which was well documented in the DRC work, to appear in a publication from research it was funding. The Center was eventually able to publish the study results under its own auspices (Adams 1970).

As far as we have been able to ascertain, all the other early studies by other groups which we have mentioned likewise were not subjected to any direct pressure or control. It may be that DRC and the other researchers escaped direct control because the contract funding typically provided for the study of very broad topics such as "organizational functioning in disaster." Another possibility is that perhaps the lack of any knowledge about the subject matter on the part of the sponsoring groups provided freedom from direct control or supervision. Our judgement is that something more important was operative which allowed considerable freedom from sponsor control. It is that the sponsored research, at least in the early days, was primarily commissioned at the highest levels of the agencies for reasons other than seeking answers to practical problems (which however may have not been the point of view of lower level officials who actually negotiated the research agreements with academic researchers). It could be argued that disaster research was initiated (and the initiation came from the agencies and not social scientists) because of internal bureaucratic pressure for agencies to be current with the post World War II phenomena of social science research being on the agenda of many government groups. Whatever was involved, the sponsoring agencies, military for the most part, and contrary to certain images which developed in the late 1960's (see, e.g., documented accusation in Fisher 1972, p. 208), directly dictated very little if anything at all in the disaster research area.

However, while the applied orientation of the research sponsors did not lead to direct control or guidance in the research that was done, there were nonetheless, a number of indirect consequences. Let us mention just three of them. Any one of them alone has had in our view important effects on the work done in the last 35 years in the disaster area.

(1) The very conception of what constitutes a disaster was strongly influenced by the applied orientation. Thus both at NORC and DRC the prototype of a disaster was visualized, sometimes explicitly, as a major earthquake. In terms of possible extensive impact over a wide area, the sudden and unwarned occurrence of an earthquake was seen as being closest to a bombing attack on a community.

It is only possible to speculate, but we feel that substantive social science work on disasters would have developed remarkably different in the last 30 years if, for example, such diffuse emergencies as famines or droughts or epidemics or even large scale riverine flooding had provided the prototype of what constituted a disaster. In the disaster research area we early implicitly accepted a conception of disaster as a particular kind of event concentrated in time and space, and for various reasons have avoided until very recently, facing up to the serious problem of not being at all clear or certain about the core and parameters of what we are studying under the label of "disaster" (see, Quarantelli 1987). As will be discussed in a forthcoming article, we do not think we can advance significantly on further studies on disasters until we move forward on the conceptual problem.

In the collective behavior area, a subspecialty of sociology, the development of the field has been handicapped by taking a very concentrated happening in time and space—primarily a crowd—as the prototype of collective behavior even though most of collective behavior phenomena is diffuse in time and space (Aguirre and Quarantelli 1983). We have implicitly done the same thing in the disaster area. We have tended to think of disasters as concentrated space-time events, even though it might be argued that most collective stress situations (to use Barton's term, 1970) are usually much more diffuse in time and space. DRC always has had more problems in deciding in its field work whether to study a widespread riverine flood than a tornado, reflecting its implicit image of disasters.

It is interesting to note the comment of the major researcher in the University of Maryland pioneering field studies. In a little known article he raises an interesting speculative question as to the kind of disasters American disaster researchers came to focus on in their work. He wrote:

> As has been suggested, American urgency about disaster study grows out of our uncertainty about how we will act if war is ever brought directly into our continent: modern war, especially atomic war. Our anxiety over our own prospective performance is. I think, demonstrated by the spotty and perhaps guilt-motivated concentration on disasters approximating atomic explosion. (If we had dropped nerve gas or a virulent toxin on Japan, what would our focus of study be now?) (Powell 1954b, p. 61)

However, it should be noted contrary to what we have heard said at meetings, the disasters which were studied by the pioneering field teams included others than those involving only natural disaster agents. All three of the field team operations studied explosions, fires, crashes, and other concentrated in time and space human created occurrences. Neither the Academy work or the early DRC work included only natural disaster agents. It is true that relatively few non-natural disaster situations were studied, but this was more a function of what occurred during the course of the research periods involved than a deliberate focus only on natural disasters. A more recent argument (e.g., Couch and Kroll-Smith 1985) that disaster researchers have neglected chronic or slow moving as over against sudden disasters, is a much more valid criticism.

Our overall point is that we have tended to accept the notion of disaster as a concen-trated time and space occurrence. This view, a constraining one on what should be researched, was developed at the time of the origin of study in the area. This conception of disaster was to a great extent implicitly and indirectly produced by the applied wartime orientation of the early research sponsors.

(2) The early focus on the emergency time period and on the emergency response in disasters is also, we think, a partial result of the early applied orientation. If war or a military situation is thought of as the generating context, it follows that emphasis in research will be on reaction, not prevention. That the field of geography came to focus on mitigation measures and such issues as land use as part of natural hazard research problems (and the difference in focus on something called "disasters" and on something called "natural hazards" is neither an accidental or unimportant matter in our view) far before sociologists addressed such matters, may be partly a function of disciplinary differences. But we suspect it also has something to do with who initially sponsored studies by sociologists on disasters and by geographers on natural hazards. The major research program in natural hazards initiated in the late 1960's by three geographers was supported by a grant from the National Science Foundation and included studies of such matters as coastal erosion, frost and high wind, humid area drought, urban snow hazard, and water quality (see White, Kates and Burton 1969). These topics would not have interested the military supporters of the initial work in the disaster area.

The almost complete neglect by the early disaster researchers of the longer run post-impact recovery activities can also be partly attributed to the interests of the funding agencies. DRC did do some longitudinal studies of organizational long run recoveries from disasters, but they had to be done independent of OCD support (e.g. Anderson 1969). It is not that there was any objection to such studies; in fact, some OCD funding was used to obtain the relevant field data, but there simply was little interest in the results. This matter, of course, is also not independent of the funding cycles and inabilities of most governmental bureaucracies to commit themselves to support for more than one fiscal year at a time. Studies of recovery would usually have to go considerably beyond one post-impact year.

(3) The related emphasis in early studies, and to this day on planning for instead of managing disasters, we also believe is an indirect consequence of the applied orientation of the early funding agencies. The early disaster researchers assumed that they needed better knowledge of what happened in disasters so that better planning for disasters could be instituted. To a considerable extent we believe this reflected the similar bias in perspective of the military or national civil defense sponsoring agencies, who spend a great deal of time, effort and resources on planning for events with low probabilities for occurrence. Management of the military in wars or of civil defense responses in disasters is not a frequent occurrence.

There is a difference between disaster planning and disaster management, a crucial distinction still little appreciated even though it took us only 30 years to grasp its significance (Quarantelli 1985). The latter does not follow automatically from the former in the same sense as that good tactics do not follow directly from a good strategy. Management, of course deals with actual happenings, and good managing is what is needed for efficient and effective response and recovery, and, while it does not and cannot replace planning, it probably needs an equivalent emphasis. Such an emphasis was not present in the early days of disaster research and it was unlikely to be to the extent researchers reflected the bias of their supporting funding sources. The emphasis on planning also partly reflects a "command and control" model for handling emergency time problems. While disaster researchers extremely early criticized "command and control" conceptions of disaster response (e.g.,

Fritz 1961), none of them essentially challenged the primary and almost exclusive focus on planning instead of managing.

We do think it is illustrative of our point that in DRC's early days, a formal DRC proposal to study the operation and management of the United States Office of Foreign Disaster Relief and an informal one to study the operation and management of the Office of Emergency Planning (OEP) were rejected out of hand. But DRC had little difficulty in obtaining funds to study community emergency planning. The matter, of course, is a complicated one, and even in the examples given, for a variety of reasons it might be understandable why research into local agencies might be seen as more acceptable than study of national organizations, apart from a preference for a focus on planning than on management. But we think the preference needs to be accounted for, and we think it partly has its roots in the early days of disaster research.

There were other indirect consequences for disaster research that, perhaps, stemmed as much from the fact that the sponsoring agencies were American as that they had an applied orientation. Thus, there was an almost necessary focus not only on the kinds of disasters which occur in American society (e.g., tornadoes rather than famines), but also on relatively small scale and minor impact disasters (compared with the massive casualties, losses and disruptions which occur in some disasters in Latin America, Asia, or Africa). Some of the funding agencies allowed and supported overseas studies by the first American researchers. The events studied, such as floods in Holland (e.g., the volumes by the Institute Voor Social Onderzoek Van Het Nederlandse Volk 1955), massive fires in Australia (e.g., Anderson and Whitman 1967), and a dam collapse in Italy (e.g., Dynes, Haas and Quarantelli 1964) seemed to be researched because of a perceived similarity or a parallel to potential wartime situations rather than because they might be a learning situation for a potential peacetime catastrophe in the United States. We leave aside that field studies outside of the country might also have been partly supported for totally nonscientific reasons— e.g., for agency officials to be able to boast in their own bureaucratic circles, they were supporting research halfway across the world of a disaster that was the focus of international mass media attention.

The general focus on American disasters also meant that only a certain kind of social structure was studied by the early disaster researchers (e.g., one with a decentralized authority structure, with relatively weak social class differences, with highly developed social institutions, such as in the mass communication area). For instance, the almost total ignoring of social class as a factor in any way in disaster phenomena is certainly partly attributable to the locus of study used (Taylor 1978). Similarly, disaster researchers tended to look at populations with certain sociocultural characteristics (e.g., norms regarding volunteering, beliefs as to governmental responsibilities, values with regard to private property, etc). From this, for example, probably has come some of the concern of American disaster researchers about the citizen's view of emergency organizations.

Our point of course is that certain topics have been either focused on or ignored in disaster studies and that this indirectly is related to the applied research funding pattern in American society. To the extent that agencies with strong applied orientations of a particular kind emerged as the research funders rather than governmental organizations supportive of basic research (it should be noted that the initial DRC proposal went to NSF not OCD), indirectly there is going to be a reflection of this in what is assumed, studied and reported on by researchers. The applied agencies did not directly dictate much of anything, but indirectly from the start they have implicitly provided much of the research agenda and, like all agendas, the one that initially sets the stage became the one that tended to be continued to be used.

## ANOTHER IMPORTANT INFLUENCE

Although the applied orientation of sponsoring agencies looms large in our accounting for much of what has happened in the development of disaster studies, to leave it at this point would be to present an incomplete picture. Probably equally as important in the development of the area, is the fact that the early students in the area were primarily sociologists. To a considerable extent they imposed much more of a sociological perspective on how and what was studied than is realized by practically anyone. In our view, the applied orientation was married to basic sociological conceptions and ideas, although neither the research supporters nor the researchers were very aware of it at the time, and most still do not recognize the situation is the same today. However, the exposition of this point can not be provided here but will be elaborated upon in a succeeding article (Quarantelli forthcoming).

### REFERENCES

Adams, David. 1970. "The Red Cross: Organizational Sources of Operational Problems." *American Behavioral Scientist* 13:392–403.

Aguirre, Benigno E. and E. L. Quarantelli. 1983. "Methodological, Ideological, and Conceptual-Theoretical Criticisms of the Field of Collective Behavior: A Critical Evaluation and Implications for Future Study." *Sociological Focus* 16:195–216.

Anderson, William. 1969. *Disaster and Organizational Change*. Book and Monograph Series #5. Columbus, Ohio: Disaster Research Center, The Ohio State University.

Anderson, William and Robert Whitman. 1967. *A Few Preliminary Observations on "Black Thursday": The February 9, 1967 Fires in Tasmania, Australia*. Research Report #19. Columbus, Ohio: Disaster Research Center, The Ohio State University.

Barton, Allen. 1970. *Community in Disaster*. Garden City, New York: Anchor.

Bucher, Rue. 1957. "Blame and Hostility in Disasters." *American Journal of Sociology* 62:467–75.

Chandessais, Charles. 1966. *La Catastrophe de Feyzin*. Paris: Centre D'Etudes Psychosociologiques Des Sinistres Et De Leur Prevention.

Couch, Stephen and J. Stephens Kroll-Smith. 1985. "The Chronic Technical Disaster: Toward a Social Scientific Perspective." *Social Science Quarterly* 66:564–75.

Drabek, Thomas E. 1970. *Laboratory Simulation of a Police Communication System Under Stress*. Columbus, Ohio: The Ohio State University.

———. 1986. *Human System Responses to Disasters. An Inventory of Sociological Findings*. New York: Springer Verlag.

Drabek, Thomas E. and J. Eugene Haas. 1969. "Laboratory Simulation or Organizational Stress." *American Sociological Review* 34:223–38.

Dynes, Russell. 1983. "Problems in Emergency Planning." *Energy* 8:653–60.

Dynes, Russell, J. E. Haas and E. L. Quarantelli. 1964. *Some Preliminary Observations on Organizational Responses in the Emergency Period After the Niigata, Japan, Earthquake of June 16, 1964*. Research Report #1. Columbis, Ohio: Disaster Research Center, The Ohio State University.

Fisher, Charles S. 1972. "Observing a Crowd: The Structure and Description of Protest Demonstrations." Pp. 187–211 in *Research on Deviance*, edited by Jack Douglas. New York: Random House.

Fritz, Charles E. 1961. "Disaster." Pp. 651–94 in *Contemporary Social Problems*, edited by Robert Merton and Robert Nisbet. New York: Harcourt.

———. 1968. "Disasters." Pp. 202–07 in *International Encyclopedia of the Social Sciences*. Volume 3. New York: Macmillan.

Fritz, Charles E. and Eli Marks. 1954. "The NORC Studies of Human Behavior in Disaster." *Journal of Social Issues* 10(No. 3):26–41.

Instituut Voor Social Onderzoek van Het Nederlandse Volk. 1955. *Studies in Holland Flood Disaster 1953*. Washington, D.C.: National Academy of Sciences.

Johnson, John. 1975. *Doing Field Research*. New York: Free Press.

Kaplan, Abraham. 1964. *The Conduct of Inquiry*. San Francisco: Chandler.

Kreps, Gary. 1981. "The Worth of the NAS-NRC (1952–1963) and DRC (1963–Present) Studies of Individual and Social Responses to Disasters." Pp. 91–121 in *Social Science and Natural Hazards*, edited by James D. Wright and Peter H. Rossi. Cambridge, Massachusetts: Abt Books.

Kuhn, Thomas. 1970. *The Structure of Scientific Revolutions*. Chicago: University of Chicago Press.

Logan, Leonard, Lewis M. Killian and Wyatt Mars. 1952. "A Study of the Effects of Catastrophe on Social Disorganization." Chevy Chase, Maryland: Operations Research Office.

Mannheim, Karl. 1936. *Ideology and Utopia*. New York: Harcourt Brace.

Marks, Eli, et al. 1954. *Human Reactions in Disaster Situations*. Chicago, Illinois: National Opinion Research Center, University of Chicago.

Okabe, K. and H. Hirose. 1985. "The General Trend of Sociobehavioral Studies in Japan." *International Journal of Mass Emergencies and Disasters* 3:7–19.

O'Neill, John. 1981. "The Literary Production of Natural and Social Sciences Inquiry." *Canadian Journal of Sociology* 6:105–120.

Powell, John. 1954a. *An Introduction to the Natural History of Disaster*. Baltimore, Maryland: Psychiatry Institute, University of Maryland.

———. 1954b. "Gaps and Goals in Disaster Research." *Journal of Social Issues* 10:61–65.

Quarantelli, E. L. 1960. "Images of Withdrawal Behavior in Disasters: Some Basic Misconceptions." *Social Problems* 9:68–79.

———. 1967. *The Disaster Research Center Simulation Studies of Organized Behavior Under Stress*. Final Project Report #6. Columbus, Ohio: Disaster Research Center, The Ohio State University.

———. 1972. "Study and Research in the United States." Pp. 17–26 in *Proceedings of Organizational and Community Responses to Disasters*. Book and Monograph #8. Columbus, Ohio: Disaster Research Center, The Ohio State University.

———. 1981. "A Response to the Description and Evaluation of the DRC Work in the Disaster Area." Pp. 122–35 in *Social Science and Natural Hazards*, edited by James D. Wright and Peter M. Rossi. Cambridge, Massachusetts: Abt Books.

———. 1985. *Organizational Behavior in Disaster and Implications for Disaster Planning*. Newark, Delaware: Disaster Research Center, University of Delaware.

———. 1987. "What Should We Study? Questions and Suggestions for Researchers About the Concept of Disasters." *International Journal of Mass Emergencies and Disasters* 5:7–32.

———. Forthcoming. "Disaster Studies: An Analysis of the Consequences of the Historical Use of a Sociological Approach in the Development of Research in the Area." *International Journal of Mass Emergencies and Disasters*.

Quarantelli, E. L. and Russell R. Dynes. 1977. "Response to Social Crisis and Disaster." *Annual Review of Sociology* 2:23–49.

Quarantelli, E. L. and Robert Roth. 1969. *Simulation Studies of Communication Behavior Under Stress: Phase One and Two*. Final Project Report #7. Newark, Delaware: Disaster Research Center, University of Delaware.

Quarantelli, E. L. and Dennis Wenger. 1985. "Disasters: An Entry for an Italian Dictionary of Sociology." Preliminary Paper #97. Newark, Delaware: Disaster Research Center, University of Delaware.

Schatzman, Leonard. 1960. "A Sequence of Pattern Disaster and Its Consequences for Community."

Ph.D. Dissertation. Bloomington, Indiana: Indiana University.

Taylor, Verta. 1978. "Future Directions for Study." Pp. 251–80 in *Disasters: Theory and Research*, edited by E. L. Quarantelli. London: Sage.

Tibbetts, Paul. 1986. "The Sociology of Scientific Knowledge: The Constructivest Theses and Relativism." *Philosophy of the Social Sciences* 16:39–57.

Tyhurst, J. S. 1950. "Individual Reactions to Community Disaster." *American Journal of Psychiatry* 107:764–69.

U.S. Strategic Bombing Survey. 1947. *The Effects of Strategic Bombing on German Morale*. Washington, D.C.: U.S. Government Printing Office.

White, Gilbert. 1964. *Choice of Adjustment of Floods*. Chicago, Illinois: Department of Geography, University of Chicago.

White, Gilbert, Robert Kates and Ian Burton. 1969. *Collaborative Research on Natural Hazards Progress Report*. Chicago, Illinois: Natural Hazards Research Program, Department of Geography, University of Chicago.

Williams, Harry. 1954. "Fewer Disasters, Better Studies." *Journal of Social Issues* 10:5–11.

# 3

# RISK ANALYSIS AND RISK MANAGEMENT:
# AN HISTORICAL PERSPECTIVE

*Vincent T. Covello* ❖ *Jeryl Mumpower*

## 1. INTRODUCTION

In the Tigris-Euphrates valley about 3200 B.C. there lived a group called the Asipu. One of their primary functions was to serve as consultants for risky, uncertain, or difficult decisions. If a decision needed to be made concerning a forthcoming risky venture, a proposed marriage arrangement, or a suitable building site, one could consult with a member of the Asipu. The Asipu would identify the important dimensions of the problem, identify alternative actions, and collect data on the likely outcomes (e.g., profit or loss, success or failure) of each alternative. The best available data from their perspective were signs from the gods, which the priest-like Asipu were especially qualified to interpret. The Asipu would then create a ledger with a space for each alternative. If the signs were favorable, they would enter a plus in the space; if not, they would enter a minus. After the analysis was completed, the Asipu would recommend the most favorable alternative. The last step was to issue a final report to the client, etched upon a clay tablet.[1]

Reprinted from *Risk Analysis* Volume 5 (2) (1985), pp. 103–20. Used with permission of the Society for Risk Analysis.

According to Grier,[2,3] the practices of the Asipu mark the first recorded instance of a simplified form of risk analysis. The similarities between the practices and procedures of modern risk analysts and those of their Babylonian forebears underscore the point that people have been dealing with problems of risk for a long time, often in a sophisticated and quantitative way.

This paper reviews the history of risk analysis and risk management giving special emphasis to the neglected period prior to the 20th century. It is hoped that this review will accomplish the following:

♦ Dampen the prevailing tendency to view present-day concerns about risk in an ahistorical context.

♦ Shed light on the intellectual antecedents of current thinking about risk.

♦ Clarify how contemporary ideas about risk analysis and societal risk management differ significantly from the past.

♦ Provide a basis for anticipating future directions in risk analysis and management.

This paper is divided into five major sections. The first discusses the early antecedents of quantitative risk analysis, with an emphasis on the development of probability theory. It would be difficult, if not impossible, to separate contemporary risk analysis from mathematical notions of probability. Yet our review indicates that probability, expressed quantitatively, is a relatively recent idea. Although precursors of contemporary risk analysis can be traced as far back as early Mesopotamia, it was not until the emergence of probability theory in the 17th century that the intellectual tools for quantitative risk analysis became available.

The second section discusses the development of scientific methods for establishing or demonstrating causal links or connections between adverse health effects and different types of hazardous activities. Such methods are as essential to modern risk analysis as is probability theory. Despite their importance, however, progress in developing such methods was exceedingly slow. Several possible explanations are considered.

The third section focuses on mechanisms for coping with risks and discusses the principal antecedents of contemporary societal risk management strategies. Four major strategies are discussed: insurance, common law, government intervention, and private sector self-regulation. In each instance, examples are cited that closely resemble but considerably predate modern practice.

The fourth section discusses nine changes between the past and the present which we consider to be among the most significant for risk analysis and risk management.

The final section attempts to anticipate some likely future directions in risk analysis and risk management.

## 2. QUANTITATIVE RISK ANALYSIS AND PROBABILITY

Unlike modern risk analysts, who express their results in terms of mathematical probabilities and confidence intervals, the Asipu of ancient Babylonia expressed their results with certainty, confidence, and authority. Since the Asipu were empowered to read the signs of the gods, probability played no part in their analyses. Faulty predictions, as in other forms of devination, were readily rationalized according to initial premises and posed no threat to the system.[4,5] The search for the origins of modern quantitative risk analysis must, therefore, look elsewhere.

An important thread leading to modern quantitative risk analysis can be traced to early religious ideas concerning the probability of an afterlife. This should hardly be surprising, considering the salience and seriousness of the risks involved (at least for true believers). Beginning with Plato's *Phaedo* in the 4th century B.C., numerous treatises have been written discussing the risks to one's soul in the afterlife based on how one conducts oneself in the here and now.

One of the most sophisticated analyses of this issue was carried out by Arnobius the Elder, who lived in the 4th century A.D. in North Africa. Arnobius was a major figure in a pagan church that was competing at the time with the fledging Christian church. Members of Arnobius' church, who maintained a temple to Venus complete with virgin sacrifices and temple prostitution, led a decadent life in comparison to the austere Christians. Arnobius taunted the Christians for their lives of pointless self-abnegation; but, after a revelatory vision, renounced his previous beliefs and attempted to convert to Christianity. The bishop of the Christian church, suspicious of Arnobius' motives and the sincerity of his conversion, refused him the rite of baptism. In an effort to demonstrate the authenticity of his conversion, Arnobius authored an eight-volume monograph entitled *Against the Pagans*. In his work, Arnobius made a number of arguments for Christianity, one of which is particularly relevant to the history of probabilistic risk analysis. After thoroughly discussing the risks and uncertainties associated with decisions affecting one's soul, Arnobius proposed a 2 ×

2 matrix. There are, he argued, two alternatives: "accept Christianity" or "remain a pagan." There are also, he argued, two possible, but uncertain, state of affairs: "God exists" or "God does not exist." If God does not exist, there is no difference between the two alternatives (with the minor exception that Christians may unnecessarily forgo some of the pleasures of the flesh enjoyed by pagans). If God exists, however, being a Christian is far better for one's soul than being a pagan.

According to Grier,[3] Arnobius' argument marks the first recorded appearance of the *dominance principle*, a useful heuristic for making decisions under conditions of risk and uncertainty. Through his student Lactinius, and later St. Jerome and St. Augustine, this argument entered the mainstream of Christian theology and intellectual thought. When Pascal introduced probability theory in 1657, one of his first applications was to extend Arnobius' matrix. Given the probability distribution for God's existence, Pascal concluded that the expected value of being a Christian outweighed the expected value of atheism.

In addition to Pascal's seminal work,[6] the late 17th and 18th centuries witnessed a remarkable spurt of intellectual activity related to probability theory.[7] In 1692, John Arbuthnot argued that the probabilities of different potential causes of an event could be calculated. In 1693, Halley proposed improved life expectancy tables. In 1728, Hutchinson examined the tradeoff between probability and utility in risky-choice situations. In the early 18th century, Cramer and Bernoulli proposed solutions to the St. Petersburg paradox. Then, in 1792, LaPlace developed a true prototype of modern quantitative risk assessment—an analysis of the probability of death with and without smallpox vaccination.[8]

What caused this unprecedented surge of activity in the mathematical theory of probability? For decades, historians of science have grappled with this question. In 1865, Isaac Todhunter wrote a work entitled *A History of the Mathematical Theory of Probability From the Time of Pascal to that of LaPlace*.[a] Only six of the 618 pages in the text deal with Pascal's predecessors. The dearth of material was not a simple omission by Todhunter. Nor was it due to a lack of historical diligence and scholarship. Instead, it appears that formal quantitative concepts of probability were not understood to any substantial degree before the time of Pascal. Prior to Pascal, there was virtually no history of probability theory. Yet after LaPlace, the laws of probability were so well understood that a bibliography of early work on the subject would cover several hundred pages.

How can this be? What makes the situation even more difficult to understand is that man's fascination with games of chance appears to be nearly as old as man himself. As David[9] has shown, games of chance may have been one of the first inventions of primitive man. In sites throughout the ancient world, archeologists have uncovered large numbers of tali, a predecessor of modern dice shaped from the "knucklebone" or heel of deer, horses, oxen, or sheep. A talus is so formed that when it is thrown on a level surface, it can come to rest in only four ways. Well polished and often engraved examples of tali are regularly found in ancient Egyptian, Sumerian, and Assyrian sites. Tomb illustrations and scoring boards make it virtually certain that these were used for gaming. During the Roman era, Marcus Aurelius was so obsessed with throwing dice that he was regularly accompanied by his own master of games. It would seem to follow that the mathematical calculation of relative frequencies and averages should be as old as the rolling of such ancient devices. Yet mathematical theories of relative frequency, betting, randomness, and probability only appear 1500 years later in the work of Pascal, Bernoulli, and others.

Several tentative explanations have been offered, none of which is entirely satisfactory.[10,11] First, it has been argued that mathematical probability theory developed in response to specific economic needs. According to this argument, the rapid development of probability theory in the 17th and 18th centuries can, in part, be traced to the rise of capitalism. This, in turn, can be related to the desire of the new mercantile class for improved methods of business calculation and for greater economic security in the form of insurance. But early probability theorists were generally not involved in commerce, nor was their work readily applicable to business.

A second argument, related to the first, argues that the development of mathematical probability theory was related to the growth of firms dealing in life annuities. This argument falters, however, when one considers that by the third century A.D., the selling of annuities was already a common practice in Rome.

A third argument is that prior to the 17th century mathematics was not sufficiently rich in concepts and ideas to generate a theory of probability. According to this argument, the mathematics of probability became possible when limit theorems became possible. This argument falters when one considers that the concept of probability requires little besides simple arithmetic. In response, supporters of this argument claim that prior to the 17th century the arithmetic symbolism needed for easy addition and multiplication did not exist, and that such a symbolism is a prerequisite for probability.

A fourth argument is that the conditions leading to the emergence of a mathematical theory of probability are the same as those leading to the emergence of modern science in the 16th and 17th centuries. While it has long been commonly accepted that theoretical and methodological developments during this period, particularly in England and France, constituted a scientific revolution, it is not at all clear why or how this came about. Numerous theories have been proposed, from the theories of Marx,[12] concerning changes in the economic means of production; to Merton's theory[13] concerning the link between religion, the Protestant reformation, and scientific developments; and to relatively recent theories which link the emergence of modern science to a complex chain of scientific, technological, political, economic, religious, institutional, and ideological changes.[14-16]

A fifth argument, offered by Grier,[3] is that the preconditions for the emergence of probability theory were established approximately a century and a half before Pascal, largely because of a change in attitude of the Catholic Church. Grier notes that in the 14th century the Catholic Church experienced a serious cash flow situation. On the one hand, money was needed to pay the massive debts arising from the Crusades—which had been extraordinarily, almost ruinously, expensive. On the other, money was also needed to pay for new church construction in response to growing population pressure. Money was, however, in short supply, since the Church prohibited usury. In order to reconsider the matter, the Church formed an advisory panel of scholars. At the same time, the financial community hired John Ecks, a university professor, to argue on their behalf. In 1518, the Laettian Council adopted a scholar's report that redefined usury in such a way that interest was permitted as long as there was risk incurred on the part of the lender. Although this definition was rescinded in 1586 (and the Church did not sanction interest again until 1830), Grier argues that the 68 years of sin-free interest rates were enough to stir up intellectual thought about probability. The real change, he argues, was not in law or morality, but in making risk a legitimate topic of discussion. He believes that much of the intellectual thinking about probability in the 17th and 18th centuries had its roots in the discussions of risk in the Church's debates over interest rates.

Whatever the true explanation or explanations, it did not take long for the new theories of probability to be applied to the human condition. Within 100 years of Pascal's discovery, several individuals were using mathematical theories of probability to examine life expectancies. Prior to this work the only life expectancy tables of note were those developed by the Roman Domitius Ulpianus in 230 A.D. Similar efforts were not made until Graunt published his life expectancy tables in 1662. In explaining this large gap, it appears that philosophical objections played a major role. As David[9] points out, there ". . . seems to have been a taboo on speculations with regard to health, philosophers implying that to count the sick or even the number of boys born was impious in that it probed the inscrutable purpose of God."

Graunt's work represents the first recorded attempt to calculate empirical probabilities on any scale. The raw data for his tables were parish records of births and deaths. In the conclusion of his work, Graunt[17] offers several remarks that might apply equally to the work of contemporary risk analysts:

"It may be now asked, to what purpose tends all this laborious puzzling and groping? . . . I might answer; That there is much pleasure in deducing so many abstruse and unexpected inferences out of these poor despised Bills of Mortality; and in building upon that ground, which hath lain waste these eighty years. And there is pleasure in doing something new, though never so little, without pestering the World with voluminious Transcriptions. But, I answer more seriously, . . . that a clear knowledge of these particulars, and many more, whereat I have shot but at rovers, is necessary in order for good, certain, and easy Government, and even to balance Parties and factions both in Church and State. But whether the knowledge thereof be necessary to many, or fit for others, than the Sovereign, and his chief Ministers, I leave to consideration."

Not long after Graunt published his tables, Halley (better remembered for the comet that bears his name) published an article[18] containing mortality tables based on the records of births and deaths at various ages in the city of Breslau. Halley's probabilistic analysis of the data convincingly disproved popular theories about the effect of phases of the moon on health. His results, as will be discussed later, provided the foundation or one of the pillars of modern societal risk management—life insurance.

One of the earliest systematic attempts to apply probability theory to a risk problem was by Von Bortkiewicz in the 19th century.[19] Von Bortkiewicz built on previous work done by Poisson to calculate theoretically the annual number of Prussian soldiers dying from kicks by horses. He studied records covering a span of 10 years to determine whether an observed rash of kicking accidents represented a random event or a change in circumstance requiring action (e.g., a rise in carelessness by soldiers or an increase in the number of wild horses). The analysis indicated that the occurrences he observed were indeed random events and that no special disciplinary actions were required.

## 3. METHODS FOR ESTABLISHING CAUSALITY

Modern risk analysis has its twin roots in mathematical theories of probability, and in scientific methods for identifying causal links between adverse health effects and different types of hazardous activities. Researchers throughout history have relied principally on methods of observation to unravel these links. The most basic form of such methods, and the most universally practiced, is experience based on trial and error. Since primitive times, human beings have upon occasion simply undertaken a new activity of interest (e.g., tasted a stange plant or launched a new boat) and have observed the adverse effects, if any. A slight variant of this method has been to distance oneself and conduct the experiment on a surrogate (e.g. feed new foods to animals). On a more complex level, researchers have used both indirect observational methods; such as the burn tests developed by Pliny the Elder in first century A.D. Rome to detect food adulteration[20]; and epidemiological observational methods (i.e., methods that seek to establish associations or cause-effect relationships through the observation of adverse health effects in clusters of cases). Although the early researchers in this second tradition did not adhere to the rigorous scientific and statistical standards of modern epidemiological studies,[21,22] the historical record is replete with examples. The association between malaria and swamps, for example, was established in the 5th century B.C. even though the precise reason for the association remained obscure. In the book *Airs, Waters, and Places*, thought to have been written by Hippocrates in the 4th or 5th century B.C., an attempt was made to set forth a causal relationship between disease and the environment. As early as the 1st century B.C., the Greeks and Romans had observed the adverse effects of exposure to lead through various mediums.[23,24] For example, the Roman Vitruvious (cited in Ref. 25), wrote:

> "We can take example by the workers in lead who have complexions affected by pallor. For when, in casting, the lead receives the current of air, the fumes from it occupy the members of the body, and burning them thereon, rob the limbs of the virtues of the blood. Therefore it seems that water should not be brought in lead pipes if we desire to have it wholesome."

Unfortunately, the observations of the ancient Greeks and Romans were soon forgotten and work did not begin anew until the 16th, 17th, and 18th centuries. Among the many studies conducted during this period, the following stand out:

◆ A study by Agricola[26] linking adverse health effects to various mining and metallurgical practices.
◆ A study by Evelyn[27] linking smoke in London to various types of acute and chronic respiratory problems.
◆ A study by Ramazzini[28] indicating that nuns living in Appennine monasteries appeared to have higher frequencies of breast cancer (Ramazzini suggested that this might be due to their

celibacy, an observation that is in accord with recent observations that nulliparous women may develop breast cancer more frequently than women who have had children—see Refs. 29 and 30).
◆ A study by Hill[31] linking the use of tobacco snuff with cancer of the nasal passage.
◆ A study by Sir Percival Pott[32] indicating that juvenile chimney sweeps in England were especially susceptible to scrotal cancer at puberty.
◆ A study by Ayrton-Paris[33] as well as by Hutchinson[34] indicating that occupational and medicinal exposures to arsenic can lead to cancer.
◆ A study by Chadwick[35] linking nutrition and sanitary conditions in English slums to various types of ailments.
◆ A study by Snow[36] linking cholera outbreaks to contaminated water pumps.
◆ Studies by Unna[37] and Dubreuilh[38] linking sunlight exposure with skin cancer.
◆ A study by Rehn[39] linking aromatic amines with bladder cancer.

Despite these studies, progress in establishing causal links between adverse health effects and different types of hazardous activities was exceedingly slow. It appears that at least two major obstacles impeded progress. The first was the paucity of scientific models of biological, chemical, and physical processes, especially prior to the 17th and 18th centuries. Related to this was the lack of instrumentation and the lack of rigorous observational and experimental techniques for collecting data and testing hypotheses. Shapiro (cited in Ref. 40) described two revolutionary intellectual developments affecting science (and law):

> "The first was the drive for systematic arrangements and presentation of existing knowledge into scientifically organized categories . . . The second . . . was the concern with degrees of certainty . . . or probability . . . By the end of the 17th century . . . traditional views . . . had been upset and new methods of determining truth and investigating the natural world had replaced those that had been accepted for centuries . . . there was a strong movement toward arranging both concepts and data into some rational ordering that could be easily communi-

cated and fitted into the materials of other fields so that a universal knowledge might emerge."

Although often taken for granted in today's world, even basic medical knowledge is a relatively recent development. It is surprisingly easy to forget that it was not until the work of Pasteur in the late 19th century that scientists first began to comprehend adequately the concept of infection or the causal relationship between the environment and biological agents of infectious disease.

The second obstacle was the belief, rooted in ancient traditions, that most illnesses, injuries, misfortunes, and disasters could best be explained in social, religious, or magical terms (e.g., by the will of the gods, by some moral transgression, or by the malevolence of an unseen enemy). In 16th and 17th century Europe, witch hunting resulted in death by fire for an estimated half-million people, as the Church attempted to eradicate a perceived source of crop failures, disease, death, and other ill fortune.[41] In 1721, an influential critic of medical experimentation in Boston insisted that smallpox is "a judgment of God on the sins of the people" and that "to avert it is . . . an encroachment on the prerogatives of Jehovah, whose right it is to wound and smite."[42] For many such critics, the direct physical agent that caused the harm was of considerably less interest than the moral status of the victim. In the mid-19th century, for example, critics opposed to health reforms in the Lowell, Massachusetts textile factories (including a large number of influential physicians) blamed symptoms of disease among factory workers on the workers' "improvident" style of life.[43]

## 4. SOCIETAL RISK MANAGEMENT

In response to identified risks, individuals and groups have historically employed a number of techniques for reducing or mitigating adverse health effects. These include the following:

- ◆ Avoiding or eliminating the risk, such as prohibiting the use of a potentially dangerous object or substance.
- ◆ Regulating or modifying the activity to reduce the magnitude and/or frequency of adverse health effects, e.g., by constructing dams, levees, and seawalls.
- ◆ Reducing the vulnerability of exposed persons and property, e.g., by requiring the use of safety devices, by elevating buildings in floodplains, by immunizing the population, by implementing quarantine laws, or by establishing disaster warning systems.
- ◆ Developing and implementing post-event mitigation and recovery procedures, e.g., by establishing search and rescue teams, stockpiling food, providing first aid training, or providing fire extinguishing equipment and services.
- ◆ Instituting loss-reimbursement and loss-distribution schemes through such mechanisms as insurance systems or incentive pay schedules for high risk activities.

Although all of these techniques are still practiced, most of our current ideas about societal risk management are rooted in four basic strategies or mechanisms of control: insurance, common law, government intervention, and private sector self-regulation. Each is discussed below.

### 4.1. Insurance

Insurance is one of the oldest strategies for coping with risks.[44] Its origins can be traced to early attempts at setting interest rates, which first arose in Mesopotamia. Records of interest rates in that area predate 3000 B.C. The practice appears to have originated when successful farmers loaned a portion of their excess production in exchange for a share of the return. At first, loaned goods were returned in kind along with the interest charge, but subsequently barley and then silver became the media of exchange. Differences in interest rates quickly emerged, ranging from zero for personal loans among friends to 33% for presumably much riskier agricultural loans. Inasmuch as interest rates reflected the perceived riskiness of the loan, they represented one of the earliest attempts to quantify and manage risk.[3]

The connection between interest rates and insurance can be traced to ancient Babylon. A natural trade center, Babylonia occupied a place as the center of the world economy in the pre-Greek period. Imports and exports flowed through Babylonia to and from both the east and the west. Traders faced numerous hazards in the form of bandits, pirates, fires, storms, and other assorted misfortunes. Loans extended on cargo in transit ordinarily included risk premiums as large as 200% in excess of interest. Because the borrower often posted all his property and sometimes his family as collateral, misfortune could prove truly catastrophic. Under these conditions trade declined, until insurance emerged as a more effective risk management technique. The Code of Hammurabi, issued about 1950 B.C.,[45] established several doctrines of risk management and laid the basis for the institutionalization of insurance. It formalized the concept of *bottomry*, which formed the basis for maritime contracts on vessels and cargoes. These bottomry contracts contained three elements: (1) the loan on the vessel, cargo, or freight; (2) an interest rate; and (3) a risk premium for the chance of loss and consequent cancellation of the debt. Bottomry quickly spread throughout the ancient world and represents one of the oldest attempts to cope with risk in a systematic, quantitative fashion.

By 750 B.C., bottomry was highly developed, particularly in Greece. At that time almost all voyages were covered by bottomry contracts, with 10–25% risk premiums, depending upon the perceived riskiness of the venture. (It is interesting to note here the etymology of the English word "risk." According to the *Unabridged Random House Dictionary*, the word "risk" comes to us through the French, Italian, and Latin, originating from the Greek word *rhiza*, in reference to the hazards of sailing around a cliff.) During this same period, the concept of *general average*, which called for all parties to share proportionately in any loss suffered during a venture, was also developed. This device established a mechanism for risk sharing, and provided a foundation for the first insurance exchange that developed and flourished in Athens.

With the decline of Greek civilization the Western development of insurance institutions also declined, although the Romans continued the practice of bottomry. The Romans did institute a rudimentary form of life and health insurance, however, in the form of *collegia*. Although burial societies had existed in Greece, the collegia of the Romans were much more highly developed. Members made regular contributions, a fund was maintained, and burial and last expenses were paid by the collegia.

Insurance almost disappeared in the West after the fall of the Roman Empire. Although the European guilds provided some protection to their members against various calamities, it was not insurance per se. Marine insurance reappeared in the Italian port cities, perhaps as early as 1000 A.D. and certainly by the 12th–14th centuries, becoming progressively more widespread and better developed. The Hanseatic League and Lombards, in particular, developed detailed sets of regulations pertaining to marine insurance, as evidenced by the Laws of Wisby (1300), the Ordinance of Barcelona (1435), and the Guidon de la Mer (circa 1600). In 1688, Lloyd's was established and London emerged as the nucleus of the global marine insurance market, later extending into other areas of insurance.

From the 17th century on, the insurance industry flourished in England. Fire insurance, for example, developed in London in response to the Great Fire of 1666 and quickly achieved great success. Life insurance in something resembling its modern form emerged during the 16th and 17th centuries in England, France, and Holland, although the first life insurance policies may date back to 1800–1200 B.C. Grier[3] points out that life insurance policies existed in Spain by about 1100 A.D. and that "tontines" were highly popular in 17th century France. (Members of tontines made payments into a general fund; if one lived long enough one received a share of the pool; and if one were the last member of the tontine to remain alive one could become quite rich.) In England, the first recorded life insurance policy was issued in 1583. Life insurance then grew rapidly under the sponsorship of the various so-called Friendly Societies.

Historical records show that the failure rates of the English Friendly Societies were initially exceedingly high. In 1867, for example, an official of the British government estimated that during the previous 75 years one-third of the Friendly Societies had failed. The reason for the high rate of failure seems clear. Methods of probabilistic assessment were either not known or not utilized, and comprehensive sets of vital statistics were not available. Without appropriate tools for quantitative thinking about risk or the requisite data base, attempts to think quantitatively about risk often went awry. It was not until professional actuaries became an integral part of the industry in the 19th century that insurance companies stood on a firmer footing.

Developments in the life insurance field in the 17th century appear to foreshadow the modern debate concerning whether it is acceptable to place a monetary value on the loss of human life. The Church in particular raised serious questions about the morality of life insurance. For the Church, life insurance was an immoral—or at least highly suspect—wager on human life. Indeed, life insurance was prohibited in France until 1820. Debate about the morality of life insurance has long since died out, but similar issues are still discussed in debates about the moral status of cost–benefit analyses addressing risks to human life.

## 4.2. Common Law

In the English and American legal systems, the common law (that is, judge-made law) of "torts" has long permitted one citizen to re-cover damages from another for harms resulting from such actions as *nuisance* (use of one's own property that unreasonably interfers with the use of another's property), *negligence* (conduct that unreasonably imposes risks on another), and the pursuit of *abnormally dangerous activities*. These grounds for lawsuits amount to risk management in the sense that people must conform to a standard of reasonable conduct (e.g., cleaning their chimneys, disposing of waste products) or face the prospect of being liable and successfully sued for damages. Common law thus provides two risk management functions—compensation and deterrence.

Hammer[46] has argued that the origins of modern liability laws can be traced back to the Code of Hammurabi and to the Old Testament, both of which stressed the notion of *strict liability* (i.e., the concept that the manufacturer of a product is liable for injuries due to defects regardless of negligence or fault). With the advent of the Industrial Revolution, however, the principle of strict liability embodied in ancient laws fell into temporary decline, and proof of negligence or other fault on behalf of the defendant became an essential requirement for recovery of damages in most areas of common law. By 1850, the law stated that "the plaintiff must come prepared with evidence to show that the intention was unlawful, or that the defendant was at fault; for if the injury was unavoidable, and the conduct of the defendant was free from blame, he will not be liable." An injured party could seek redress only if there were proof of negligence and, according to the principle of *privity*, only from the party contracted to supply the product.

In the United States, it was not until 1916—in the MacPherson vs. Buick Motor Company[b] case—that this historically new and narrow concept of liability was partially broadened. In this case, it was ruled that the manufacturer had a responsibility to inspect products for defects, and that the lack of privity should not affect a plaintiff's right to recover damages for his injuries. It was not until the 1960s that the ancient notion of strict liability began to be reinstated through a series of court decisions and the passage of legislation that weakened the necessity to prove negligence in order to collect damages.

## 4.3. Direct Government Intervention

Since ancient times, government authorities have directly intervened to reduce, mitigate, or control risks. As Handler[47] pointed out, it ". . . has long been a function of government to shield the citizenry from those dangers against which it cannot readily protect itself; hence police and fire departments, armies and navies."

Many of the earliest efforts by government authorities relied heavily on magico-religious practices. In 5th century B.C. China, for example, provincial officials and priests required the annual sacrifice of a maiden to propitiate the Yellow River gods and thereby control the ravages of annual flooding, as is described in the following passage:

"Adorned in ceremonial regalia, the victim was flung into the stream, where she was swiftly dragged beneath the surface by her heavy accouterments. Needless to say, the maiden was invariably selected from a peasant family rather than from the local gentry, and Chinese historians record that as the years passed, farmers who had eligible daughters deserted the district in increasing numbers. Eventually, around 400 B.C., a magistrate named Hsimen Pao stepped forth and put an end to the practice with one final, highly appropriate, sacrifice: He had the priests and officials hurled to their deaths in the swirling yellow waters."[48]

Paralleling, and sometimes alternating with, these magico-religious techniques were direct government interventions based on firmer ground. The following section provides several examples of early interventions that predate present-day practices.

### 4.3.1. Natural Disasters

Virtually all of the great ancient civilizations (e.g., China, Maya, Egypt, and Mesopotamia) directly intervened to mitigate the effects of natural disasters. Historical records indicate, for example, that throughout history, governments have played a major role in developing and financing elaborate systems of flood control, including dams, dikes, and canals. One of the first efforts of this kind was recorded by the Roman historian, Pliny the Elder. Pliny noted that the Egyptian authorities had successfully devised an elaborate system for dealing with the risk of famine due to the periodic overflow of the Nile. Pliny reported the system as follows:

"The Nile begins to rise at the next new moon after midsummer, the rise being gradual and moderate while the Sun is passing through the Crab and at its greatest height when it is in the Lion; and when in Virgo it begins to fall by the same degrees as it rose. It subsides entirely within its banks, according to the account given by Herodotus, on the hundreth day, when the sun is in the Scales . . . Its degrees of increase are detected by means of wells marked with a scale. The province takes careful note of both extremes; in a rise of 18 ft. it senses famine, and even at one of 19 1/2 ft. it begins to feel hungry, but 21 ft. brings cheerfulness, 22 1/2 ft. complete confidence, and 24 ft. delight. The largest rise up to date was one of 27 ft. in the principate of Claudius (1st century A.D.) and the smallest 7 1/2 ft. in the year of the war of Pharsalus (48 B.C.), as if the river were attempting to avert the murder of Pompey by a sort of portent. When the rise was to a standstill, the floodgates are opened and irrigation begins; and each strip of land is sown as the flood relinquishes it."

As protection against a bad year, Pliny noted that Egyptian authorities used data on flooding and crop surpluses to adjust the taxes leveled on crops harvested in the current season.

In addition to attempts to prevent or control disasters, government authorities have also responded by providing relief after disasters occur. In 1803, for example, the U.S. Congress passed legislation to assist victims of a fire in Portsmouth, New Hampshire. In following years, Congress approved on an *ad hoc* basis more than 100 separate acts granting relief after specific disasters had occurred.[49] It was not, however, until the 20th century that the first U.S. agency was authorized to make loans to the private sector for rehabilitation, repair of damage, and alleviation of hardship caused by natural disasters.

### 4.3.2. Epidemic Disease

Throughout history, government authorities have attempted to mitigate the effects of epidemic disease. The magnitude of the problem was in many cases staggering. The 1348 to 1349 epidemic of the Black Death (bubonic plague), for example, killed over a quarter of the population of Europe—approximately 25 million people.[50-52] Given the lack of knowledge about the causes of diseases such as plague and typhus, government authorities often adopted one of the oldest and most direct strategies of disease control—quarantine and isolation. Fear of leprosy, for example, has throughout history caused wide-scale adoption of the practice of isolating the infected and cleansing or burning of their garments. Fear of infection also prompted healthy communities to adopt strict measures in preventing the entry of goods and persons from infected communities. In the 7th century A.D., for example, armed guards were stationed between plague-stricken Provence and the diocese of Cahors. One thousand years later, in 1720, when Marseilles was suffering a severe epidemic of the plague, a ring of sentries was placed around the city to prevent any person from escaping.[53,54]

In addition to quarantines and public health efforts (discussed below), the development of vaccines in the 18th and 19th centuries had a major impact on the problem. Although governments played only a minor role in these developments, it is interesting to note that the first federal regulatory health statute in the United States was the Federal Vaccine Act of

1813.[55] The act gave the President the power to appoint a federal vaccine agent to test the safety of the newly discovered smallpox vaccine. The law was repealed, however, in 1822 on the ground of "states rights."

### 4.3.3. Pollution

Pollution of the air, water, and land has long been recognized as a problem, but efforts at pollution control have been highly sporadic. Air pollution (due to dust and smoke from wood and coal fires) has been a ubiquitous problem in congested urban areas since ancient times.[25] The first act of government intervention did not occur until 1285, however, when King Edward I of England responded to a petition from members of the nobility and others concerning the offensive coal smoke in London. Smoke arising from the burning of soft coal had long been a problem in London.[14,56] Edward's response to the petition was one that is now commonly practiced by government risk managers—he established a commission in 1285 to study the problem. In response to the commission's report, several private sector actions were taken, including a voluntary decision by a group of London smiths in 1298 not to ". . . work at night on account of the unhealthiness of coal and damage to their neighbors."[56] These voluntary efforts were not sufficient, however, and in 1307 Edward issued a royal proclamation prohibiting the use of soft coal in kilns. Shortly after this, Edward was forced to establish a second commission, the main function of which was to determine why the royal proclamation was not being observed.

The history of water and land pollution control has been equally sporadic. Over three thousand years ago, the governments of Minoa and Crete built community sewage-drainage systems, and at least some citizens enjoyed the benefits of flush toilets and indoor plumbing.[57,58] Athens and other Greek cities also built sewage-disposal systems and enacted laws requiring that waste matter be carried outside the walls for a certain distance before it was dumped. Fines were frequently levied and pollution of the city water supply could merit the death penalty.[25] The ancient Romans, however, are credited with developing the most extensive system, consisting of paved streets, gutters, and a complex of tunnels and aqueducts.[52,59] Roman authorities also enacted strict laws to control foul smells and the disposal of waste products.[57] After the fall of the Roman Empire, many of these laws were unfortunately forgotten and the structures fell into disrepair. A resurgence of interest did not appear again until the 14th and 15th centuries when, in response to the spread of contagious diseases, public officials in Europe created a rudimentary system of pollution and sanitary control. The system included the development of pure water supplies, garbage and sewage disposal, observation stations, hospitals, disinfection procedures, and food inspection. The extent and effectiveness of these efforts should not, however, be overestimated. As several authors have noted (e.g., Bettman[60]), prior to the 19th and 20th centuries:

". . . filth, squalor, and disease of community life were apparently accepted as a usual and normal state of affairs. The crude attempts to alleviate the conditions of those days were almost always local efforts. The situation was aggrevated by the Industrial Revolution, when hordes of men, women, and children flocked to the cities seeking employment in the new factories. The cities, utterly unprepared to meet the influx, had no means of housing the newcomers except in areas where living conditions were already wretched. To make matters worse, flimsy tenements were improvised without proper provision for ventilation, light, water, and waste disposal. Streets were dark, narrow, and barely passable owing to filth, stagnant pools, and the stench arising from them. Inevitably, the drinking water became contaminated, and as a result typhoid, dysentery, and cholera took a large toll of lives."[61]

In dealing with these problems, public health efforts were seldom effective. Despite the passage of laws by many localities, such as the 17th century ordinance enacted by colonial New Amsterdam prohibiting ". . . the throwing of rubbish and filth into the streets and canals, . . ."[62] and a law passed in 1671 requiring that each peasant coming into Berlin had to leave with a load of filth,[59] little change took place until the 19th century. One factor contributing to the changes that occurred in the 19th century were a number of government-sponsored reports documenting the abominable conditions in European and American cities. In England, Edwin Chadwick published his classic work *Report on an Inquiry into the Sanitary Conditions of the Laboring Population of Great Britain in 1842.* This report, which was commissioned by the British Parliament four years earlier, played a major role in the creation of the General Board of Health for England in 1848. Similarly, in the United States, Lemuel Shattuck's publication of his *Report of the Sanitary Commission of Massachusetts* (1850) led to the establishment of the State's Board of Health in 1869. The act creating the Board directed it to:

". . . make sanitary investigations and inquiries in respect to the people, the causes of disease, and especially of epidemics, and the source of mortality and the effects of localities, employments, conditions, and circumstances on the public health; and they shall gather information in respect to those matters as they may deem proper, for diffusion among the people".[57]

Over the next few decades, several localities in the U.S. and Europe created similar bodies, leading to major improvements in street paving, refuse collection, water purification, water distribution, and sewage disposal. Several important laws were also passed including the English Nuisance Removal Act of 1855 which attempted to regulate gross pollution of the Thames River. Unfortunately these laws were seldom heeded (see Ref. 63), and effective pollution controls occurred only after major outbreaks of infectious diseases.

### 4.3.4. Food Contamination and Adulteration

As the basic sustenance of life, virtually all societies have been concerned about the safety of the food supply. The Biblical abominations of Leviticus, particularly the prohibition against the eating of pork, are often cited as an early attempt at controlling food safety. Douglas[64] has argued, however, that it would be a mistake to view all such attempts as simple forerunners of modern food and drug regulations. She observes that food prohibitions often serve a variety of purposes, including: the affirmation of ethical norms, a means of distinguishing one group from another, and a symbolic mechanism for bringing order into a chaotic world by classification and category. (see also Douglas and Wildavsky[65]) In her discussion of Leviticus, she asks:

> "Why should the camel, the hare, and the rock badger be unclean? Why should some locusts, but not all, be unclean? Why should the frog be clean and the mouse and hippotamus unclean? What have chameleons, snakes, and crocodiles got in common?"[64]

In response, Douglas suggests that the abomination of Leviticus can be seen as a mix of pragmatic-classificatory rules and the threefold classification of Genesis which divided creation into the earth, sea, and sky. "Clean" animals fully conform to the archetypes of their class: cloven-hoofed ruminants; four-legged animals of the earth that hop, jump, or walk; scaly fish of the sea that swim with fins; and two-legged fowls that fly with wings. Species that are "unclean" are those that are imperfect members of their class, or whose class itself violates the Biblical system.

Aside from the Biblical prohibitions, it appears that the first important law to be enacted regulating food was the English Assize of Bread (1263), which made it unlawful to sell any food "unwholesome for man's body." Interestingly, Hutt[55] has argued that this statute is practically indistinguishable from the current U.S. standard prohibiting additives which "may render food injurious to health." For nearly six hundred years, the Assize of Bread and later statutes covering other food products were in effect until many were repealed in the early 1800s at the height of the industrial revolution and a *laissez faire* philosophy of government. By the late 19th century, however, the medieval laws were reinstated, culminating in federal legislation such as the U.S. Biologies Act of 1902, the Federal Pure Food and Drug Act of 1906, and the Federal Meat Inspection Act of 1906. (The earliest food adulteration act enacted by a state had been passed more than a century earlier in 1785 by Massachusetts.)

### 4.3.5. Building and Fire Codes

In what is perhaps the first recorded attempt to manage risks through government regulation, the Code of Hammurabi (circa 1950 B.C.) decreed that should a house collapse and kill the occupants, the builder of the house must forfeit his own life.[45,66] Although not quite as strict, the Romans also enacted laws regulating the quality of building constructing.[57] Aside from construction risks, virtually all societies have been concerned with the risks of fire. Despite this concern, however, it appears that a concerted effort by government to deal with the problem did not occur until the 17th century. In 1626, for example, the Plymouth colony enacted a law directing that new houses not be thatched, but roofed with board or other materials.[67] In 1648, New Amsterdam prohibited the construction of wooden or plaster chimneys on new homes, and required that chimneys on existing homes be inspected regularly. An even stricter abridgement of individual freedom occurred in 1740 when the city of Charleston required that ". . . all buildings should be of brick or stone, that all 'tall' wooden houses must be pulled down by 1745, and that the use of wood . . . be confined to window frames, shutters, and to interior work."[68] The event of perhaps the greatest significance in stimulating government authorities to action was the Great London fire of 1666, which destroyed over three-quarters of the city's buildings. Largely as a result of this disastrous fire, nearly all large cities in Europe and America established municipal fire extinguishing companies during the next hundred years.

### 4.3.6. Transportation Accidents

Regulation of the transportation system, in the interest of safety, substantially predates modern mechanized transportation technologies. Traffic safety regulations, for instance, date back at least to ancient Rome. According to Hughs,[25] a municipal law under Julius Caesar prohibited all wheeled vehicles to operate in Rome between sunrise and 2 hours before sunset, except for essential public service traffic. This regulation was largely for the benefit of pedestrians, for whom the combination of narrow streets and heavy traffic created a genuine hazard.

The highly regulated character of the modern transportation system was foreshadowed by responses to earlier technological developments. Indeed, the first regulation of a technological risk in the United States occurred in 1838, when Congress passed legislation governing boiler testing, inspection, and liability.[69,70] This legislation was enacted in response to a series of boiler explosions on steamboats that led to thousands of injuries and fatalities during the early 19th century. The initial legislation was too lax to foster effective risk reduction, but was replaced by stricter legislation in 1858. This law specified engineering safety criteria, gave inspectors authority to examine boats and refuse licenses, and created a regulatory agency–the Board of Directors of Inspectors.

The steamboat remained the dominant form of transportation technology in the United States until the latter part of the century, when it was replaced by the railroad. Both in the United States and Europe, disputes over the risks of railroads clearly reflected broader so-cial values. The major concerns in Britain regarding this new transportation technology were described by Cohen[71] as "horror at attaining speeds over 40 kilometers an hour, concern about the capacity of new kinds of organizations to run large operations safely, fears of social consequences of change, worries about the desecration of the Sabbath, and even concern at the ease with which dangerous radicals might travel about the country." At the turn of the 20th century, disputes about the automobile also reflected social concerns broader than those associated with risk.[70] For both the railroads and automobiles (and later airplanes), the substantial intrinsic risks associated with these transportation modes led quickly to the development of a regulatory scheme that, while much stricter today than in its earliest versions, is not essentially different in concept.

### 4.3.7. Occupational Injuries

Prior to the 18th and 19th centuries, occupational health and safety issues were apparently of only minor concern to government authorities. Although working conditions in most industries were generally abominable (e.g., see Agricola[26] and Engels,[72] it was not until the Industrial Revolution that government officials took note. Most of the first efforts by government authorities were focused on the conditions of child labor.[73] As Samuelson has noted,

". . . No Dickens novel did full justice to the dismal conditions of child labor, length of working day, and conditions of safety and sanitation in the early nineteenth century factories. A workweek of 84 hours was the prevailing rule, with time out at the bench for breakfast and sometimes supper, as well as lunch. A good deal of work could be got out of a six year old child, and if a man lost two fingers in a machine, he still has eight left."[74]

In 1842, a British Parliamentary commission estimated that about one-third of the mine workers in Britain were less than 13 years old.

The commission's report noted that many of these children were employed as "trappers," who manned the air doors that separated the various sections of the mines. Their life consisted of sitting ". . . in the pit the whole time it is worked, frequently above 12 hours a day. They sit, moreover, in the dark, often with a damp floor to stand on, and exposed necessarily to drafts . . . ."[75]

Most efforts at reform were initially strongly resisted by mine and factory owners, although there were a few notable exceptions. The 19th century British millowner Robert Owen, for example, played an important role in bringing about change through the way he operated his mills and through his writings about the responsibility of employers toward their employees.[76]

At the same time that improvements in working conditions were being made, significant changes were also taking place in the way societies dealt with work-related accidents and occupational diseases. In the late 19th century, workers' compensation statutes were enacted in Germany under Bismarck. Within 20 years, similar laws were passed in England and by a number of states in the United States.[77] Under these laws, requirements to demonstrate employers' negligence or fault were waived for most occupational injuries and an employee was entitled to compensation based on a percentage of lost wages.

## 4.4. Private Sector Self-Regulation

Insurance, common law, and government intervention are not, of course, the only societal strategies for managing risks. Voluntary, private self-regulation aimed at preventing or reducing potential adverse health effects has always played an important part in societal risk management efforts. In virtually all societies, there have been strong incentives for the private sector to refrain from actions that would recklessly endanger the health of the public. Such incentives range from moral and altruistic norms and values to simple self-interest based on fear of monetary loss, possible civil or criminal litigation, or punitive or restrictive government action.

Private risk management activities are intrinsically less publicly obvious than other risk management strategies. Two of the more visible forms of this strategy are industrial self-regulation and licensure and certification.[78] Both these types of voluntary self-regulation, however, appear to have few clear historical precedents prior to the late 19th and early 20th centuries.

### 4.4.1 Industrial Self-Regulation

Reliance on privately developed standards is particularly widespread at the local level and in areas such as fire safety and the provision of electrical, building, boiler, plumbing, and similar services. Baram[78] points out that such reliance is virtually a necessity given the characteristically limited technical and financial resources available at local governmental levels. He observes that historical experience suggests two essential conditions for the successful use of this type of strategy in risk management: (1) the involved risks and technologies must be well understood, (2) the potential liability must be significant enough to force a responsible industrial approach to risk reduction.

Perhaps the most important institutional mechanism for industrial self-regulation are the standard-setting organizations, professional and technical societies, trade societies, and testing laboratories that set consensus-based standards covering a wide variety of products, materials, systems, services, processes, and practices. Such organizations were, for the most part, founded during the late 19th and early 20th centuries in growing recognition of the hazards associated with increased industrialization. Major standard-setting organizations include the American Society of Mechanical Engineers, founded in 1880; the Underwriters Laboratory, founded in 1894; the National Fire Protection Association, founded in 1896; the American Society for Testing and Materials, founded in 1898; and the American National Standards Institute, founded in 1918.

### 4.4.2 Licensure and Certification

Although over 550 occupations are currently licensed in the United States, licensing and certification appear to have been little used as a form of risk management prior to the turn-of-the-century. Surprisingly, this appears to have been true even for such clearly risky and currently heavily regulated areas as medicine. The control of physicians in the U.S. by licensure first began in the eighteenth century but was abandoned from 1820 to 1850. Our present form of physician licensure did not really begin until the late 1800s.[78] The system of licensure that evolved, however, has often been criticized as serving economic self-interests (e.g., by excluding competition) as much as protecting public health and safety.

## 5. NINE IMPORTANT CHANGES BETWEEN PAST AND PRESENT

It should not be surprising that contemporary ways of thinking about, and coping with, risks are different in many respects from earlier times. In this century, especially in the last few decades, major changes have taken place in the nature of the risks that society faces, as well as in the social and political context for risk analysis and risk management efforts. Nine changes between past and present that we consider among the most important for risk analysis and risk management are discussed below.

### 5.1. Shift in the Nature of Risks

In the United States, the leading causes of death in 1900 were infectious diseases—pneumonia, influenza, and tuberculosis.[79] By 1940, infectious diseases had been displaced by two chronic degenerative diseases of adulthood—heart disease and cancer. Although

there has been no substantial change in the rank of accidents as another leading cause of death, there has been a shift in the types of accidents to which human beings are subject. The rate of fatal accidents in British coal mines, for example, fell from 4 per 1000 workers in the mid-19th century to less than 1 per 1000 workers in recent decades. Similarly, the average annual rate of fatal accidents in British factories fell from 17.5 deaths per 100,000 employees seventy years ago to a recent rate of less than 4.5 deaths.[80] Natural hazards still cause substantial property damage, but in industrialized nations such events account for only a small number of annual fatalities. While these types of accidents have been declining in significance, other types have increased. In 1900, the number of automobile accidents in the United States was, understandably, insignificant; however, 1980 automobile accidents accounted for over 50,000 deaths.[81]

### 5.2. Increase in Average Life Expectancies

A female born in the United States in 1900 could expect to live, on the average, 51 years; a male born in the same year could expect to live 48 years.[79] But a female born in 1975 could expect to live for 75 years, and a male born in the same year could expect to live to 66. Looking further back in history, the average life expectancy was about 33 years in the Middle Ages, 20–30 years during the Roman Empire[82] and 18 years in prehistoric times.[83,82,84] The factors leading to these increases are complex and not entirely understood, but certainly include substantial improvements in nutrition, hygiene, sanitation, working conditions, education, standards of living, and medical services.

### 5.3. Increase in New Risks

There has been an increase in new risks fundamentally different in both character and magnitude from those encountered in the past.

These include nuclear war, nuclear power plant accidents, radioactive waste, exposure to synthetic pesticides and chemicals, supertanker oil spills, chemical plant and storage accidents, recombinant DNA laboratory accidents, ozone depletion due to emissions of fluorocarbons, and acid rain. The magnitude of many of these risks cannot easily be estimated because historical or actuarial data do not exist or are extremely difficult to collect. Moreover, cause–effect relationships are often highly problematic for these risks. Of perhaps greatest importance is that many of these new risks are latent, long-term, involuntary, and irreversible. At least some are conceivably globally catastrophic, and most are derived from science and technology (in contrast to risks from "acts of nature or God").

## 5.4. Increase in Ability of Scientists to Identify and Measure Risks

These improvements include major advances in laboratory tests (e.g., animal bioassays and *in vitro* tests), epidemiological methods, environmental modelling, computer simulations, and engineering risk assessment (e.g., fault trees and event trees). Because of these advances, scientists are now routinely able to detect design faults in extremely complex engineering systems; even weak causal links between hazards and deleterious outcomes; and infinitesimally small amounts (e.g., parts per trillion) of potentially harmful carcinogenic or mutagenic substances.

## 5.5. Increase in the Number of Scientists and Analysts Whose Work is Focused on Health, Safety, and Environmental Risks

In recent years risk analysis has emerged as an identifiable discipline and profession, with its own societies, annual meetings, journals, and practitioners. In the last decade alone,

the risk analysis literature has grown from a handful of articles and books to a formidable collection of material.[85]

## 5.6. Increase in Number of Formal Quantitative Risk Analyses that are Produced and Used

In the past, risk management decisions were based primarily on common sense, ordinary knowledge, trial and error, or nonscientific knowledge and beliefs. In recent years risk management decisions have been increasingly based on highly technical quantitative risk analyses. Increased reliance on such analyses reflect a related trend—a growing societal preference for planning, forecasting, and early warning in contrast to *ad hoc* responses to crisis.

## 5.7. Increase in Role of Federal Government in Assessing and Managing Risks

There have been dramatic increases in: (1) the number of health, safety, and environmental laws, with over 30 major pieces of federal legislation passed within the last two decades; (2) the number of federal agencies charged with managing health, safety, and environmental risks; including the Environmental Protection Agency, the Occupational Safety and Health Administration, the Consumer Product Safety Commission, the National Highway Traffic Safety Administration, and the Nuclear Regulatory Commission; and (3) the number of health, safety, and environmental cases adjudicated by the courts both in the tort-liability system and in judicial review of agency decisions.[86–88] Although attempts have recently been made to reverse the trend toward growth in federal regulatory involvement, several factors have contributed to its continuation, including the increasing health, safety, and environmental consciousness of the nation; a decline in the level of public confidence in business; the emergence of the public interest movement; and the growth of a complex, interdependent, highly technological society.[86]   Additional factors leading toward continued federal regulatory involvement include the following:

♦ An accelerating rate of technological change, resulting in enormous increases in the physical and temporal scale and complexity of risks (for example, approximately 70,000 chemicals are in current use, with perhaps 1000 new chemicals being introduced each year).[79]
♦ An increase in the speed of scientific and technological developments, so that there are shorter and shorter time lags between scientific experimentation, technological development, and entrepreneurial production.
♦ The increasing role of government as a producer of risks through its sponsorship of scientific and technological research and development.
♦ The rising cost of technological risk control and damages—estimated by one research group[89] to be 179–283 billion dollars a year.

## 5.8.  Increase in Participation of Special Interest Groups in Societal Risk Management

Risk analysis and risk management activities have become increasingly politicized, with virtually every major health, safety, and environmental decision subject to intense lobbying by interest groups representing industry, workers, environmentalists, scientific organizations, and other groups.[70]   Not only has there been a substantial increase in the number of such groups and their members, but also substantial growth in their scientific sophistication and modes of operation.   These changes have contributed to at least two others: (1) It has become increasingly necessary for government decision makers to consult representatives from these groups and to make risk analysis information publicly available. (2) The adversarial nature of most contemporary risk debates appears to be causing increasing confusion among the public (due in part to the inscrutability for the layperson of competing technical risk analyses and the widely publicized and often heated debates between scientists).

## 5.9.  Increase in Public Interest, Concern, and Demands for Protection

Despite increases in average life expectancies, reductions in the frequency of catastrophic events, and assurances that "the health of the American people has never been better,"[90] surveys indicate that most Americans believe that life is getting riskier.   A recent Louis Harris[91] poll found that approximately four-fifths of those surveyed agreed that ". . . people are subject to more risk today than they were 20 years ago."   Only 6% thought there was less risk (although it should be noted that the definition of *risk* implied in the survey questions may have been considerably broader than the meaning used in this paper).   Research has suggested that the primary correlates of public concern are not mortality or morbidity rates, but characteristics such as potentially catastrophic effects, lack of familiarity and understanding, involuntariness, scientific uncertainty, lack of personal control by the individuals exposed, risks to future generations, unclear benefits, inequitable distribution of risks and benefits, and potentially irreversible effects.[92,93]   Many of the most salient contemporary risks—nuclear power plant accidents, nuclear waste, airplane crashes, exposure to toxic chemicals, ozone depletion, exposure to low level radiation, recombinant DNA, acid rain—possess precisely these characteristics.   Additional factors contributing to heightened public concern include a better-informed public, the seemingly weekly scientific discovery of previously unknown risks, advances in communication technologies leading to widespread and intensified media coverage of risk problems, rising levels of affluence accompanied by expectations of decreasing risks, rising expectations about the ability of science and technology to control risks, and loss of confidence in the major risk management institutions in contemporary industrialized societies—particularly, business and government.[86]

## 6. IMPLICATIONS FOR THE FUTURE

Making projections about the future is always a risky enterprise, especially in an area as complex as risk analysis and risk management. Nonetheless, a historical perspective suggests certain trends that can reasonably be expected to be important in the foreseeable future.

We expect that public concern about risk will continue to increase, and we expect this to occur in spite of the simultaneous trend toward longer, healthier lives. Part of this is due to the changing nature of the risks faced by modern society, including increases in the number of "mysterious" technological hazards offering prospects of dread, ill-understood, or potentially catastrophic consequences. But the more profound change may be the increasing prevalence of the idea that injuries, deaths, and diseases are not acts of God to be fatalistically accepted, but avoidable events subject to some degree of human control. This change in perspective implies that something *can* be done about most risks. Paralleling this is a change in perspective implying that something *should* be done—derived in part from changing ideas about the rights of individuals to live their lives free of risks imposed on them by others and about the role of government in protecting individuals from such risks.

Improved scientific, technical, and engineering capabilities should lead to steady improvements in our ability to control, reduce, or eliminate risks. The same set of capabilities are also expected to lead, however, to steady increases in the number of identified risks. In the near term, we suspect that improved risk management capabilities will be outstripped by improved risk identification capabilities. Although improved risk management will be welcome, improved abilities to identify and measure risks will not necessarily lead to feelings of greater understanding or control. Indeed, we expect just the opposite. Already, improved science has raised more questions than it has settled about the possible risks of both new and familiar objects, substances, and activities.[94,95] This phenomena might be dubbed the "Hydra effect—for every risk problem that is resolved, two new ones are raised in its place (Baram, personal communication). It is quite likely that the probabilistic and uncertain world created by modern science and technology will seem to many an increasingly risky and uncomfortable place, even in the face of overall improved prospects for a longer, healthier life.

*Acknowledgments* We would like to extend special thanks to Brown Grier, Department of Psychology, Northern Illinois University, for allowing us to draw on his two unpublished conference papers cited in the references. Thanks are also due to Arthur Norberg, The Charles Babbage Institute, University of Minnesota, for his contribution to our section on the control of natural disasters, and to Michael Baram, Ward Edwards, Baruch Fischhoff, Patrick Johnson, Ralph Keeney, Howard Kunreuther, Lester Lave, Joshua Menkes, Jiri Nehnevasja, Paul Slovic, Jack Sommer, Detlof von Winterfeldt, and Chris Whipple for their helpful comments on earlier drafts.

The views expressed in this paper are exclusively those of the authors and do not necessarily represent the views of the National Science Foundation.

### NOTES

[a] Reprinted by G. E. Stechert and Company, New York, 1931.

[b] See MacPherson vs. Buick, 217 New York 382, 111 N.E. 1050, 1916.

### REFERENCES

1. L. Oppenheim, *Ancient Mesopotamia* (University of Chicago Press, Chicago, 1977).
2. B. Grier, *One Thousand Years of Mathematical Psychology* (Paper presented at Society for Mathematical Psychology Convention, Madison, Wisconsin, 1980).
3. B. Grier, *The Early History of the Theory and Man-*

*agement of Risk* (Paper presented at the Judgment and Decision Making Group Meeting, Philadelphia, Pennsylvania, 1981).

4. K. Thomas, *Religion and the Decline of Magic* (Weidenfeld and Nicolson, London, 1971).

5. V. Turner, *Ndembu Divination* (University of Manchester Press, Manchester, 1961).

6. O. Ore, "Pascal and the Invention of Probability Theory," *American Mathematical Monthly* **67**, 409–19 (1960).

7. A. Lightman, "Weighing the Odds," *Science '83* **4**, 21–22 (1983).

8. P. Laplace, *Theorie Analytique de Probabilities* (Paris, 1812).

9. F. N. David, *Games, Gods, and Gambling* (Griffin and Co., London, 1962).

10. I. Hacking, *The Emergence of Probability* (Cambridge University Press, Cambridge, 1975).

11. O. Sheynin, "On the Prehistory of the Theory of Probability," *Archive for the Hisotry of Exact Science* **12**, 97–141 (1974).

12. T. Bottomore and M. Rubel, (Eds.) *Karl Marx: Selected Writings* (McGraw-Hill, New York, 1956).

13. R. K. Merton, *Science, Technology, and Society in Seventeenth Century England* (Saint Catherine Press, Bruges, Belgium, 1938).

14. L. White, "The Historical Roots of Our Ecological Crisis," *Science* **155**, 1203–7 (1967).

15. J. Ben-David and T. Sullivan, "Sociology of Science," in A. Inkeles, J. Coleman, and N. Smelser (eds.), *Annual Review of Sociology* (Annual Reviews, Palo Alto, 1975), pp. 203–22.

16. J. Needham, *Science and Civilization in China* (Cambridge University Press, Cambridge, 1956).

17. J. Graunt, *Natural and Political Observations Made Upon the Bills of Mortality* (1662).

18. E. Halley, "An Estimate of the Degrees of Mortality of Mankind, Drawn from Curious Tables of the Births and Funerals at the City of Breslau, with an Attempt to Ascertain the Price of Annuities Upon Lives," *Philosophical Transactions of the Royal Society of London* **17**, 596–610 (1693).

19. I. Campbell, *Accident Statistics and Significance* Occasional Paper No. 34. (Safety Accident Compensation Commission, Wellington, New Zealand, 1980).

20. P. B. Hutt, "The Basis and Purpose of Government Regulation of Adulteration and Misbranding of Food," *Food and Drug Cosmetic Law Journal* **33**, (10) (1978), pp. 2–74.

21. M. Shimkins, *Contrary to Nature* NIT 79-720, (Dept of Health and Human Services, Washington, D.C., 1979).

22. M. Shimkins, *Some Classics of Experimental Oncology, 1775–1965*, NIH 80-2150, (U.S. Dept of Health and Human Services, Washington, D.C., Oct. 1980).

23. J. Nriagu, *Lead and Lead Poisoning* (Wiley Interscience, New York, 1983).

24. S. Gilfillan, "Roman Culture and Dysgenic Lead Poisoning," *Mankind Quarterly* **5**, 3–20 (January–March, 1965).

25. J. Hughs, *Ecology in Ancient Civilizations* (University of New Mexico Press, Albuquerque, 1975).

26. G. Agricola, (1556), *De re metallica*. 1st ed. Reprint, translated by H. C. Hoover and L. C. Hoover, (Dover Publications, New York, 1950).

27. J. Evelyn (1661), "Fumifugium or the Inconvenience of the Aer and Smoake of London Dissipated," Reprinted in: *The Smoke of London* (Maxwell Reprint Co., Fairview Park, Elmsford, New York, 1969).

28. B. Ramazzini (1700), *De Morbia artificium*, (Chapter XX Capponi, Italy). "Ueber biasentumoren bei Fuchsinarbeitern,"*Arch. Clin. Chur.* **50**, 588. Also see B. Ramazzini, *Diseases of Workers*, 1713 Edition,Wilner Wright (Trans.), Classics of Medicine Library, (Birmingham, Alabama, 1940).

29. B. MacMahon and P. Cole, "Endocrinology and Epidemiology of Breast Cancer," *Cancer* **24**, 1146–51 (1969).

30. B. M. Sherman and S. G. Korenman, "Inadequate Corpus Luteum Function: A Pathophysiological Interpretation of Human Breast Cancer Epidemiology," *Cancer* **33**, 1306–12 (1974).

31. J. Hill, *Cautious Against the Immoderate Use of Snuff* (Baldwin and Jackson, London, 1781).

32. P. Pott (1775), *Cancer Scroti: The Chirurgical Works of Percival Pott* (Clark and Collins, London, 1975).

33. J. A. Ayrton-Paris, *Pharmaecologia* (1822).

34. J. Hutchinson, "Arsenic cancer," *British Medical J.* **2**, 1280 (1887).

35. E. Chadwick, (1842), *Report on the Sanitary Condition of the Labouring Population of Gt. Britain* (ed. with introduction by M. W. Flinn), (Edinburgh University Press, Edinburgh, 1965).

36. J. Snow, *On the Mode of Communication of Cholera* (Churchill, London, 1855).

37. P. G. Unna, *Die Histopathologie der Hautkrankheiten* (A. Hirschwald, Berlin, 1894).

38. W. Dubreuilh, "Des Hyperkeratoses Circonscrites," in *Ann. Dermatol. Syphilig.*, 3rd Series, 1158–1204 (1896).

39. L. Rehn, "Blasengeschwulste bei Fuchsin-Arbeitern," *Arch. Klin. Chir.* 1895, **50**, 588–800 (1895).

40. M. Baram, "Technology Assessment and Social Control," *Science* **180**, 465–73 (1973).

41. W. C. Clark, "Witches, Floods, and Wonder Drugs: Historical Perspectives on Risk Management," in R. Schwing and W. Albers (eds.), *Societal Risk Assessment* (Plenum, New York, 1980), pp. 287–313.

42. A. White (1895). *A History of the Warfare of Science with Theology in Christendom* (George Brazillier, New York, 1955).

43. G. Rosen, "The Medical Aspects of the Controversy Over Factory Conditions in New England, 1840–1850," *Bulletin of the History of Medicine* **XV**, 483–97 (1944).

44. I. Pfeffer and D. Klock, *Perspectives on Insurance* (Prentice Hall, Englewood Cliffs, 1974).

45. C. H. Johns, *Babylonian and Assyrian Laws Contracts and Letters* (Charles Scribner's Sons, New York, 1904).

46. W. Hammer, *Product Safety Management and Engineering* (Prentice-Hall, Englewood Cliffs, New Jersey, 1980).

47. P. Handler, "Some Comments on Risk" in *The National Research Council in 1979: Current Issues and Studies* (National Academy of Sciences, Washington, D.C., 1979).

48. W. Clark, *Flood* (Time-Life Books, Alexandria, Virginia, 1982).

49. H. Kunreuther, *Recovery from Natural Disasters: Insurance or Federal Aid?* (American Enterprise Institute, Washington, D.C., 1973).

50. K. Helleiner, "The Population of Europe from the Black Death to the Eve of the Vital Revolution," in E. E. Rich and C. H. Wilson, (eds.), *The Cambridge Economic History of Europe, Vol. 4, the Economy of Expanding Europe in the Sixteenth and Seventeenth Centuries* (Cambridge, 1967).

51. J. Nohl, *The Black Death* (Ballantine Books, Cambridge, 1960).

52. Philip Ziegler, *The Black Death* (Penguin Books, Middlesex, England, 1969).

53. C. Winslow, *The Evolution and Significance of the Modern Public Health Campaign* (Yale University Press, New Haven, 1923).

54. H. Zinssler, *Rats, Lice, and History* (Atlantic Monthly Press, New York, 1935).

55. P. Hutt, "Legal Considerations in Risk Assessment Under Federal Regulatory Statutes," in J. Rodricks and R. Tardiff (eds.), *Assessment and Management of Chemical Risks* (American Chemical Society, Washington, D.C. 1984), pp. 84–95.

56. W. Te Brake, "Air Pollution and Fuel Crisis in Pre-Industrial London, 1250–1650," *Technology and Culture* **16**, 337–59 (July 1975).

57. J. Hanlon, *Principles of Public Health Administration* (C. V. Mosby Company, St. Louis, 1969).

58. G. Rosen, *A History of Public Health* (M. D. Publications, New York, 1958).

59. H. F. Gray, "Sewage in Ancient and Medieval Times," *Sewage Works Journal* **12**, 939–46 (1940).

60. O. Bettman, *The Good Old Days—They Were Terrible* (Random House, New York, 1974).

61. L. A. Dublin, A. J. Lotka, and M. Spiegelman, *Length of Life: A Study of the Life Table* (The Ronald Press Company, New York, 1949).

62. J. Ford, *Slums and Housing, Vol. 1* (Harvard University Press, Cambridge, Massachusetts, 1936).

63. D. Kidd, "The History and Definition of Water Pollution," *Bulletin of Science, Technology, and Society* **3**, 121–26 (1983).

64. M. Douglas, *Purity and Danger* (Routledge & Kegan Paul, London, 1966).

65. M. Douglas and A. Wildavsky, *Risk and Culture: An Essay on the Selection of Technological and Environmental Dangers* (University of California Press, Berkeley and Los Angeles, 1982).

66. H. Webster, *History of Civilization, Ancient and Medieval* (D. C. Heath and Company, Boston, 1947).

67. G. Beyer, *Housing and Society* (The MacMillan Company, New York, 1968).

68. T. J. Wertenbaker, *The Old South: The Founding of American Civilization* (Charles Scribner's Sons, New York, 1942).

69. T. G. Burke, "Bursting Boilers and the Federal Power," *Technology Culture* **7**, 1–23 (1965).

70. W. Edwards and D. von Winterfeldt, "Public Disputes about Risky Technologies: Stakeholders and Arenas," in V. Covello, J. Menkes, and J. Mumpower, (eds.), *Risk Evaluation and Management* (Plenum Press, New York, 1984).

71. A. Cohen, *Overview and Definition of Risk* Research Paper. (United Kingdom Health and Safety Executive, London, 1983).

72. F. Engels (1845), *The Condition of the Working Class in England* (Blackwell, Oxford) Translated and edited by W. O. Henderson and W. H. Chaloner (Macmillan Co., New York, 1958).

73. British Parliamentary Papers (1816–17), *Report of the Minutes of Evidence on the State of Children Employed in the Manufactories of the United Kingdom, Together with a Report of the Employment of Boys in Sweeping Chimneys with Minutes of Evidence and Appendix* House of Commons, (Reprinted by Irish University Press, Shannon, Ireland, 1968).

74. P. A. Samuelson, *Economics* 8th Edition. (McGraw-Hill Book Company, New York, 1970).

75. British Parliamentary Papers (1842), *First Report of the Commissioners-Mines*; *Children's Employment Commission* (Reprinted by Irish University Press, Shannon, Ireland, 1968).

76. R. Owen, *A New View of Society and Other Writings* (J. M. Dent and Sons, London/Toronto; E. P. Dutton and Co., New York, 1927).

77. H. Weiss, "Employers' Liability and Workmen's Compensation," in J. Commons, (eds.), *History of Labor in the United States 1896–1932* (Macmillan, New York, 1935).

78. M. Baram, *Alternatives to Regulation: Managing Risks to Health, Safety and the Environment* (Lexington Books, Lexington, Massachusetts, 1982).

79. National Academy of Sciences, *Science and Technology: A Five Year Outlook* (W. H. Freeman, San Francisco, 1979).

80. L. Rubens, *Safety and Health at Work* Report of the Committee, (HMSO, London, 1972).

81. J. Claybrook, "Motor Vehicle Occupant Restraint Policy, in V. Covello, W. G. Flamm, J. Rodricks, and R. Tardiff, (eds.), *The Analysis of Actual Versus Perceived Risks* (Plenum, New York, 1983), pp. 21–47.

82. J. Durand, "Mortality Estimates from Roman Tombstone Inscriptions," *American Journal of Sociology* (January, 1960).

83. M. Spiegelman, *Health Progress in the United States: A Survey of Recent Trends in Longevity* (American Enterprise Association, Inc., New York, 1950).

84. A. Atkisson, W. Petak, and J. Fuller, *An Examination of Premature Death As a Target of U.S. Occupational Health Policies* Report No. 81–82/54. (Public Policy Institute, University of Southern California, Los Angeles, 1981).

85. V. Covello and M. Abernathy, "Risk Analysis and Technological Hazards: A Policy-Related Bibliography," in P. Ricci, L. Sagan, and C. Whipple, (eds.), *Technological Risk Assessment* (Martinus Nijhoff Publishers, Boston, 1984), pp. 283–363.

86. National Academy of Sciences, *Risk and Decision Making: Perspectives and Research* (National Academy Press, Washington, D.C., 1982).

87. A. Oleinick, L. Disney, and K. East, "Institutional Mechanisms for Converting Sporadic Agency Decisions into Systematic Risk Management Strategies: OSHA, the Supreme Court and the Court of Appeals for the District of Columbia," in V. Covello, J. Menkes, and J. Mumpower, (eds.), *Risk Evaluation and Management* (Plenum, New York, 1984).

88. V. Covello and J. Menkes, "Issues in Risk Analysis," in C. Hohenemser and J. Kasperson, (eds.), *Risk in the Technological Society* (Westview Press, Boulder, 1982), pp. 287–301.

89. C. Hohenemser, R. Kasperson, and R. Kates, "Casual Structure: A Framework for Policy Formulation" in C. Hohenemser and J. Kasperson: (eds.), *Risk in the Technological Society* (Westview Press, Boulder, 1982), pp. 109–39.

90. U.S. Surgeon General, *Healthy People* (U.S. Government Printing Office, Washington, D.C., 1979).

91. Louis Harris and Associates, *Risk in a Complex Society*, Public opinion survey conducted for Marsh and McLennan, Inc., (Marsh and McLennan, New York, 1980).

92. P. Slovic, B. Fischhoff, and S. Lichtenstein, "Facts and Fears: Understanding Perceived Risk," in R. Schwing and W. Albers, Jr., (eds.), *Societal Risk Assessment: How Safe is Safe Enough?* (Plenum Press, New York, 1980), pp. 181–216.

93. V. Covello, "Social and Behavioral Research on Risk: Uses in Risk Management Decisionmaking," *Environmental International* 4 (December, 1984).

94. B. Ames, "Dietary Carcinogens and Anticarcinogens," *Science* 21, 1256–63 (1983).

95. S. Epstein and J. Swartz, "Letter to Science on Cancer and Diet," *Science* 18 May, 660–66 (1984).

# SOCIAL BENEFIT VERSUS
# TECHNOLOGICAL RISK

*Chauncey Starr*

The evaluation of technical approaches to solving societal problems customarily involves consideration of the relationship between potential technical performance and the required investment of societal resources. Although such performance-versus-cost relationships are clearly useful for choosing between alternative solutions, they do not by themselves determine how much technology a society can justifiably purchase. This latter determination requires, additionally, knowledge of the relationship between social benefit and justified social cost. The two relationships may then be used jointly to determine the optimum investment of societal resources in a technological approach to a social need.

Technological analyses for disclosing the relationship between expected performance and monetary costs are a traditional part of all engineering planning and design. The inclusion in such studies of *all* societal costs (indirect as well as direct) is less customary, and obviously makes the analysis more difficult and less definitive. Analyses of social value as a function of technical performance are not only uncommon but are rarely quantitative. Yet we know that implicit in every nonarbitrary national decision on the use of technology is a trade-off of societal benefits and societal costs.

In this article I offer an approach for establishing a quantitative measure of benefit relative to cost for an important element in our spectrum of social values—specifically, for accidental deaths arising from technological developments in public use. The analysis is based on two assumptions. The first is that historical national accident records are adequate for revealing consistent patterns of fatalities in the public use of technology. (That this may not always be so is evidenced by the paucity of data relating to the effects of environmental pollution.) The second assumption is that such historically revealed social preferences and costs are sufficiently enduring to permit their use for predictive purposes.

In the absence of economic or sociological theory which might give better results, this empirical approach provides some interesting insights into accepted social values relative to personal risk. Because this methodology is based on historical data, it does not serve to distinguish what is "best" for society from what is "traditionally acceptable."

Reprinted from *Science* Volume 165 (1969), pp. 1232–38. Copyright 1969 by the AAAS. Used with the permission of the author and the AAAS.

# MAXIMUM BENEFIT
# AT MINIMUM COST

The broad societal benefits of advances in technology exceed the associated costs sufficiently to make technological growth inexorable. Shef's socioeconomic study (1) has indicated that technological growth has been generally exponential in this century, doubling every 20 years in nations having advanced technology. Such technological growth has apparently stimulated a parallel growth in socioeconomic benefits and a slower associated growth in social costs.

The conventional socioeconomic benefits—health, education, income—are presumably indicative of an improvement in the "quality of life." The cost of this socioeconomic progress shows up in all the negative indicators of our society—urban and environmental problems, technological unemployment, poor physical and mental health, and so on. If we understood quantitatively the causal relationships between specific technological developments and societal values, both positive and negative, we might deliberately guide and regulate technological developments so as to achieve maximum social benefit at minimum social cost. Unfortunately, we have not as yet developed such a predictive system analysis. As a result, our society historically has arrived at acceptable balances of technological benefit and social cost empirically—by trial, error, and subsequent corrective steps.

In advanced societies today, this historical empirical approach creates an increasingly critical situation, for two basic reasons. The first is the well-known difficulty in changing a technical subsystem of our society once it has been woven into the economic, political, and cultural structures. For example, many of our environmental-pollution problems have known engineering solutions, but the problems of economic readjustment, political jurisdiction, and social behavior loom very large. It will take many decades to put into effect the technical solutions we know today. To give a specific illustration, the pollution of our water resources could be completely avoided by means of engineering systems now available, but public interest in making the economic and political adjustments needed for applying these techniques is very limited. It has been facetiously suggested that, as a means of motivating the public, every community and industry should be required to place its water intake downstream from its outfall.

In order to minimize these difficulties, it would be desirable to try out new developments in the smallest social groups that would permit adequate assessment. This is a common practice in market-testing a new product or in field-testing a new drug. In both these cases, however, the experiment is completely under the control of a single company or agency, and the test information can be fed back to the controlling group in a time that is short relative to the anticipated commercial lifetime of the product. This makes it possible to achieve essentially optimum use of the product in an acceptably short time. Unfortunately, this is rarely the case with new technologies. Engineering developments involving new technology are likely to appear in many places simultaneously and to become deeply integrated into the systems of our society before their impact is evident or measurable.

This brings us to the second reason for the increasing severity of the problem of obtaining maximum benefits at minimum costs. It has often been stated that the time required from the conception of a technical idea to its first application in society has been drastically shortened by modern engineering organization and management. In fact, the history of technology does not support this conclusion. The bulk of the evidence indicates that the time from conception to first application (or demonstration) has been roughly unchanged by modern management, and depends chiefly on the complexity of the development.

However, what *has* been reduced substantially in the past century is the time from first use to widespread integration into our social system. The techniques for *societal diffusion* of a new technology and its subsequent exploitation are now highly developed. Our ability to organize resources of money, men, and materials to focus on new technological programs has reduced the diffusion-exploitation time by roughly an order of magnitude in the past century.

Thus, we now face a general situation in which widespread use of a new technological development may occur before its social impact can be properly assessed, and before any empirical adjustment of the benefit-versus-cost relation is obviously indicated.

It has been clear for some time that predictive technological assessments are a pressing societal need. However, even if such assessments become available, obtaining maximum social benefit at minimum cost also requires the establishment of a relative value system for the basic parameters in our objective of improved "quality of life." The empirical approach implicitly involved an intuitive societal balancing of such values. A predictive analytical approach will require an explicit scale of relative social values.

For example, if technological assessment of a new development predicts an increased per capita annual income of $x$ percent but also predicts an associated accident probability of $y$ fatalities annually per million population, then how are these to be compared in their effect on the "quality of life"? Because the penalties or risks to the public arising from a new development can be reduced by applying constraints, there will usually be a functional relationship (or trade-off) between utility and risk, the $x$ and $y$ of our example.

There are many historical illustrations of such trade-off relationships that were empirically determined. For example, automobile and airplane safety have been continuously weighed by society against economic costs and operating performance. In these and other cases, the real trade-off process is actually one of dynamic adjustment, with the behavior of many portions of our social systems out of phase, due to the many separate "time constants" involved. Readily available historical data on accidents and health, for a variety of public activities, provide an enticing stepping-stone to quantitative evaluation of this particular type of social cost. The social benefits arising from some of these activities can be roughly determined. On the assumption that in such historical situations a socially acceptable and essentially optimum trade-off of values has been achieved, we could say that any generalizations developed might then be used for predictive purposes. This approach could give a rough answer to the seemingly simple question "How safe is safe enough?"

The pertinence of this question to all of us, and particularly to governmental regulatory agencies, is obvious. Hopefully, a functional answer might provide a basis for establishing performance "design objectives" for the safety of the public.

## VOLUNTARY AND INVOLUNTARY ACTIVITIES

Societal activities fall into two general categories—those in which the individual participates on a "voluntary" basis and those in which the participation is "involuntary," imposed by the society in which the individual lives. The process of empirical optimization of benefits and costs is fundamentally similar in the two cases—namely, a reversible exploration of available options—but the time required for empirical adjustments (the time constants of the system) and the criteria for optimization are quite different in the two situations.

In the case of "voluntary" activities, the individual uses his own value system to evaluate his experiences. Although his eventual trade-off may not be consciously or analytically determined, or based upon objective knowledge, it nevertheless is likely to represent, for that individual, a crude optimization appropriate to his value system. For example, an urban dweller may move to the suburbs because of a lower crime rate and better, schools, at the cost of more time spent traveling on highways, and a higher probability of accidents. If, subsequently, the traffic density increases, he may decide that the penalties are too great and move back to the city. Such an individual optimization process can be comparatively rapid (because the feedback of experience to the individual is rapid), so the statistical pattern for a large social group may be an important "real-time" indicator of societal trade-offs and values.

"Involuntary" activities differ in that the criteria and options are determined not by the individuals affected but by a controlling body. Such control may be in the hands of a government agency, a political entity, a leadership group, an assembly of authorities or "opinion-makers," or a combination of such bodies. Because of the complexity of large societies, only the control group is likely to be fully aware of all the criteria and options involved in their decision process. Further, the time required for feedback of the experience that results from the controlling decisions is likely to be very long. The feedback of cumulative individual experiences into societal communication channels (usually political or economic) is a slow process, as is the process of altering the planning of the control group. We have many examples of such "involuntary" activities, war being perhaps the most extreme case of the operational separation of the decision-making group from those most affected. Thus, the real-time pattern of societal trade-offs on "involuntary" activities must be considered in terms of the particular dynamics of approach to an acceptable balance of social values and costs. The historical trends in such activities may therefore be more significant indicators of social acceptability than the existent trade-offs are.

In examining the historical benefit-risk re-lationships for "involuntary" activities, it is important to recognize the perturbing role of public psychological acceptance of risk arising from the influence of authorities or dogma. Because in this situation the decision-making is separated from the affected individual, society has generally clothed many of its controlling groups in an almost impenetrable mantle of authority and of imputed wisdom. The public generally assumes that the decision-making process is based on a rational analysis of social benefit and social risk. While it often is, we have all seen after-the-fact examples of irrationality. It is important to omit such "witch-doctor" situations in selecting examples of optimized "involuntary" activities, because in fact these situations typify only the initial stages of exploration of options.

## QUANTITATIVE CORRELATIONS

With this description of the problem, and the associated caveats, we are in a position to discuss the quantitative correlations. For the sake of simplicity in this initial study, I have taken as a measure of the physical risk to the individual the fatalities (deaths) associated with each activity. Although it might be useful to include all injuries (which are 100 to 1000 times as numerous as deaths), the difficulty in obtaining data and the unequal significance of varying disabilities would introduce inconvenient complexity for this study. So the risk measure used here is the statistical probability of fatalities per hour of exposure of the individual to the activity considered.

The hour-of-exposure unit was chosen because it was deemed more closely related to the individual's intuitive process in choosing an activity than a year of exposure would be, and give substantially similar results. Another possible alternative, the risk per activity, involved a comparison of too many dissimilar units of measure; thus, in comparing the risk for various modes of transportation, one could use risk per hour, per mile, or per trip. As this study was directed toward exploring a methodology for determining social acceptance of risk, rather than the safest mode of transportation for a particular trip, the simplest common unit—that of risk per exposure hour—was chosen.

The social benefit derived from each activity was converted into a dollar equivalent, as a measure of integrated value to the individual. This is perhaps the most uncertain aspect of the correlations because it reduced the "quality-of-life" benefits of an activity to an overly simplistic measure. Nevertheless, the correlations seemed useful, and no better measure was available. In the case of the "voluntary" activities, the amount of money spent on the activity by the average involved individual was assumed proportional to its benefit to him. In the case of the "involuntary" activities, the contribution of the activity to the individual's annual income (or the equivalent) was assumed proportional to its benefit. This assumption of roughly constant relationship between benefits and monies, for each class of activities, is clearly an approximation. However, because we are dealing in orders of magnitude, the distortions likely to be introduced by this approximation are relatively small.

In the case of transportation modes, the benefits were equated with the sum of the monetary cost to the passenger and the value of the time saved by that particular mode relative to a slower, competitive mode. Thus, airplanes were compared with automobiles, and automobiles were compared with public transportation or walking. Benefits of public transportation were equated with their cost. In all cases, the benefits were assessed on an annual dollar basis because this seemed to be most relevant to the individual's intuitive process. For example, most luxury sports require an investment and upkeep only partially dependent upon usage. The associated risks,

of course, exist only during the hours of exposure.

Probably the use of electricity provides the best example of the analysis of an "involuntary" activity. In this case the fatalities include those arising from electrocution, electrically caused fires, the operation of power plants, and the mining of the required fossil fuel. The benefits were estimated from a United Nations study of the relationship between energy consumption and national income; the energy fraction associated with electric power was used. The contributions of the home use of electric power to our "quality of life"—more subtle than the contributions of electricity in industry—are omitted. The availability of refrigeration has certainly improved our national health and the quality of dining. The electric light has certainly provided great flexibility in patterns of living, and television is a positive element. Perhaps, however, the gross-income measure used in the study is sufficient for present purposes.

Information on acceptance of "voluntary" risk by individuals as a function of income benefits is not easily available, although we know that such a relationship must exist. Of particular interest, therefore, is the special case of miners exposed to high occupational risks. In Fig. 1, the accident rate and the severity rate of mining injuries are plotted against the hourly wage (2, 3). The acceptance of individual risk is an exponential function of the wage, and can be roughly approximated by a third-power relationship in this range. If this relationship has validity, it may mean that several "quality of life" parameters (perhaps health, living essentials, and recreation) are each partly influenced by any increase in available personal resources, and that thus the increased acceptance of risk is exponentially motivated. The extent to which this relationship is "voluntary" for the miners is not obvious, but the subject is interesting nevertheless.

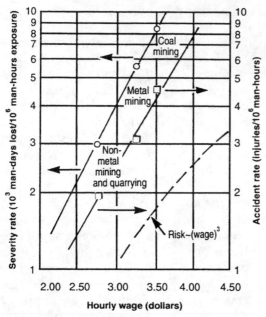

**FIGURE 1.** *Mining accident rates plotted relative to incentive.*

## RISK COMPARISONS

The results for the societal activities studied, both "voluntary" and "involuntary," are assembled in Fig. 2 (For details of the risk-benefit analysis, see the appendix.) Also shown in Fig 2 is the third-power relationship between risk and benefit characteristic of Fig. 1 For comparison, the average risk of death from accident and from disease is shown. Because the average number of fatalities from accidents is only about one-tenth the number from disease, their inclusion is not significant.

Several major features of the benefit-risk relations are apparent, the most obvious being the difference by several orders of magnitude in society's willingness to accept "voluntary" and "involuntary" risk. As one would expect, we are loathe to let others do unto us what we happily do to ourselves.

The rate of death from disease appears to play, psychologically, a yardstick role in de-

termining the acceptability of risk on a voluntary basis. The risk of death in most sporting activities is surprisingly close to the risk of death from disease—almost as though, in sports, the individual's subconscious computer adjusted his courage and made him take risks associated with a fatality level equaling but not exceeding the statistical mortality due to involuntary exposure to disease. Perhaps this defines the demarcation between boldness and foolhardiness.

In Fig. 2 the statistic for the Vietnam war is shown because it raises an interesting point. It is only slightly above the average for risk of death from disease. Assuming that some long-range societal benefit was anticipated from this war, we find that the related risk, as seen by society as a whole, is not substantially different from the average nonmilitary risk from disease. However, for individuals in the military service age group (age 20 to 30), the risk of death in Vietnam is about ten times the normal mortality rate (death from accidents or disease). Hence the population as a whole and those directly exposed see this matter from different perspectives. The disease risk pertinent to the average age of the involved group probably would provide the basis for a more meaningful comparison than the risk pertinent to the national average age does. Use of the figure for the single group would complicate these simple comparisons, but that figure might be more significant as a yardstick.

The risks associated with general aviation, commercial aviation, and travel by motor vehicle deserve special comment. The latter originated as a "voluntary" sport, but in the past half-century the motor vehicle has become an essential utility. General aviation is still a highly voluntary activity. Commercial aviation is partly voluntary and partly essential and, additionally, is subject to government administration as a transportation utility.

Travel by motor vehicle has now reached a benefit-risk balance, as shown in Fig. 3. It is interesting to note that the present risk level is only slightly below the basic level of risk

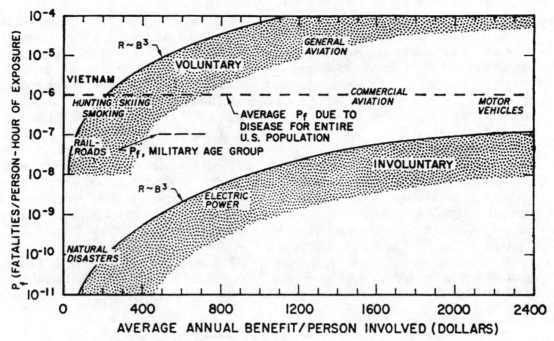

**FIGURE 2.** Risk (R) plotted relative to benefit (B) for various kinds of voluntary and involuntary exposure.

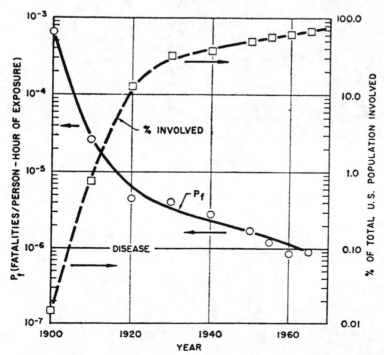

**FIGURE 3.** Risk and participation trends for motor vehicles.

from disease. In view of the high percentage of the population involved, this probably represents a true societal judgment on the acceptability of risk in relation to benefit. It also appears from Fig. 3 that future reductions in the risk level will be slow in coming, even if the historical trend of improvement can be maintained (4).

Commercial aviation has barely approached a risk level comparable to that set by disease. The trend is similar to that for motor vehicles, as shown in Fig. 4. However, the percentage of the population participating is now only 1/20 that for motor vehicles. Increased public participation in commercial aviation will undoubtedly increase the pressure to reduce the risk, because, for the general population, the benefits are much less than those associated with motor vehicles. Commercial aviation has not yet reached the point of optimum benefit-risk trade-off (5).

For general aviation the trends are similar, as shown in Fig. 5. Here the risk levels are so high (20 times the risk from disease) that this activity must properly be considered to be in the category of adventuresome sport. However, the rate of risk is decreasing so rapidly that eventually the risk for general aviation may be little higher than that for commercial aviation. Since the percentage of the population involved is very small, it appears that the present average risk levels are acceptable to only a limited group (6).

The similarity of the trends in Figs. 3–5 may be the basis for another hypothesis, as follows: the acceptable risk is inversely related

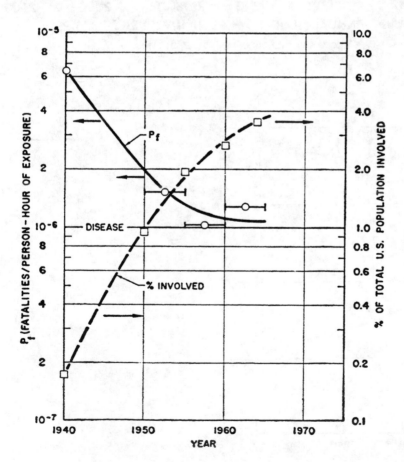

**FIGURE 4.** Risk and participation trends for certified air carriers.

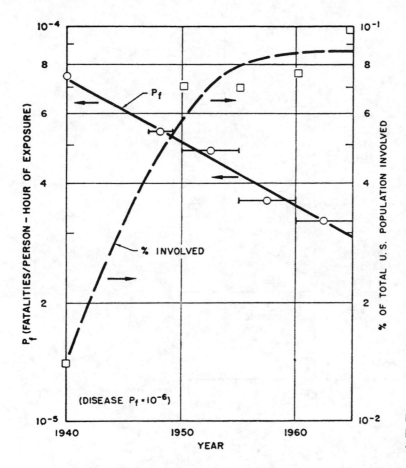

**FIGURE 5.** Risk and participation trends for general aviation.

to the number of people participating in an activity.

The product of the risk and the percentage of the population involved in each of the activities of Figs. 3–5 is plotted in Fig. 6. This graph represents the historical trend of total fatalities per hour of exposure of the population involved (7). The leveling off of motor-vehicle risk at about 100 fatalities per hour of exposure of the participating population may be significant. Because most of the U.S. population is involved, this rate of fatalities may have sufficient public visibility to set a level of social acceptability. It is interesting, and disconcerting, to note that the trend of fatalities in aviation, both commercial and general, is uniformly upward.

## PUBLIC AWARENESS

Finally, I attempted to relate these risk data to a crude measure of public awareness of the associated social benefits (see Fig. 7). The "benefit awareness" was arbitrarily defined as the product of the relative level of advertising, the square of the percentage of population involved in the activity, and the relative usefulness (or importance) of the activity to the individual (8). Perhaps these assumptions are too crude, but Fig. 7 does support the reasonable position that advertising the benefits of an activity increases public acceptance of a greater level of risk. This, of course could subtly produce a fictitious benefit-risk ratio— as may be the case for smoking.

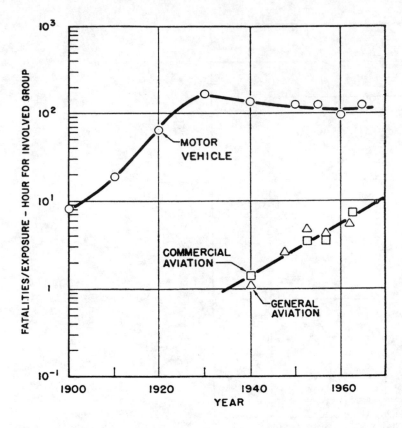

**FIGURE 6.** Group risk plotted relative to year.

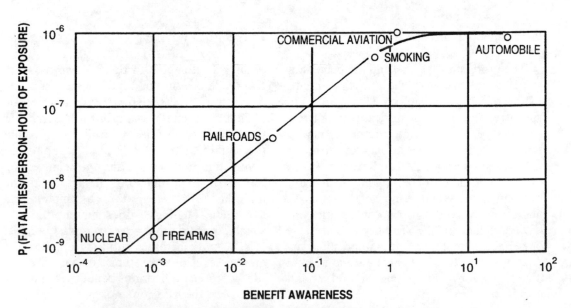

**FIGURE 7.** Accepted risk plotted relative to benefit awareness (see text).

## ATOMIC POWER PLANT SAFETY

I recognize the uncertainty inherent in the quantitative approach discussed here, but the trends and magnitudes may nevertheless be of sufficient validity to warrant their use in determining national "design objectives" for technological activities. How would this be done?

Let us consider as an example the introduction of nuclear power plants as a principal source of electric power. This is an especially good example because the technology has been primarily nurtured, guided, and regulated by the government, with industry undertaking the engineering development and the diffusion into public use. The government specifically maintains responsibility for public safety. Further, the engineering of nuclear plants permits continuous reduction of the probability of accidents, at a substantial increase in cost. Thus, the trade-off of utility and potential risk can be made quantitative.

Moreover, in the case of the nuclear power plant the historical empirical approach to achieving an optimum benefit-risk trade-off is not pragmatically feasible. All such plants are now so safe that it may be 30 years or longer before meaningful risk experience will be accumulated. By that time, many plants of varied design will be in existence, and the empirical accident data may not be applicable to those being built. So a very real need exists now to establish "design objectives" on a predictive-performance basis.

Let us first arbitrarily assume that nuclear power plants should be as safe as coal-burning plants, so as not to increase public risk. Figure 2 indicates that the total risk to society from electric power is about $2 \times 10^{-9}$ fatality per person per hour of exposure. Fossil fuel plants contribute about $\frac{1}{5}$ of this risk, or about 4 deaths per million population per year. In a modern society, a million people may require a million kilowatts of power, and this is about the size of most new power stations.

So, we now have a target risk limit of 4 deaths per year per million-kilowatt power station (9).

Technical studies of the consequences of hypothetical extreme (and unlikely) nuclear power plant catastrophes, which would disperse radioactivity into populated areas, have indicated that about 10 lethal cancers per million population might result (10). On this basis, we calculate that such a power plant might statistically have one such accident every 3 years and still meet the risk limit set. However, such a catastrophe would completely destroy a major portion of the nuclear section of the plant and either require complete dismantling or years of costly reconstruction. Because power companies expect plants to last about 30 years, the economic consequences of a catastrophe every few years would be completely unacceptable. In fact, the operating companies would not accept one such failure, on a statistical basis, during the normal lifetime of the plant.

It is likely that, in order to meet the economic performance requirements of the power companies, a catastrophe rate of less than 1 in about 100 plant-years would be needed. This would be a public risk of 10 deaths per 100 plant-years, or 0.1 death per year per million population. So the economic investment criteria of the nuclear plant user—the power company—would probably set a risk level 1/200 the present socially accepted risk associated with electric power, or 1/40 the present risk associated with coal-burning plants.

An obvious design question is this: Can a nuclear power plant be engineered with a predicted performance of less than 1 catastrophic failure in 100 plant-years of operation? I believe the answer is yes, but that is a subject for a different occasion. The principal point is that the issue of public safety can be focused on a tangible, quantitative, engineering design objective.

This example reveals a public safety consideration which may apply to many other activities: The economic requirement for the protection of major capital investments may often be a more demanding safety constraint than social acceptability.

## CONCLUSION

The application of this approach to other areas of public responsibility is self-evident. It provides a useful methodology for answering the question "How safe is safe enough?" Further, although this study is only exploratory, it reveals several interesting points. (i) The indications are that the public is willing to accept "voluntary" risks roughly 1000 times greater than "involuntary" risks. (ii) The statistical risk of death from disease appears to be a psychological yardstick for establishing the level of acceptability of other risks. (iii) The acceptability of risk appears to be crudely proportional to the third power of the benefits (real or imagined). (iv) The social acceptance of risk is directly influenced by public awareness of the benefits of an activity, as determined by advertising, usefulness, and the number of people participating. (v) In a sample application of these criteria to atomic power plant safety, it appears that an engineering design objective determined by economic criteria would result in a design-target risk level very much lower than the present socially accepted risk for electric power plants.

Perhaps of greatest interest is the fact that this methodology for revealing existing social preferences and values may be a means of providing the insight on social benefit relative to cost that is so necessary for judicious national decisions on new technological developments.

## Appendix: Details of Risk-Benefit Analysis

*Motor-Vehicle Travel.* The calculation of motor-vehicle fatalities per exposure hour per year is based on the number of registered cars, an assumed 1½ persons per car, and an assumed 400 hours per year of average car use [data from 3 and 11]. The figure for annual benefit for motor-vehicle travel is based on the sum of costs for gasoline, maintenance, insurance, and car payments and on the value of the time savings per person. It is assumed that use of an automobile allows a person to save 1 hour per working day and that a person's time is worth $5 per hour.

*Travel by Air Route Carrier.* The estimate of passenger fatalities per passenger-hour of exposure for certified air route carriers is based on the annual number of passenger fatalities listed in the *FAA Statistical Handbook of Aviation* (see 12) and the number of passenger-hours per year. The latter number is estimated from the average number of seats per plane, the seat load factor, the number of revenue miles flown per year, and the average plane speed (data from 3). The benefit for travel by certified air route carrier is based on the average annual air fare per passenger-mile and on the value of the time saved as a result of air travel. The cost per passenger is estimated from the average rate per passenger-mile (data from 3), the revenue miles flown per year (data from 12), the annual number of passenger boardings for 1967 ($132 \times 10^6$, according to the United Air Lines News Bureau), and the assumption of 12 boardings per passenger.

*General Aviation.* The number of fatalities per passenger-hour for general aviation is a function of the number of annual fatalities, the number of plane hours flown per year, and the average number of passengers per plane (estimated from the ratio of fatalities to fatal crashes) (data from 12). It is assumed that in 1967 the cash outlay for initial expenditures and maintenance costs for general aviation was $1.5 \times 10^9$. The benefit is expressed in terms of annual cash outlay per person, and the estimate is based on the number of passenger-hours per year and the assumption that the average person flies 20 hours, or 4000 miles, annually. The value of the time saved is based on the assumption that a person's time is worth $10 per hour and that he saves 60 hours per year through traveling the 4000 miles by air instead of by automobile at 50 miles per hour.

*Railroad Travel.* The estimate of railroad passenger fatalities per exposure hour per year is based on annual passenger fatalities and passenger-miles and an assumed average train speed of 50 miles per hour (data from *11*).The passenger benefit for railroads is based on figures for revenue and passenger-miles for commuters and noncommuters given in *The Yearbook of Railroad Facts* (Association of American Railroads, 1968). It is assumed that the average commuter travels 20 miles per workday by rail and that the average noncommuter travels 1000 miles per year by rail.

*Skiing.* The estimate for skiing fatalities per exposure hour is based on information obtained from the National Ski Patrol for the 1967–68 southern California ski season: 1 fatality, 17 days of skiing, 16,500 skiers per day, and 5 hours of skiing per skier per day. The estimate of benefit for skiing is based on the average number of days of skiing per year per person and the average cost of a typical ski trip [data from "The Skier Market in Northeast North America," *U.S. Dep. Commerce Publ.* (1965)]. In addition, it is assumed that a skier spends an average of $25 per year on equipment.

*Hunting.* The estimate of the risk in hunting is based on an assumed value of 10 hours' exposure per hunting day, the annual number of hunting fatalities, the number of hunters, and the average number of hunting days per year [data from *11* and from "National Survey of Fishing and Hunting," *U.S. Fish Wildlife Serv. Publ.* (1965)]. The average annual expenditure per hunter was $82.54 in 1965 (data from *3*).

*Smoking.* The estimate of the risk from smoking is based on the ratio for the mortality of smokers relative to nonsmokers, the rates of fatalities from heart disease and cancer for the general population, and the assumption that the risk is continuous [data from the *Summary of the Report of the Surgeon General's Advisory Committee on Smoking and Health* (Government Printing Office, Washington, D.C., 1964)]. The annual intangible benefit to the cigarette smoker is calculated from the American Cancer Society's estimate that 30 percent of the population smokes cigarettes, from the number of cigarettes smoked per year (see *3*), and from the assumed retail cost of $0.015 per cigarette.

*Vietnam.* The estimate of the risk associated with the Vietnam war is based on the assumption that 500,000 men are exposed there annually to the risk of death and that the fatality rate is 10,000 men per year. The benefit for Vietnam is calculated on the assumption that the entire U.S. population benefits intangibly from the annual Vietnam expenditure of $30 \times 10^9$.

*Electric Power.* The estimate of the risk associated with the use of electric power is based on the number of deaths from electric current; the number of deaths from fires caused by electricity; the number of deaths that occur in coal mining, weighted by the percentage of total coal production used to produce electricity; and the number of deaths attributable to air pollution from fossil fuel stations [data from *3* and *11* and from *Nuclear Safety* **5**, 325 (1964)]. It is assumed that the entire U.S. population is exposed for 8760 hours per year to the risk associated with electric power. The estimate for the benefit is based on the assumption that there is a direct correlation between per capita gross national product and commercial energy consumption for the nations of the world [data from Briggs, *Technology and Economic Development* (Knopf, New York, 1963)]. It is further assumed that

35 percent of the energy consumed in the U.S. is used to produce electricity.

*Natural Disasters.* The risk associated with natural disasters was computed for U.S. floods ($2.5 \times 10^{-10}$ fatality per person-hour of exposure), tornadoes in the Midwest ($2.46 \times 10^{-10}$ fatality), major U.S. storms ($0.8 \times 10^{-10}$ fatality), and California earthquakes ($1.9 \times 10^{-10}$ fatality) (data from *11*). The value for flood risk is based on the assumption that everyone in the U.S. is exposed to the danger 24 hours per day. No benefit figure was assigned in the case of natural disasters.

*Disease and Accidents.* The average risk in the U.S. due to disease and accidents is computed from data given in *Vital Statistics of the U.S.* (Government Printing Office, Washington, D.C., 1967).

### References and Notes

1. A. L. Shef, "Socio-economic attributes of our technological society," paper presented before the IEEE (Institute of Electrical and Electronics Engineers) Wescon Conference, Los Angeles, August 1968.

2. *Minerals Yearbook* (Government Printing Office, Washington, D.C., 1966).

3. *U.S. Statistical Abstract* (Government Printing Office, Washington, D.C., 1967).

4. The procedure outlined in the appendix was used in calculating the risk associated with motor-vehicle travel. In order to calculate exposure hours for various years, it was assumed that the average annual driving time per car increased linearly from 50 hours in 1900 to 400 hours in 1960 and thereafter. The percentage of people involved is based on the U.S. population, the number of registered cars, and the assumed value of 1.5 people per car.

5. The procedure outlined in the appendix was used in calculating the risk associated with, and the number of people who fly in, certified air route carriers for 1967. For a given year, the number of people who fly is estimated from the total number of passenger boardings and the assumption that the average passenger makes six round trips per year (data from *3*).

6. The method of calculating risk for general aviation is outlined in the appendix. For a given year, the percentage of people involved is defined by the number of active aircraft (see *3*); the number of people per plane, as defined by the ratio of fatalities to fatal crashes; and the population of the U.S.

7. Group risk per exposure hour for the involved group is defined as the number of fatalities per person-hour of exposure multiplied by the number of people who participate in the activity. The group population and the risk for motor vehicles, certified air route carriers, and general aviation can be obtained from Figs. 3–5.

8. In calculating "benefit awareness" it is assumed that the public's awareness of an activity is a function of $A$, the amount of money spent on advertising; $P$, the number of people who take part in the activity; and $U$, the utility value of the activity to the person involved. $A$ is based on the amount of money spent by a particular industry in advertising its product, normalized with respect to the food and food products industry, which is the leading advertiser in the U.S.

9. In comparing nuclear and fossil fuel power stations, the risks associated with the plant effluents and mining of the fuel should be included in each case. The fatalities associated with coal mining are about ¼ the total attributable to fossil fuel plants.[1] As the tonnage of uranium ore required for an equivalent nuclear plant is less than the coal tonnage by more than an order of magnitude, the nuclear plant problem primarily involves hazard from effluent.

10. This number is my estimate for maximum fatalities from an extreme catastrophe resulting from malfunction of a typical power reactor. For a methodology for making this calculation, see F. R. Farmer, "Siting criteria—a new approach," paper presented at the International Atomic Energy Agency Symposium in Vienna, April 1967. Application of Farmer's method to a fast breeder power plant in a modern building gives a prediction of fatalities less than this assumed limit by one or two orders of magnitude.

11. "Accident Facts," *Nat. Safety Counc. Publ.* (1967).

12. *FAA Statistical Handbook of Aviation* (Government Printing Office, Washington, D.C., 1965).

# 5

# PATHS TO RISK ANALYSIS

*Gilbert F. White*

## 1. INTRODUCTION

Anyone taking the Sunshine Canyon road into the mountain foothills west of Boulder on a cold, rainy night two weeks ago and turning up a steep dirt side road to a muddy corral at the top of the hill would have found four figures, huddled around a gravely ill but still cantankerous donkey, engaged in risk assessment and management analysis. The situation was that the donkey, with a record of behavior well known to us all, was suffering from some undiagnosed complaint—possibly a pasture toxin, possibly a human error—and that narrowing the uncertainty by further painful examination had a low probability of success but a high probability of a flying hoof injuring one of us in attempting restraint or sedation in the cold, slippery corral. Our veterinarian concluded that whatever the correct diagnosis for the ailment, there was no known sure-fire treatment, and we knew without doubt that any failure to take all feasible steps would be regarded by loving young neighbors and grandchildren as a dereliction of duty.

Reprinted from *Risk Analysis* Volume 8 (2) (1988), pp. 171–75. Used with permission of the Society for Risk Analysis.

Later, having done what we perceived to be our risky duty, as we were shedding our rain gear, my wife posed a question in her usual calm, incisive manner: "What would those risk experts gathering in Houston think of our analysis, and would they call this a natural or a technological hazard?" With the annual meeting abstracts and biographical sketches in mind, this set me to thinking about the paths by which its members come to the Society, and the roles of natural and technological hazards in its activities. In consequence, I would like to share with you some of my ruminations on my own path, and a few hypotheses as to what has been learned from studies of natural hazards that may have broader significance.

## 2. ONE PERSON'S PATH

As a boy working during the summers on a ranch along the upper Tongue River in Wyoming I first became acutely aware of the risks inherent in precipitation as it affected pastures and irrigation water supply. Going as a student to the University of Chicago in 1928, I found a faculty stirred by interests in newly emerging fields of ecology—geographical,[1] botanical, social—in which extreme events had an important role. A wiry economics professor occasionally asked penetrating questions in geography seminars and had written a book that I now feel might be recommended reading for SRA members: his name was Knight and the title was *Risk, Uncertainty, and Profit*.[2] I invested my 1929 summer earnings in that most promising of all vehicles, the stock market, and learned agriculture was not the only risky enterprise.

With the ensuing economic depression and New Deal came a period of experimentation with technical and social measures to reduce the deleterious impacts of drought, flood, and soil erosion. A young geographer fresh out of graduate school was caught up in those efforts as they were channeled through the Executive Office of the President.[3] The probabilities and uncertainties attaching to events in nature and their consequences for human lives were inherent in much of the analysis of water management and broader environmental issues in which I was involved in later years.

Thus, my path originated in the hazardous earth and looked out for social processes and their interactions with biological and physical processes. The continuing emphasis was on hazards in which a principal or primary component was a perturbation in a natural system.

## 3. NATURAL HAZARDS IN A BROADER CONTEXT

As I look back at the course of developments in thinking about extreme events in nature since the late 1920s, I am impressed by the number of them carrying import for broader fields of risk analysis. For example, the great flood of 1927 in the alluvial valley of the Mississippi river spurred Federal legislation[4] to prevent a repetition of that catastrophe but also authorized a series of river basin studies in which systematic attempts were made to estimate probabilistically the costs and benefits of water management projects. Those

so-called "308" findings became the basis for the first Federal policy in the resource field to specify benefit–cost analysis in determining the feasibility of projects to cope with unpredictable extremes of too much or too little water. The flood control acts of 1936 and 1938, each stimulated by flood disasters, established B/C and multiobjective optimization procedures that were progressively refined and extended.[5]

Efforts to understand why farmers responded as they did to soil erosion hazards and land management and farm security program laid the groundwork for the first government opinion polls, initiated by the Department of Agriculture in the 1940s and later dismantled. Some of the early attempts to investigate the ways in which decisionmakers depart from probabilistic estimates of consequences were made in the hazards field.[6] Comparative examination of the costs of natural disasters led to more sophisticated analysis and to the specification of the tradeoffs among shareholders wherever risk is managed.[7] The role of perception of natural phenomena in drought and flood decisions was investigated at an early time.[8,9] The need for some kind of clearinghouse for exchange of experience among research workers and administrators of risk management was recognized in the creation of the Natural Hazards Research Applications and Information Center.[10]

By the late 1980s, most of the genuinely global environmental problems were seen to entail the risk of consequences of technical interventions: the environmental effects of nuclear war[11]; permanent disposal of high-level radioactive waste[12]; alterations in biogeochemical cycles of carbon, sulfur, and other greenhouse gases as they affect climate and vegetation[13,14]; and development of a sustainable agriculture.[15] Looking ahead on the national scale, major social disturbances of large magnitude and unpredictable timing are to include a great earthquake in mid-America or the Pacific Coast, a change in the lower channel of the Mississippi River, or acidification of forest soils and water.

## 4. PARALLEL STREAMS OF CONCERN

One of the notable aspects of this period of development of risk analysis has been that studies of natural and of technological hazards have flowed like two streams in roughly parallel courses in an alluvial valley. They touch here or there, and sometimes join each other during high water, but for the most part they are separate, with few direct connections. Chauncey Starr and Chris Whipple organized an early exercise to examine both streams.[16] Clark University was the first to establish a research center addressing technological as well as natural hazards problems,[17] and we know of others like those at the University of Pennsylvania that have followed. *Risk Abstracts* deals with both broad classes of risk. Paul Slovic in his pioneering analyses of risk perception embraced hazards of all types.[18] Such instances are not numerous in the publications and meetings of the Society.

It is not clear why the two streams have mingled so infrequently. Their drawing from different scientific disciplines may have been a factor, but while geophysicists study earthquakes and biologists study toxic substances, the study of how those phenomena are perceived and how society may choose to evaluate them are matters of common concern. Vince Covello ought to illuminate some of the reasons. They may also be clarified by a comparison recently undertaken at the University of Colorado. Perhaps the risk managers in a field such as flood loss mitigation feel little empathy for risk managers dealing with toxic waste disposal. Perhaps those who seek to estimate the probable frequency of a management failure or the survival rate from a toxic exposure have nothing to learn from hydrologists who for 50 years have struggled with suitable methods of estimating the frequency

of a maximum flood.[19] Perhaps there is, indeed, insignificant carryover from one field to another. In any event, it seems likely that more careful examination of similarities and differences and what promotes them might be beneficial to both streams.

## 5. OBSERVATIONS FROM NATURAL HAZARDS RESEARCH

It is dangerous to seek exact analogies, and it may be more helpful to pick out a few characteristics that may possibly have broad application. To advance such an exploration, I offer seven observations from natural hazards research.

### 5.1

Almost all hazards have both natural and technological components, with the mix differing from one place to another. A flash flood episode always has its human components of land use and warnings or lack thereof. A chemical spill's consequences are heavily influenced by factors of terrain, weather, vegetation. It is extremely rare that any natural event fails to have human consequences and that the magnitude and extent of such consequences are not the product of natural–social interactions.

### 5.2

There is a widespread and persistent paucity of postaudits or risk assessment and management that might be helpful in further research and mitigation. It has long been painfully apparent that although reports on floods and ways of mitigating their effects on society are voluminous and numerous, there are only a few studies of what happened after the study and protective action, and that those findings are not necessarily useful.[20] The need for postaudits should be heeded, and the troublesome and generic problem of how to make a useful evaluation should receive far more attention in both natural and technological fields.

### 5.3

The use of benefit–cost analysis to appraise the efficacy of proposed methods of handling risk has severe limitations and may be misleading rather than helpful in providing tools for decision. The elements of such analysis are readily susceptible to manipulation advancing political goals. More serious, the limitations imposed by the types of readily available data distort the analysis so that it may judge what is readily measured rather than what is important for social choice. Increasing sophistication in methods of benefit–cost analysis for technological measures has much to learn from the long history of applications of such analysis, including multiobjectives in management of water surplus and deficiency.[21]

### 5.4

Several decades of study of droughts, earthquakes, and floods shows that any analysis of a risk needs to take account of how it is perceived by the people directly affected as well as by individuals and organizations involved in responding to information and experience. Reliance solely on the perceptions of scientific and technical analysts may give a false notion of the actual situation. And students of earthquakes or floods have learned the importance of examining how perception of one hazard may be related to perception of other hazards in the same place and time (Refs. 22 and 23; see also Ref. 24).

### 5.5

An important set of hypotheses about messages and channels of risk communication have been derived from extensive investigations of how people in a variety of circumstances of age, gender, income, education, and experience respond to flash flood warnings, hurricane coastal zone evaluation advice, and notification of earthquake hazard as well as to other extreme events in nature.[25,26] To speculate about risk communication without awareness of that rich body of experience may ignore significant findings and cautions.[27]

### 5.6

Unless a risk analysis comprehends the social structure within which individual decisions are made, it may fall far short of understanding either the process or the consequences of those decisions. In the natural hazard field there has been discontent with what some workers regard as inadequate study of how the prevailing social systems restrict or guide the decisions of individuals.[28] There is growing recognition of how the social fabric in which decisions take place may influence the outcome profoundly (Ref. 29; see also Ref. 30).

### 5.7

The separation of analysis of risk from appraisal of risk management may be administratively convenient but an analytic trap. Each stands to gain from recognition of the methodological and policy issues encountered by the other. It is messy to mix the two, but neatness may lead to unrealistic findings as to action options and social response. Increasingly the field of flood hazard, for example, embraces analysis of the events with means of mitigating them.

These illustrate observations from the body of natural hazards studies and action that may be seen as relevant to technological hazards. Few efforts have been made to help the two streams learn from each other. Looking ahead, I hope there will be more explicit attempts to do so as systematic research advances on the widening array of social and environmental risks and as the results from both streams of study find wider acceptance in political arenas.

## 6. IN CONCLUSION

While working with colleagues in Moscow last summer I was told the five essential stages of any new intellectual movement: (1) universal enthusiasm, (2) general confusion, (3) searching for the guilty, (4) punishing the innocent, and (5) decorating the uninvolved. With that sobering perspective for a number who has not been involved directly in its management, I salute those members who, coming along quite different paths, have given the Society its vision and initiative in developing a burgeoning field of analysis, and I am grateful for the opportunity to share in the effort as well as for your gracious recognition of past work. In looking to the Society's continuing activity, I covet for it a further and mutually healthy mingling of natural and technological currents of analysis to produce a growing, occasionally turbulent, and more influential stream of thought about ways of coping with an increasingly uncertain and complex world.

*Acknowledgments* I thank the following, who have kindly offered comments on an earlier draft: Howard Kunreuther, William E. Riebsame, and Paul Slovic.

### REFERENCES

1. H. H. Barrows, "Geography as Human Ecology," *Annals AAG* **13**, 1–14 (1923).
2. F. H. Knight, *Risk, Uncertainty, and Profit*. New York: (Harper Torchbooks, New York, 1921).
3. R. W. Kates and I. Burton (eds.) *Geography, Resources, and Environment*, Vol. I (University of Chicago Press, Chicago, 1986).
4. H. Rosen and M. Reuss (eds.) *The Flood Control Challenge: Past, Present, and Future* (Public Works Historical Society, Chicago, 1988).
5. R. H. Platt, "Floods and Man: A Geographer's Agenda," in R. W. Kates and I. Burton (eds.), *Geography, Resources, and Environment*, Vol. II, pp. 28–68 (University of Chicago Press, Chicago, 1986).
6. H. Kunreuther and P. Slovic, "Decision Making in

7. H. Kunreuther *et al.* "An Interactive Modeling System for Disaster Policy Analysis," Monograph No. 26, University of Colorado, Institute of Behavioral Science, Boulder (1978).
8. T. F. Saarinen, "Perception of the Drought Hazard on the Great Plains," Research Paper No. 106, University of Chicago, Department of Geography (1966).
9. R. W. Kates, "Hazard and Choice Perception in Flood Plain Management," Research Paper No. 78, University of Chicago, Department of Geography (1962).
10. G. F. White and J. E. Haas, *Assessment of Research on Natural Hazards* (MIT Press, Cambridge, 1975), pp. 229–31.
11. B. Pittock *et al.* *Environmental Consequences of Nuclear War*, SCOPE Report No. 28 (John Wiley and Sons, Chichester, England, 1986), 2 vols.
12. R. Bryant, G. Brown, L. Carter, and J. Gervers, *Environment* **29**(8), 4–35 (1987).
13. B. Bolin *et al.* *The Greenhouse Effect, Climatic Change, and Ecosystems*, SCOPE Report No. 29 (John Wiley and Sons, Chichester, England, 1986).
14. R. W. Kates, J. H. Ausabel, and M. Berberian (eds.), *Climate Impact Assessment*, Scope Report No. 27 (John Wiley and Sons, Chichester, England, 1985).
15. W. C. Clark and R. E. Munn, *Sustainable Development of the Biosphere* (Cambridge University Press, Cambridge, 1986).
16. Committee on Public Engineering, *Policy Perspectives on Benefit–Risk Decision Making* (National Academy of Engineering, Washington, D.C., 1972).
17. R. W. Kates *et al.* "Managing Technological Hazard: Research Needs and Opportunities," Monograph No. 25 (University of Colorado, Boulder, Institute of Behavioral Science (1977).
18. P. Slovic, B. Fischhoff, and S. Lichtenstein, "Facts Versus Fears: Understanding Perceived Risk," in D. Kahneman, P. Slovic, and A. Tversky (eds.), *Judgment Under Uncertainty: Heuristics and Biases* (Cambridge University Press, Cambridge, 1982).
19. Committee on Techniques for Estimating Probabilities of Extreme Floods, *Estimating Probabilities of Extreme Floods: Methods and Recommended Research* (National Research Council, Washington, D.C., 1988).
20. G. F. White, "When May a Postaudit Teach Lessons?" in H. Rosen and M. Reuss (eds.), *The Flood Control Challenge: Past, Present, and Future* (Public Works Historical Society, Chicago, 1988).
21. D. D. Baumann and Y. Haimes (eds.), *The Roles of the Social and Behavioral Sciences in Water Resources Planning and Management* (American Society of Civil Engineers, New York, in press).
22. I. Burton, R. W. Kates, and G. F. White, *The En-

vironment as Hazard (Oxford University Press, New York, 1978).

23. T. F. Saarinen, D. Seamon, and J. L. Sell (eds.), "Environmental Perception and Behavior: An Inventory and Prospect," Research Paper No. 209, University of Chicago, Department of Geography (1984).

24. A. V. T. Whyte, "From Hazard Perception to Human Ecology," in R. W. Kates and I. Burton (eds.), Geography, Resources, and Environment, Vol. II (University of Chicago Press, Chicago, 1986), pp. 240–71.

25. The Natural Hazards Observer (The Natural Hazards Research Applications and Information Center, University of Colorado, Boulder, Coloardo 80309-0482). (The same center publishes an annual collection of annotations of major publications.)

26. T. E. Drabek, Human System Responses to Disaster: An Inventory of Sociological Findings (Springer Verlag, New York, 1986).

27. T. O'Riordan, "Coping with Environmental Hazards," in R. W. Kates and I. Burton (eds.), Geography, Resources, and Environment, Vol. II (University of Chicago Press, Chicago, 1986), pp. 272–309.

28. K. Hewitt (ed.), Interpretations of Calamity: From the Viewpoint of Ecology (Allen and Unwin, London, 1983).

29. J. F. Short, Jr., "Social Dimensions of Risk: The Need for a Sociological Paradigm and Policy Research," American Sociologist, 22, 167–72 (1987).

30. J. F. Short, Jr., "The Social Fabric at Risk: Toward the Social Transformation of Risk Analysis," American Sociological Review 49, 711–25 (1984).

# PART II

# THEORETICAL INNOVATIONS

One of the long-standing criticisms of hazards research is its seeming lack of theoretical developments. Natural hazards research, for example, has a wealth of case study data on a range of specific hazards, yet we are often puzzled by the most basic questions of how and why people respond to hazards the way they do. Throughout its fifty-year history, the natural hazards paradigm has not wavered from its five thematic areas:

◆ To identify and map the human occupance of the hazard zone
◆ To identify the full range of human adjustments to the hazard
◆ To study how people perceive and estimate the occurrence of hazards
◆ To describe the processes whereby mitigation measures are adopted, including their social contexts
◆ To identify the optimal set of adjustments to hazards and their social consequences.

In his pioneering article, Kates (Reading 6) proposes one of the earliest models of human adjustments to natural hazards. Using a general systems framework, he characterizes the human use system as well as the natural events system as separate entities. It is the interaction of these two that produces the hazard in question. Once the hazard is identified and its effects made known, emergency responses are made initially, with longer term adjustments and recovery following. These adjustments provide feedback into the human use systems and the natural events systems, thus altering their basic characteristics and initiating subsequent changes in human adjustments.

Dissatisfied with the natural event–centered model, other researchers argued that the increasing number of natural disasters was a function of socioeconomic and political forces, not natural factors. O'Keefe, Westgate, and Wisner (Reading 7) advocate a political economic view of disasters, claiming that increased vulnerability to hazards and disasters by many of the world's people results more from political and economic struggles than from any profound changes in the natural systems themselves. According to this view, societal processes govern the extent and occurrence interval of natural disasters. Hazards are intensified and populations made more vulnerable by social, political, and economic constraints on their responses. The inability of individuals and nations to cope with hazards in terms of political systems, emergency response, and wealth results in the increased impact of disasters on society.

This radical critique of natural hazards work led to a reinterpretation of nature-society interactions. Still using the human-ecological approach, hazards in context emerged as an improvement on hazards theory. Hazards in context claims that the nature-society interaction entails a dialogue between the physical setting, the political-economic context, and the role of individuals and agents in effecting change. This interactive model utilizes many traditional hazards elements, such as physical processes, population vulnerability, adjustments, and losses, but imbeds them in larger political, cultural, social, economic, or historical frameworks. Thus collective responses to natural hazards are but one of many social problems that confront nations. In an example of the windstorms that struck Great Britain in the fall of 1987 Mitchell, Devine and Jagger (Reading 8) describe the transformation of natural hazards from human-ecological phenomena to sociopolitical phenomena. Despite its ferocity and heavy losses, the storm's impact was muted by more pressing problems, illustrating the importance of considering hazards in context.

Combining traditional views of hazards with the risk assessment literature, Kasperson et al. (Reading 9) provide yet another innovative approach to hazards theory. Technical assessments of risk are often at odds with socially defined perceptions of the same risk. Why, then, does a minor risk (defined by technical assessments) produce such massive public reactions? In proposing the social amplification of risk model, Kasperson and colleagues contend that risks interact with psychological, social, cultural, and institutional processes, which in turn either amplify or attenuate the risks. The amplification of risk or its attenuation influences how we respond to it.

Hazards, defined as socially constructed phenomena, is one widely accepted perspective in the hazards community. Social theory has helped to refine our understanding of how social structures initiate human responses, how these social structures are interpreted, how the interpretations are linked to the responses in both time and space, and finally the importance of historical knowledge. Bogard (Reading 10) argues for the primacy of social context as the foremost contributor to the increased impacts of environmental hazards and disasters, taking up where the political economy view left off.

Environmental risks, originating from both natural and technological events, are products of failures in political, social, and economic systems that govern the use of the environment and technology. The highly politicized nature of risks and hazards is a consequence of differing perceptions and uneven burdens, as the social amplification and social theory models suggest. The highly politicized nature of risk and hazards constrains the choice of adjustments and the management of such hazards and risks. In the final article in this section, Cutter, Tiefenbacher, and Solecki (Reading 11) question the traditional view of science, environment, and technology in hazard perception. They illustrate how a feminist analysis of hazards offers an alternative view of the identification and assessment of risks and hazards, increases the public debate on the uneven burdens of those risks and hazards, questions acceptability judgments that are based solely on "scientific truth" rather than social discourse, and how social activism and resistance offer the potential for reducing vulnerability to hazards and disasters.

# DISCUSSION POINTS

1. How would you describe the differences between the major theories (natural hazards paradigm, political economy, hazards in context, social theory, and social amplification of risk) in hazard research?
2. What five themes characterize the traditional human-ecological approach in hazards work?
3. How might each of the major theories or perspectives address the problem of global climate change as a hazard? How would they address hunger? Nuclear proliferation?
4. Traditionally, the political economy perspective has been applied to hazards in developing country contexts. Why is this? How might this view be applied to hazards in the industrialized world?
5. Gender differences in hazards research are virtually nonexistent in the literature. How might a feminist or gender-based approach to hazards research improve our understanding of how and why people and institutions respond to disasters?

# ADDITIONAL READING

Blaikie, P. M. and H. C. Brookfield, 1987. *Land Degradation and Society*. London: Methuen.

Bogard, W. C., 1989. *The Bhopal Tragedy: Language, Logic, and Politics in the Production of Hazard*. Boulder: Westview Press.

Cutter, S. L., 1993. *Living with Risk: The Geography of Technological Hazards*. London: Edward Arnold.

Hewitt, K. (Ed.), 1983. *Interpretations of Calamity*. Winchester, MA: Allen and Unwin.

Kirby, A. (Ed.), 1990. *Nothing to Fear: Risks and Hazards in American Society*. Tucson: University of Arizona Press.

Palm, R. I., 1990. *Natural Hazards: An Integrative Framework for Research and Planning*. Baltimore: Johns Hopkins University Press.

Perrow, C., 1984. *Normal Accidents: Living with High Risk Technologies*. New York: Basic Books.

Watts, M., 1983. *Silent Violence: Food, Famine, and the Peasantry in Northern Nigeria*. Berkeley: University of California Press.

Wijkman, A. and L. Timberlake, 1984. *Natural Hazards: Acts of God or Acts of Man*. London: Earthscan.

# 6

# Natural Hazard in Human Ecological Perspective: Hypotheses and Models

*Robert W. Kates*

The lives and affairs of men constantly interact with the natural world. Elaborate technical and social mechanisms enable men to seek in nature that which is useful and to buffer that which is harmful to man. To cope with the harmful effects of nature, complex sets of human adjustments are found in all human use systems. By chance, or even by design, these adjustments can prove insufficient to cope with a given set of natural events, and serious and detrimental effects may ensue. Thus a natural hazard is an interaction of man and nature, governed by the coexistent state of adjustment in the human use system and the state of nature in the natural events system. In this context, it is those extreme events of nature that exceed the capabilities of the system to reflect, absorb, or buffer that lead to the harmful effects, ofttimes dramatic, that characterize our image of natural hazards. But it is also the continuous process of adjustment that enables men to survive and indeed benefit from the natural world. Therefore, the burden of hazard is twofold: a continuing effort to make the human use sys-

tem less vulnerable to the vagaries of nature, and specific impacts on man and his works arising from natural events that exceed the adjustments incorporated into the system.

For the past dozen years, the collaborators in Natural Hazard Research have sought to study this process of adjustment.[1] Beginning with floods, these studies were extended to coastal storm, earthquake, drought, and snow hazard. Subsequently, the list has been enlarged by colleagues and students to include tsunami, frost, coastal erosion, and water pollution hazards. These varied studies employed all or part of a research paradigm which sought to: 1) assess the extent of human occupance in hazard zones; 2) identify the full range of possible human adjustment to the hazard; 3) study how men perceive and estimate the occurrence of the hazard; 4) describe the process of adoption of damage reducing adjustments in their social context; and 5) estimate the optimal set of adjustments in terms of anticipated social consequences.

But it is only now that we can begin to structure a primitive general framework of human adjustment to natural hazard, in which we try to preserve its human ecological perspective. In this perspective, with its focus on man as the ecological dominant, the inter-

Reprinted from *Economic Geography* Volume 47 (3) (1971), pp. 438–51. Used with permission of Clark University.

actions between men and nature tend, over the short run, to be stable, homeostatic, and self-regulating and, over the long run, to be dynamic, adaptive, and evolutionary in the direction of increasing control over nature's resources and buffering from nature's hazards.[2]

A rudimentary model of the short-run process of adjustment constitutes the major focus of this paper. Our present understanding of this process, particularly in North America, is considerably greater than our comprehension of the long-run adaptive process in the global context. Nevertheless, some hypotheses, having as their core the man-nature interaction and an evolutionary sequence of techno-social stages of adjustment, have been developed from the body of hazard-specific and place-specific research.

# GLOBAL HYPOTHESES OF NATURAL HAZARDS

The present state of global understanding of natural hazard phenomena may be stated as a series of linked, succinct, but complex hypotheses as to the nature of natural hazard, adjustments to it, and the choice thereof made by the human occupants of hazard areas.[3] They purport to explain major sources of variation in human behavior, as between great techno-social stages, specific hazards, specific classes of decisions and decision makers, and between individuals within a specific group of managerial decision makers (these are linked as in Figure 1).

## Man-Nature Interaction

Natural hazard is an aspect of the interaction of man and nature arising from the common process in which men seek in nature that which is useful, and attempt to buffer that which is harmful to man. This process, whether employing elaborate technical and social mechanisms or simple ones, makes possible human occupance of areas of even frequent and recurrent natural hazard.

Thus it is rare in such areas to discover individuals in substantial ignorance of the hazard or unaware of alternative locations. Rather, locations either offer opportunities of relative or absolute superiority, or appear less threatening from the unique, terminal perspective of the individual than from the longer-run view of the external observer.

## Techno-Social Stages

Human response to natural hazard is organized into three distinctive techno-social patterns or stages of adjustment: folk or preindustrial; modern technological or industrial; and comprehensive or postindustrial. Each stage is marked by a preferred cluster of adjustments, a distinctive process of choice, and characteristic patterns of damage occurrence.

Folk or preindustrial adjustments, for example, are often mystical, arational, or imbedded in the broader cultural context of life and livelihood; require more modification of human behavior in harmony with nature than reliance on the control of nature; are flexible and easily abandoned; are low in capital requirements; require action only by individuals or small groups; and can vary drastically over short distances. Damage-causing natural events appear to be frequent; the average loss per event is low; but the ratio of deaths-to-damage is high.

Modern technological, or industrial, adjustments: involve more or less conscious decisions from a limited range of technological actions emphasizing control of nature; are inflexible and difficult to change; are high in capital requirements; require interlocking and interdependent social organization; and tend to be uniform. Damage-causing natural events become less frequent; death rates diminish drastically; but average damage loss per event is extremely high.

Comprehensive, or postindustrial, adjustments combine features of both earlier stages so as to involve a larger range of adjustments, greater flexibility and variety of capital and organizational requirements, and the institu-

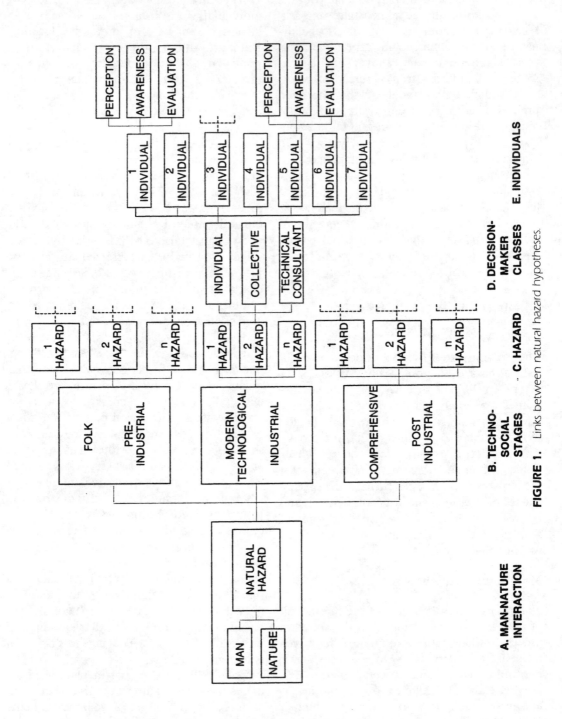

**FIGURE 1.** Links between natural hazard hypotheses.

A. MAN-NATURE
   INTERACTION

B. TECHNO-
   SOCIAL
   STAGES

C. HAZARD

D. DECISION-
   MAKER
   CLASSES

E. INDIVIDUALS

tionalization of broadened choice from the array of potential actions. Damage-causing natural events increase slightly, death rates further diminish; and average damage losses per event decrease by up to half the maximum potential damage. Nevertheless, absolute damages and deaths may remain high as a function of increase in population and wealth.

## Hazard Differences

Within the context of any of these three patterns of response to natural hazard, considerable variation exists. There are noticeable differences in the choice of adjustments between various hazards, and differences as well between decision makers. These decision makers include both collectivities such as communities, public bodies, and corporations, together with their technical consultants and individuals who occupy or use hazard areas.

Four critical features of natural hazards give rise to different choices of adjustments. Three are features of the natural events: the frequency of occurrence, the magnitude of energy release, and the suddenness of onset. A fourth feature arises from the ecological setting, namely, whether the hazard is intrinsic to the use characteristics or locational advantage of the site (e.g., drought in rain-fed agriculture, coastal storms at scenic locations), or is not intimately related to occupance activity (e.g., earthquake, tornado).

## Decision Maker Differences

In the context of a single hazard, the characteristics of the choice process vary with the nature of the decision maker: some choices being collective actions, others, individual actions, and many actions are sequentially constrained by previous collective or individual choices. While differing in detail and setting, our reading of the community, organization, and administration literature does not suggest a fundamental discrepancy between individual and collective behavior. Thus, while the

appropriate managerial unit may differ, the ways in which the choice of adjustment is made does not fundamentally differ.

## Individual Differences

Thus, all men who choose—user of a hazard area, public guardian, technical consultant, or single individuals or committees—seem to perceive hazard and are aware of a range of adjustments. They evaluate these adjustments with reference to their environmental fit, technical feasibility, economic gainfulness, and social conformity. But again in the context of a single hazard, considerable variation can be found in the perception of hazard, the knowledge of adjustments, and the evaluation criteria applied to these adjustments.

## Perception of Hazard

Variation in the perception of a specific natural hazard (expectation of future occurrence and of personal vulnerability) can be accounted for by a combination of: the way in which characteristics of the natural event are perceived, the nature of personal encounters with the hazard, and factors of individual personality. Such perception appears to be independent of common socioeconomic indicators. Of the many possible characteristics of natural events, the perception of magnitude, duration, frequency, and temporal spacing of the natural event appears to be most significant. For personal experience, it is the recency, frequency, and intensity of such an experience that appears most critical with intermediate frequency generating greatest variation in hazard interpretation and expectation. Of the many possible personality factors, fate control, differential views of nature, and tolerance of dissonance-creating information seem most relevant. Risk-taking propensity, which appeared logically relevant, has not been shown to be a consistent trait and has proved operationally difficult to measure.

## Awareness of Adjustments

Awareness of adjustments, of their number and type, and the quality of knowledge thereof, is a function in the main of the casual access to communication networks and, to a lesser degree, of motivation to search for new modes of adjustment. Variation in awareness might be accounted for by factors controlling access to information. Intensity of personal experience or role-related responsibility might provide motivation for increased knowledge of adjustments when encouraged by positive views of fate control and the efficacy of action.

## Evaluation of Adjustments

Evaluation of known adjustments, with reference to environmental fit involves the conformity of the adjustment to an appraisal of site or situation for certain activities. Technical feasibility involves an assessment as to the efficacy of the adjustment, the availability of skills, tools, and materials, and the indivisibility of the activity from related processes. Economic gain involves an estimate of anticipated costs and gains in the light of the perceived time horizon, the ratio of reserves to anticipated loss, and the degree to which the choice is required. Social conformity involves a judgment of the degree of conflict or conformity with law, tradition, or expected mores of behavior.

The foregoing criteria for evaluating adjustments are not of equal importance and vary as between major stages, hazards, and individuals. For preindustrial adjustments, criteria of environmental fit and social conformity seem most important, while those of technological feasibility and economic gainfulness appear more prominent in considering industrial adjustments. The entire set of criteria appear relevant for postindustrial adjustments.

In the context of a single hazard and stage of response, variation in the importance of criteria appears related both to the perception of the hazard and the role training and responsibility of the decision maker. For ex-

ample, in modern industrial adjustments, for decision makers with high hazard perception, technological feasibility should dominate questions of economic gainfulness. In cases of moderate to low hazard perception, role inclinations towards technological or economic considerations dominate.

The foregoing hypotheses range from those of great culture realms of nations and history to those explaining the diversity of behavior of individual farmers on the shores of Lake Victoria or residents of the flood plain of La Follette, Tennessee. To move from a set of hypotheses to a theory of hazard behavior requires the careful refinement of questions and the extensive research for answers in the series of comparative cross-cultural studies presently underway in over fifteen countries.[4] Models can contribute in a special way to the refinement of good questions.

## ON MODELS

A good model of a system is a theory of that system. It purports to identify major elements of the system, describe the strengths and direction of the linkages between those elements, and to simulate dynamically the processes that underlie the elements and linkages. Good models serve also as practical laboratories for social scientists in which the consequences of changes in process elements or linkages can be examined for their practical import.

Most models fail to do either function well. Lacking a theoretical understanding of process, the model builder resorts to black boxes, frequently in the form of some probability distribution. A working model may ensue, even one useful for prediction; but unless one subscribes to the fiction that equates prediction with understanding, the model itself does not necessarily enhance the state of theory. Nor do most models succeed very well in their practical simulations. It is common to find in the literature authors who bemoan the absence of certain critical data, the size of computer memory, or the fact that by the time the elaborate simulation is completed, the real world policies, towards which the model was intended to contribute, have changed several times over.

Nevertheless, we do learn many things from models, even in their failure, and that is why we turn to them again and again in our research strategies.[5] Faced with the need to model processes that we do not understand, we are given pause to determine whether we should seek to understand them before proceeding further. Then if we resort to a black box, it may be because we have found that the process is not intrinsic to understanding the phenomenon directly under study. Or when faced with the absence of critical data, we now might be encouraged to try to obtain it, but with increased confidence, having now established that in truth the data are critical. Thus we can emerge with what is most helpful for science, a statement not of gross ignorance but of highly specific ignorance, a veritable agenda for research needs.

A model may not only humble us in our ignorance but give us courage as well. So complex is the world, so many the events that occur, so simultaneous their occurrence, that the mind boggles at ever hoping to capture any complex process in all its dimensions. When we model a system, we reduce it to a mosaic, with distinguishable elements, boundaries, and single characteristics which combine, nevertheless, to give a representation greater than the sum of its parts. To make it dynamic, we can animate the mosaic and if its representation is still recognizable, we have some reason to be encouraged.

Based on these general observations, let me suggest some specific qualities for a model for hazard research: it should be parsimonious, conservative, flexible, useful, and aesthetically pleasing. The model should strip the adjustment process to its barest bones, it should minimize detail, subject only to the constraint of some verisimilitude toward the real world. It should be conservative with what has been done over the last decade utilizing wherever possible accumulated materials rather than demanding fresh constructs or data. It should be flexible in its ability to accept new findings and insight, such as may be derived from the extensive cross-cultural studies now underway. It should be capable of providing practical answers by duplicating desired patterns of adjustment and evaluating unambiguously the advantages and disadvantages of each. To do so it must be programmable, suitable to the simplistic linear thinking (loops notwithstanding) of the modern computer. Finally, and perhaps most importantly, it should prove aesthetic. None of those involved in hazards research finds model building a pleasant avocation for its own sake. If it is to be justified it must have aesthetic scientific appeal, namely that the final product has genuinely enhanced our understanding in such a way as to provide some sense of pleasure.

These qualities, desirable in themselves, seriously compromise reality. Human society or natural process are simplified in ways that seldom meet the approval of specialists in a specific area of study. Subtle but cumulatively important processes are ignored. A complex, subtle, variegated, almost infinite process of adjustment is reduced to a sequence of crude, iterative steps. But even in its crude form a model of human adjustment involves a formidable understanding of systems, collection of data, and knowledge of functional relationships.

## MODELING THE ECOLOGICAL PERSPECTIVE WITHIN A GENERAL SYSTEMS FRAMEWORK

The model shown in outline in Figure 2 is only a small slice of the global system for which the above hypotheses represent the first step towards a theoretical formulation. The system modeled is a single cross-section of space and time: an area with a relatively homogenous expectation of a single hazard and a duration in time appropriate to the temporal

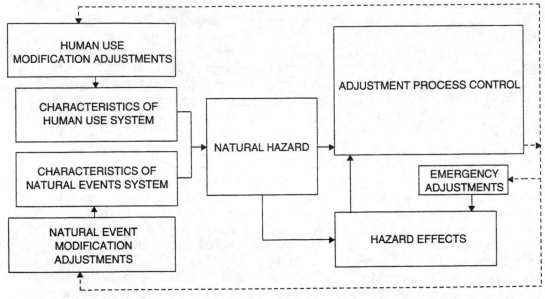

**FIGURE 2.** Human adjustment to natural hazards: outlines of a general systems model.

character of the natural events and the related human activity.

For some bit of the earth's surface, for some small moment in time, man and nature, in the form of a *human use system* and a *natural events system*, interact to pose a *natural hazard*. The existence of such hazard generates a specific set of *hazard effects* and its own homeostatic control in the guise of various adjustments.[6] The *adjustment process control* governs the adoption of adjustments that modify the human use system, modify the natural events system, and modify the hazard effects through *emergency adjustments*. The characteristics of each of the elements are specific and the linkages are functional between elements. These are shown in greater detail in Figure 3.

### Characteristics of the Human Use System

What constitutes a minimal but sufficient description of the human use characteristics of a relatively small, homogeneously hazardous area? To evade a classic question in regional description (and in all behavioral and social science), we define sufficiency in terms of adjustment capability and hazard effects. Thus we describe the human use in terms of managerial units; the smallest units of occupance capable of independent and indivisible decision making relative to adjustment adoption. For each managerial unit we describe those characteristics which capture the most significant of hazard effects.

With these general but imprecise guides, smaller managerial units would consist of households in most societies, but include as well all sorts of commercial, industrial, and governmental units, and on the highest level constitute an aggregate based on the nation-state. Unfortunately, one cannot simply aggregate the smaller units to arrive at the higher levels of hierarchy, where very different managerial responsibilities are found.

For each unit, a minimal set of descriptive data would include: 1) the specific human occupance in such terms as the number, age, sex, and diurnal or seasonal occupance of the hazard area; 2) activities, with material or service-productive activities described simply

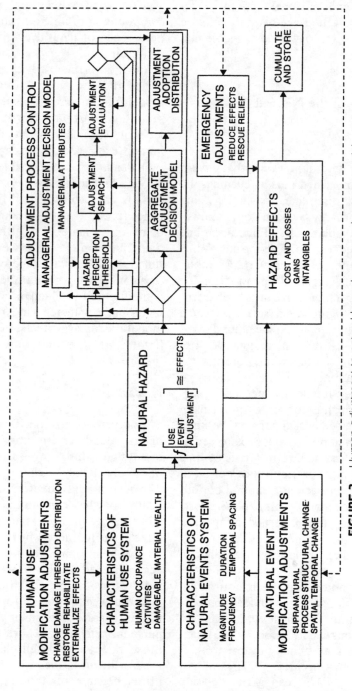

**FIGURE 3.** Human adjustment to natural hazards: a general systems model.

85

in terms of their outputs and factor or process inputs, and non material-productive but important social and personal activities, described in terms of their age-sex participation rates; and 3) an inventory of damageable material wealth.

## Characteristics of the Natural Event System

The study of the natural processes that govern the generation of hazard-causing events is the subject of entire disciplines: seismology, hydrology, meteorology, vulcanology, parasitology, just to name a few. Each discipline develops key indexes to describe its events and these are not necessarily transferable. Nevertheless, in a current attempt to describe all types of environmental hazards twelve critical indexes are being used, seven of which are primarily characteristics of the natural event system: spatial distribution, magnitude, frequency, duration, areal extent, forecast capability, and warning time.[7]

In the model further reduction is suggested. Events can be described in terms of *magnitude* expressed as a dimension, volume, or energy expression; *frequency* expressed as a probability of occurrence in a unit of time or an average return or recurrence period of time; *duration* expressed as temporal periods ranging from seconds to years; and *temporal spacing* describing the patterned occurrence of the event in time—random, even (seasonal or regular periodic), or clustered (serially correlated).

It should be noted that the measurement of these characteristics are indeed perceptions, those of the scientist and engineer. Other perceptions exist, those of the manager, and these may employ different characteristics and measurements. These enter the model in terms of managerial decision making; for the description of events we seek the best technical, albeit subjective, appraisal.

## Natural Hazard

A natural hazard is a threatening state to man, compounded of an expectation of the future occurrence of natural events which impinge on a human use system that is provided, through adjustments, with a certain capacity to absorb these events. In the context of the model, natural hazard takes on meaning as a set of functional statements that relate for each level of assumed adjustment, for each set of human uses, and for each pattern of event occurrence, a set of possible hazard effects.

Such functional statements are available for certain characteristics of flood plain occupance (productive and residential activities and damageable material wealth) under differential adjustments.[8] Thus we have generalized functions that relate stage of flood (a dimensional measure of magnitude) to the structure, contents, and productive activities of manufacturing, commerce, and residence to yield monetary damage effects. But for most other hazards, relationships with predictive potential are rare.

## Hazard Effects

The occurrence of a specific natural event may or may not have any impact on the human use system, this being a function of the size of the event and the character of adjustment. The cost of adjustment, which can vary drastically, may be a continuing levy on the wealth and energies of the managerial units. These effects are registered in their direct impacts on the health and well-being of the human occupance, in the loss of wealth from curtailment of productive activities and damage to material wealth, and in losses in the opportunity to participate in important social and personal activities. There are gains as well, the work or location well-placed to profit from a hazard: the farmer who benefits in higher prices because of another's loss, the welldigger in time of drought. Finally, there are intangible impacts, some unidentifiable, and others, though identifiable, for which the consequences are not easily assessed. The model needs to identify these contrasting impacts, to cumulate and store them, and to follow the resulting change in the human system and its level of adjustment.

## Adjustment Process Control: Managerial Adjustment Decision Model

The presence of a natural hazard encourages human action to minimize its threat and mitigate its effects. For any individual managerial unit the decision process is a complex but interesting one, and it has been a focus of hazard research for many years.

A model of decision making applicable both to the choice of resource and natural hazard adjustment has been developed. This model by White [10] is heavily influenced by the work of Simon particularly in the notions of "bounded rationality" and "satisficing" [9]. The work also parallels the complex model of resource use developed by Firey [4].

Over the years, variants of this approach have been tested in different hazard and resource use situations. Two emphases can be found in this work: to develop a sharper, more predictive decision making model and to incorporate individual personality characteristics into it. This is a continuing task, providing new challenges in the cross-cultural context.

The sub-model presented in Figure 4, then, is really the current state of our decision making theory, strung together in an operative sequence. The sequence is as follows: for the manager of each unit there is a threshold of hazard perception below which he does not seek nor evaluate adjustments. This threshold is in turn a function of the way in which the manager perceives natural events, his personal hazard experience, and specific personality characteristics that include attitudes towards fate and the efficacy of action, differential views of nature, tolerance of dissonance, and risk-taking propensity. The perception of

events and personal hazard experience can change at each iteration; personality traits are fixed for the duration of a model run.

The initial set of known adjustments is also a function of an individual manager's attributes, specifically his casual and specialized access to communication networks. General access can be approximated by socioeconomic indications of age, education, income, and travel, specialized access by unique role responsibilities and training.

When the hazard perception threshold reaches a certain value, a search of known alternatives begins, and each is evaluated in turn by reference to four basic questions.

1. Is it suitable for the environmental setting?
2. Is is technically efficacious and feasible given the available tools, skills, materials and the indivisibility of activity?
3. Is it economically gainful in the context of the managerial unit's time horizon, reserve-loss ratio, and constraints on choice?
4. Does it conform to social guides of law, tradition, or expected norms of behavior?

The questions are not of equal priority however, and the model would allow sequencing them or giving the evaluation criteria different worth. It is clear for example that in much of the world engineers faced with a problem of adjustment use technical feasibility—Will it work?—as their prime criterion and employ considerations of cost, social conformity, and the like as constraints. Similarly, social conformity—to do as my father did—is a basic guide in many areas. In the context of a specific area, the order of criteria is probably a function of hazard perception and role responsibility and training.

The actual application of the evaluation criteria is a function primarily of the human use characteristics for the managerial unit. Based on the evaluation, a decision to adopt or not is made. If rejected, provision is made to feed this back into the process. We know from the concept of satisficing and from cognitive consistency theories, that in the face of

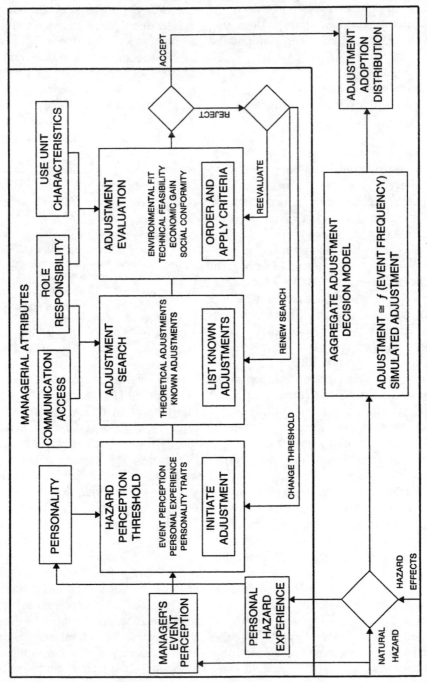

**FIGURE 4.** Adjustment process control.

either ease or difficulty in developing problem solutions, men will often change their evaluation of the problem, seek new alternatives or modify their standards of acceptance of a solution. Provision is made for each or all of these modes of cognition.

This is truly a complex sub-model reflecting the complexity of real-world processes. But the overall model does not require successful simulation of each individual decision; an alternative path of adjustment control is provided.

## Adjustment Process Control: Aggregate Adjustment Decision Model

In any hazard area, the frequency of adoption of adjustments appears to be a function of the hazard frequency. The variation in adoption between individual managerial units is also related to frequency, but variation is highest in areas of intermediate frequency.

In areas of low frequency, most people adopt few, if any, adjustments. In areas of high frequency, widespread adoption is found. These relationships are modeled by simple functional relationships in the aggregate adjustment decision model, which is also capable of accepting simulated adjustment distributions.

## Adjustments to Natural Hazard

Three distinct sets of adjustments are postulated in the model: those that seek to *modify the natural events system*, those that attempt to *modify the human use system*, and a set of post-event *emergency* adjustments.

The most common adjustment of the set designed to modify the natural events system is an appeal to, or activity for, some supranatural power. In the face of calamity or to prevent it, men everywhere appeal to "whatever Gods may be." Nor are such practices limited to nonliterate peoples; witness the widespread practice of water well dowsing in North America.

For some hazards, an attempt can be made to affect the natural process as in fog or hail dispersal by seeding or other forms of weather modification. More common is to seek to shift the spatial or temporal distribution of the natural events to a more favorable one. Barriers of all sorts seek to limit the spread of a hazardous event, and water storage and retardation structures are familiar means of dealing with floods and droughts. Finally, some adjustments affect the internal or chemical structure of the event as in the case of snow melting adjustments.

An even wider array of adjustments affect the human use system. Some seek to raise the damage threshold, the point at which damage begins, while others tend to change the entire damage potential distribution. Examples of the former would be the flood-proofing of basement areas, and of the latter, the use of a drought-resistant crop variety. Evacuation and related adjustments may lead to eliminating entirely some damage potential, while conversely the process of restoration and rehabilitation of previous hazard effects may add to or diminish the damage potential. The most common of all adjustments is the bearing of losses when they occur. But this burden can be shared. Indeed all hazards effects can be externalized, spreading them over a wider space, greater time, or broader society by insurance, relief, extended family relations, and the like.

Finally, after a specific event ensues, emergency adjustments can diminish its effects. Rescue and relief operations can save lives and reduce the burden of the hazard; flood-fighting, fire-fighting, evacuation, and emergency repairs prevent greater losses from occurring.

## THE GENERAL MODEL APPLIED TO EAST AFRICAN AGRICULTURAL DROUGHT

By way of illustration, let me briefly note the progress being made in modeling drought in a specific context—East African smallholder agriculture.[9] This agriculture is characterized by markedly seasonal rainfall, a hoe-based cultivation system with little or no capital inputs, which mixes crops and cattle, family food and commercial crops, and perennial and annual crops, depending on locale. For all but the most well watered mountain areas, drought and adjustments to drought are essential features of the agricultural system. As the rains come late, are sparse, or fail altogether, varied and widespread suffering is reported. If we would seek to model this system, based on our present knowledge, what would its components look like?

## The Human Use System

The managerial unit of the human use system is the household in almost all cases (Figure 5). The productive activities of such a household are very complex. Up to 25 different crops or trees will be grown, many occupying the same plot of land. Nevertheless, much

progress has been made in describing the nature of the agricultural system; a dozen or more detailed studies exist today, where there were but one or two five years ago.[10] In these recent studies in Tanzania, it has been found possible to describe crudely the major factors of production, land, and labor, with some precision; but accurate description of crop yields still seems to escape us. In terms of nonproductive activities we know little of their relationship, if any, to drought; and of material wealth, only animal wealth seems particularly sensitive to the phenomenon.

## The Natural Events System

It now appears possible to provide a rather sophisticated description of the natural events system in terms of the water balance. Rainfall data, estimates of potential evapotranspiration, and computer programs exist capable of providing probability estimates of soil

**FIGURE 5.** *Human adjustment to natural hazards: an application to East African agricultural drought.*

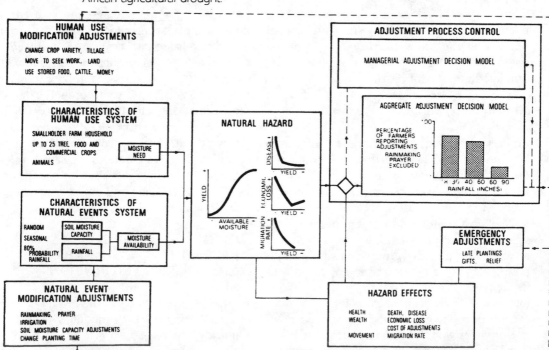

moisture surplus or deficit by ten-day or monthly periods for selected areas of East Africa [1, 3]. A monthly period appears to be adequate for capturing major variations in weather, the growth cycle, and the human use system during the growing season, but for certain growth processes a daily or weekly period would be desirable. With little carryover for soil moisture, random temporal spacing or the independence of growing season rainfall can be assumed (intra-seasonal rainfall, however, is serially correlated).

## Natural Hazard

The essential features of the natural and human use system can be captured in the water-yield relationship, and this can be used to describe the natural hazard. In practice, the most significant range of hazard lies between the wilting point and potential evapotranspiration. We know little about this range of the relationship although we do know something about the optimal and minimal water needs at each point of the plant growth cycle. A distinctive water-yield relationship is needed for each variation in crop, location, cultivation, and tillage practice, essentially for each variation in adjustment. It would be perhaps possible to obtain such relationships for a single place, the East Africa Agricultural and Forest Research Organization station at Muguga, Kenya, but extremely difficult to obtain such data elsewhere. An additional problem with these experimental data is the excellence of the agriculture; the level of practice employed at the station far exceeds that of the general level of agricultural practice.

## Hazard Effects

Three kinds of drought effects are considered: effects on health, wealth, and movement. We know very little about the relationship between variation in crop yield and death and disease, economic gains and losses, or move-ments of population; estimating the impacts of these effects is a major research task.

## Adjustments

Pilot studies of adjustment involving interviews with 460 Tanzanian farmers have indicated the range of adjustment and have provided a rudimentary conceptualization of the process.[11] There is widespread resort to supranatural appeal, to bearing losses or externalizing them in time by using stored food or money in the form of cattle, or by moving to seek work or land. More rarely employed are on-farm adjustments that manipulate the human use system by changing crop varieties, cultivation practices, planting dates, and the like.

## Adjustment Process Control

The adoption of adjustments appears, as elsewhere, to be a function of the hazard frequency, but not a simple one, and much more needs to be learned in this respect. We think that we now know how to better ask farmers for the data required to simulate managerial decisions and with out collaborators are collecting such data in Australia, Brazil, Mexico, and Tanzania.

In operation of such a model, different adjustments affect either the available moisture, the water-yield relationship, or the hazard effects. With a simulated or historical trace of precipitation employed as an independent variable, it is possible to evaluate the longer term effects of changes in adjustment or the decision process itself. For at least one or two places, where the complex data assembly needs can be met, the model can provide an agenda of research needs, serve as a test of our decision theory when compared with observed behavior, or provide a simulation for the potential outcome of our policy suggestions.

## AFTERTHOUGHTS

The general and specific models herein have focused on the interaction of man and nature, as a continuous process where certain extreme concurrences are identifiable as hazards of natural origin and harmful to man. In addition to modeling this process from a human ecological perspective, the models seek to fashion major hypotheses of hazard behavior into structures capable of computer simulation. At this point in their development, their heuristic value seems established but the capability for relatively efficient and meaningful simulation is still in doubt.

But more important and still debatable is the desirability of the general perspective. A case can be made that many of the real determinants of human behavior related to natural hazard lie outside the interface of the natural and human systems modeled here. For example, the simultaneous occurrence of the droughts and floods with economic depression in the 1930s surely led to development of policies different from those that might have prevailed in the absence of the depression. And the encourgement of cash cropping and the prohibitions on migration by the colonial administration in East Africa probably intensified the effects of the disastrous droughts occurring there in the thirties. Critical events such as these, seen as important with hindsight, are not easily handled by the model.

An alternative would be to model in much greater detail the specific human use system related to a particular hazard: smallholder agriculture, urban residential development, municipal water supply, and the like. In this context natural hazard would be but one of a series of concerns facing the manager, and the ways in which he dealt with other forms of uncertainty would surely affect the ways in which he deals with the natural hazard. A greater fidelity of the system to reality would be obtained at the cost of comparative generalization.

This is, of course, an old dilemma and one familiar to many. How quickly have models of systematic systems become regional systems? One begins to model the transportation and ends up modeling the city, the region, its activities and growth. In my earliest introduction to modeling, in working with the Harvard Water Program, I learned that decision comes from the Latin verb "to cut." The decision as to where to make the cut that severs the decision-model from the matrix of reality comes no more easily than do the decisions of even the most knowledgeable in the face of an uncertain natural world.

## NOTES

[1] See Burton, Kates, and White [2].

[2] Given present, rapid rates of change, the long run increasingly shortens, and it remains to be seen whether that which is seemingly adaptive will not prove maladaptive in the future.

[3] Walker Banning and Carlos Alsina helped develop an initial list of hazard hypotheses, subsequently refined in many discussions with project collaborators.

[4] These studies, involving the collaborative effort of many colleagues, have been organized by the Commission on Man and Environment of the International Geographical Union, and consist of comparative field observations of hazards including drought, earthquake, flood, frost, hurricane, landslide, pollution, and volcanic activity, and national appraisal of drought, flood, hurricane, and air pollution.

[5] That is, quite aside from fads or fun in research, which also contribute to the spate of model building.

[6] The notion of natural hazard as a joint probability of states of natural events and human adjustments to them was developed with Russell and Arey in a study of humid area drought in Massachusetts [7]. A fuller appreciation of the ecological perspective comes from the work of Hewitt and Burton [5].

[7] The remaining five indexes being used for a study designed for UNESCO are: damage potential, adjustments, adoption of adjustments, perception of hazard, and perception of adjustment.

[8] See White [11] and Kates [6].

[9] Work undertaken in conjunction with Len Berry, Director of the Bureau of Resource Assessment and Land Use Planning, University College, Dar es Salaam, Tanzania and Philip Porter of the University of Minnesota.

[10] For a review of many of these studies, see Ruthenberg [8].

[11] Reported on in detail in Berry [1].

# REFERENCES

1. Berry, L., T. Hankins, R. W. Kates, L. Maki, and P. Porter. "Human Adjustment to Agricultural Drought in Tanzania: Pilot Investigations," Working Paper, Natural Hazard Research, Department of Geography, University of Toronto, 1971.

2. Burton, I., R. Kates, and G. White. "The Human Ecology of Extreme Geophysical Events," Working Paper No. 1, Natural Hazard Research, Department of Geography, University of Toronto, 1968.

3. Dagg, M. "A Rational Approach to the Selection of Crops for Areas of Marginal Rainfall in East Africa," *East African Agricultural and Forestry Journal*, 30 (1965), pp. 296–300.

4. Firey, W. *Man, Mind, and Land: A Theory of Resource Use*. Glencoe: The Free Press, 1960.

5. Hewitt, K. and I. Burton. *The Hazardousness of a Place: Extreme Events in London, Ontario*. Toronto: University of Toronto Press. (Forthcoming).

6. Kates, R. W. "Industrial Flood Losses: Damage Estimation in the Lehigh Valley," Research Paper No. 98, Department of Geography, University of Chicago, 1965.

7. Russell, C. S., D. Arey, and R. W. Kates. *Drought and Water Supply: Implications of the Massachusetts Experience for Municipal Planning*. Baltimore: Johns Hopkins Press for Resources for the Future, 1970.

8. Ruthenberg, H. *Smallholder Farming and Smallholder Development in Tanzania*. Munich: Weltforum Verlag, 1968.

9. Simon, H. A. *Models of Man: Social and Rational*. New York: John Wiley and Sons, 1957.

10. White, G. F. "The Choice of Use in Resource Management," *National Resources Journal*, 1 (March, 1961), pp. 30–36.

11. White, G. F. "Choice of Adjustment to Floods," Research Paper No. 43, Department of Geography, University of Chicago, 1964.

# 7

# TAKING THE NATURALNESS
# OUT OF NATURAL DISASTERS

*Phil O'Keefe* ❖ *Ken Westgate*
*Ben Wisner*

The media continually present us with graphic accounts of natural disasters—the Bangladesh cyclone, the Nicaraguan earthquake and the African drought are just some of the recent examples of catastrophes that caused much death and destruction. Since the beginning of 1976, we have already had detailed accounts of floods in Venezuela, Australia and Indonesia, famine in Niger, landslides in Ecuador, drought in Malaysia and the mammoth earthquake in Guatemala. It is difficult to gather global information on the frequency, and, more importantly, the impact of these disasters. But a set of global statistics from all available resources has recently been compiled.

The Disaster Research Unit at the University of Bradford has collected data from international organisations, government departments, academic institutions and insurance companies. The information was confused because most institutions recording disaster data had an implicit role in the disasters and their aftermath which coloured their recording. For example, the international and governmental organisations were chiefly concerned with disasters in which aid was being donated, academic institutions were primarily interested in recording unusual phenomena which might not necessarily be disastrous, while insurance companies only recorded information directly related to their business.

In spite of the data's unreliability, several tendencies can be observed.

The most important tendency is an increase in the occurrence of disasters over the last 50 years. Figure 1 shows this increase from 1947–70 of large-scale disasters, that is, those disasters covering more than a ten-degree square on a world map and where damage exceeds $1 million. This tendency is paralleled by an increasing loss of life per disaster. The greatest loss of life per disaster is observed in underdeveloped countries, and there are general indications that the vulnerability of these countries in particular is increasing. Such conclusions clearly require further explanation.

Disaster marks the interface between an extreme physical phenomenon and a vulnerable human population. It is of paramount importance to recognise both of these elements. Without people there is no disaster. The two elements are basic to an explanation of an increase in the occurrence of disasters. No major geological or climatological changes over the last 50 years adequately explain the rise. There is little argument about geological change, but there has been much mystifying argument about climatic change, especially following the prolonged drought over the African and Asian continents. But no firm conclusion can be drawn about changing climatic conditions from available evidence. Randall Baker at the Development Studies School of the University of East

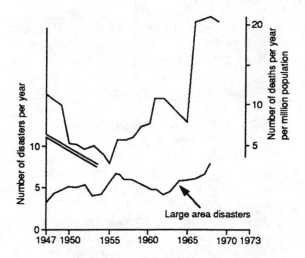

**FIGURE 1** Global disasters 1947–1973: Five-year moving average.

Anglia recently reviewed all the evidence of climatic change in Africa, and offered the Scottish judgment of "case not proven". Even if some long-term change was observable it would not explain the increase in disaster occurrence observed in the data.

If it is accepted that there has been no major geological and climatological change in recent years, then it can be assumed that the probability of the extreme physical occurrence is constant. If the probability is constant, then logically the explanation of the increasing numbers of disasters must be sought in an explanation of the growing vulnerability of the population to extreme physical events. Ongoing research suggests that some radical rethinking on the nature of "natural" disasters is necessary.

It is known that the frequency of natural disasters is increasing especially in underdeveloped countries. Indeed, the increased vulnerability of people to extreme physical events can be seen as intimately connected with the continuing process of underdevelopment recorded throughout the world. As population continues to expand, and as resources continue to be controlled by a minority, the real standard of living drops for much of the world's population. This population is increasingly vulnerable to environmental variation as the process continues. Paul Richards of the Environmental Unit in the International African Institute

recently emphasized this in his introduction to *African Environment: Problems and Perspectives*: ". . . just as natural processes such as lack of rainfall affect social structures", he argued, "so social processes such as economic 'development' can affect natural systems, 'causing' famine and soil erosion for example". He went on:

"[I]n a continent where international ties of dependency, massive international labour migration and multinational companies prevail and in a world where growth does not necessarily mean development, and development does not necessarily bring enrichment or an increase in personal happiness, the ultimate cause of environmental problems may well be traceable to the structural imbalances between rich and poor countries and we would be right to replace the term *natural* with the more appropriate term *social* or *political* disaster".

These suggestions would strike the Guatemalan peasant as commonsense. The recent earthquake there is no longer identified as a natural event—the local inhabitants who survived are referring to the event as a "class-quake". It is a view which reflects their broad experience. That experience is the basis for an explanation of their general plight in terms of a process, not of development, but of underdevelopment—a process which their increasing vulnerability reflects. Instead of moving independently (as one school of thought argues) from a state of undevelopment to development along the lines that the now-developed west was already done, Third World countries (says an alternative, radically opposed school) have in fact been moving retrogressively from a state of undevelopment to one of underdevelopment, in a process of "marginalisation" which is not so much separate from as the price of the west's development in an increasingly interdependent world. As this has happened, the relatively deprived sectors of the populations within these countries have become even worse off.

The reality of this situation is shown in Figure 2. It is the poorest countries which tend to suffer most disaster strikes. This point ought to encourage precautionary planning to mitigate the effect of future disaster. Such precautionary planning needs to be totally integrated into planning for real development, which means the necessary concentration on the vulnerability of a population to future disaster can only be done successfully through an understanding of the marginalisation process. Emphasis on precautionary planning does not, of course, make unnecessary the valuable action invariably mounted in the aftermath of a disaster: in fact, the formation of the London Technical Group, which supplies accurate technical reports on the position after disasters and generally provides expertise on relief measures,

is to be warmly welcomed. In the long run, however, precautionary planning would be more beneficial than relief work, since it would aim to consider and alleviate the causes and not merely the symptoms of disaster.

Successful precautionary planning, then, in focusing on the population's vulnerability, depends upon the identification of cultural attitudes towards the use of indigenous resources at local and regional levels, and the incorporation into development planning of strategies to mitigate disasters; precautionary planning should be seen as the insurance chapter in any development plan. The aim would be to raise the standards of life of people currently ill-placed to resist disasters because those standards are too low. The average "cost" of a disaster strike, a rather meaningless concept when the range of impact is so great, is about $20 million. It is sufficient to say that more than $1,100 million was spent on disaster assistance in 1973; and 96 countries of the world had less than this amount of resource capital as GNP in 1973. The time is ripe for some form of precautionary planning which considers vulnerability of the population as the real cause of disaster—a vulnerability that is induced by socio-economic conditions that can be modified by man, and is not just an act of God. Precautionary planning must commence with the removal of concepts of naturalness from natural disasters.

**FIGURE 2** *Number of disaster strikes by per capita income of disaster-strike area: the large number of disasters recorded in developed countries is a reflection of the closer monitoring of disaster events.*

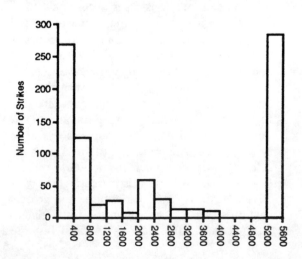

# 8

# A Contextual Model of Natural Hazard

*James K. Mitchell* ❖ *Neal Devine* ❖ *Kathleen Jagger*

On the night of 15 October 1987, a severe cyclonic storm moving across southern England generated the strongest winds that have affected the region in centuries. At least thirteen persons died, and millions of trees were blown down. Property damage was heavy in many communities. Disruption of daily life was greater than at any time since 1945. The ensuing discourse was noteworthy for the rapidity with which issues of hazard management were eclipsed by other public-policy concerns. A debate about the adequacy of the hazard-management system became a referendum about the merits of British economic and governmental restructuring during the administration of Prime Minister Margaret Thatcher. After the worldwide stock-market crash of 19 October, the storm also became a metaphor for financial catastrophe.

Human responses to this storm suggest that many Britons view natural hazard as a sociopolitical phenomenon rather than a human-ecological one. They also cast light on the contextual character of natural hazards, a little-

explored subject that has recently attracted considerable attention in the hazards-research community. Context is multilayered. It includes not only factors of space and time that confer uniqueness on specific events, independent processes that affect exposure, vulnerability, and other components of natural-hazards systems, but also the salience of hazards on public and private agendas. Based on the evidence provided by this storm, an expanded contextual model of natural hazard is proposed.

## PHYSICAL CHARACTERISTICS OF THE STORM

On 14 October 1987 a shallow depression (988 mb central pressure) formed over the Atlantic Ocean approximately 500 miles southwest of England. By the following evening this cyclonic storm had intensified (970 mb) and was accelerating toward the coast of Cornwall, which it reached at about midnight. During the next six hours the storm traversed 300 miles across England on a northeasterly track and passed out to sea over the estuary of the River Humber around dawn on 16 October (Fig. 1).

Three aspects of cyclonic storms are po-

Reprinted from *Geographical Review* Volume 79 (4) (1989), pp. 391–409. Used with permission of the American Geographical Society.

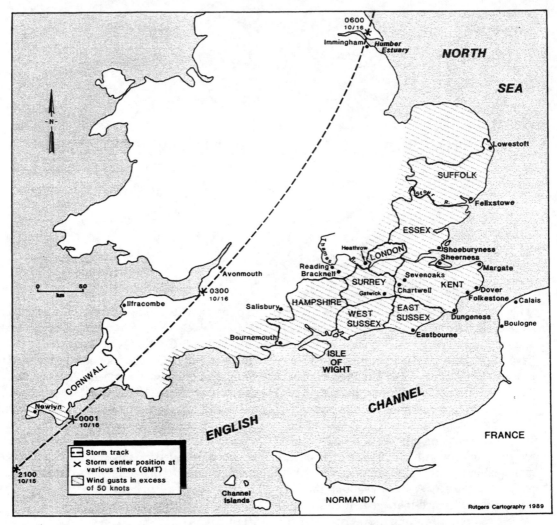

**FIGURE 1** Track of cyclonic storm across England 15–16 October 1987.

tentially damaging: winds, rain, and surge. In this case wind was the most important factor. The highest reported gusts were at the Channel Islands (110 mph) and at Pointe du Roc on the adjacent coast of Normandy (134 mph). These areas were brushed by the southern edge of the storm, and only a few places suffered heavy damage. In southern England, sustained winds of 40 to 60 knots (46–69 mph) were common during the early morning hours of 16 October. Many gusts exceeded 80 knots (92 mph), and some surpassed 90 knots (104 mph) on the Isle of Wight,

at Eastbourne and Dover, and in Essex.[1] Observers remarked on the hurricane-strength winds, and media references to this extratropical storm erroneously labeled it a hurricane.

Little rain fell during the storm, possibly because it moved so rapidly. Maximum recorded rainfall during the 24 hours preceding 9 a.m. GMT on 16 October was approximately 30 millimeters, less than on many days in southern Great Britain. However, the first half of October was unusually wet, perhaps the third-wettest fortnight since reliable rain-gauge records began in 1727.[2] As a result, the ground was well saturated before the storm arrived, and there were reports of significant flooding. Autumn had lingered, and trees bore heavy foliage that would normally have been shed before the onset of year-end gales. A combination of saturated ground, wind-resistant foliage, and thin soils on exposed hill slopes reduced the stability and resilience of trees to strong winds and contributed to the heavy losses that were incurred.[3]

A small storm surge occurred at a few places on the coast of southwestern England, for example, at Newlyn, Ilfracombe, and Avonmouth. Given the predominantly rocky upland character of this area, the surge posed little threat of flooding to people and property. Elsewhere, on the more vulnerable low-lying coasts of southern and eastern England, prevailing winds tended to suppress normal high tides, and a negative surge was recorded, as at Dover, Sheerness, Lowestoft, and Immingham.[4] Water levels were lower than normal along much of this coastline.

In contrast with other recent natural extremes in the United Kingdom, this storm was a major event. Most of the affected areas recorded their highest-ever wind speeds; in many places gusts were twice the previous record. London experienced its highest winds since systematic measurements began there in 1940, and at several sites on the southern coast wind velocity exceeded the measurement capacity of anemometers. In London and the Thames estuary the return period for winds of the strength experienced during this storm was estimated at 200 to 500 or more years (Table I).

## IMPACTS AND LOSSES

At least thirteen persons and possibly as many as twenty-five died as a direct result of the storm. Most deaths occurred singly in south-coast communities where victims were struck by falling trees and collapsing masonry. Unknown numbers of people were injured. In-sured property losses were estimated at between $450 million and $750 million.[5] Damage to facilities owned by county governments alone reached $85 million.[6] The adjusted economic losses may have been the largest associated with any natural disaster in Great Britain since 1945. Nonetheless, the losses were lighter than they could have been. Fortunately for local residents, the band of peak wind velocities was relatively narrow. The storm occurred during night hours when few people were out-of-doors or otherwise at risk and moved rapidly across the country. The heaviest losses were inflicted by falling trees. They blocked transportation routes, sometimes for more than ten days. Uprooted trees struck adjacent buildings, crushed parked vehicles, and brought down electricity-distribution lines throughout the affected region. In addition to the associated economic loss, the destruction of trees was important because of their symbolic role in British society.

## DAMAGE TO TREES AND FORESTS

Some fifteen million trees, containing approximately 3.9 million cubic meters of timber, were blown down.[7] The number of trees eventually salvaged for commercial timber is unknown, but amounts that could be processed were restricted by the high cost of moving logs to mills, by the variable quality of the wind-felled timber, and by the rapid onset of rot in the fallen trees. Tree loss assumed great significance in the United Kingdom, one of the least forested countries in western Europe. Approximately 10 percent of the total land area is classified as forest, in contrast with 28 percent in France and 30 percent in West Germany. Only Ireland, with 6 percent, is substantially less forested.[8] Losses of the scale that occurred in this storm are clearly visible on the landscape.[9] The economic impact is harder to assess because much of the

**TABLE I**  Maximum Wind Gusts at Selected Locations

| Location | Highest Reported Gusts (mph) | Return Period (years) |
|---|---|---|
| Salisbury | 81 | 20 |
| Bournemouth | 71 | 10 |
| Isle of Wight | 104 | na |
| Jersey | 98 | 15 |
| Eastbourne | 104 | na |
| Dover | 104 | na |
| Margate | 99 | 200+ |
| Gatwick Airport | 99 | 300+ |
| London Airport | 76 | 40 |
| London Weather Centre | 94 | 120 |
| Shoeburyness (Essex) | 100 | 500+ |
| Great Yarmouth | 90 | na |

*Source:* Preliminary Summary, text footnote 1.
na = not available.

affected woodland was not commercially managed.

Widespread concern about loss of woodland was also attributable to the types of trees that were damaged, their location, and public attitudes toward tree preservation. In the past forty years most severe storms have affected rural parts of northern England and Scotland, where large, publicly owned conifer plantations predominate. By contrast, the storm of 15 October damaged many small, privately owned, broad-leaved, ornamental woodlots and orchards in urban, suburban, and exurban areas throughout heavily populated sections of southeastern England.

Slightly more than half of the total losses, 1,908,000 cubic meters, were sustained by broad-leaved woodlands that constitute approximately one-quarter of all British forests.[10] The storm felled almost twice the volume of broad-leaved trees that are harvested commercially in any typical year. The volume of conifers downed, which was 1,930,000 cubic meters, was less than half the average annual commercial harvest of these species. Pines were the most seriously affected conifer.[11] The National Farmers' Union reported that 250,000 fruit trees were uprooted and a further 500,000 were badly damaged.[12] Impacts were especially large in East Sussex, West Sussex, and Kent, where 24 percent, 19 percent, and 18 percent of the original standing volume of trees were blown down (Table II).[13]

Broad-leaved trees are key landscape features in southern England. They are strongly preferred to conifers for aesthetic, amenity, and recreational reasons. Additionally, individual stands are venerable and well-known features of local landscapes that have been carefully managed for centuries. Trees and their management have long had symbolic and practical importance in England. Many British nature-conservation organizations were formed to encourage the replanting of deforested areas and to protect existent trees and wooded landscapes against unwanted changes.[14] Many of these efforts were intended to protect broad-leaved species, not conifers. Indeed, there is significant public opposition to large-scale planting of conifers, because of their allegedly monotonous appearance, their perceived ecological impoverishment, and their inimical habitat for favored forms of wildlife. Landscaped parks and gardens, created during the eighteenth and nineteenth centuries, are numerous in southern England and contain rare or valuable trees and cherished wooded vistas. Of

**TABLE II** *Fallen Timber as a Percentage of Original Standing Volume*

| County | Publicly owned | | Privately Owned | |
|---|---|---|---|---|
| | Conifers | Broad-leaved | Conifers | Broad-leaved |
| Suffolk | 27 | 3 | 20 | 8 |
| Essex | 5 | 0 | 7 | 4 |
| Kent | 44 | 10 | 25 | 18 |
| East Sussex | 51 | 23 | 35 | 18 |
| West Sussex | 20 | 22 | 30 | 18 |
| Surrey | 8 | 0 | 10 | 5 |
| Hampshire | 3 | 3 | 4 | 2 |

*Source:* Forest Windblow Action Committee, text footnote 7.

1,200 gardens listed in the Register of Parks and Gardens of Specific Historical Interest in England, 350 were affected by the storm. One hundred of them were badly damaged, and another thirty to forty were very badly damaged, that is, entirely devastated.[15] At the Royal Botanic Gardens in Kew, 10 percent of the trees were lost, including many irreplaceable specimens gathered from around the world.

Throughout England, especially in the southeast, pruning and removal of trees are strictly controlled by the local-government mechanism of tree-preservation orders. After the storm property owners in the small villages of the London greenbelt and adjoining areas alleged that these regulations exacerbated damage because they prevented the replacement of old trees that were vulnerable to wind damage with young, resilent ones.[16] Nonetheless, the urge to minimize human interference with trees ran deep, even in the wake of the storm. One ecological-interest group, Common Ground, counseled homeowners to leave some fallen trees untouched because "a fallen tree is not a dead tree . . . with only a quarter of its roots left in the ground, it may survive and become a fascinating old character."[17] Newspapers and magazines carried solicitations for donations to replant and restore damaged trees and woodlots, in contrast with the few appeals for funds to assist human victims. Sponsoring organizations included Men of the Trees, the National Trust, the Woodland Trust, English Heritage, and the Conservation Foundation, to mention only a few.

When the storm leveled woodlands in southern England, it did more than alter the physical environment; it removed highly valued cultural artifacts that served as historical points of reference, tourist attractions, and centerpieces of local conservation programs. Moreover, it affected a disproportionately large number of communities where luxurious, well-tended deciduous woodlands had long been an expression of preferred social status. Damage to these trees was not mere physical loss; it was a symbolic wound to the national heritage.

## DAMAGE TO STRUCTURES AND INFRASTRUCTURE

The damage to buildings by the storm was less than might have been expected. Recently constructed industrial plants, houses, and office buildings survived intact, with the exception of some broken windows and missing sections of roofs. An expert at the Building Research Station judged that even the highest wind gusts did not exert forces that were much in excess of modern design standards.[18] Certain types of structures sustained heavy damage—old farm outbuildings, commercial nursery glasshouses, mobile homes, flat-roofed structures with overhangs, seafront amusement arcades, schools, churches, and light-aircraft hangars. Schools and other publicly owned buildings were especially hard hit. Damage reports for 700 county-owned buildings in East Sussex approximated the average annual total for wind damage to all buildings in England. Nevertheless, private property owners incurred losses at least four times those of public bodies.

Commercial shipping experienced relatively light losses. Strong winds stalled all seven of the rotary radar scanners monitoring traffic on the Thames estuary, but few ships were in motion during the night. English Channel ferries were delayed; one was driven aground but later refloated. Two lives were lost when a bulk carrier capsized in Dover harbor; the area around Felixtowe harbor was evacuated after a freighter with a toxic cargo broke loose from moorings and rammed an oil jetty. A detention ship housing illegal Tamil immigrants went adrift on the River Stour and ran into a sandbank. A supply vessel broke down and threatened to collide with offshore gas-drilling platforms, and at least one platform was evacuated as a result. Hundreds of moored small craft were wrecked along the southern and eastern coasts.

For most people the primary impact of the storm was the disruption of electrical service that followed loss of transmission lines from power stations. The cross-channel transmission link to France was severed as were lines from the Dungeness nuclear power station and from fossil-fuel stations in the Thames valley. Electrical-power transmission lines were not adequately protected against wind damage, and a countrywide overload was averted only when the electrical-system managers isolated southeastern England from the rest of the grid. Blackouts occurred in 1.7 million houses and businesses throughout the Greater London area and surrounding counties, but there was an orderly and rapid restoration of power to all areas within nine hours, except where local

distribution lines had been lost. Ten days after the storm almost 50,000 customers were still without electricity. Additionally, approximately 3,000 miles of telephone lines, together with 3,000 poles, were blown down.

Loss of electrical power set in motion a cascade of other effects. The Underground system stopped. Communities dependent on electrically pumped wells were forced to obtain water from emergency tankers. Banks, the London Stock Exchange, and other financial institutions closed or suspended operations because computers and telecommunications ceased to function. Refrigeration equipment was inoperable, automated traffic controls failed, and domestic cooking, heating, and lighting were seriously impaired.

The storm was a signal event in modern Great Britain. Actual damage and threat of losses reminded residents that southern England remains vulnerable to sudden, disruptive natural hazards. The supposedly benign, if damp, weather became more than a ritual topic of conversation. People who had previously not questioned the existence of adequate safeguards against disasters began to reassess existent public policies.

## POLICY ISSUES AFTER THE STORM

The unanticipated arrival of the storm took most people, even weather experts, by surprise. Not until midnight of 15 October did forecasters begin to inform people in southern England, including emergency services, that a severe storm was almost upon them. Later the Meteorological Office drew strong criticism for failing to alert the public. As late as a television news broadcast at 6 p.m. on 15 October, a forecaster, responding to a viewer's query, flatly discounted the possibility of unusual weather that night. After the storm, weather officials pointed out that advisories of Beaufort Force 9 winds had been issued for the English Channel early on 15 October and that the warning had been upgraded to Force 10 later in the day. Advisories of Force 9 winds in the Thames estuary were issued at 1:15 p.m., and British Rail, which operates cross-channel ferry services, received a similar advisory at 5:30 p.m. The actual wind velocities were much higher. Forecasters acknowledged that they had not sufficiently stressed the possibility of high winds over land.[19]

Precise forecasting of midlatitude storms is difficult in the United Kingdom. North Atlantic weather systems are highly changeable. Great Britain is a relatively small island, and slight changes in the path of oncoming storms can produce strikingly different weather conditions onshore. Also there are no large land areas immediately to the west from which detailed reports of developing weather might be obtained. British forecasters depend on satellite data, government weather ships, and commercial ships and aircraft for most information. These sources proved inadequate. Some weather professionals alluded to the limitations of forecasting, but others reminded critics that the Meteorological Office had no statutory responsibility for issuing public warnings of severe storms. They also questioned whether warnings would have made a significant difference to human safety because most people were asleep in relatively secure structures.

More importantly, from the viewpoint of its long-term implications for public policy, few informed observers believe that storm warnings would be beneficial in the United Kingdom at any time of day. For example, most emergency managers have concluded that British public and private institutions are unaware of appropriate protective actions and have almost no responsibility to implement such measures. Nor did the public demand improved warnings in the weeks after the storm. Most people accepted the official view that it was a rare event that did not warrant major expenditures on improved protection. By erroneously labeling the storm a hurricane, the mass media placed it in a category of events that Britons do not need to plan for and thereby may have helped curtail an extended debate about upgrading emergency protection.[20]

Although there was only limited analysis of storm-protection policy, there was widespread discussion of reductions in British public spending that forced the withdrawal of weather ships from the area in which the storm developed. Controversy over warnings also contributed to a debate about the relative merits of rival computer-based weather forecasting technologies preferred by the Meteorological Office at Bracknell, the European Centre for Medium Range Forecasting at Reading, and the U.S. National Meteorological Center in Maryland.[21] After making allowances for the limitations of forecasting models used by these agencies and by their own Meteorological Service, French forecasters issued severe storm warnings for northwestern France on 15 October. British forecasters accepted the models' predictions without modification and decided that a major storm was unlikely to occur. Recently it has been recognized that differences in forecasts provided by the various models should act as a caution against exclusive reliance on any single source.[22]

It is possible to interpret both the argument about lack of weather ships and the debate about computer systems as outcomes of attempts to diagnose failure in hazard-management systems and to allocate blame. But they are more than that: they reflect a common tendency to favor technological fixes for hazards over nonstructural alternatives. In England during the late 1980s they are also outgrowths of a continued debate about reduced spending and reshuffled spending priorities of the government of the United Kingdom. To a significant degree contention about economic policy subsumed and truncated the debate about hazard management. The need for improved warning, evacuation, and sheltering systems was not seriously assessed by central-government policy makers. Neither were the choices among available alternatives. In other words, economic concerns effectively displaced the topic of disaster preparedness from the policy agenda. Similar pressures affected the public discourse about postdisaster emergency response and recovery.

## ADEQUACY
## OF EMERGENCY
## RESPONSES

Because of their infrequency, natural disasters are not widely feared in Great Britain.[23] There is little research on disaster management and only a limited tradition of automatic social support in the wake of disasters. In view of these circumstances, it is not surprising that British governments appear to give natural-disaster planning a low priority. Moreover, civil emergencies are a security-conscious subject in the British government, which is itself an institution practicing much secrecy.[24]

British civil-defense policies are primarily designed to provide protection in the event of nuclear war, but since 1982 the central government has encouraged county governments to develop a comprehensive approach that includes protection against natural hazards.[25] Some individuals interpret this change as an attempt to assuage widespread public uneasiness about the entire concept of preparedness by linking preparedness against nuclear annihilation with protection against peacetime emergencies.[26]

Little has been published about many aspects of central-government plans for emergencies. But it is known that the emphasis is on private and local actions directed toward postdisaster emergency response and recovery. The bulk of disaster-management responsibilities falls on the shoulders of local police, fire and hospital services, and sixty-nine county and regional emergency-planning officers who devise and coordinate local-authority services.[27]

Although recent changes in the structure of governmental financing have diverted emergency funds from localities to national and county emergency-preparedness agencies, most of the immediate response to the storm was provided by local governments, voluntary organizations, and private individuals. Army units and other groups were also involved in the late stages of tree clearance and cleanup. Evidence of self-help, mutual aid among neighbors, and community cooperation was widely reported in the mass media, but apart from the professional civil-defense community and a handful of external observers, much less attention was paid to flaws in civil-protection and preparedness planning that were exposed by the storm.

One criticism leveled at British emergency-response plans is that civil-defense organizations were not ready to deal with the impact of major natural disasters. Based on information from a selection of county and local emergency managers who were interviewed by the authors of this article, it is clear that central preparedness plans and county contingency plans had negligible effect on response to the storm. Instead, local governments and private groups such as utilities, businesses, town councils, parishes, voluntary bodies, and individuals bore the brunt of storm response and recovery efforts.

Communities with well-established volunteer programs responded better than those without, but there was widespread realization that civil-defense services were inadequate for a major emergency. Disruption of communications was an especially troublesome problem. Three days after the storm 30 percent of potable water was being pumped by mobile generators, and 270 sewage works lacked power. Schools were expected to have important roles as shelters and providers of postdisaster services, yet most were inadequately designed to withstand high winds and many were severely damaged. It was also argued that the storm raised questions about the value of existent nuclear-warfare response plans.

Despite a considerable volume of evidence that all was not well with the emergency-response system, very little of the public discourse in the aftermath of the storm focused on its difficulties. Although a debate about the adequacy of emergency management might have occurred eventually, it was forestalled by the emergence of other problems, the first of which happened three days after the storm.

## A STOCK-MARKET CRASH

The international stock-market crisis that occurred on 19 October 1987 brought about the single largest loss of value in shares traded on the world's stock exchanges since the crash of 1929, which ushered in the Great Depression.[28] The 1987 crash quickly struck most of the principal stock exchanges throughout the world and thereby confirmed the emergence of a technologically driven, global financial network.[29]

The crash started at the New York Stock Exchange but affected all principal equity markets within the next twenty-four hours. One possible link between the storm and the crash is suggested by the closure of the London Stock Exchange on 16 October because of electricity cuts attributable to the storm. Without power for computers and other electronic equipment, it was almost impossible for traders to communicate either among themselves or with markets elsewhere. It can be argued that the disruption in London helped to trigger even larger subsequent market losses there and in New York.[30] Over the storm weekend, British residents had time to absorb information about the volatile state of the market during the previous week and the magnitude of the storm damage. Both pent-up selling pressure and the prospect of substantial storm-related losses among British insurance companies and other British investors may have encouraged immediate selling when the London exchange reopened the morning of 19 October.[31] This pattern might have contributed to the perception of an impending global financial crisis that encouraged accelerated selling later in New York. The circumstantial and suppositional evidence raises intriguing possibilities of this sort of linkage.

Between 16 October and 19 October the mass media devoted considerable attention to the storm. Some comments were prescient about future economic disasters. However, the stock-market crash shifted attention to economic and financial topics. The storm was not entirely ignored, but it became a widely used metaphor for the economic upheaval in financial markets. Echoing the criticisms of weather forecasts during the storm, cartoons depicted dazed investors who blamed the Meteorological Office for its failure to warn them about falling share prices. Under the title Plant a Yuppie Week, one satirical magazine drew parallels between the plight of young entrepreneurs who had suffered financial reverses and the trees that lay scattered across the countryside. Images like these show how mass-media reporting of the storm and public discourse about it changed rapidly after 19 October. The speed of this transition seems remarkable; in the process the policy implications received only cursory examination.

Clearly the market crash altered the attention that was given to the storm and to natural-disaster management in general. At a time when public interest might have been focused on the shortcomings of British domestic policies for disaster preparedness and response, attention was abruptly refocused on global financial trading and international policy. Nevertheless, one domestic issue proved durable.

## WHO PAYS THE COST?

Of the several issues that arose in the aftermath of the storm, payment of disaster-recovery costs attracted the most sustained public comment. Although insurance reimbursements covered many private expenses, they did not cover the costs of public or quasi-public responsibilities: clearing debris from roads, repairing public structures, restoring communications and other infrastructure, and replanting trees. Initially the central government was determined to resist attempts to involve it in paying costs of disaster relief and recovery. Instead local municipalities were encouraged to be self-reliant. To government leaders, repayment of storm costs became a test of sweeping new policies initiated under the administration of Prime Minister Thatcher.

An emphasis of this government has been to return many services and activities to the private sector. The powers of local government have generally been reduced, and those of some regional-government units, like the Greater London County Council, have been abolished. There has been a concomitant increase in the role of the central government and concentration of authority in it. The reorganization altered the financing of local municipalities and thereby affected the pace and the effectiveness of recovery after the storm. Most of the affected counties lacked contingency reserves and sought financial assistance from the central government.

The dispute over cost reimbursements took on added notoriety because voters in most districts that suffered heavy storm damage had staunchly supported the Conservative government. Its leaders faced the politically embarrassing choice of denying assistance to their supporters or admitting that the funds necessary to carry out effective recovery services and activities were being withheld. The demonstration of need for emergency reserves cast doubt on the wisdom of the general cutbacks in public spending. The central government initially refused assistance, but a compromise was eventually reached that required local governments to pay storm-related costs up to the sum that would be raised by a 1 percent local-tax increase. The central government agreed to reimburse 75 percent of costs in excess of this figure but also announced that some

of these funds would be recouped by reducing future contributions to local-government budgets. Public discourse on reimbursement gradually was replaced by three other concerns: the implications of the stock-market crash, new proposals for local-government funding, and reemergence of unresolved questions about the construction of the cross-channel tunnel.

During 1987 the Thatcher ministry unveiled plans to change the system of financing local government in England and Wales. Traditionally local governments had been funded by levies that are similar to real-estate taxes levied in the United States. The Thatcher proposal called for their replacement by a poll tax, based on the numbers of voters residing in a local municipality. Population rather than property would become the basis of local-government funding, and the capability of local units to meet demands from their constituents would be affected. Natural-hazard management and disaster policy might be changed, but so too would almost every aspect of local government. It was no surprise that the hazard issues were submerged in the uncertainties generated by this proposal. Additionally, at least for residents of Kent and adjacent counties, eclipse of the hazard issues was paralleled by another wave of concern about the possible impacts of the channel tunnel.

Even before the storm the physical, economic, and social impacts of the channel tunnel were a concern of many residents in Kent, East Sussex, and West Sussex. Awareness of this issue was heightened by the announced sale of shares in the Eurotunnel venture in November 1987 and the publicity campaign that opened in October. Work commenced in 1986 on a system of three tunnels to accommodate railroad trains that will carry automobiles, trucks, and people under the English Channel. Scheduled for completion in 1993, the tunnel project is expected to handle 13.8 to 15.7 million people and 7 million metric tons of freight annually.[32] Journey time from shore to shore will be reduced from 140 minutes to less than an hour, and the trip between London and Paris will take four and one-quarter hours rather than seven.

In the past, British opposition to this type of cross-channel link had centered on fears of military invasion and cultural penetration, but recent concerns have focused on economic and environmental impacts. Most of the current transchannel services and their communities will be severely affected, and ferry sites like Dover, Folkestone, Calais, and Boulogne may face serious unemployment and business closures. Places near the tunnel entrances and access routes will experience large increases in road and railroad traffic together with land use changes and undesirable environmental impacts. New railroad lines may be necessary between London and the tunnel, and for many people in this crowded section of England uncertainties about such changes constitute additional environmental and economic stresses. As the storm receded into memory, the tunnel reemerged as a continuing threat to livelihoods and lifestyles.

## THE STORM AND MODELS OF NATURAL HAZARD

After more than forty years of research, geographers have achieved considerable understanding of the human ecology of natural hazards.[33] The concept of hazards as external events impinging on unsuspecting people has been shed in favor of the interpretation that they emerge from interactions between people and environments. Natural hazards involve different combinations of physical processes and human activities that create a variety of risks. Exposure and vulnerability to hazard are uneven within societies, and willingness and ability to undertake protective measures vary widely. Moreover, cognition of risk influences behavior in the face of hazard, and risk perceptions are often amplified or atten-

uated by sociocultural and psychological processes.[34]

A natural-hazard system comprises two parts: a subsystem of hazard components, and a subsystem of hazard contexts (Fig. 2). The first subsystem contains four separate but interacting components: physical processes, human populations, adjustments to hazard, and net losses. Important measures of these components include risk (physical processes), exposure and vulnerability (human populations), responses (adjustments to hazard), and costs (net losses). Components modify each other through seven endogenous feedback relationships. Physical processes affect human activities (link 1), but the latter can also inadvertently change the former (link 2). Adjustments may deliberately modify physical processes (link 5) and human exposure or vulnerability (link 4), but they rarely eliminate all losses (link 6). Net losses are monitored by society (link 7), and new adjustments may be adopted if a threshold of tolerance is exceeded (link 3).

The hazard-context subsystem includes exogenous factors that interact with and modify components of hazard but are largely independent of them. Exogenous factors also change through time (dotted arrows). Many examples might be identified, but three suffice here for purposes of illustration. First, megascale changes in global environmental systems affect physical processes of hazard; these changes include movement of tectonic plates and long-term fluctuations in the concentration of atmospheric gases. Second, human populations, among others, are affected by exogenous demographic processes that fuel population expansion and migration, which sometimes encourage invasion of areas at risk. Third, other exogenous factors influence the range of possible adjustments to hazard. For example, the number and the variety of adjustments tend to increase over time as scientific knowledge increases and new technologies are developed. Much hazard research currently under way seeks to measure the rel-

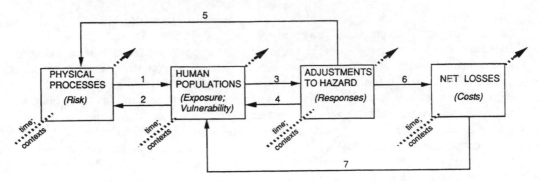

**FIGURE 2**  A natural-hazard system.

ative contributions of hazard components and to explore links between them. More is known about endogenous aspects of the hazard system than about exogenous factors. Specifically there is a dearth of research on contexts of hazard that are created by the interaction of exogenous factors and hazard components.

## CONTEXTS

Natural disasters occur in specific, often unique, contexts. As suggested above, contexts are large problem sets that include or overlap with natural-hazard components. Conversely, those components are dependent on or partly determined by subsets of hazard contexts. Contexts may be spatial, temporal, organizational, environmental, sociocultural, economic, political, or of some other form. Because hazard contexts are largely independent of natural-hazard systems and are subject to a host of exogenous factors, they are highly changeable. This characteristic poses a dilemma for a hazards analyst. It is very difficult to derive broadly applicable conclusions from studies of specific natural disasters, because contexts are likely to vary. But evidence about the performance of structures in disasters and the behavior of humans before, during, and after disasters is often used to validate hypotheses about natural hazards and human responses to them.[35]

It has been argued elsewhere that the study of hazard contexts requires the development of new investigative strategies.[36] Three alternative approaches have been proposed: analysis in a broad hazard context, analysis of components and contexts using different methods and concepts of each, and examination of linkages connecting hazard components and contexts. The third alternative is explored below.

The 15 October 1987 storm illustrates a variety of hazard contexts composed of the structures and processes of physical and human systems that exist at specific places and times. A detailed analysis of all hazard contexts is beyond the scope of this article, but several examples are warranted. First, storm damage was not solely a function of high wind velocities. It was greatly influenced by local environmental conditions at the time the storm struck. The old deciduous woodlands that dominate upland parts of southern England were very susceptible to wind damage after a period of heavy rains that occurred before the onset of leaf fall. Second, the willingness of the British government to deviate from its established policy by providing limited assistance to stricken communities derived from the disproportionate impact of the storm on the constituencies of party-in-power supporters. Finally, public debate about ways of improving hazard-management policies was truncated because a storm of this magnitude was perceived to be very rare, and people were more concerned about the possible effects of a major economic depression.

In the first example, the storm acted as a forcing mechanism for physical processes that were temporarily close to failure. Timing was crucial: if the storm had occurred a month earlier or later, it would have probably caused much less physical destruction. In the second example, the spatial incidence of the storm had central importance. If it had affected more northerly parts of Great Britain, the likelihood of any change in natural-hazards policy would have been negligible. Populations that live in such places are better adjusted to high winds, and in contemporary Great Britain they have relatively limited political influence on central-government policy. Instead the storm affected areas that were ripe for disaster and people who were capable of bringing political pressure to bear. Thus the physical magnitude and human impact of a specific hazard event were amplified by temporal and spatial contexts in which it occurred, and the potential for far-reaching changes of hazard-management policy was increased. If the storm had occurred during daylight, losses would have been even greater. However, few, if any, significant policy changes resulted, because the storm was quickly overshadowed by the stock-market collapse. Both events can be labeled emergencies or crises, but the stock-market upset was widely perceived as being more portentous than the storm. In this case, a threat to countrywide economic security muted the significance of a natural disaster. In the public calculus, one crisis outranked the other.

The speed and completeness with which the storm disappeared from the policy arena cannot be attributed simply to the perceived primacy of the economic crisis. The disappearance was also aided by a widespread perception that the storm was mainly a sociopolitical phenomenon, not a human-ecological one. Most commentators assumed that developed countries like Great Britain should be able to manage natural hazards effectively. Losses were typically attributed to the inadequacies of protective systems brought about by cuts in governmental spending. Other complex scientific, technical, and human issues, such as the changing character of natural-hazard and alternative hazard-management policies, were glossed over. Public discourse became largely an effort to affix blame, chiefly by indicting national economic policy. An opportunity for a searching examination of hazard-management policy was lost.

Both the storm and the public response to it may be examined in terms of the model proposed here. The sequence of developments that followed the storm fits well within the model's hazard-component subset. A low-probability physical event affected a vulnerable population, exceeded coping capacity, inflicted heavy losses, and generated pressures for the adoption of new adjustments. Turning to the hazard-context subset, it is clear that several components were amplified by contextual factors. These include physical processes (failure-susceptible soils, slopes, and vegetation), human populations (ability to mobilize political power), and net losses (prolonged disruption in key political constituencies). Although there was an embryonic debate about the adequacy of adjustments to hazard, the search for improved ones was confined to a canvas of technological fixes and was stalled by other contextual factors. They included a perception that the storm was a rare event, unlikely to be repeated; a judgment that lack of economic investment was primarily responsible for failures of protection and that there was little need for a thorough overhaul of hazard-management policy; and a belief that storm-caused losses would pale to insignificance compared with a major collapse of global trading relations. In short, the stimulus for reconsideration of existent hazard-management policy that was provided by an unusually heavy and strategically concentrated burden of losses was offset by other pressures for inaction. These pressures arose both within the natural-hazard system itself, through the mechanism of risk cognition of low-probability events, and from outside this system in the context of other crises that competed for public attention.

## CONCLUSION

The British experience confirms that an extreme physical event can inflict a major disaster without prompting significant changes in hazard-management programs or policies. This familiar circumstance is exemplified by the continuing plight of many people in third-world countries, where poverty and lack of other resources often impede effective hazard management. But the British storm inflicted heavy losses on one of the most developed countries in the world without leading to significant changes in hazard management. Even in developed countries, favorable conditions for improved management of hazards are not readily found.

The creation of a stimulus for the introduction of new programs and policies requires more than the temporary coming together of great and unusual physical risks, vulnerable and resourceful populations, failed responses to hazard, and heavy losses. As the British storm demonstrates, hazard components are embedded in a variety of contexts that may or may not favor the introduction of new policies. Measures to reduce disasters are more likely to be taken when contexts are changing so as to increase the likelihood of future disasters. Even then, it is possible that other crises will outrank hazards on policy agendas.

To increase the prospects for successful reduction of disasters it is crucial for analysts and managers to understand the dynamics of hazard contexts, to chart their trends, and to broaden the process of adjusting to include hazard contexts. Where societies are beset by different types of crises, it is also desirable to design new public policies in light of their interrelated character.

The year 1990 marks the beginning of the International Decade for Natural Disaster Reduction, a global enterprise sponsored by the United Nations.[37] The observations presented here suggest that projects undertaken during this decade will benefit from close collaboration among many types of people. They should not be confined only to scientists and managers whose work focuses specifically on hazards. Instead they should include the broad spectrum of organizations and interest groups that affect the many contexts of hazard.

*Acknowledgments* This research was funded by the Natural Hazards Research and Applications Information Center, University of Colorado, under a grant from the National Science Foundation. We acknowledge the help received from many informants in the United Kingdom, especially Dennis Parker, John McWhirter, Alan Heasman, M. S. Shaw, and W. M. Powell.

NOTES

[1] A Preliminary Summary of Information on the Exceptionally Strong Winds of 16 October 1987 over the South of England, Meteorological Office Advisory Services, Bracknell, 1987.

[2] Summary of Weather over England, Wales and Northern Ireland: 7 to 22 October 1987, Meteorological Office Advisory Services, Bracknell, 1987.

[3] Mike Sykes, In the Wake of the Storm: Planting Amenity Trees, *NERC News* (January 1988): 17–19.

[4] Personal communication from M. S. Shaw, conveying tidal data analyzed by Proudman Oceanographic Laboratory, National Environment Research Council, Birkenhead, England, 19 February 1988.

[5] *Times* (London), 17 October 1987; *Sunday Times* (London), 18 October 1987; *Independent* (London), 20 October 1987.

[6] County Council Management Report, London, 30 October 1987.

[7] Forest Windblow Action Committee: Report to the Forestry Commissioners, Forestry Commission, Edinburgh, January 1988.

[8] Forestry Facts and Figures 1986–87, Forestry Commission, Edinburgh, 1987.

[9] *Country Life* (19 November 1987): 82–87.

[10] Forestry Facts and Figures, footnote 8 above.

[11] Sykes, footnote 3 above.

[12] *Times* (London), 3 November 1987.

[13] Forest Windblow Action Committee, footnote 7 above.

[14] H. L. Edlin, Trees, Woods and Man (London: Collins, 1956), 96–131.

[15] *Observer* (London), 22 November 1987.

[16] *Times* (London), 24 October 1987; *Guardian* (London), 24 October 1987.

[17] *Observer* (London), 1 November 1987.

[18] Personal communication from Nicholas Cook, Building Research Station, 2 November 1987.

[19] *Civil Protection* (Winter 1987): 3–5.

[20] *Country Life* (5 November 1987, 19 November 1987).

[21] *Guardian* (London), 19 October 1987; *Independent* (London), 20 October 1987; Richard A. Kerr, Whom to Blame for the Great Storm?, *Science* 239 (11 March 1988): 1238–39.

[22] Richard A. Kerr, Telling Weathermen When to Worry, *Science* 244 (9 June 1989): 1137–39.

[23] John C. Chicken, Hazard Control Policy in Britain (Oxford: Pergamon Press, 1975); A. H. Perry, Environmental Hazards in the British Isles (London: Allen and Unwin, 1981); S. John Harrison, Climatic Hazards in Scotland (Norwich: Geo Books, 1985); Mike Walsh, Disasters: Current Planning and Recent Experience (London: Edward Arnold, 1989).

[24] Timothy O'Riordan, The Politics of Environmental Regulation in Great Britain, *Environment* 30, No. 8 (October 1988): 5–9, 39–44.

[25] Duncan Campbell, War Plan UK: The Truth about Civil Defence in Britain (London: Burnett Books, 1982).

[26] Bruce Kent, *Civil Protection* (Autumn 1987): 6–7.

[27] Walsh, footnote 23 above, 86–90.

[28] *Guardian* (London), 24 October 1987.

[29] Geoffrey Dobilas, Information Technology and Simultaneous Financial Markets: The Crash of October 1987, *Geography Discussion Papers New Series 23*, London School of Economics, Graduate School of Geography, London, 1988.

[30] *Observer* (London), 22 November 1987.

[31] *Times* (London), 24 October 1987; *Daily Telegraph* (London), 31 October 1987.

[32] Murray Hughes, Racehorse on Rails, *Geographical Magazine* (April 1988): 36–43.

[33] James K. Mitchell, Hazards Research, *in* Geography in America (edited by Gary Gaile and Cort Willmott; Columbus, Ohio: Merrill, 1989), 410–24.

[34] Roger E. Kasperson and others, Social Amplification of Risk: A Conceptual Framework, *Risk Analysis* 8, No. 2 (1988): 177–87.

[35] James K. Mitchell, Human Dimensions of Environmental Hazards: Complexity, Disparity and the Search for Guidance, *in* Nothing to Fear: An Examination of Risks and Hazards in American Society (edited by Andrew Kirby; Tucson: University of Arizona Press, forthcoming).

[36] Mitchell, footnote 35 above.

[37] James K. Mitchell, Confronting Natural Disasters: An International Decade for Natural Hazards Reduction, *Environment* (30 March 1988): 25–29.

# THE SOCIAL AMPLIFICATION OF RISK: A CONCEPTUAL FRAMEWORK

*Roger E. Kasperson* ❖ *Ortwin Renn* ❖ *Paul Slovic*
*Halina S. Brown* ❖ *Jacque Emel* ❖ *Robert Goble*
*Jeanne X. Kasperson* ❖ *Samuel Ratick*

## 1. RISK IN MODERN SOCIETY

The investigation of risks is at once a scientific activity and an expression of culture. During the twentieth century, massive governmental programs and bureaucracies aimed at assessing and managing risk have emerged in advanced industrial societies. Despite the expenditure of billions of dollars and steady improvements in health, safety, and longevity of life, people view themselves as more rather than less vulnerable to the dangers posed by technology. Particularly perplexing is that even risk events with minor physical consequences often elicit strong public concern and produce extraordinarily severe social impacts, at levels unanticipated by conventional risk analysis.

Several difficult issues require attention:

♦ The technical concept of risk focuses narrowly on the *probability* of events and the *magnitude* of specific consequences. Risk is usually defined by multiplication of the two terms, assuming that society should be indifferent toward a low-consequence/high-probability risk and a high-consequence/low-probability risk with identical expected values. Studies of risk perception have revealed clearly, however, that most persons have a much more comprehensive conception of risk. Clearly, other aspects of the risk such as voluntariness, personal ability to influence the risk, familiarity with the hazard, and the catastrophic potential shape public response.[1,2] As a result, whereas the technical assessment of risk is essential to decisions about competing designs or materials, it often fails to inform societal choices regarding technology.[3]

♦ Cognitive psychologists and decision researchers have investigated the underlying patterns of individual perception of risk and identified a series of heuristics and biases that govern risk perceptions.[4,5] Whereas some of these patterns of perception contrast with the results of formal reasoning, others involve legitimate concern about risk characteristics that are omitted, neglected, or underestimated by the technical concept of risk. In addition, equity issues, the circumstances surrounding the process of generating risk, and the timeliness of management response are considerations, important to people, that are insufficiently addressed by formal probabilistic risk analysis.[6,7]

♦ Risk is a bellwether in social decisions about technologies. Since the resolution of social conflict requires the use of factual evidence for assessing the validity and fairness of rival claims, the quantity and quality of risk are major points of contention among participating social groups.

Reprinted from *Risk Analysis* Volume 8 (2) (1988), pp. 177–87. Used with permission of the Society for Risk Analysis.

As risk analysis incorporates a variety of methods to identify and evaluate risks, various groups present competing evidence based upon their own perceptions and social agenda. The scientific aura surrounding risk analysis promotes the allocation of substantial effort to convincing official decision makers, and the public, that the risk assessment performed by one group is superior in quality and scientific validity to that of others. Controversy and debate exacerbate divergences between expert and public assessment and often erode confidence in the risk decision process.[8,9]

In short, the technical concept of risk is too narrow and ambiguous to serve as the crucial yardstick for policy making.

Public perceptions, however, are the product of intuitive biases and economic interests and reflect cultural values more generally. The overriding dilemma for society is, therefore, the need to use risk analysis to design public policies on the one hand, and the inability of the current risk concepts to anticipate and explain the nature of public response to risk on the other. After a decade of research on the public experience of risk, no comprehensive theory exists to explain why apparently minor risk or risk events,[a] as assessed by technical experts, sometimes produce massive public reactions, accompanied by substantial social and economic impacts and sometimes even by subsequently increased physical risks. Explaining this phenomenon, and making the practice of risk analysis more sensitive to it, is one of the most challenging tasks confronting the societal management of risk. This paper takes up that challenge.

The explanations that have emerged, while affording important insights, have been partial and often conflicting. The past decade has witnessed debates between the "objectivist and subjectivist" schools of thought, between structuralistic and individualistic approaches, between physical/life scientists and social scientists. Even within the social sciences, psychologists see the roots of explanation in individual cognitive behavior,[10] a claim extensively qualified by anthropologists, who insist that social context and culture shape perceptions and cognition,[11,12] and by analysts of technological controversies, who see "stakeholder" interaction and competing values as the keys.[13] The assumption underlying these debates is that the interpretations are mutually invalidating. In fact, we shall argue, the competing perspectives illuminate different facets of the public experience of risk.

A comprehensive theory is needed that is capable of integrating the technical analysis of risk and the cultural, social, and individual response structures that shape the public experience of risk. The main thesis of this article is that risk events interact with psychological, social, and cultural processes in ways that can heighten or attenuate public perceptions of risk and related risk behavior. Behavioral patterns, in turn, generate secondary social or economic consequences but may act also to increase or decrease the physical risk itself. Secondary effects trigger demands for additional institutional responses and protective actions, or, conversely (in the case of risk attenuation), impede needed protective actions. The social structures and processes of risk experience, the resulting repercussions on individual and group perceptions, and the effects of these responses on community, society, and economy compose a general phenomenon that we term *the social amplification of risk*. This article sets forth an initial conceptualization of the elements, structure, and processes that make up this phenomenon.

## 2. BACKGROUND

The technical assessment of risk typically models the impacts of an event or human activity in terms of direct harms, including death, injuries, disease, and environmental damages. Over time, the practice of characterizing risk by probability and magnitude of harm has drawn fire for neglecting equity issues in relation to time (future generations), space (the so-called LULU or NIMBY issue), or social groups (the proletariat, the highly vulnerable, export of hazard to developing countries). It also has become apparent that the consequences of risk events extend far beyond direct harms to include significant indirect impacts (e.g., liability, insurance costs, loss of confidence in institutions, or alienation from community affairs).[14] The situation becomes even more complex when the analysis also addresses the decision-making and risk-management process. Frequently, indirect impacts appear to be dependent less on the direct outcomes (i.e., injury or death) of the risk event than on judgments of the adequacy of institutional arrangements to control or manage the risk, the possibility of assigning blame to one of the major participants, and the perceived fairness of the risk-management process.

The accident at the Three Mile Island (TMI) nuclear reactor in 1979 demonstrated dramatically that factors besides injury, death, and property damage can impose serious costs and social repercussions. No one is likely to die from the release of radioactivity at TMI, but few accidents in U.S. history have wrought such costly societal impacts. The accident devastated the utility that owned and operated the plant and imposed enormous costs—in the form of stricter regulations, reduced operation of reactors worldwide, greater public opposition to nuclear power, and a less viable role for one of the major long-term energy sources—on the entire nuclear industry and on society as a whole.[15] This mishap at a nuclear power plant may even have increased public concerns about other complex technologies, such as chemical manufacturing and genetic engineering.

The point is that traditional cost—benefit and risk analyses neglect these higher-order impacts and thus greatly underestimate the variety of adverse effects attendant on certain risk events (and thereby underestimate the overall risk from the event). In this sense, social amplification provides a corrective mechanism by which society acts to bring the technical assessment of risk more in line with a fuller determination of risk. At the other end of the spectrum, the relatively low levels of interest by the public in the risks presented by such well-documented and significant hazards as indoor radon, smoking, driving without seat belts, or highly carcinogenic aflatoxins in peanut butter serve as examples of the social attenuation of risk. Whereas attenuation of risk is indispensable in that it allows individuals to cope with the multitude of risks and risk events encountered daily, it also may lead to potentially serious adverse consequences from underestimation and underresponse. Thus both social amplification and attenuation, through serious disjunctures between expert and public assessments of risk and varying responses among different publics, confound conventional risk analysis.

In some cases, the societal context may, through its effects on the risk assessor, alter the focus and scope of risk assessment. A case in point is the series of actions taken in 1984 by the Environmental Protection Agency with regard to a soil and grain fumigant, ethylene dibromide (EDB).[16] An atmosphere charged with intense societal concern about protecting the nation's food and groundwater supplies from chemical contaminants prompted the Agency to focus primarily on these two pathways of population exposure to EDB, although it was well aware that emissions of EDB from leaded gasoline were a significant source of population exposure. Consequently, the first-line receivers of the risk information—the risk managers, the mass media, the politicians, and the general public—heard from the start about cancer risks from tainted water and food, but not from ambient air. This example illustrates how the filtering of information about hazards may start as early as in the risk assessment itself and may profoundly alter the form and content of the risk information produced and conveyed by technical experts.[16]

Other researchers have noted that risk sources create a complex network of direct and indirect effects that are susceptible to change through social responses.[9,17] But because of the complexity and the transdisciplinary nature of the problem, an adequate conceptual framework for a theoretically based and empirically operational analysis is still missing. The lack of an integrative theory that provides guidelines on how to model and measure the complex relationships among risk, risk analysis, social response, and socioeconomic effects has resulted in a reaffirmation of technical risk assessment, which at least provides definite answers (however narrow or misleading) to urgent risk problems.

The concept of social amplification of risk can, in principle, provide the needed theoretical base for a more comprehensive and powerful analysis of risk and risk management in modern societies. At this point, we do not offer a fully developed theory of social amplification of risk, but we do propose a fledgling conceptual framework that may serve to guide ongoing efforts to develop, test, and apply such a theory to a broad array of pressing risk problems. Since the metaphor of amplification draws upon notions in communications theory, we begin with a brief examination of its use in that context.

## 3. SIGNAL AMPLIFICATION IN COMMUNICATIONS THEORY

In communications theory, amplification denotes the process of intensifying or attenuating signals during the transmission of information from an information source, to intermediate transmitters, and finally to a receiver.[18] An information source sends out a cluster of signals (which form a message) to a transmitter, or directly to the receiver. The signals are decoded by the transmitter or receiver so that the message can be understood. Each transmitter alters the original message by intensifying or attenuating some incoming signals, adding or deleting others, and sending a new cluster of signals on to the next transmitter or the final receiver where the next stage of decoding occurs.

The process of transmitting is more complex than the electronic metaphor implies. Messages have a meaning for the receiver only within a sociocultural context. Sources and signals are not independent entities but are perceived as a unit by the receiver who links the signal to the sources or transmitters and draws inferences about the relationship between the two. In spite of the problems of the source–receiver model, the metaphor is still powerful enough to serve as a heuristic framework for analyzing communication processes. In a recent literature review of 31 mass-communication textbooks, the source–receiver metaphor was, along with the concept of symbolic meaning, the predominant theoretical framework.[19]

Each message may contain factual, inferential, value-related, and symbolic meanings.[20] The factual information refers to the content of the message (e.g., the emission of an air pollutant is $X$ mg per day) as well as the source of the message (e.g., EPA conducted the measurement). The inferential message refers to the conclusions that can be drawn from the presented evidence (e.g., the emission poses a serious health threat). Then those conclusions may undergo evaluation according to specific criteria (e.g., the emission exceeds the allowable level). In addition, cultural symbols may be attached that evoke specific images (e.g., "big business," "the military-industrial complex," "high technology," etc.) that carry strong value implications.

Communication studies have demonstrated that the symbols present in messages are key factors in triggering the attention of potential receivers and in shaping their decoding processes.[21] If, for example, the communication source is described as an independent scientist, or a group of Nobel laureates, the content of the message may well command public attention. Messages from such sources may successfully pass through the selection filters of the transmitters or receivers and be viewed as credible. A press release by the nuclear industry, by contrast, may command much less credibility unless other aspects of the message compensate for doubts about the impartiality of the source.

Transmitters of signals may detect amplification arising from each message component.[22] A factual statement repeated several times, especially if by different sources, tends to elicit greater belief in the accuracy of the information. An elaborate description of the inference process may distract attention from the accuracy of the underlying assumptions. Reference to a highly appreciated social value may increase the receiver's tolerance for weak evidence. And, of course, a prestigious communication source can (at least in the short run) compensate for trivial factual messages. But adding or deleting symbols may well be the most powerful single means to amplify or attenuate the original message.

Amplification of signals occurs during both transmission and reception. The transmitter structures the messages that go to a receiver. The receiver, in turn, interprets, assimilates, and evaluates the messages. But a transmitter, it should be noted, is also a new information source—one that transcribes the original message from the source into a new message and sends it on to the receiver, according to institutional rules, role requirements, and anticipated receiver interests. Signals passing through a transmitter may therefore, be amplified twice—during the reception of information and in recoding.

Signal amplification in communications, then, occupies a useful niche in the overall structure of the social amplification of risk. A discussion of the proposed conceptional framework takes up the next section of this paper.

## 4. A STRUCTURAL DESCRIPTION OF THE SOCIAL AMPLIFICATION OF RISK

Social amplification of risk denotes the phenomenon by which information processes, institutional structures, social-group behavior, and individual responses shape the social experience of risk, thereby contributing to risk consequences (Fig. 1). The interaction between risk events and social processes makes clear that, as used in this framework, risk has meaning only to the extent that it treats how people think about the world and its relationships. Thus there is no such thing as "true" (absolute) and "distorted" (socially determined) risk. Rather the information system and characteristics of public response that compose social amplification are essential elements in determining the nature and magnitude of risk. We begin with the information system.

Like a stereo receiver, the information system may amplify risk events in two ways:

♦ By intensifying or weakening signals that are part of the information that individuals and social groups receive about the risk;
♦ By filtering the multitude of signals with respect to the attributes of the risk and their importance.

Signals arise through direct personal experience with a risk object or through the receipt of information about the risk object.[18] These signals are processed by social, as well as individual, amplification "stations," which include the following:

♦ The scientist who conducts and communicates the technical assessment of risk;
♦ The risk-management institution;
♦ The news media;
♦ Activist social organizations;
♦ Opinion leaders within social groups;
♦ Personal networks of peer and reference groups;
♦ Public agencies.

Social amplification stations generate and transmit information via communications channels (media, letters, telephones, direct conversations). In addition, each recipient *also* engages in amplification (and attenuation) processes, thereby acting as an amplification station for risk-related information. We hypothesize that the key amplification steps consist of the following:

♦ Filtering of signals (e.g., only a fraction of all incoming information is actually processed);
♦ Decoding of the signal;

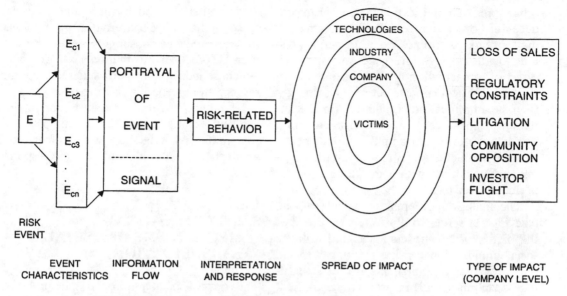

| RISK<br>EVENT | EVENT<br>CHARACTERISTICS | INFORMATION<br>FLOW | INTERPRETATION<br>AND RESPONSE | SPREAD OF IMPACT | TYPE OF IMPACT<br>(COMPANY LEVEL) |

**FIGURE 1.** *Highly simplified representation of the social amplification of risk and potential impacts on a corporation.*

♦ Processing of risk information (e.g., the use of cognitive heuristics for drawing inferences);

♦ Attaching social values to the information in order to draw implications for management and policy;

♦ Interacting with one's cultural and peer groups to interpret and validate signals;

♦ Formulating behavioral intentions to tolerate the risk or to take actions against the risk or risk manager;

♦ Engaging in group or individual actions to accept, ignore, tolerate, or change the risk.

A full-fledged theory of the social amplification of risk should ultimately explain why specific risks and risk events undergo more or less amplification or attenuation. Whether such a theory will carry the power to predict the specific kinds of public responses and the anatomy of social controversy that will follow the introduction of new risks must await the test of time. It may prove possible to identify and classify attributes of the risk source and of the social arena that heighten or attenuate the public response to risk.

Social amplifications of risk will spawn behavioral responses, which, in turn, will result in *secondary impacts*. Secondary impacts include such effects as the following:

♦ Enduring mental perceptions, images, and attitudes (e.g., antitechnology attitudes, alienation from the physical environment, social apathy, stigmatization of an environment or risk manager);

♦ Local impacts on business sales, residential property values, and economic activity;

♦ Political and social pressure (e.g., political demands, changes in political climate and culture);

♦ Changes in the physical nature of the risk (e.g., feedback mechanisms that enlarge or lower the risk);

♦ Changes in training, education, or required qualifications of operating and emergency-response personnel;

♦ Social disorder (e.g., protesting, rioting, sabotage, terrorism);

♦ Changes in risk monitoring and regulation;

♦ Increased liability and insurance costs;

♦ Repercussions on other technologies (e.g., lower levels of public acceptance) and on social institutions (e.g., erosion of public trust).

Secondary impacts are, in turn, perceived by social groups and individuals so that another stage of amplification may occur to produce third-order impacts. The impacts thereby may spread, or "ripple," to other parties, distant locations, or future generations. Each order of impact will not only disseminate social and political impacts but may also trigger (in risk amplification) or hinder (in risk attenuation) positive changes for risk reduction. The concept of social amplification of risk is hence dynamic, taking into account the learning and social interactions resulting from experience with risk.

The analogy of dropping a stone into a pond (see Fig. 1) serves to illustrate the spread of the higher-order impacts associated with the social amplification of risk. The ripples spread outward, first encompassing the directly affected victims or the first group to be notified, then touching the next higher institutional level (a company or an agency), and, in more extreme cases, reaching other parts of the industry or other social arenas with similar problems. This rippling of impacts is an important element of risk amplification since it suggests that amplification can introduce substantial temporal and geographical extension of impacts. The same graphic representation demonstrates the possibility that social amplification may, quantitatively and qualitatively, increase the direct impacts. In this case the inner circle changes it shape with each new round of ripples. Figure 2 depicts in greater detail the hypothesized stages of social amplification of risk and its associated impacts for a hypothetical corporation.

Several examples illustrate the ripple effect of risk events. Following the Three Mile Island accident, nuclear plants worldwide were shut down and restarted more frequently for safety checks, although these phases of operations (as with aircraft takeoffs and landings) are by far the riskiest operational stages. In a more recent case of risk amplification, Switzerland recalled and ordered the incineration of 200 tons of its prestigious Vacherin Mont d'Or cheese because of bacterial contamination. Rival French cheesemakers at first celebrated their good fortune until it became apparent that public concern over the event had caused worldwide consumption of the cheese, from all producers, to plummet by over 25%. An entire industry, in short, suffered economic reversal from a specific risk event.[23]

Social amplification of risk, in our current conceptualization, involves two major stages (or amplifiers)—the transfer of information about the risk or risk event, and the response mechanisms of society.

## 5. INFORMATIONAL MECHANISMS OF SOCIAL AMPLIFICATION

The roots of social amplification lie in the social experience of risk, both in direct personal experience and in indirect, or secondary, experience, through information received about the risk, risk events, and management systems. Direct experience with risky activities or events can be either reassuring (as with automobile driving) or alarming (as with tornadoes or floods). Generally, experience with dramatic accidents or risk events increases the memorability and imaginability of the hazard, thereby heightening the perception of risk.[24] But direct experience can also provide feedback on the nature, extent, and manageability of the hazard, affording better perspective and enhanced capability for avoiding risks. Thus, whereas direct personal experience can serve as a risk amplifier, it can also act to attenuate risk. Understanding this interaction for different risks, for different social experiences, and for different cultural groups is an important research need.

But many risks are not experienced directly. When direct personal experience is lacking or minimal, individuals learn about risk from other persons and from the media. Information flow becomes a key ingredient in public response and acts as a major agent of amplification. Attributes of information that may influence the social amplification are *vol-*

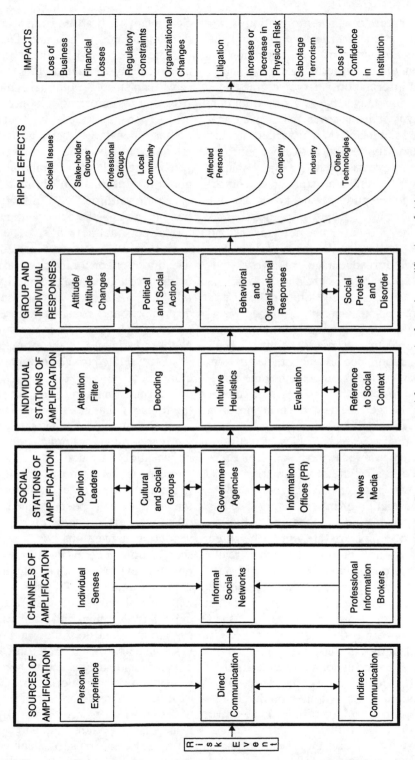

**FIGURE 2.** Detailed conceptual framework of socal amplification of risk.

119

*ume*, the degree to which information is *disputed*, the extent of *dramatization*, and the *symbolic connotations* of the information.

Independent of the accuracy and particular content of information, a large volume of information flow may serve as a risk amplifier. In an analysis of media coverage of Love Canal and Three Mile Island, Mazur argued that the massive quantity of media coverage not only reported the events but defined and shaped the issues.[25] Repeated stories, of course, direct public attention toward particular risk problems and away from competing sources of attention. Moreover, the news media tend to become battlegrounds where various participants vie for advantage. However balanced the coverage, it is unclear that reassuring claims can effectively counter the effects of fear-arousing messages.[26] In Alvin Weinberg's metaphor, it is much harder to "unscare" people than to scare them.[27] High volumes of information also mobilize latent fears about a particular risk and enhance the recollection of previous accidents or management failures or enlarge the extent to which particular failures, events, or consequences can be imagined. In this way, technologies or activities may come to be viewed as more dangerous.[2,28]

The second attribute of information is the degree to which individuals or groups dispute factual information or inferences regarded as credible by interested members of the public. Debates among experts are apt to heighten public uncertainty about what the facts really are, increase doubts about whether the hazards are really understood, and decrease the credibility of official spokespersons.[29] If the risks are already feared by the public, then increased concern is the likely result.

Dramatization, a third attribute, is undoubtedly a powerful source of risk amplification. The report during the Three Mile Island accident that a hydrogen bubble inside the reactor could explode within the next two days, blow the head off the reactor, and release radioactive material into the atmosphere certainly increased public fears near the nuclear plant (and around the world). Sensational headlines ("Thousands Dead!") following the Chernobyl accident increased the memorability of that accident and the perceived catastrophic potential of nuclear power. If erroneous information sources find ready access to the mass media without effective antidotes, then large social impacts, even for minor events, become entirely possible.

The channels of information are also important. Information about risks and risk events flows through two major communication networks—the news media and more informal personal networks. The news media as risk articulators have received the bulk of scientific attention for their critical role in public opinion formation and community agenda setting.[29,30] Since the media tend to accord disproportionate coverage to rare or dramatic risks, or risk events, it is not surprising that people's estimates of the principal causes of death are related to the amount of media coverage they receive.[31]

Informal communication networks involve the linkages that exist among friends, neighbors, and co-workers, and within social groups more generally. Although relatively little is known about such networks, it is undoubtedly the case that people do not consider risk issues in isolation from other social issues or from the views of their peers. Since one's friends or co-workers provide reference points for validating perceptions but are also likely to share a more general cultural view or bias, the potential exists for both amplifying and attenuating information. If the risk is feared, rumor may be a significant element in the formation of public perceptions and attitudes. Within social group interaction, these interpretations of risks will tend to be integrated into larger frames of values and analysis and to become resistant to new, conflicting information. It should be expected, therefore, that interpersonal networks will lead to divergent risk perceptions, management preferences, and levels of concern. Since experts also exhibit cultural biases in their selections of theories, methods, and data, these variable public perceptions will also often differ as a group from those of experts.

Finally, specific terms or concepts used in risk information may have quite different meanings for varying social and cultural groups. They may also trigger associations independent of those intended.[32] Such symbolic connotations may entail "mushroom clouds" for nuclear energy, "dumps" for waste disposal facilities, or feelings of "warmth and comfort" for solar power technologies.

## 6. RESPONSE MECHANISMS OF SOCIAL AMPLIFICATION

The interpretation and response to information flow form the second major stage of social amplification of risk. These mechanisms involve the social, institutional, and cultural contexts in which the risk information is interpreted, its meaning diagnosed, and values attached. We hypothesize four major pathways to initiate response mechanisms:

♦ *Heuristics and Values*. Individuals cannot deal with the full complexity of risk and the multitude of risks involved in daily life. Thus people use simplifying mechanisms to evaluate risk and to shape responses. These processes, while permitting individuals to cope with a risky world, may sometimes introduce biases that cause distortions and errors.[28] Similarly, the application of individual and group values will also determine which risks are deemed important or minor and what actions, if any, should be taken.

♦ *Social Group Relationships*. Risk issues enter into the political agenda of social and political groups. The nature of these groups will influence member responses and the types of rationality brought to risk issues.[3] To the extent that risk becomes a central issue in a political campaign or in a conflict among social groups, it will be vigorously brought to more general public attention, often coupled with ideological interpretations of technology or the risk-management process.[11,12] Polarization of views and escalation of rhetoric by partisans typically occur and new recruits are drawn into the conflicts.[29] These social alignments tend to become anchors for subsequent interpretations of

risk management and may become quite firm in the face of conflicting information.

♦ *Signal Value*. An important concept that has emerged from research on risk perception is that the seriousness and higher-order impacts of a risk event are determined, in part, by what that event signals or portends.[4] The informativeness or "signal value" of an event appears to be systematically related to the characteristics of the event and the hazard it reflects. High-signal events suggest that a new risk has appeared or that the risk is different and more serious than previously understood (see Table I). Thus an accident that takes many lives may produce relatively little social disturbance (beyond that experienced by the victims' families and friends) if it occurs as part of a familiar and well-understood system (such as a train wreck). A small accident in an unfamiliar system (or one perceived as poorly understood), such as a nuclear reactor or a recombinant-DNA laboratory, however, may elicit great public concern if it is interpreted to mean that the risk is not well understood, not controllable, or not competently managed, thus implying that further (and possibly worse) mishaps are likely. In sum, signals about a risk event initiate a process whereby the significance of the event is examined. If found to be ominous, these implications are likely to trigger higher-order social and economic impacts.

♦ *Stigmatization*. Stigma refers to the negative imagery associated with undesirable social groups or individuals.[33] But environments with heavy pollution, waste accumulation, or hazardous technology may also come to be associated with negative images. Love Canal, the Valley of the Thousand Drums, Times Beach, and the Nevada Test Site evoke vivid images of waste and pollution. Since the typical response to stigmatized persons or environments is avoidance, it is reasonable to assume that risk-induced stigma may have significant social and policy consequences.[34] Research is needed to define the role of risk in creating stigma, the extent of aversion that results, and how durable such stigma become.

In addition to these four mechanisms, *positive feedback to the physical risk itself* can occur due to social processes. If a transportation accident with hazardous materials were to occur close to a waste-disposal site, for example, protests and attempted blockage of the transportation route could result. Such ac-

**TABLE I** Risk Events with Potentially High Signal Value

| Events | Messages |
|--------|----------|
| Report that chlorofluorocarbon releases are depleting the ozone layer | A new and possibly catastrophic risk has emerged |
| Resignation of regulators or corporate officials in "conscience" | The managers are concealing the risks: they cannot be trusted |
| News report of off-site migration at a hazardous waste site | The risk managers are not in control of the hazard |
| Scientific dispute over the validity of an epidemiological study | The experts do not understand the risks |
| Statement by regulators that the levels of a particular contaminant in the water supply involve only very low risks as compared with other risks.[35] | The managers do not care about the people who will be harmed; they do not understand long-term cumulative effects of chemicals |

tions could themselves become initiating or coaccident events, thereby increasing the probabilities of future accidents or enlarging the consequences should an accident occur. Or, alternatively, an accident in waste handling at the facility could lead opponents, or a disgruntled worker, to replicate the event through sabotage. Especially where strong public concern exists over a technology or facility, a wide variety of mechanisms is present by which health and safety risks may be enlarged through social processes.[35]

## 7. NEXT STEPS

Only partial models or paradigms exist for characterizing the phenomenon we describe as the social amplification of risk. Understanding this phenomenon is a prerequisite essential for assessing the potential impacts of projects and technologies, for establishing priorities in risk management, and for setting health and environmental standards. We put forth this conceptual framework to begin the building of a comprehensive theory that explains why seemingly minor risks or risk events often produce extraordinary public concern and social and economic impacts, with rippling effects across time, space, and social institutions. The conceptualization needs scrutiny, elaboration, and competing views.

Empirical studies, now beginning, should provide important tests and insights for the next stage of theory construction.

*Acknowledgment* This work was supported by the Nevada Nuclear Waste Project Office and by NSF grant No. SES 8796182 to Decision Research. We wish to thank Brian Cook, Christoph Hohenemser, Nancy Kraus, Sarah Lichtenstein, Steve Rayner, and three anonymous reviewers for their constructive comments on earlier drafts of the manuscript.

NOTE

[a] In this article, the term "risk event" refers to occurrences that are manifestations of the risk and that initiate signals pertaining to the risk. Risk events thus include routine or unexpected releases, accidents (large and small), discoveries of pollution incidents, reports of exposures, or adverse consequences. Usually such risk events are specific to particular times and locations.

REFERENCES

1. P. Slovic, B. Fischhoff, and S. Lichtenstein, "Why Study Risk Peception?" *Risk Analysis* **2**, 83–94 (1982).

2. O. Renn, "Risk Perception: A Systematic Review of Concepts and Research Results," in *Avoiding and Managing Environmental Damage from Major Industrial Accidents*, Proceedings of the Air Pollution Control Association International Conference in Vancouver, Canada, November 1985 (The Association, Pittsburgh, 1986), pp. 377–408.

3. S. Rayner and R. Cantor, "How Fair is Safe Enough? The Cultural Approach to Societal Technology Choice," *Risk Analysis* **7**, 3–13 (1987).

4. P. Slovic, "Perception of Risk," *Science* **236**, 280–90 (1987).

5. C. A. Vlek and P. J. M. Stallen, "Judging Risks and Benefits in the Small and the Large," *Organizational Behavior and Human Performance* **28**, 235–71 (1981).

6. J. M. Doderlein, "Understanding Risk Management," *Risk Analysis* **3**, 17–21 (1983).

7. R. E. Kasperson (ed.), *Equity Issues in Radioactive Waste Management* (Cambridge, Oelgeschlager, Gunn and Hain, 1983).

8. H. J. Otway and D. von Winterfeldt, "Beyond Acceptable Risk: On the Social Acceptability of Technologies," *Policy Sciences* **14**, 247–56 (1982).

9. B. Wynne, "Public Perceptions of Risk," in *The Urban Transportation of Irradiated Fuel*, J. Surrey (ed.), (Macmillan, London, 1984), pp. 246–59.

10. B. Fischhoff, P. Slovic, S. Lichtenstein, S. Read, and B. Combs, "How Safe is Safe Enough?: A Psychometric Study of Attitudes Towards Technological Risks and Benefits," *Policy Sciences* **8**, 127–52 (1978).

11. M. Douglas and A. Wildavsky, *Risk and Culture: An Essay on the Selection of Technological and Environmental Dangers* (University of California Press, Berkeley, 1982).

12. B. Johnson and V. Covello (eds.), *Social and Cultural Construction of Risk* (Reidel, Boston, 1987).

13. D. von Winterfeldt and W. Edwards, *Understanding Public Disputes about Risky Technologies*, technical report (Social Science Research Council, New York, 1984).

14. M. T. Katzman, *Chemical Catastrophies: Regulating Environmental Risk Through Pollution Liability Insurance* (R. D. Irwin, Springfield, Illinois, 1985).

15. C. D. Heising and V. P. George, "Nuclear Financial Risk: Economy Wide Cost of Reactor Accidents," *Energy Policy* **14**, 45–52 (1986).

16. H. I. Sharlin, *EDB: A Case Study in the Communication of Health Risk* (Office of Policy Analysis, Environmental Protection Agency, Washington, D.C., 1985).

17. I. Hoos, "Risk Assessment in Social Perspective," in *Perceptions of Risk* (National Council on Radiation Protection and Measurement, Washington, 1980), pp. 37–85.

18. M. L. DeFleur, *Theories of Mass Communication* (D. McKay, New York, 1966).

19. P. J. Shoemaker, "Mass Communication by the Book: A Review of 31 Texts," *Journal of Communication* **37**(3), 109–31 (1987).

20. H. D. Lasswell, "The Structure and Function of Communication in Society," in L. Bryson (ed.), *The Communication of Ideas: A Series of Addresses* (Cooper Square Publishers, New York, 1948), pp. 32–35.

21. C. J. Hovland, "Social Communication," in *Proceedings of the American Philosophical Society* **92**, 371–75 (1948).

22. J. H. Sorensen and D. S. Mileti, "Decision-Making Uncertainties in Emergency Warning System Organizations," *International Journal of Mass Emergencies and Disasters* (in press).

23. S. Grunhouse, "French and Swiss Fight about Tainted Cheese," *New York Times* (1 January 1988), p. 2.

24. P. Slovic, "Informing and Educating the Public about Risk," *Risk Analysis* **6**, 403–15 (1986).

25. A. Mazur, "The Journalist and Technology: Reporting about Love Canal and Three Mile Island," *Minerva* **22**, 45–66 (1984).

26. J. Sorensen *et al.*, *Impacts of Hazardous Technology: The Psycho-Social Effects of Restarting TMI* (State University of New York Press, Albany, 1987).

27. A. Weinberg, "Is Nuclear Energy Acceptable?" *Bulletin of the Atomic Scientists* **33**(4), 54–60 (1977).

28. D. Kahneman, P. Slovic, and A. Tversky (eds.), *Judgment under Uncertainty: Heuristics and Biases* (Cambridge University Press, New York, 1982).

29. A. Mazur, *The Dynamics of Technical Controversy* (Communication Press, Washington, D.C., 1981).

30. National Research Council, *Disasters and the Mass Media* (National Academy of Sciences Press, Washington, D.C., 1980).

31. B. Combs and P. Slovic, "Newspaper Coverage of Causes of Death," *Journalism Quarterly* **56**, 837–43, 849 (1979).

32. H. Blumer, *Symbolic Interactionism: Perspective and Method* (Prentice Hall, Englewood Cliffs, New Jersey, 1969).

33. E. Goffman, *Stigma* (Prentice Hall, Englewood Cliffs, New Jersey, 1963).

34. P. Slovic, "Forecasting the Adverse Economic Effects of a Nuclear Waste Repository," in R. G. Post (ed.), *Waste Management '87* (Arizona Board of Regents, University of Arizona, Tucson, 1987).

35. R. E. Kasperson, J. Emel, R. Goble, C. Hohenemser, J. X. Kasperson, and O. Renn, "Radioactive Wastes and the Social Amplification of Risk," in R. G. Post (ed.), *Waste Management '87* (Arizona Board of Regents, University of Arizona, Tucson, 1987).

# 10

# BRINGING SOCIAL THEORY
# TO HAZARDS RESEARCH

### Conditions and Consequences of the Mitigation
### of Environmental Hazards

*William C. Bogard*

While the social mitigation of environmental hazards has been the subject of sociological study for some time now, there has been relatively little concern with tying this subject to general contemporary sociological theory. Theoretical works that do exist in this area are more likely to deal with the concept of "disaster" than the concepts of "hazard" or "mitigation" (see Dynes, 1970; Quarantelli, 1974; Fritz, 1961; Barton, 1970; Kreps, 1982), while those that do explicitly deal with the latter usually do so only at the midrange level of organizational theory (White, 1974; White and Haas, 1975; Saarinen, 1982; Rossi, Wright, and Weber-Burdin, 1982; Mileti, 1980; Mileti, Drabek, and Haas, 1975).

Organizational accounts of the hazards mitigation process are undoubtedly important to our theoretical understanding of the relations between society and the environment. They constitute one explanation of how features of the environment come to be perceived as threats and how social organizations subsequently emerge to counter those threats. Most organizational hazards research, however, is never related to a general sociological theory of action. Consequently, the ties that do exist to general theory remain implicit. Frequently, these implicit ties are to some variety of sociological functionalism or equilibrium theory that stresses the reactive character of social mitigation in bringing hazardous situations back to "normal." Within this framework, mitigation is defined as any action— collective or individual, public or private— taken to reduce the potential harm posed by an environmental hazard. In light of contemporary theories of action, such conceptions are both somewhat dated and no longer entirely adequate.

## UNACKNOWLEDGED CONDITIONS AND UNANTICIPATED CONSEQUENCES OF MITIGATION

One of the central projects of Western sociology has been to trace the causes and consequences of rational action. Habermas (1984, pp. 3–7) has gone so far as to claim this project, generally speaking, as the primary legitimate focus of all sociological theory. While

Reprinted from *Sociological Perspectives* Volume 31(2) (1988), pp. 147–68. Used with permission of JAI Press Inc.

substantive problems of rational action are found in all the humanistic disciplines, theoretical sociology's interest historically has been in the diverse modes of rational action per se.

Rational action is traditionally conceptualized as goal directed, utilizing means chosen predominantly for their efficiency or efficaciousness in reaching those goals. Theoretical considerations of rational action are thus necessarily implicit in debates over public policy, economic choice, and political strategy in general, and for how these debates affect the mitigation of environmental hazards in particular (Slovic, Kunreuther, and White, 1974; Dacy and Kunreuther, 1969; Rossi, Wright, and Weber-Burdin, 1982).

Early sociological theorizing took account of the fact that the social effects of rational action may themselves be irrational (or nonrational). Both Marx (1963) and Weber (1958) frequently noted that the rational organization of society and the economy was capable of producing just such effects—alienation, bureaucratic red tape, dehumanization, contradiction, and conflict. This theme has continued in varied forms in subsequent theoretical work. Sociological functionalism, for instance, maintains that rational action can be explained by reference to its unintended, latent effects on social organization (Merton, 1957; Coser, 1971). Game theorists observe that strategic action, utilizing rational choice criteria, may lead to suboptimal or counterfinal solutions (Elster, 1978, Chap. 5). Such theories stress the fact that uncertainty and risk permeate the entire process of decision making. Recent research has also noted nonrational social effects resulting from the exclusion of policy alternatives from political agendas, or "non-decision-making" (Bachrach and Baratz, 1973). Finally, a whole literature has developed concerning the theoretical and practical limitations of rational choice models that must assume perfect information, homogeneous goods, and equal access to markets (see Simon, 1954; Lindblom, 1964).

In attempting to tie such diverse theories together at the level of action, Giddens (1979, Chap. 2) has argued that any adequate general sociological theory must account for the fact that all forms of rational action operate within a dynamic framework of unacknowledged conditions and unanticipated consequences. Giddens refers to this general process as "structuration," a schematic representation of which is given in Figure 1.

Giddens terms the representation in Figure 1 a "stratification" model of action, insofar as to be adequately specified it must include general structural considerations about the distribution of power and resources in society. For the limited purposes of this article, however, the simplified outline of this model will suffice.

In Figure 1, the reflexive monitoring, rationalization, and motivation of action collectively refer to the fact that human activity is purposive, intentional, and feedback oriented. Actors are conceived as having the ability to account theoretically for their behavior utilizing the same social knowledge that is implicated in the production of that behavior. The structures of social knowledge are, to use Giddens's terminology, both the media and the outcomes of rational social action.

Social knowledge, however, as the medium of rational action is neither complete nor always readily available in explicitly codified forms. What is known is often uncritically taken for granted (demarking the limits imposed by practical as opposed to theoretical knowledge), while what is not known forms a background of uncertainty. In either case, rational action is bounded by its unacknowledged conditions.

Precisely because of this bounded character, the outcomes of rational action are often unforeseen and unintended, that is, they can "escape . . . the scope of the purposes of the actor" (Giddens, 1979, p. 59). Within social theory, functionalism, with its concept of latent functions, has been the dominant perspective to express this viewpoint. The past

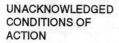

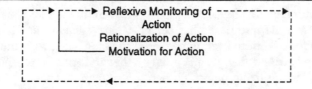

| UNACKNOWLEDGED CONDITIONS OF ACTION | Reflexive Monitoring of Action / Rationalization of Action / Motivation for Action | UNANTICIPATED CONSEQUENCES OF ACTION |

SOURCE: From A. Giddens (1979). **Central Problems In Social Theory.** London: MacMillan. Used by permission of publisher.

**FIGURE 1:** *A General Model of Action*

years, however, have seen functionalism come under severe criticism in sociology.[1] This criticism has centered not so much on the notion of unintended effects of rational action, which clearly identifies a valid empirical concern, but on the problems of teleological explanation, anthropomorphism in the case of collective entities such as the state or society, and a tendency toward theoretical conservatism, reactionism, and equilibrium explanations of social action. In the latter case, sociological functionalism tends to overstress the *beneficial* unintended consequences of action for society rather than consequences that may be disvalued or disruptive (see Elster, 1983, p. 55).

In contrast to this, the theory of structuration developed by Giddens is antifunctionalist (Giddens, 1976). It does not rely on teleological forms of explanation, nor does it hypostatize societal "needs" that social action functions beneficially to maintain in some form of equilibrium. How this is critical to a new conception of hazard mitigation will be explored in the next section.

In the literature on hazards, unacknowledged conditions of mitigation can refer to two things. A fair amount of work has been done to document the many unexamined popular myths that individuals hold (usually erroneously) about crisis behavior (e.g., the notion that persons tend to panic, loot, or generally behave egoistically in times of crisis). Much of this work on the unacknowl-

edged practical consciousness of persons in disaster has yet to be theoretically systematized, although Wenger (1985) and Wenger and Friedman's (1986) work is noteworthy in this respect. Despite the obvious importance of this area of study, in this article I limit myself to a second concern more amply addressed in the literature, namely, the fact that mitigation irremediably operates against constraints imposed by uncertainty or the lack of information. Due to the variable predictability of disasters and their social impacts, uncertainty is endemic to hazards research and operates at many different levels. For practical purposes, however, most uncertainties can be classified under two broad headings to be examined below—uncertainty regarding the physical parameters of a future environmental event (e.g., onset, duration, or location), and uncertainty regarding that event's social impacts, for example, short- or long-term organizational responses to the event.

Unanticipated consequences of hazards mitigation on the other hand refers to those effects that modify the hazard potential in unintended and perhaps disvalued ways. The notion of a "hazard potential" is complex and somewhat resistant to definition. I shall do no more in this article than to characterize this notion as the probability of an object, event, or location to cause personal harm either as a feature of an individual or group's vulnerability or as an objective characteristic of the environment itself.[2] In general, the beneficial, intended effects of mitigation (reduction of losses to life and property) have been fairly well investigated (see Anderson, 1969; Perry, Lindell, and Greene, 1981). Potential negative consequences of mitigation (increased vulnerability to a threat, increased hazardousness of a location, or cost shifts to particular social groups or classes, and so on) have been less systematically examined or tied to an overall theory.

White (1974, p. 3), one of the pioneers in this latter area, has noted the rather disturbing fact that placing flood mitigation in the hands of the federal government and increasing expenditures for such mitigation has actually appeared to increase vulnerability to losses from flooding. Although White's research was confined to the United States and indicated that losses were generally increasing only with respect to property and not to life, the implication that mitigation as a rational action could have disvalued effects for hazard potential was not lost to those investigating fatality rates from diasters in the Third World. Over the last twenty years, deaths from natural disasters in Third World countries have increased sixfold despite no evidence of any increase in the rate of actual physical events over this period (Wijkman and Timberlake, 1984, pp. 23–24; Hagman, 1984; Tinker, 1984). Much of the research in this area suggests that people's large-scale intervention into the environment, ostensibly to mitigate the potential for future disasters, has in part been responsible for this rise in casualties.

In theory, unanticipated, disvalued consequences of social actions such as mitigation can generate new structural parameters of knowledge for further action. This can be translated into saying that the range of uncertainties involved in mitigation changes with each mitigation effort and the effects it produces. Negative effects unavoidably feed back to alter the bounded knowledge the decision maker has of his or her environment, perhaps occasionally reducing uncertainty by allowing for the possibility of learning from mistakes,

but never eliminating uncertainty altogether. At the midrange level, the general model of structuration thus becomes as follows (Figure 2).

The variable predictability of hazardous events and their potentially negative social impacts makes the study of mitigation strategies a useful illustration of the general theory of unacknowledged conditions and unanticipated consequences. Additionally, focusing on the unintended, disvalued aspects of mitigation serves as a needed counterbalance to implicit functional theories of social action found in much hazards research. Finally, by examining the structure of informational constraints on the hazards mitigation process (and its potentially disvalued impacts), the possibility arises for bringing an element of critique to a literature not generally known for its criticism of social hazards policy. One such criticism concerns the traditionally accepted wisdom of conceiving mitigation as a response or reaction to hazards in the environment.

## PROBLEMS IN CONCEIVING MITIGATION AS A RESPONSE

It is necessary at the outset to reexamine some commonly held assumptions about the role of mitigation in hazards research. The theory of structuration suggests that mitigation cannot be conceived as a simple response, reaction, or adjustment to an actual or potential threat from the environment. Rather, mitigation is better conceptualized as a set of strategic actions that actively reshape and redistribute the social parameters of hazards. This view is not new but it has become more prominent in the hazards mitigation literature over the last decade (Turner, 1979; Wijkman and Timberlake, 1984; Tinker, 1985; Hagman, 1984). The contention is that this view represents an advance and the legitimate focus for a genuinely sociological investigation of hazards.

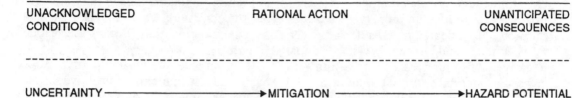

| UNACKNOWLEDGED CONDITIONS | RATIONAL ACTION | UNANTICIPATED CONSEQUENCES |

UNCERTAINTY ⟶ MITIGATION ⟶ HAZARD POTENTIAL

**FIGURE 2:** *Midrange Model of Mitigation*

Sociologists have long recognized the need to include social factors in any adequate conceptualization of the hazards process. In using the older vocabulary of "disasters," Prince (1920) noted that disaster could be functional in uniting a community for the task of recovery and rebuilding. Kutak (1938) observed that in disaster social role and status distinctions were overcome and that new normative patterns emerged in the process of coping with events. Studies such as these were the first to be concerned with the effects of disaster on social organization, particularly changes in the rules governing that organization. At a higher level of abstraction, these studies reproduced our commonsense idea of disasters as temporally bounded events in the environment necessitating a response. The content of that response, however, was perceived as constrained by certain essential features of disaster itself, as an event in the environment over which little or no control could be exercised. The unpredictability of disaster, its perceived externality to the routine of social life, its characterization as an "act of God," all entered into and reinforced this idea.

It is difficult for sociologists to extricate themselves from the tendency to see disasters as unique and temporally bounded, although unpredictable, natural events, and to some extent this older view has confused current research on hazards. The social aspects of disaster, when considered theoretically, have stubbornly remained temporally unbounded or at least indeterminate (e.g., the long-term effects of psychological stress or economic dis-

organization). Later research has recognized some of these problems by making a conceptual distinction between the physical aspects of disaster (utilizing such terms as "environmental extreme," "trigger event," or disaster "agent") and its social impacts (Mileti, 1980, p. 327; Dynes, 1970, pp. 50–51; Wijkman and Timberlake, 1984, p. 29). Much of this research was crucial in making the transition from the study of disasters to study of hazards, although in terms of establishing temporal bounds for the social impacts of disasters much work remains to be done (see Rossi, Wright, and Weber-Burdin, 1982).

With these problems in mind, the commonsense view of disaster that is taken over from the earlier disaster literature can oversimplify the complex relations between human and natural environments, positing these relations as external and contingent rather than internal and necessary. Some of the current literature on hazards has attempted to remedy this simplification by forwarding a variety of strategies through which social organizations routinely interact with environmental extremes or triggers to mitigate their effects in terms of potential harm to human populations (Mileti, 1980; Dynes, 1970; Dacy and Kunreuther, 1969; Rossi, Wright, and Weber-Burdin, 1982; Quarantelli and Tierney, 1979; Anderson, 1969). There has been some progress, if not total agreement, in the extent to which these studies have been able to identify and categorize the multitude of mitigation strategies. At a general classificatory level, Mileti (1980, p. 329) has developed a typology of strategies that include settlement constraints and the reduction or redistribution of losses. Rossi, Wright, and Weber-Burdin (1982, pp. 4–9) identify typical strategies of mitigation as falling into four broad categories: (1) free market approaches, (2) relief and rehabilitation assitance, (3) technological fixes, and (4) land use management. Milliman (1982, p. 5) cites eight different strategies: (1) siting decisions, (2) land use regulations, (3) construction codes, (4) insurance, (5) warning and prediction, (6) evacuation and relocation, (7) emergency planning, and (8) relief and reconstruction aid.

While important for orienting our attention to various forms of mitigation, these typologies leave us with theoretical questions of how and why mitigation actually redistributes patterns of harm in disasters. One source of this problem is that mitigation has continued to be unreflectively defined in commonsense terms as a response to an actual or potential disruption of social life by an environmental event.

The hazards process itself is usually conceived as a simple linear causal process such as the following:

ENVIRONMENTAL EXTREME → SOCIAL
DISRUPTION → SOCIAL RESPONSE

As theory and research on social interaction with the environment has become more sophisticated, it has been recognized that theoretically it is misleading to place the environment extreme variable at the beginning of the causal chain. It is more common today to find models that locate the physical parameters of extreme events in the middle of a temporal continuum and that thus enable one to speak or preimpact and postimpact stages in disaster (Mileti, Drabek, and Haas, 1975, pp. 14–23). Within these stages, analysis can then proceed to model strategic mitigation activities with increasing specificity and that vary with regard to their temporal relation to the trigger event.

Research during the 1970s and early 1980s made the first tentative steps toward the notion that such strategic activities constitute not only a simple response to extreme events, but also influence these events' potential for imposing losses (Haas and Mileti, 1976; White, 1974; Sorenson and White, 1980; Burton, Kates, and White, 1978; White and Haas, 1975). Much of this effort has been classified under the label "hazards research" as opposed to "disaster research." While the latter tends to emphasize explanations for various modes of response to environmental extremes, the former seeks explanations for adjustment to the risk of future disasters (Mileti, 1980, p. 328).

Emphasizing the idea of adjustment, however, has compromised the key insight of this research, namely, the idea of an *active* role for mitigation in altering the potential for harm from disaster. Essentially, there are at least three criticisms of utilizing the language of adjustment for hazards research, all of which are directed at its implicit ties to sociological functionalism.

(1) Problems of teleological explanation. The idea of adjustment in current hazards research still assigns causal primacy to the environmental extreme or trigger event itself in determining the content of mitigation, even if this event lies only in the future. Adjustment to future environmental extremes, however, if it is to make any sense at all, cannot be grounded on any *actual* features of these extremes, which are in any case known only uncertainly, but on our *expectations* concerning these extremes. These expectations are formed by experience with past events, whose actual features can be known and taken into account in the choice of mitigation. Simply stated, the expectations upon which choices are made do indeed concern future events, but they are not caused by them.

(2) Closely related to this is the idea that decision makers concerned with hazards somehow react or respond to future environmental events. But again, it is the concern

itself that forms the basis for decision and not any actual physical feature of the potential event itself. Mitigation, if it can be said to be reactive or responsive at all, has this character only in relation to such concerns, which can only be formed on the basis of past experience with hazards. Characterizing mitigation as a reaction or response to a future event has the theoretical result of making mitigation an *effect*, when the proper concern should be with how mitigation itself *produces* certain effects, that is, actively changes the parameters of the future environmental event with which it is concerned.

(3) Finally, the notion of adjustment to environmental extremes assumes that the relation of society to the environment can be directed through mitigation to achieve an equilibrium state. From this it is a short step to the assumption that society "needs" to be in balance with its environment, that is, needs to relieve the stress or strain resulting from potential environmental disruption. Society, however, has no such "needs," and to posit them is to transpose metaphorically features of individuals onto society itself. In this case, it makes far better sense theoretically to say that individuals simply take measures to protect themselves (or others) against hazards and to drop implicit ties to societal equilibrium notions. Again, the *effect* of such individual measures may or may not be a state of equilibrium, but equilibrium itself is not a prior theoretical reason or justification for taking such measures in the first place.

When individuals take measures to protect themselves or others against hazards, they do so with imperfect knowledge. And imperfect knowledge, whether in the form of popular myths about disasters or genuine uncertainties, leads to imperfect outcomes for adopted measures. These are outcomes that may, in game-theoretical terms, be suboptimal (everyone gets something but no one gets as much as they could) or even counterfinal (everyone gets just the opposite of what was intended). The interplay between uncer-

tainty and its unintended consequences for hazard potentials is a complex phenomenon, and in the following section I can provide no more than a few examples. These examples do, however, illustrate the usefulness of a theory of structuration to the area of hazards research and how diverse observations from various research pieces can be tied together within a comprehensive action framework.

## THE RELATION OF UNCERTAINTY AND HAZARD POTENTIAL TO MITIGATION

Habermas (1984, p. 8) has noted that an essential feature of all forms of rational action is their fallibility:

> What does it mean to say that persons behave "rationally" in a certain situation or that their expressions can count as "rational"? Knowledge can be criticized as unreliable. The close relation between knowledge and rationality suggests that the rationality of an expression depends on the reliability of the knowledge embedded in it. Consider two paradigmatic cases: an assertion with which A in a communicative attitude expresses a belief and a goal directed intervention in the world with which B pursues a specific end. Both embody fallible knowledge; both are attempts that can go wrong.

The failure of action may occur both in the process of rational communication and in the technical rationality of means-end relationships. In both cases, failure reflects the bounded, that is, uncertain character of our knowledge. In both cases the action does not "come off"—it produces effects that may be clearly unintended or irrational.

Presumably, the intention of mitigation is to reduce the hazard potential within someone's environment, that is, to lessen the chance for harm to persons from future environmental events. If, however, all mitigations (as forms of rational action) are fallible, that is, if the knowledge embedded in mitigation is essentially uncertain, then the theoretical possibility is opened for mitigation to produce just the opposite effect of what it intends. In other words, mitigation may actually increase the hazard potential within the environment due to outcomes that are unanticipated at the time choices concerning mitigation are made.

There are ready examples of this process. Relief supplies of food sent to regions stricken by drought are routinely seen as a necessary mitigation against famine, and they undoubtedly save many lives. But if too much aid is forthcoming (i.e., if too many nations act as outside suppliers), the result may be bottlenecks in distribution, economic disruption, or long-term dependence of populations on outside help (which may dry up as quickly as it materialized) (Hagman, 1984). In such cases prevention may actually be worse than the disease.

The content of aid itself, while both rational and well intentioned, may be totally inappropriate, as Wijkman and Timberlake (1984, p. 108) humorously yet grimly note:

> Stories abound in the relief field of completely inappropriate aid: the British charity that sent packs of tea, tissues and Tampax; the European Community sending powdered milk into an earthquake area where few cows had perished, but there was no water; and the West German charity which constructed 1,000 polystyrene igloos which proved too hot to live in. But the igloos could not be dismantled or moved. They had to be burned down, and when burning gave off toxic fumes. Tins of chicken cooked in pork fat have been sent to Moslem countries which do not eat pork. Blankets donated to India were donated by India to Nepal, which donated them back to India; the blankets were never needed or used. Turkey after a 1983 earthquake asked donors not to send any medicine or second-hand clothes, but a Northern donor flew in a few days later with a planeload of precisely these items—and a TV crew to cover the distribution.

When everyone becomes involved in the relief game, the collective result can be chaos. But similar processes are at work in other forms of mitigation. If everyone tries to evacuate a location at once, the result may be that no one can get out in time (the classic illustration is the fire in a crowded theater). A warning siren that is tested only at 12 P.M. on Sundays will be useless for a disaster that strikes at 12 P.M. on Sunday, but if many *different* tests of a warning system are conducted at many different times no one might be able to distinguish a test from the real thing.

The phenomena referred to in the above examples illustrate the theoretical concept of counterfinality.[3] Counterfinality occurs when a set of actions, each of which is rational in the individual case, produces unanticipated and disvalued effects for the aggregate (Elster, 1978, pp. 106–122).

Perhaps the best known illustration of counterfinality in sociological theory derives from Marx's explanation for the falling rate of profit in capitalist societies (Marx, 1967, Chaps. XIII–XIV). Individual capitalists, seeking to increase their profits, mechanize their operations in an effort to reduce wage costs. But if all capitalists behave in this manner (which is rational for the individual capitalist), the outcome is an increase in unemployment and a lessening of the overall system's capacity to extract surplus labor. The result is a decline rather than a rise in the rate of profit. Individual rationality produces collective irrationality.

In the literature on hazards, examples of the operation of counterfinality are far from rare. Several authors have noted the desensitizing effect resulting from too many disaster warnings that prove to be unfounded (Anderson, 1969; Mack and Baker, 1961; Breznitz, 1984). When a real warning comes, no one believes it; features of the warning process actually inhibit action to counter the threat. To take perhaps a more far-reaching example, the clearing of forests in the Third World countries, rational for the individual or corporate landowner seeking to increase available land for agriculture, poses a greater threat of flooding for the collective by reducing the capacity of the land to absorb or divert the flow of water (Wijkman and Timberlake,

1984, pp. 57–62). The process of counterfinality is also evident in the case of pesticide use, a rational mitigation for the individual farmer seeking to increase agricultural yields, but that may engender long-term problems resulting from the development of resistant strains of insects, or "super pests" (Norris, 1982). It can be argued that most general hazards, such as those rising from the over-exploitation of resources, pollution, and rapid population growth also exhibit features of counterfinality (Elster, 1978, p. 110).

Counterfinal outcomes for collective mitigation processes rest on two common and taken-for-granted assumptions: (1) that not everyone will behave in the same fashion at the same time, and (2) that the actions of relevant others at time $t_1$ will be roughly the same at time $t_2$. If each individual acts on one or another of these (supposedly rational) assumptions, the mitigative actions that ensue may be collectively irrational, that is, they may produce the opposite of their individually intended results. Action on the first assumption—that not everyone will do the same thing simultaneously—could result in the adoption of mitigation strategies that overtax a system's resources for coping with disaster. Mass evacuations, for example, have the possibility of placing such a strain on the available transportation system that evacuation is itself slowed down and perhaps more people exposed to a threat than if they had stayed at home (see Ruch, 1981). The aforementioned incapacity of states to deal with an oversupply of well-intentioned relief donors is also a result of action on this first assumption. In contrast, action on the second assumption—that actors will behave similarly between time $t_1$ and time $t_2$—could result in mitigation strategies that produce adequate results initially but develop problems later (e.g., the "cry wolf" syndrome noted above). Potential victims (including those who issue warnings) often become desensitized to repeated false warnings.

Potential failures in mitigation arise not only from rational assumptions about the behavior of others, but also from the inability to form any expectations whatsoever regarding the outcomes of mitigations, that is, from classical uncertainty. Over the last 30 years, the theory of decision under uncertainty has generally come to replace classical conceptions of how rational choices are made.

Simon (1959, p. 253) has noted that classical decision analysis, which presupposes rational choice by actors regarding pregiven alternatives with known outcomes, is inadequate as a model for explaining how decisions are actually made:

> The classical theory is a theory of a man choosing among fixed and known alternatives, to each of which is attached known consequences. But when perception and cognition intervene between the decision-maker and his objective environment, this model no longer proves adequate. We need a description of the choice process that recognizes that alternatives are not given but must be sought; and a description that takes into account the arduous task of determining what consequences will follow on each alternative.

With regard to the selection and implementation of mitigations, uncertainties may for simplicity be grouped into two broad categories—those involving the physical parameters of an environmental extreme, and those dealing with social outcomes of the extreme event (Bogard, 1986). Uncertainties regarding the former include such factors as the time of occurrence (Graham and Brown, 1983), the magnitude of the event (Saarinen et al., 1985; Perry, Lindell, and Green, 1981; Moore, 1964; Savage et al., 1984; Sorenson and Gersmehl, 1980), the proximity of the threat (Diggory, 1956), and the recognition of the event itself (Anderson, 1969; Erikson, 1976). Uncertainties regarding social outcomes include the possibility of panic (Savage et al., 1984; Scanlon et al., 1976; Quarantelli, 1974), pressure from outside groups (Sorenson and Gersmehl, 1980; Gray, 1981), evacuation problems (Greene and Lindell, 1981), role ambiguity and conflict (Perry and Mushkatel, 1984;

Erikson, 1976), and many more (Mileti, Sorenson, and Bogard, 1985).

Uncertainty generally arises from difficulties in obtaining or processing information (Turner, 1979, p. 55). One way these difficulties may increase a mitigation's potential for failure is by altering the timing of its implementation. Problems in obtaining or processing information may delay the notification of relevant authorities, the latter's notification of the public, or the decision to evacuate hazardous areas (Mileti, Sorenson, and Bogard, 1985). A warning or evacuation order issued too late may actually be worse than no warning being given at all, especially if it is hastily worded, conveys the uncertainty of officials, or generates confusion.

To return to the example of foreign aid, relief supplies often arrive too late to aid disaster victims substantially. Wijkman and Timberlake (1984, p. 110) have noted that the most pressing needs of victims are usually within 48 hours of a disaster. Information processing difficulties, however—often political in nature but also those generated by the disaster itself—can cause delays. Aid arriving after this time can often interfere with victims' own attempts at recovery:

> Human nature dictates that when a plane arrives with free goods, people will stop what they are doing and queue for hours to get whatever is available, whether they need it or not. Thus relief can actually slow down reconstruction and interfere with self-help.

Conversely, uncertainty may force mitigation to be implemented too early. Premature warnings may be forgotten, ignored, or unwisely acted on if a disaster does not materialize within a reasonably expected time frame. In the Bhopal tragedy, there was at least one individual who explicitly warned residents of the dangers of the Union Carbide plant years in advance, but as time went on with few serious accidents these warnings lost their salience (Bhandari, 1984, p. 104). The same process is also evident for natural hazards. Much of the debate concerning the issuing of warnings for earthquakes is based on the wide time frames that may separate the warning and the actual event. Action on warnings of this type may produce clearly unintended effects. Mileti, Hutton, and Sorenson (1981, pp. 112–114) have noted that such warnings have the potential for economic dislocations in the communities in which they are issued. Corporate enterprises may be dissuaded from making investments in areas that have been declared earthquake zones, this despite the fact of the tremendous uncertainties in pinpointing the precise time, location, or magnitude of the forecast event.

Finally, uncertainty can be a result of the complexity of mitigation itself. Difficulties in obtaining or processing information can be exacerbated by mitigations that are tightly coupled or interactive (Perrow, 1984). When multiple-function mitigations are involved (e.g., backup systems in nuclear deterrence), it becomes increasingly difficult (and takes increasing amounts of time) to pinpoint the exact location of particular mitigation failures. In extreme cases the result may be catastrophic. The complexity of the system designed to detect an incoming nuclear attack, for example, is so great that a false warning may not be corrected in time to halt a decision for a "retaliatory" strike. Here the potential failure of the system negates any possible benefits that could derive from its implementation.

## CONCLUSION

The list of examples illustrating potential disvalued effects of the mitigation process can be easily extended. It has not been the purpose of this article, however, to suggest that *all* unanticipated consequences of mitigation are negative or disvalued. This is certainly far from the case, even though accounts of the negative effects of mitigation are quite common in the literature. The purpose, rather, has been to suggest a general sociological theory that, because of its generality, is able to unite diverse research observations on mitigation by providing them with a common theme, that is, the unacknowledged conditions and unanticipated consequences of rational action.

I have concentrated primarily on negative, disvalued outcomes of the mitigation process in this article because some theoretical balance needs to be achieved against the claims of functional theories of this process. These theories, if anything, tend to overstress the *beneficial* functions of rational action for society. The theory of structuration developed by Giddens does not deny that the effects of rational action may be beneficial for society, only that it is somehow their *function* to achieve such effects.

Functional theories of action implicitly underlie many organizational accounts of the hazards process. This is particularly evident in the use of the terms *response* and *adjustment* to characterize organizational relations to environmental events. These terms have the unfortunate feature of positing a reactive rather than active role for mitigation in changing the hazard potential of the environment.

The motivation for this article has been the well-noted fact that uncertainties are present in all forms of social mitigation. As conditions of all rational action, these uncertainties often result in counterfinal outcomes for the mitigation process. Empirically, we may find that these outcomes do or do not increase the overall hazard potential of the environment. The point is that (a) they may in fact do so, and (b) this fact can be explicitly accounted for by a general theory of action.

## NOTES

[1] However, see Alexander (1983) for what has been termed a revival of functional thought in sociology.

[2] Most research on hazards contains at a minimum some reference to the idea of potential harm to persons, whether this harm is bodily, economic, or social-psychological. For work in this area see Burton, Kates, and White (1978), Hewitt and Burton (1971), and White (1974).

[3] The game theory notion of suboptimality is also important in this context, but will not be addressed here. For a good discussion, see Elste (1978, pp. 122–34).

## REFERENCES

Alexander, Jeffery. 1983. *Theoretical Logic in Sociology*. Berkeley: University of California Press.

Anderson, William A. 1969. "Disaster Warning and Communication Processes in Two Communities." *Journal of Communication* 19(June):92–104.

Bachrach, Peter and Morton S. Baratz. 1973. "Two Faces of Power." In *The Bias of Pluralism*, edited by W. E. Connoly. New York: Lieber-Atherton.

Barton, Allen H. 1970. *Communities in Disaster*. New York: Anchor.

Bhandari, Arvind. 1984. "The Avaricious Giants." *Tribune* (India) 19(December).

Bogard, William. 1986. "Unacknowledged Conditions and Unanticipated Consequences of Hazards Mitigation." Paper presented at the 1986 Western Social Science Meetings, Reno, Nevada, April.

Breznitz, Shlomo. 1984. *Cry Wolf: The Psychology of False Alarms*. Hillsdale, NJ: Lawrence Erlbaum.

Burton, Ian, Robert W. Kates, and Gilbert F. White. 1978. *The Environment as Hazard*. New York: Oxford University Press.

Coser, Lewis. 1971. "Social Conflict and the Theory of Social Change." In *Conflict Resolution: Contributions of the Behavioral Sciences*, edited by C. E. Smith. Notre Dame: University of Notre Dame Press.

Dacy, Douglas C. and Howard Kunreuther. 1969. *The Economics of Natural Disasters*. New York: Free Press.

Diggory, James C. 1956. "Some Consequences of Proximity to a Disaster Threat." *Sociometry* 19(March):47–53.

Dynes, Russell R. 1970. *Organized Behavior in Disaster*. Lexington, MA: Lexington.

Elster, Jon. 1978. *Logic and Society: Contradictions and Possible Worlds*. Chichester: John Wiley.

———. 1983. *Explaining Technical Change*. Cambridge: Cambridge University Press.

Erikson, Kai T. 1976. *Everything In its Path*. New York: Simon & Schuster.

Fritz, Charles. 1961. "Disasters." pp. 651–94 in *Contemporary Social Problems*, edited by R. Merton and R. A. Nisbet. New York: Harcourt.

Giddens, Anthony. 1976. "Functionalism: Apres la lutte." *Social Research* 43(2):325–66.

———. 1979. *Central Problems in Social Theory*. Berkeley: University of California Press.

Graham, W. J. and C. A. Brown. 1983. *The Lawn Lake Dam Failure: A Description of the Major Flooding Events and an Evaluation of the Warning Process*. Denver: Bureau of Reclamation.

Gray, Jane. 1981. "Characteristic Patterns of and Variations in Community Response to Acute Chemical

Emergencies." *Journal of Hazardous Materials* 4:357–65.

Greene, M., R. Ferry, and M. Lindell. 1981. "The March 1980 Eruptions of Mt. St. Helens: Citizen Perceptions of Volcano Threat." *Disasters* 5(1):49–66.

Haas, J. Eugene and Dennis S. Mileti. 1976. "Socioeconomic Impact of Earthquake Prediction on Government, Business and Community." *California Geology* 30(7):147–57.

Habermas, Jurgen. 1984. *The Theory of Communicative Action: Reason and Rationalization of Society*. Vol. 1. Boston: Beacon.

Hagman, Gunnar. 1984. *Prevention Better Than Cure*. Stockholm: Swedish Red Cross.

Hewitt, Kenneth and Ian Burton. 1971. "The Hazardousness of a Place." Department of Geography Research Publication No. 6. Toronto: University of Toronto.

Kreps, Gary A. 1982. "A Sociological Theory of Organized Disaster Response." Paper presented at the Tenth World Congress of Sociology, Mexico City, August.

Kutak, Robert L. 1938. "The Sociology of Crisis." *Social Forces* 17(2):66–72.

Lindblom, C. E. 1964. "The Science of Muddling Through." In *The Making of Decisions*, edited by W. J. Gore and J. W. Dyson. New York: Free Press.

Mack, Raymond W. and George W. Baker. 1961. *The Occasion Instant*. Washington, DC: National Academy of Sciences.

Marx, Karl. 1963. *The 18th Brumaire of Louis Bonaparte*. New York: International Publishers.

——— 1967. *Capital*. Vol. 3. New York: International Publishers.

Merton, Robert. 1957. *Social Theory and Social Structure*. New York: Free Press.

Mileti, Dennis S. 1980. "Human Adjustment to the Risk of Environmental Extremes." *Sociology and Social Research* 64(3):327–47.

——— T. Drabek, and J. Haas. 1975. *Human Systems in Extreme Environments: A Sociological Perspective*. Boulder: University of Colorado, Institute of Behavioral Sciences.

——— Janice R. Hutton, and John H. Sorenson. 1981. *Earthquake Prediction Response and Options for Public Policy*. Boulder: University of Colorado, Institute of Behavioral Sciences.

——— John H. Sorenson, and William C. Bogard. 1985. "Evacuation and Decision-Making: Process and Uncertainty." Prepared for the Office of Nuclear Safety. Unpublished.

Milliman, Jerome W. 1982. "Economic Issues in Formulating Policy for Earthquake Hazard Mitigations." Prepared for NSF Workshop on Hazards Research, Policy Development, and Implementation Incentives, University of Redlands, California.

Moore, Harry Estill. 1964. *And The Winds Blew*. Austin: University of Texas, Hogg Foundation for Mental Health.

Norris, Ruth (Ed.) 1982. *Pills, Pesticides & Profits: The International Trade in Toxic Substances*. Croton-on-Hudson, N.Y.: North River Press.

Pelanda, Carlo. 1982. "Disaster and Sociosystemic Vulnerability." pp. 67–91 in *The Social and Economic Aspects of Earthquakes*, edited by B. G. Jones and M. Tomazevic. Proceedings of the 3rd International Conference, Bled, Yugoslavia.

Perrow, Charles. 1984. *Normal Accidents: Living With High Risk Technologies*. New York: Basic Books.

Perry, Ronald W., Michael K. Lindell, and Marjorie R. Greene. 1981. *Evacuation and Emergency Management*. Lexington, MA: Lexington Books.

Perry, Ronald W. and Alvin Mushkatel. 1984. *Disaster Management: Warning Response and Community Relocation*. Westport, CT: Quorum.

Prince, Samuel H. 1920. "Catastrophe and Social Change." Ph.D. dissertation, Columbia University, New York.

Quarantelli, E. L. 1974. *Disasters: Theory and Research*. Beverly Hills, CA: Sage.

——— and Kathleen Tierney. 1979. *Disaster Preparation Planning*. Columbus: Ohio State University, Disaster Research Center.

Rossi, Peter H., James D. Wright, and Eleanor Weber-Burdin. 1982. *Natural Hazards and Public Choice: The State and Local Politics of Hazard Mitigation*. New York: Academic Press.

Ruch, Carlton. 1981. *Hurricane Relocation Planning for Brazoria, Galveston, Harris, Fort Bend, and Chambers Counties*. College Station: Texas A&M University, Center for Strategic Technology, Texas Engineering Experiment Station.

Saarinen, F., Victor R. Baker, Robert Durrenberger, and Thomas Maddock. 1985. "The Tucson, Arizona, Flood of October 1983." National Research Council, Washington, DC: National Academy Press.

Saarinen, Thomas F. (Ed.) (1982). *Perspectives on Increasing Hazard Awareness*. Boulder: University of Colorado, Institute of Behavioral Science.

Savage, Rudolph P., Jay Baker, Joseph Golden, Ashan Kareem, and Billy Manning. 1984. "Hurricane Alicia: Galveston and Houston, Texas, August 17–18, 1983." National Research Council on Natural Disasters, National Academy Press.

Simon, Herbert. 1954. "A Behavioral Theory of Rational Choice." *Quarterly Journal of Economics* 69:99–118.

——— 1959. "Theories of Decision-making in Economics and Behavioral Science." *American Economic Review* 49:253–83.

Slovic, Paul, Howard Kunreuther, and Gilbert F. White. 1974. "Decision Processes, Rationality, and Adjustment to Natural Hazards." pp. 187–206 in *Natural Hazards; Local, National, Global*, edited by G. F. White. New York: Oxford University Press.

Sorenson, John H. and Philip J. Gersmehl. 1980. "Volcanic Hazard Warning System: Persistence and Transferability." *Environmental Management* 4(March):125–36.

Sorenson, John and Gilbert F. White. 1980. "Natural Hazards: A Cross-Cultural Perspective." In *Human Behavior and the Environment*, edited by I. Altman, A. Papaport, and J. Wohwill. New York: Plenum.

Tinker, Jon. 1984. "Are Natural Disasters Natural?" *Socialist Review* 78(14):7–25.

Turner, Barry. 1979. "The Social Aetiology of Disasters." *Disasters* 3(1):53–59.

Weber, Max. 1958. *The Protestant Ethic and the Spirit of Capitalism*. New York: Scribners.

Wenger, Dennis. 1985. "Collective Behavior and Disaster Research." Preliminary Paper 104. University of Delaware, Newark, Disaster Research Center.

——— and Barbara Friedman. 1986. "Local and National Media Coverage of a Disaster: A Content Analysis of the Print Media's Treatment of Disaster Myths." Preliminary Paper 99. University of Delaware, Newark, Disaster Research Center.

White, Gilbert F. (Ed.) (1974). *Natural Hazards: Local, National, Global*. New York: Oxford University Press.

——— 1975. *Flood Hazard in the United States: A Research Assessment*. Boulder: University of Colorado, Institute of Behavioral Science.

——— and J. Eugene Haas. 1975. *Assessment of Research on Natural Hazards*. Cambridge: MIT Press.

Wijkman, Anders and Lloyd Timberlake. 1984. *Natural Disasters: Acts of God or Acts of Man?* London: Earthscan.

# EN-GENDERED FEARS: FEMININITY
# AND TECHNOLOGICAL RISK PERCEPTION

*Susan L. Cutter* ❖ *John Tiefenbacher*
*William D. Solecki*

## INTRODUCTION

More than a decade has passed since the original research by Fischhoff, Lichtenstein and Slovic on perception of technological risks and benefits (Fischhoff et al., 1978; Lichtenstein et al., 1978). Critiques are now emerging leading to an overall reappraisal of the psychometric paradigm in particular (Renn and Swaton, 1984; Gardner and Gould, 1989; Hansson, 1989) and risk perception research in general (Douglas and Wildavsky, 1982; Johnson and Covello, 1987; Kasperson et al., 1988). While there is some work on cross-cultural comparisons of risk perceptions (Vlek and Stallen, 1981; Englander et al., 1986; Teigen et al., 1988; Bastide et al.,1989; Keown, 1989; Kleinhesselink and Rosa, 1989), research on social distinctions (race, ethnicity, gender) is virtually nonexistent. Yet race, ethnicity and gender may be crucial in understanding the social acceptability of risk at both the local and national levels. In particular, the issues of social equity and the burdens of technological risk and hazards are now being raised (Commission for Social Justice, 1987; Bullard, 1990).

We argue for a feminist analysis of technological risk that addresses these social distinctions. Specifically, a pilot study of gender differences in the perception of technological risk is used to illustrate the shortcomings of the psychometric paradigm and the need for alternative approaches to extract gender-based differences in risk perception and behavior toward technological hazards.

## RE-EVALUATING NATURE, SOCIETY AND TECHNOLOGY

If viewed on a continuum, the underlying industrialized world philosophies that govern nature–society interactions range from ecocentrism at one pole to technocentrism at the other. Ecocentrism views the world according to natural laws. The Earth's functioning involves a complex and sensitive balance between all parts of nature striving for stability through the principles of diversity and homeostasis (O'Riordan, 1981). Ecocentrism preaches responsibility and care, where humans are a part of the larger system, not apart

Reprinted from *Industrial Crisis Quarterly* Volume 6 (1992), pp. 5–22. Used with permission of the authors and Elsevier Science Publishers B.V.

from it. Smallness and low-impact technology are also governing principles.

Technocentrism, the dominant view since the Industrial Revolution, argues that nature exists to be used by societies through the application of rationality, science and professionalized managerial techniques. Progress, efficiency and control form the ideology that underpins technocentrism (O'Riordan, 1981). Human domination of nature through technology can be traced historically to the development of Western culture (White, 1967; Moncrief, 1970). Merchant (1980), for example, has shown that the domination mentality co-evolved with rationality and science during and since the Renaissance.

At the beginning of the 18th century, nature was the primary target of rationalism as Baconian thinking encouraged order, hierarchy, linearity and logic. These were traits of men, not nature. By distinguishing the duality of nature and man, and placing man at the pinnacle, the anthropocentric world view was born. Rationalism also changed the belief that nature was the nurturing mother. Nature became threatening—a foe that needed conquering. Chaos, disorder, uncontrollability and unpredictability were anathema to this rationalized world view. One way that man could overcome the antagonisms of nature was through technological innovation. Earth was no longer passive and "her" wrath and fury justified the use of technology to tame and subdue her (Plant, 1990). Women were closely identified with nature because of their nurturing qualities and ability to reproduce (Merchant, 1980). Their provision of these services (mother, wife, homemaker, gardener) relegated them to a more passive role than men (hunter, provider, protector). Patriarchal systems soon developed, ultimately resulting in the domination of women. Many of the technologies that liberated men and empowered them over nature have subjugated women.

Natural hazards are often characterized as "acts of God" or products of "the wrath of nature." Technology is employed to control nature and protect people from harm or mitigate adverse impacts. Not until very recently has an alternative framework been suggested (Hewitt, 1983). This political-economic perspective suggests that natural hazards are prompted by natural events, but their impacts and society's responses to them are largely governed by the failures of the political, economic and social systems. Therefore, technological fixes will not simply solve the problem or render the wrath of nature immutable.

Technological hazards, on the other hand, are viewed as acts of people. Technological systems fail because of their complexities and because of human fallibility. Accidents should be seen as normal, not extreme occurrences, at least in complex systems (Perrow, 1984). The idea that technological systems can be improved through bigger and better science and technology, better risk assessments and more management clearly represents the technocentric view.

## MASCULINITY AND TECHNOLOGY

Masculinity manifests itself clearly in nature-society interactions. Its effect on the technological world is less apparent. As Paehlke (1989) suggests, men in decision making positions have traditionally undervalued the nurturing dimension that forms the central tenet of ecocentrism or what is now called environmentalism. Instead, these "experts" feel more comfortable with a rational, scientific view embodied in such disciplines as engineering. Technology is a creation of science and engineering, where women traditionally have had limited access and professional acceptance. The growing sophistication of resource management meant the separation of the resource manager from the public. Policy makers became dependent on "scientific advisers" from a narrow range of specialities, most of whom were men. Women, if they were involved in resource decision making, were most likely the citizen activists, or the "bird, bug and bunny lovers" according to the professional managerial elite. In accordance with the moniker, women's views were ridiculed and discounted.

The separation of informed judgements (by experts) from those of an uninformed public fuels public debates on the acceptability of technological risks. Some feel the lay public does not have a valid understanding of, or constructive opinions on, technological risks (Lewis, 1990), and thus should simply defer judgements and management to the experts. It is, in fact, the creation and maintenance of the aura of expertise that establishes hierarchies that ultimately reduce the value of the opinion of the "other," be they women, minorities or the poor (Hynes, 1989, 1991). Risk perceptions based upon anti-technocentric, anti-masculine or anti-progress modes of thought are not only ignored but disabled. Clearly, what the public views as acceptable risk is as important to the use of technology as what the expert thinks. Often the public view is more persuasive with lawmakers.

Numerous cases can be mentioned that emphasize the disparity between the acceptability of a technology or activity to experts and its unacceptability to the public. For example, the use of Alar (daminozide) on apples became an unacceptable activity to mothers based on evidence of the likely carcinogenic effects on children—the most apple-consuming cohort. Experts insisted that levels ingested even in this most vulnerable population were not enough to induce cancer. Likewise, the apple industry asserted their product was safe. Nevertheless, millions of parents boycotted red apples, sales plummeted, and apple growers withdrew the products from market as a sign of concern, losing millions of dollars in the process.

In contrast, the control of ozone-depleting substances like CFCs and halons is championed more by the experts and less by the general public, although it is the latter who would need to change their lifestyles to support the goal. In fact, most of the concern for global warming, greenhouse gases and ozone depletion is coming from scientists, large environmental organizations and other institutions that focus on global concerns. Thus far, the public has demonstrated little direct concern about this issue.

## ECOFEMINISM AND THE POLITICS OF TECHNOLOGY

Within the last decade, feminists and environmentalists have discovered one another, recognizing commonalities in their goals. The pragmatism of radical environmentalists (Foreman and Haywood, 1987; Manes, 1990; Scarce, 1990) and the critical nature of radical ecologies (like social ecology) provide support to feminist critiques of the anthropocentrism of technology and society. The nearest kindred spirits, deep ecology (Devall and Sessions, 1985) and bioregionalism (Sale, 1985) argue for a biocentric world view, but it is a biocentralism built upon the foundations of masculinity-based science, the cause of the problem. The interaction between feminism and environmentalism provides a wide range of ecofeminisms that challenge the social, cultural and political zeal of masculine modern culture. The message is clear: when women suffer through social discrimination and the domination of nature, the planet is also threatened (King, 1989). There are a number of central tenets to ecofeminism that are pertinent to our discussion. Ecosystem malaise and abuse is rooted in androcentric (man-centered) values and institutions. Human liberation and its relationship to non-human nature requires a new ethic about the use of technology since technology is the tool by which humans manipulate nature. Relations between society and nature, and between men and women, need re-assessing to become more harmonizing and less dominating. Ecofeminists praise diversity and view it as enriching not confining, a strength rather than a weakness. The problems of relations and connections (human and non-human, male and female) will not be resolved until androcentric values and institutions are restructured (Oelschlaeger, 1991). Finally, as a social movement, ecofeminism supports harmony, diversity, decentralized communities and technologies based on sound ecological principles (King, 1989).

The politics of social control over the use of the environment is governed by public opinion. These collective views represent a wide range of individual perspectives that are ultimately masked by the creation of averages. The obscurity of the views of socially alienated groups is managed by their omission from telephone surveys, or their amalgamation into the average statistic to derive a "typically American" view. Indeed, what is a typical American? A case in point is the social construction of sex roles. Sexual stereotypes often dictate the activities that may be performed, given the vagaries of social approvals and disapprovals—acceptance, ostracism, security, fear. Sex roles will, for example, create guidelines for acceptable activities revolving around ideas concerning body/mind strength, body/mind agility, body image and body size—that is, ideas merely based on sex or biological difference.

When individual personality is examined, perception can be regarded as a product of non-physical, "emotional" development. "Gender" is a term that describes the extent to which traits that are socially defined as masculine or feminine are present within an individual's personality. The dynamics of gender may be unseen and unknown, yet it conditions the individual's method of processing information, and ultimately his or her view of nature. Tracing the work of psychologists, Merchant (1981) says femininity stems from the woman's role as reproducer in Western culture and is imprinted upon offspring from birth. The female psyche is based on empathy, identification and wholeness, while the male psyche is rooted in separation,

dualism and distinction. Each leads to a different view of nature and our place in it. This argument is further enhanced by Gilligan's (1982) different voice thesis. She argues that there are distinct differences in the psychological development of men and women that arise from the social construction of identity. Power, social status and reproductive biology all interact to shape the individual experiences of males and females as well as the interactions between them. Thus, if women perceive and experience the world differently than men, then it stands to reason that they would also perceive risks differently as well.

An examination of the perceptions of women versus those of men can provide insights into understanding the role gender plays within a society and the political processes occurring within that society. Is there a difference, for instance, between the levels of risk perceived by men and women automobile drivers? Would the presence of a nuclear power plant in a neighborhood elicit more concern in a woman than in a man? Would the potential of handgun use produce the same perceptual and emotional response in men as in women?

## GENDER PERCEPTIONS

Some general distinctions between men's and women's responses to risks have already been identified (McStay and Dunlap, 1983; Stallen and Thomas, 1988). For example, studies have variously described women as more nurturing and concerned with health and well-being, interested in the fate of society as a whole, and desiring to limit destructive technologies. As such, they are more concerned about industrial hazards and more opposed to military spending, military intervention and war, especially nuclear war (Boulding, 1984; Silverman and Kumka, 1987; Bastide et al., 1989). Men are described as more self-interested, prone to more aggressive behaviors, and more willing to gamble and take risks (Levin et al.,

1988).  In other words, men may be viewed as risk-takers and women as risk-avoiders.

The above statements partially reflect some truisms and oversimplifications about gender differences.  Unfortunately, we have very little empirical evidence to support or reject them, and the evidence that is available is inconclusive.  Specific studies in the feminist literature, for example, examined gender differences in the perception of nuclear war and found that women were more concerned about the risk and more willing to work for peaceful solutions to conflict (Boulding, 1984; Jensen, 1987; Silverman and Kumka, 1987; Hamilton et al., 1988).  In a study of acid rain, another technological hazard, men were found to be more concerned with the problem, and more knowledgeable than women (Arcury et al., 1987).  Currently, it is virtually impossible to determine what kind of gender perception differences might exist.

Understanding the role of gender in risk perception is critical for several reasons.  First, it is important to understand the socialization process and how sex roles influence risk perception.  Women may bear a disproportionate share of the risks, thus they may have completely different views of the risks and avenues for ameliorating them.  Second, it is crucial to understand differences for the improvement of risk communication.  The reliance on risk perception studies based solely on a male or mixed population points to the potential failure of risk communication, where the interests and beliefs of women, who might be disproportionately more involved in a series of risk management issues, are largely ignored.  Finally, the social acceptability of risk may vary widely between different gender, racial and ethnic groups.  The influence of women in the political and economic arena is increasing along with their representation among policy analysts and decision makers (Holcomb et al., 1990; Regulska et al., 1991).  Women in government, for example, may favor radically different laws and regulatory strategies for technological hazards than their male counterparts, particularly if they feel that the laws will differentially affect women and men.  A political gender gap does exist, and it is related not only to traditional women's concerns, but also to many other issues, including risk assessment and the acceptability of such risks.

## PILOT STUDY: GENDER-BASED DIFFERENCES IN RISK PERCEPTION

Many methods for studying technological risk perception have been used over the years.  They range from attitude theory and change (Ajzen and Fishbein, 1980; Verplanken, 1989) to affective responses and judgements (Johnson and Tversky, 1983, 1984).  The psychometric paradigm is the most widely used method and the one selected for this pilot study.  The survey instrument used in this research is similar in design to the one described by Slovic et al. (1985).  The questionnaire had four major sections: a rank-ordering of the perceived level of risk for 33 technologies and activities; a judgement on the acceptability of risk from each of these (e.g. could it be made more or less risky?); an evaluation of the characteristics of risk for each technology or activity; and finally the current and desired level of regulation for each technology.

The survey was administered in April 1989 to a group of undergraduate students in five separate sections of an introductory geography course.  A sample of college students was also used in the original psychometric studies (Fischhoff et al., 1978; Slovic et al., 1985) in addition to a smaller group of respondents from the League of Women Voters.  Students were asked to complete the survey during a regular 80-minute class period.  A brief introduction and description of the survey and its use in a research project was given.  A total of 294 completed questionnaires were returned.

## Respondent Characteristics

The average age of our respondents was 21. The sample had more men (56%) than women (40%) (4% failed to identify themselves). 32% of the students were in their senior year, with the remaining evenly distributed between the other years (Table 1). About 72% of the sample were Caucasian, making racial and ethnic comparisons impossible because of the small sample size and unreliability of the self-reporting. 15% of the respondents were economics majors, followed by psychology and undecided. Finally, more than three-fourths of the students were native to New Jersey. There were very few differences between males and females in referent characteristics. Psychology and economics majors were prevalent among the females in the sample, while economics was the most often listed major among males.

**TABLE 1** Sample Population (in %)

|  | Total | Female | Male |
|---|---|---|---|
| *Number* | 294 | 119 | 166 |
| *Mean age* | 21.0 | 20.9 | 21.0 |
| *Year* |  |  |  |
| First year | 21.8 | 20.2 | 24.1 |
| Sophomore | 20.1 | 25.2 | 17.5 |
| Junior | 20.1 | 15.1 | 23.5 |
| Senior | 32.0 | 36.1 | 30.7 |
| No response | 6.0 | 3.4 | 4.2 |
| *Race/ethnicity* |  |  |  |
| African American | 4.8 | 5.0 | 4.8 |
| Asian | 6.5 | 7.6 | 6.0 |
| Caucasian | 72.4 | 76.5 | 72.3 |
| Hispanic | 3.1 | 4.2 | 2.4 |
| Native American | 3.1 | 0.8 | 4.8 |
| Other | 3.7 | 2.5 | 4.8 |
| No response | 6.4 | 3.4 | 4.9 |
| *US citizen* |  |  |  |
| Yes | 89.5 | 92.4 | 91.0 |
| No | 5.8 | 4.2 | 7.2 |
| No response | 4.7 | 3.4 | 1.8 |
| *New Jersey native* |  |  |  |
| Yes | 75.2 | 77.3 | 76.5 |
| No | 17.3 | 17.6 | 18.1 |
| Non-citizen | 4.1 | 5.0 | 3.6 |
| No response | 3.3 | 0.0 | 1.8 |

## Risk Rankings

Respondents were asked to rank the relative importance of each of the 33 items in terms of the risk of dying as a consequence of the technology, substance or activity (using 11 as the least risky and 43 as the most). There are a few statistically significant differences between males and females in their ranking of technological risks, although there is considerable agreement on the top five (Table 2). Females listed nuclear weapons, handguns, illegal drugs, smoking and sexually transmitted

**TABLE 2** Relative Ranking of Risks

|  | Arithmetic Means | | |
|---|---|---|---|
| Activity/Technology | Total | Men | Women |
| 1. Illegal drugs | 36.1 | 36.8 | 35.3* |
| 2. Nuclear weapons | 35.8 | 35.3 | 36.6 |
| 3. Handguns | 35.6 | 35.5 | 35.8 |
| 4. Smoking | 34.3 | 34.1 | 34.4 |
| 5. Motor vehicles | 34.2 | 34.1 | 34.2 |
| 6. Sexual diseases | 32.9 | 32.0 | 34.2* |
| 7. Police work | 32.5 | 32.6 | 32.1 |
| 8. Nuclear power | 32.3 | 31.4 | 33.4 |
| 9. Alcoholic beverages | 31.5 | 31.7 | 31.4 |
| 10. Motorcycles | 31.3 | 31.6 | 30.6 |
| 11. Firefighting | 30.9 | 31.4 | 30.4 |
| 12. CFCs | 29.4 | 28.7 | 30.4 |
| 13. Pesticides | 28.9 | 29.0 | 28.8 |
| 14. Surgery | 28.7 | 28.6 | 29.0 |
| 15. Burning fossil fuels | 27.4 | 27.6 | 27.1 |
| 16. Large construction | 26.7 | 26.9 | 26.7 |
| 17. Commercial aviation | 25.8 | 24.9 | 27.0* |
| 18. Non-nuclear electric | 25.1 | 24.4 | 26.1 |
| 19. Mountain climbing | 25.0 | 25.8 | 24.2 |
| 20. Hunting | 24.4 | 25.0 | 23.8 |
| 21. X-rays | 23.6 | 23.7 | 23.5 |
| 22. Railroads | 22.8 | 22.5 | 23.0 |
| 23. Prescription drugs | 22.8 | 22.0 | 23.6 |
| 24. Food preservatives | 19.4 | 18.9 | 20.1 |
| 25. Skiing | 19.4 | 19.3 | 19.6 |
| 26. Power mowers | 19.3 | 19.2 | 19.3 |
| 27. Swimming | 19.3 | 19.3 | 19.2 |
| 28. Bicycles | 18.9 | 19.2 | 18.2 |
| 29. Home appliances | 17.8 | 18.3 | 17.1 |
| 30. Contraceptives | 17.7 | 17.7 | 17.3 |
| 31. Vaccinations | 17.2 | 17.6 | 16.5 |
| 32. High school football | 16.4 | 17.2 | 15.2* |
| 33. Food coloring | 16.1 | 15.7 | 16.5 |

*Difference between male/female means (*t*-statistic) significant at $p > 0.05$.

diseases as the most risky technologies/activities. Males, on the other hand, listed illegal drugs, handguns, nuclear weapons, motor vehicles and smoking. There were no differences in the evaluation of the least risky items; food coloring, high school/college football, vaccinations, contraceptives and home appliances were given a very low ranking of perceived riskiness.

## Risk Characteristics

Students were asked to evaluate risk characteristics for each of the 33 items using a seven-point bi-polar scale. The nine dimensions, based on the psychometric approach, included: voluntary/involuntary; immediate effect/delayed effect; risk level known to the exposed/unknown to the exposed; risks known to science/unknown to science; risks new and novel/old and familiar; chronic (kills one at a time)/catastrophic (kills large numbers at once); learn to live with or common/dreaded or feared; fatal or severe consequences/no effects or not severe; risks are controllable/uncontrollable. To identify those risk characteristics that best illustrate and distinguish among the different types of technological hazards, a factor analysis was done (Slovic et al., 1985). This statistical procedure identified two factors that help define the dimensions and commonalities of risk among the 33 items.

The first factor (*Known*) describes a group of technological hazards whose risks are familiar, where people engage in the activities on a voluntary basis, whose effects are chronic, and the risks are more or less acceptable. This is in sharp contrast to the second factor (*Dreaded*), which highlights technologies that are perceived as newer, where risks are imposed involuntarily, with potential catastrophic consequences, resulting in technology that is feared. Technological risks from sources such as pesticides, CFC use, nuclear power and nuclear weapons involuntarily expose people to risks and are relatively new

hazards according to our sample. In addition, these risks are known to science and, to people exposed, catastrophic in their potential to cause death, and therefore highly feared.

Based on this statistical analysis and two-dimensional mapping, two different types of technological hazards emerge. The first grouping of hazards are what we might consider societal in origin. These are hazards that are new, to which we are involuntarily exposed, have an enormous catastrophic potential (i.e. may kill large numbers of people at one time), and where individuals have very little control over the personal risk, thus making the consequences quite dreaded. Examples include the commercial use of nuclear power, nuclear weapons, pesticides, CFC use and food additives (preservatives and coloring). The second grouping of technological hazards are more individual in nature. This group is characterized by more voluntary exposures to familiar hazards, where the risks are well known to science, and where mishaps result in fatalities. Examples of these hazards include handguns, illegal drugs, smoking and motor vehicles.

While the overall classification of technological risks is similar between men and women, there are some clear distinctions in the evaluation of each of the nine attributes of risk. For example, we found that women were more likely than men to evaluate risks as more catastrophic in potential, more dreaded, and more fatal (Table 3). Women consistently rated hunting, motor vehicles, firefighting, construction projects, mountain climbing, contraceptives and motorcycles as posing a more catastrophic risk potential than did men. Similarly, women were more inclined to evaluate motor vehicles, nuclear power, pesticides, X-rays, sexually transmitted diseases, commercial aviation and the burning of fossil fuels as dreaded risks. Women were also more pessimistic than men in their evaluations of the severity of risks and the fatal consequences of alcoholic beverages, skiing, sex-

**TABLE 3** Differences in Female/Male Perceptions Based on Selected Risk Characteristics

| Activity/Technology | Risk Characteristics | | |
| --- | --- | --- | --- |
| | Catastrophic | Dread | Fatality/Severity |
| 1. Illegal drugs | 0.25 | 0.12 | 0.06 |
| 2. Nuclear weapons | −0.01 | 0.14 | 0.08 |
| 3. Handguns | 0.32 | 0.29 | 0.03 |
| 4. Smoking | 0.27 | 0.22 | 0.17 |
| 5. Motor vehicles | 0.48* | 0.51* | 0.11 |
| 6. Sexual diseases | 0.38 | 0.49* | 0.76** |
| 7. Police work | 0.37* | 0.32 | 0.26 |
| 8. Nuclear power | 0.40* | 0.43* | 0.28 |
| 9. Alcoholic beverages | 0.31 | 0.20 | 0.63** |
| 10. Motorcycles | 0.36* | 0.17 | 0.00 |
| 11. Firefighting | 0.73* | 0.12 | 0.05 |
| 12. CFCs | 0.15 | 0.44* | 0.45* |
| 13. Pesticides | 0.43* | 0.65** | 0.35* |
| 14. Surgery | 0.14 | 0.29 | 0.29 |
| 15. Burning fossil fuels | 0.19 | 0.56* | 0.12 |
| 16. Large construction | 0.61* | 0.23 | −0.13 |
| 17. Commercial aviation | 0.32 | 0.53* | 0.06 |
| 18. Non-nuclear electric | 0.39 | 0.36 | 0.33* |
| 19. Mountain climbing | 0.36* | 0.14 | −0.21 |
| 20. Hunting | 0.18* | 0.04 | −0.00 |
| 21. X-rays | 0.24 | 0.54 | 0.27 |
| 22. Railroads | −0.11 | 0.35* | 0.15 |
| 23. Prescription drugs | 0.28 | −0.08 | 0.36* |
| 24. Food preservatives | 0.49* | 0.28 | 0.50* |
| 25. Skiing | 0.17 | −0.03 | 0.48* |
| 26. Power mowers | 0.35* | 0.04 | 0.20 |
| 27. Swimming | 0.33* | 0.03 | 0.01 |
| 28. Bicycles | 0.23 | 0.09 | 0.37* |
| 29. Home appliances | 0.15 | −0.01 | 0.25 |
| 30. Contraceptives | 0.23 | 0.26 | 0.57** |
| 31. Vaccinations | 0.31 | −0.06 | 0.08 |
| 32. High school football | 0.26* | −0.01 | 0.36* |
| 33. Food coloring | 0.43* | 0.32 | 0.22 |

*Difference between female/male means ($t$-statistic) significant at $p > 0.05$.

**Difference between female/male means ($t$-statistic) significant at $p > 0.001$.

ually transmitted diseases, CFCs, contraceptives and food additives.

## Risk Regulation

Students were asked to evaluate the current and desired level of regulation for each of the 33 different technologies and activities utilizing the same 0–5 scale employed by Slovic et al. (1985). In that original work, nuclear power and spray cans (CFCs in our survey) were viewed as the activities/technologies where the most stringent regulations were desired,

followed by pesticides, handguns and hunting. Our sample is less interested in regulation, perhaps a result of the deregulation fervor of the 1980s. The mean score for desired regulation in our sample was consistently lower than the Slovic et al. sample for all items with the exception of handguns, smoking and alcoholic beverages. Our respondents felt the most stringent regulation should be placed on illegal drugs (mean score 4.54) and nuclear weapons (mean score 4.34) (Table 4). Severe restrictions on product use were desired for handguns, smoking, nuclear power, CFCs, pes-

**TABLE 4**  *Current and Desired Levels of Regulation (mean scores)*

| Activity/Technology | Regulation | | M/F Differences | |
|---|---|---|---|---|
| | Current | Desired | Current | Desired |
| 1. Illegal drugs | 3.62 | 4.54 | 0.316 | 0.002 |
| 2. Nuclear weapons | 3.22 | 4.34 | 0.135 | −0.198 |
| 3. Handguns | 2.76 | 3.80 | 0.065 | −0.249 |
| 4. Smoking | 1.98 | 3.43 | −0.073 | 0.075 |
| 5. Motor vehicles | 2.31 | 2.62 | −0.141 | −0.250 |
| 6. Sexual diseases | 1.71 | 2.42 | −0.284 | −0.109 |
| 7. Police work | 1.76 | 2.18 | −0.003 | 0.140 |
| 8. Nuclear power | 3.19 | 3.81 | −0.023 | −0.317* |
| 9. Alcoholic beverages | 2.41 | 2.90 | −0.270* | −0.121 |
| 10. Motorcycles | 1.70 | 2.24 | 0.013 | 0.095 |
| 11. Firefighting | 1.61 | 1.85 | 0.118 | −0.015 |
| 12. CFCs | 2.16 | 3.66 | −0.112 | −0.231 |
| 13. Pesticides | 2.15 | 3.40 | 0.064 | −0.283* |
| 14. Surgery | 2.20 | 2.48 | −0.003 | 0.236 |
| 15. Burning of fossil fuels | 1.97 | 3.00 | −0.120 | −0.038 |
| 16. Large construction | 1.76 | 2.28 | 0.117 | 0.060 |
| 17. Commercial aviation | 2.31 | 3.01 | −0.059 | −0.151 |
| 18. Non-nuclear electric power | 1.97 | 2.35 | −0.224 | 0.070 |
| 19. Mountain climbing | 0.76 | 1.03 | 0.120 | 0.026 |
| 20. Hunting | 1.79 | 2.26 | 0.217 | 0.022 |
| 21. X-rays | 1.98 | 2.50 | 0.021 | −0.026 |
| 22. Railroads | 1.77 | 2.25 | 0.127 | −0.028 |
| 23. Prescription drugs | 2.15 | 2.65 | −0.142 | −0.118 |
| 24. Food preservatives | 1.73 | 2.65 | −0.096 | −0.210 |
| 25. Skiing | 0.67 | 0.84 | 0.015 | 0.150 |
| 26. Power mowers | 0.85 | 1.13 | 0.035 | 0.023 |
| 27. Swimming | 0.83 | 0.93 | 0.048 | 0.127 |
| 28. Bicycles | 0.76 | 1.04 | 0.028 | 0.002 |
| 29. Home appliances | 1.12 | 1.40 | 0.116 | −0.038 |
| 30. Contraceptives | 1.25 | 1.59 | −0.290* | −0.327 |
| 31. Vaccinations | 1.72 | 2.06 | 0.123 | 0.189 |
| 32. High school football | 0.91 | 1.22 | 0.104 | 0.085 |
| 33. Food coloring | 1.62 | 2.46 | −0.022 | −0.202 |

Scale:
0 = do nothing
1 = monitor the risk and/or inform those exposed
2 = place mild restrictions on who, how, when and where something can be done or used
3 = place moderate restrictions on who, how, when and where something can be done or used
4 = place severe restrictions on who, how, when and where something can be done or used
5 = ban the product or activity
*Differences between male/female means is significant at $p > 0.05$.

ticides, commercial aviation and the burning of fossil fuels—all issues in the news at the time.

There are only a few important distinctions between men and women in their views of current levels of regulation. Men felt that alcoholic beverages were mildly restricted at present, whereas women felt that more stringent restrictions existed. Similarly, men re-

sponded that contraceptive regulations simply informed the users, whereas women felt there were some minor restrictions imposed on their use. These differences may be partly explained by each group's access to and participation in the activity. As for desired regulation, there was considerable congruity between the two groups. The two main areas of disagreement were in the regulation of nuclear power and pesticides. In each instance, women wanted more restrictive regulations, whereas men only desired moderate regulation.

## DISCUSSION

This pilot study corroborates many of the generalized findings in the risk perception literature. The pattern of risk rankings, associated risk characteristics and desire for risk regulation in our sample reflect the findings of earlier studies utilizing the psychometric technique. Some of the changes in risk ranking and characteristics can be explained by the differences in the sample populations (location and background) and in the decade difference in the timing of the surveys (1976 versus 1989).

We expected to find greater differences between males and females than we actually did. Although there are some areas of disagreement between the two subpopulations in their views of risk and risk regulation, the magnitude of difference is not great. Perhaps age played a role, as students are not sufficiently aware of these technologies or activities to have formed an opinion on them. Perhaps the lack of difference was a function of the survey instrument we used, which did not explicitly address male/female differences. More likely, we can speculate that differences in risk perceptions are not as important as how these translate into overt actions or behaviors. Specifically, how does gender influence the ways in which people respond to these perceived risks?

A number of questions arise from this case study. First, how can we improve risk perception studies to more effectively reflect the perceptions of a heterogenous sample population? Second, how can we improve the methodology to more effectively interpret the impact of spatial, temporal and compositional differences in the perception of risk? Finally, how can a feminist analysis help in understanding risk perception and behavior?

Future modifications to the psychometric design and other instruments must be more sensitive to subgroup differences and more adept at distinguishing between the perceptions of men and women. Many of the risks included in the original survey represent traditionally male-oriented activities such as motorcycling, firefighting, construction, hunting and football, while only two technologies/activities (birth control devices and home appliances) can be considered as more female-oriented. Redesigned surveys should either include a more balanced number of male- and female-oriented risks or exclude all those that are gender-specific.

Classification of technologies/activities as male- or female-oriented will improve our understanding of how males or females will respond to these risks, but better identification of gender also must be done. In order to correlate perception with gender (not sex), the survey must provide a measure of the femininity or masculinity for each individual. How one feels about certain activities and situations (not necessarily risks) may provide enough insight to enable placement of the respondent along a femininity–masculinity continuum. It is necessary to determine how much masculinity and femininity is present in the individual's psychology to better understand the effect of gender on risk perception.

Comparative questions (gender, race, ethnicity) on risk can also be included, such as: Are women or men more aware of the risk from this activity/technology? Are women more voluntarily exposed to this risk than men? Who is more likely to be exposed to this risk, men or women? Do men take part in this activity or technology more than women? Similar questions could be developed for racial or ethnic comparisons.

Risk communication strategies attempt to incorporate the findings of the risk perception research and develop approaches to allay the fears of a concerned public. How effective can these strategies be when the public is defined as a homogeneous group rather than one with distinct gender, ethnic and racial characteristics that influence the interpretation of risk and its ultimate acceptability? This is particularly important in understanding citizen rejection of governmental and industrial risk management and communication policies. Since we found that women in our survey indicated a slightly stronger preference for less risky technologies and more stringent regulation on existing technologies than men, it is obvious that the efforts of *Mothers Against Pesticides*, *Mothers Against Drunk Driving*, and Lois Gibbs at *Citizens' Clearing House for Toxic Waste* are not aberrations, and may in fact be more representative of the general public than previously thought.

Risk perception surveys can be made more robust by including measures of behavioral responses to these risks. A matched sample of respondents—one drawn from activist groups, the other drawn from a more random public—might help to elicit the relative importance of perceived risk and its relation to proactive behavior. Clearly, more understanding of the relationship between perception and behavior is warranted as well.

Most importantly, a feminist analysis allows us to frame questions in a different light and to ultimately ask different questions. Rather than an absolute judgement of risks, perhaps a more comparative approach is warranted. How women cope with the myriad of risks they face in everyday life and their role in hazards decision making needs to be further addressed.

Finally, an ecofeminist analysis can provide a critique of the ethics of technological innovation and invention. Once a clear determination has been made regarding gender division of risk, questions focusing on the unevenness of risk exposure, the disparity in the control of risks or the provision of operational and safety information to the involuntarily exposed can be examined with a moral and ethical ear turned toward the dominated. A reevaluation of current technologies, not based on the "average American," but on a range of gender, class and ethnic distinctions, will serve to enhance the assessment of current and future risks and policies to manage them.

## CONCLUSIONS

Our pilot survey demonstrates that more regulation of technology and activities is desired by our respondents. This is consistent with current public opinion that also wants more environmental regulation and environmental laws. Our sample certainly expresses a strong desire to have many environmentally harmful technologies, such as nuclear power, CFCs, pesticides and the burning of fossil fuels, severely restricted.

Before we can definitely conclude the significance of gender differences in the perception of technological risks, however, much more work is needed. Gender differences certainly exist, as suggested by feminist analysis, but the empirical proof is weak. Furthermore, the causes of gender differences in risk perception and behavior (individual personality, sex roles, socialization) are relatively unknown and not addressed in this study or any others. While this paper provides some very preliminary data on the role of gender in the perception of technological risk, its value is in highlighting the need for more localized and population-specific analyses of risk perception, a continued reappraisal of the psychometric paradigm, a broadening of risk perception research to include behavior, and a reassessment of the prevailing masculine-based view of nature, science and technology.

### REFERENCES

Ajzen, I., and Fishbein, M., 1980. Understanding Attitudes and Predicting Social Behavior. Prentice-Hall, Englewood Cliffs, N.J.

Arcury, T.A., Scollay, S.J. and Johnson, T.P., 1987. Sex differences in environmental concern and knowledge: the case of acid rain. Sex Roles, 16: 463–72.

Bastide, S., Moatti, J.P., Pages, J.P. and Fagnani, F., 1989. Risk perception and social acceptability of technologies: the French case, Risk Analysis, 9: 215–23.

Boulding E., 1984. Focus on: the gender gap. Journal of Peace Research, 21: 1–3.

Bullard, R.D., 1990. Dumping in Dixie: Race, Class, and Environmental Quality. Westview Press, Boulder, CO.

Commission for Social Justice, 1987. Toxic Waste and Race in the United States. United Church of Christ, New York.

Devall, B. and Sessions, G., 1985. Deep Ecology: Living As If Nature Mattered. Gibbs Smith, Salt Lake City, UT.

Douglas, M. and Wildavsky, A., 1982. Risk and Culture: an Essay on the Selection of Technological and Environmental Dangers. University of California Press, Berkeley, CA.

Englander, T., Farago, K., Slovic, P. and Fischhoff, B., 1986. A comparative analysis of risk perception in Hungary and the United States. Social Behavior, 1: 55–66.

Fischoff, B., Slovic, P., Lichtenstein, S., Read, S. and Combs, B., 1978. How safe is safe enough? A psychometric study of attitudes toward technological risks and benefits. Policy Sciences, 9: 127–52.

Foreman, D. and Haywood B., eds., 1987. Ecodefense: a Field Guide to Monkeywrenching (2nd ed.). Ned Ludd Book, Tucson, AZ.

Gardner, G.T. and Gould, L.C., 1989. Public perceptions of the risks and benefits of technology. Risk Analysis, 9: 225–42.

Gilligan, C., 1982. In a Different Voice: Psychological Theory and Women's Development. Harvard University Press, Cambridge, MA.

Hamilton, S.B., Van Mouwerik, S., Oetting, E.R., Beauvais, F. and Keilin, W.G., 1988. Nuclear war as a source of adolescent worry: relationships with age, gender, trait emotionality, and drug use. The Journal of Social Psychology, 128: 745–63.

Hansson, S.O., 1989. Dimensions of risk. Risk Analysis, 9: 107–12.

Hewitt, K., ed., 1983. Interpretations of Calamity. Allen & Unwin, Boston, MA.

Holcomb, H.B., Kodras, J.E. and Brunn, S.D., 1990. Women's issues and state legislation: fragmentation and inconsistency. In: J.E. Kodras and J.P. Jones III, eds. Geographic Dimensions of United States Social Policy. Arnold, London: 178–99.

Hynes, H.P., 1989. Recurring Silent Spring. Pergamon, New York.

Hynes, H.P., ed., 1991. Reconstructing Babylon: Essays on Women and Technology. Indiana University Press. Bloomington, IN.

Jensen, M.P., 1987. Gender, sex roles, and attitudes toward war and nuclear weapons. Sex Roles, 17: 253–67.

Johnson, B. and Covello, V., eds., 1987. Social and Cultural Construction of Risk. Reidel, Boston, MA.

Johnson, E.J. and Tversky, A., 1983. Affect, general-ization and the perception of risk. Journal of Personality and Social Psychology, 45: 20–31.

Johnson, E.J. and Tversky, A., 1984. Representations of perceptions of risk. Journal of Experimental Psychology: General, 113: 55–70.

Kasperson, R.E., Renn, O., Slovic, P., Brown, H.S., Emel, J., Goble, R., Kasperson, J.X. and Ratick, S., 1988. The social amplification of risk: a conceptual framework. Risk Analysis, 8: 117–87.

Keown, C.F., 1989. Risk perceptions of Hong Kongese vs. Americans. Risk Analysis, 9: 401–5.

King, Y., 1989. The ecology of feminism and the feminism of ecology. In: J. Plant, ed. Healing the Wounds: the Promise of Ecofeminism. New Society, Philadelphia, PA: 18–28.

Kleinhesselink, R.R. and Rosa, E.A., 1989. Cognitive representation of risk perceptions: a comparison of Japan and the United States. Paper presented at annual meeting of the Society for Risk Analysis, San Francisco, CA.

Levin, I.P., Snyder, M.A. and Chapman, D.P., 1988. The interaction of experiential and situational factors and gender in a simulated risky decision-making task. The Journal of Psychology, 122: 173–81.

Lewis, H.W., 1990. Technological Risk. Norton, New York.

Lichtenstein, S., Slovic, P., Fischhoff, B., Layman, M. and Combs, B., 1978. Judged frequency of lethal events. Journal of Experimental Psychology: Human Learning and Memory, 4: 551–81.

Manes, C., 1990. Green Rage: Radical Environmentalism and the Unmaking of Civilization. Little, Brown, Boston, MA.

McStay, J.R. and Dunlap, R.E., 1983. Male-female differences in concern for environmental quality. International Journal of Women's Studies, 6: 291–307.

Merchant, C., 1980. The Death of Nature: Women, Ecology and the Scientific Revolution. Harper & Row, New York.

Merchant, C., 1981. Earthcare. Environment, 23: 6–13, 38–40.

Moncrief, L.W., 1970. The cultural basis for our environmental crisis. Science, 170: 505–12.

Oelschlaeger, M., 1991. The Idea of Wilderness. Yale University Press, New Haven, CT.

O'Riordan, T., 1981. Environmentalism. Pion, London.

Paehlke, R.C., 1989. Environmentalism and the Future of Progressive Politics. Yale University Press, New Haven, CT.

Perrow, C., 1984. Normal Accidents: Living with High-Risk Technologies. Basic Books, New York.

Plant, J., 1990. Searching for common ground: ecofeminism and bioregionalism. In: I. Diamond and G.F Orenstein, eds. Reweaving the World: the Emergence of Ecofeminism. Sierra Club Books, San Francisco, CA: 155–61.

Regulska, J., Fried, S. and Tiefenbacher, J., 1991. Women, politics, and place: spatial patterns of representation in New Jersey. Geoforum, 22(2): 203–21.

Renn, O. and Swaton, E., 1984. Psychological and sociological approaches to the study of risk perception. Environment International, 10: 557–75.

Sale, K., 1985. Dwellers in the Land: the Bioregional Vision. Sierra Club Books, San Francisco, CA.

Scarce, R., 1990. Eco-Warriors: Understanding the Radical Environmental Movement. Noble Press, Chicago, IL.

Silverman, J.M. and Kumka, D.S., 1987. Gender differences in attitudes toward nuclear war and disarmament. Sex Roles, 16: 189–203.

Slovic, P., Fischhoff, B. and Lichtenstein, S., 1985. Characterizing perceived risk. In: R.W. Kates, C. Hohenemser and J.X. Kasperson, eds. Perilous Progress: Managing the Hazards of Technology. Westview Press, Boulders, CO: 91–124.

Stallen, P.J.M. and Thomas, A., 1988. Public concern about industrial hazards. Risk Analysis, 8: 237–45.

Teigen, K.H., Brun, W. and Slovic, P., 1988. Societal risks as seen by a Norwegian public. Journal of Behavioral Decision Making. 1: 111–30.

Verplanken, B., 1989. Beliefs, attitudes, and intentions toward nuclear energy before and after Chernobyl in a longitudinal within-subjects design. Environment and Behavior, 21: 371–92.

Vlek, C. and Stallen, P.J., 1981. Judging risks and benefits in the small and in the large. Organizational Behavior and Human Performance, 28: 235–77.

White, L., 1967. The historical roots of our ecological crisis. Science, 155: 1203–7.

# PART III

# RESPONDING TO THREATS

Uncertainty regarding the risks and hazards of modern living has produced such anxiety that people are routinely barraged with media reports on the latest scare of the week. Why is there excessive fear of irradiated food when scientific evidence doesn't warrant it or excessive complacency about risks such as second-hand smoke? Why do individuals perceive similar risks differently, and how do these perceptions influence their responses to risks and hazards?

Early hazard perception studies surveyed residents about events going on in the places they lived, events such as tornadoes, drought, and floods. These field studies produced a number of interesting findings. For example, people often do not recognize that they live in hazardous environments. Individuals ignore uncertainties when making judgments about the likelihood of events and reduce low probabilities to zero. People also pass the blame on to others (e.g., God, nature, government). In responding to the threat, past experience with the hazard clearly influences the adoption of mitigation measures.

Risk perception research focuses on experimental studies under controlled laboratory conditions, where risks and characteristics of risk are quantified. The psychometric paradigm developed by Slovic, Fischhoff, and colleagues (Reading 12) is a technique that explains how people make sense of uncertainty and complexity in their perceptions of risk. By developing a cognitive map of risk characteristics, two dimensions of risk stand out: dread and the unknown. Risks that fall into the highly dreaded category because the effects are potentially catastrophic or unknown include nuclear war, nuclear power plant accidents, and DNA technology. As a consequence, these risks are viewed as less acceptable than the risks associated with bicycles, alcohol, and power mowers—risks with known, less dreaded, and more tolerated effects.

Critiques of the psychometric paradigm in assessing risk perceptions are numerous. Wildavsky and Dake (Reading 13) provide a review of some of the other social science theories that help explain risk perceptions. These include knowledge theory (people worry most about what directly threatens them at the moment), personality theory (individual personalities account for risk-taking or risk-averse behaviors), economic theory (the rich are more willing to take risks especially with technology because they benefit more, or have more resources to insulate them from the risks; for the poor it is just the opposite), political theory (risk perceptions are simply struggles over interests—government versus industry, public versus industry, Democrats versus Republicans), and finally cultural theory (individuals choose what they fear based on cultural norms and the preservation of a way of life).

Many factors influence how people perceive risks and make decisions about how acceptable these risks are to the individual, groups, or society. In addition to individual personality, social influences, and culture, a number of other factors are important. These include experience, environmental ideology (or worldview), race, gender, socioeconomic status, and distance from the source of the threat. So how are these perceptions turned into actions?

Dennis Mileti's work (Reading 14) on human adjustments to risks from environmental extremes provides a few answers. His typology of risk reduction or risk mitigation measures illustrates some key points. First, the types of strategies one employs generally fall into three broad categories: (1) choose or change the situation or site (abandon floodplains, relocate); (2) reduce the loss through modifying the event or affecting the cause (generally engineering solutions); (3) preventing the loss in the first place (warning systems, building codes), or redistribute the losses (pre-event through insurance, or post-event via disaster relief). As Mileti points out, the choice of adjustments crosses technological, cultural, and regulatory controls. For example, social units adjust when they feel there is a reason to do so (heightened risk perception) and when the costs and bother associated with the action are viewed as worthwhile. Also, social values and goals influence the adoption of mitigation measures by individual social units, because social units are part of larger social systems. This latter factor accounts for the role of peer pressure and social norms in undertaking risk reduction actions.

One example of a risk reduction strategy is to redistribute the losses through insurance. Risa Palm (Reading 15) examines the mandated disclosure of environmental hazards information and its role in influencing the adoption of earthquake insurance. Her analysis questions the prevailing assumption by many hazards managers that the mere provision of hazard information is enough to elicit the appropriate behavioral response. Their conventional wisdom holds that, because most people are risk averse, when faced with undisputed information about the risk in question, they will adapt accordingly by purchasing earthquake insurance or moving. Palm found that the mandated disclosure of the earthquake risk had little impact on whether or not home buyers in northern California decided to purchase earthquake insurance. First, many questioned the reliability of the source of the environmental information (real estate agents). Second, many home buyers felt stuck and lacking in alternatives, rationalizing that since all of California is earthquake country anyway, why bother? Finally, many home buyers viewed their purchase as a financial investment overriding all other considerations, including some future threat to their nest egg. While fire insurance is required as a condition for obtaining a mortgage, earthquake insurance is not. Home buyers did not behave as lawmakers had predicted they would, illustrating once again the divergence between hazards managers and public acceptance of management strategies.

Another example of hazard mitigation is evacuation, which is designed to prevent the losses from occurring in the first place. Evacuations are a form of preimpact protective responses that temporarily remove people from threatened areas. In this country, most evacuations are voluntary because local police officials have no statutory authority to forcibly remove people from their homes. Despite the danger, many people refuse to adhere to warning messages, preferring to ride out disasters on their own. Evacuation research began in the early 1950s, with the civil defense program, in response to threats of bomb attacks. It was subsequently expanded to include natural hazard situations.

Perry (Reading 16) provides a conceptual model of evacuation behavior that is still relevent today. Those factors that influence a positive evacuation decision include the presence of an adaptive plan (prior experience with the hazard, detailed warning messages), an individual view that the threat is real (warning messages from credible sources, number of warnings), and a heightened level of perceived risk, which increases the likelihood of evacuation. If households and family units are together, evacuations are more likely. Also the actions of neighbors and friends influence evacuation decisions. Finally, sociocultural factors, such as race, ethnicity, gender, and age, are also important influences in the decision to evacuate. For example, older people are less likely to evacuate than are households with small children.

In applying these findings from evacuations in response to natural hazard to nuclear power plant accidents, Zeigler and Johnson (Reading 17) found many commonalities, but also some striking differences. In studying the Three Mile Island nuclear power plant accident, they found that evacuations from nuclear threats are unique. Unlike natural disasters where it is difficult to get people to evacuate, there is an overresponse in reaction to nuclear threats, in which many people evacuate despite limited advisories (targeted to specific groups of people or specialized areas). Furthermore, nuclear evacuees fled to greater distances, using distance as a means of threat reduction. Also evacuees used self-selected routes instead of designated routes and chose to stay with friends and relatives rather than in shelters. In applying these findings to a hypothetical accident at the Shoreham, New York plant, Zeigler and Johnson found that people have their own ideas about how to behave during nuclear emergencies that often defy conventional wisdom and preimpact planning. The researchers also found that the spatial dimensions of evacuations were greater than anticipated.

Despite this wealth of knowledge about how risks and hazards are perceived and how people respond to threats, we still cannot say definitely what people will do when a toxic cloud wafts by, when the big earthquake strikes, or when the deluge hits. Hazard managers continue to be challenged to develop public policies to reduce risks.

## DISCUSSION POINTS

1. What are some factors that influence the ways in which social units adjust to the risk of environmental extremes?
2. Based on Slovic's cognitive map of risk characteristics, where might the following hazards be located? Drought? Hurricanes? Earthquakes? Volcanic eruptions? Lightning? Air pollution? Urban Flooding? AIDS?
3. How and why is evacuation behavior different in the case of natural hazards

versus nuclear power plant accidents? Can you think of other instances in which there would be wide variations in response?

4. Which theory of risk helps explain why you personally fear some risks and not others?

5. Is it important when developing public policies to consider how people might behave in response to hazards? If so why? If not, why not?

## ADDITIONAL READING

Barton, A., 1970. *Communities in Disaster.* New York: Anchor Books.

Douglas, M. and A. Wildavsky, 1982. *Risk and Culture: An Essay on the Selection of Technological and Environmental Dangers.* Berkeley: University of California Press.

Dynes, R., 1970. *Organized Behavior in Disaster.* Lexington, MA: D.C. Heath.

Foster, H. D. 1980. *Disaster Planning: The Preservation of Life and Property.* New York: Springer-Verlag.

Haas, J. E., R. Kates, and M. Bowden (Eds.), 1977. *Reconstruction Following Disaster.* Cambridge, MA: MIT Press.

Johnson, B. B. and V. T. Covello (Eds.), 1989. *The Social and Cultural Construction of Risk: Essays on Risk Selection and Perception.* Dordrecht, Netherlands: D. Reidel Publishing Company.

Kunreuther, H., 1978. *Disaster Insurance Protection: Public Policy Lessons.* New York: John Wiley & Sons.

Perry, R. W., M. K. Lindell, and M. R. Greene, 1981. *Evacuation Planning in Emergency Management.* Lexington, MA: Lexington Books.

Perry, R. W. and A. H. Mushkatel, 1986. *Minority Citizens in Disasters.* Athens: The University of Georgia Press.

Sorenson, J. H., J. Soderstrom, E. Copenhaver, S. Carnes, and R. Bolin, 1987. *Impacts of Hazardous Technology: The Psycho-Social Effects of Restarting TMI-1.* Albany: SUNY Press.

Weinstein, N. D. (Ed.), 1987. *Taking Care: Why People Take Precautions.* Cambridge: Cambridge University Press.

# 12

# PERCEPTION OF RISK

*Paul Slovic*

The ability to sense and avoid harmful environmental conditions is necessary for the survival of all living organisms. Survival is also aided by an ability to codify and learn from past experience. Humans have an additional capability that allows them to alter their environment as well as respond to it. This capacity both creates and reduces risk.

In recent decades, the profound development of chemical and nuclear technologies has been accompanied by the potential to cause catastrophic and long-lasting damage to the earth and the life forms that inhabit it. The mechanisms underlying these complex technologies are unfamiliar and incomprehensible to most citizens. Their most harmful consequences are rare and often delayed, hence difficult to assess by statistical analysis and not well suited to management by trial-and-error learning. The elusive and hard to manage qualities of today's hazards have forced the creation of a new intellectual discipline called risk assessment, designed to aid in identifying, characterizing, and quantifying risk (1).

Whereas technologically sophisticated an-

alysts employ risk assessment to evaluate hazards, the majority of citizens rely on intuitive risk judgments, typically called "risk perceptions." For these people experience with hazards tends to come from the news media, which rather thoroughly document mishaps and threats occurring throughout the world. The dominant perception for most Americans (and one that contrasts sharply with the views of professional risk assessors) is that they face more risk today than in the past and that future risks will be even greater than today's (2). Similar views appear to be held by citizens of many other industrialized nations. These perceptions and the opposition to technology that accompanies them have puzzled and frustrated industrialists and regulators and have led numerous observers to argue that the American public's apparent pursuit of a "zero-risk society" threatens the nation's political and economic stablity. Wildavsky (3, p. 32) commented as follows on this state of affairs.

How extraordinary! The richest, longest lived, best protected, most resourceful civilization, with the highest degree of insight into its own technology, is on its way to becoming the most frightened.

Is it our environment or ourselves that have changed? Would people like us have had this

Reprinted from *Science* Volume 236 (1987), pp. 280–85. Copyright © 1987 by the AAAS. Used with the permission of the author and the AAAS.

sort of concern in the past? . . . Today, there are risks from numerous small dams far exceeding those from nuclear reactors. Why is the one feared and not the others? Is it just that we are used to the old or are some of us looking differently at essentially the same sorts of experience?

During the past decade, a small number of researchers has been attempting to answer such questions by examining the opinions that people express when they are asked, in a variety of ways, to evaluate hazardous activities, substances, and technologies. This research has attempted to develop techniques for assessing the complex and subtle opinions that people have about risk. With these techniques, researchers have sought to discover what people mean when they say that something is (or is not) "risky," and to determine what factors underlie those perceptions. The basic assumption underlying these efforts is that those who promote and regulate health and safety need to understand the ways in which people think about and respond to risk.

If successful, this research should aid policy-makers by improving communication between them and the public, by directing educational efforts, and by predicting public responses to new technologies (for example, genetic engineering), events (for example, a good safety record or an accident), and new risk management strategies (for example, warning labels, regulations, subtitute products).

## RISK PERCEPTION RESEARCH

Important contributions to our current understanding of risk perception have come from geography, sociology, political science, anthropology, and psychology. Geographical research focused originally on understanding human behavior in the face of natural hazards, but it has since broadened to include technological hazards as well (4). Sociological (5) and anthropological studies (6) have shown

that perception and acceptance of risk have their roots in social and cultural factors. Short (5) argues that response to hazards is mediated by social influences transmitted by friends, family, fellow workers, and respected public officials. In many cases, risk perceptions may form afterwards, as part of the ex post facto rationale for one's own behavior. Douglas and Wildavsky (6) assert that people, acting within social groups, downplay certain risks and emphasize others as a means of maintaining and controlling the group.

Psychological research on risk perception, which shall be my focus, originated in empirical studies of probability assessment, utility assessment, and decision-making processes (7). A major development in this area has been the discovery of a set of mental strategies, or heuristics, that people employ in order to make sense out of an uncertain world (8). Although three rules are valid in some circumstances, in others they lead to large and persistent biases, with serious implications for risk assessment. In particular, laboratory research on basic perceptions and cognitions has shown that difficulties in understanding probabilistic processes, biased media coverage, misleading personal experiences, and the anxieties generated by life's gambles cause uncertainty to be denied, risks to be misjudged (sometimes overestimated and sometimes underestimated), and judgments of fact to be held with unwarranted confidence. Experts' judgments appear to be prone to many of the same biases as those of the general public, particularly when experts are forced to go beyond the limits of available data and rely on intuition (8, 9).

Research further indicates that disagreements about risk should not be expected to evaporate in the presence of evidence. Strong initial views are resistant to change because they influence the way that subsequent information is interpreted. New evidence appears reliable and informative if it is consistent with one's initial beliefs; contrary evidence tends to be dismissed as unreliable, erroneous, or unrepresentative (*10*). When people lack strong prior opinions, the opposite situation exists—they are at the mercy of the problem formulation. Presenting the same information about risk in different ways (for example, mortality rates as opposed to survival rates) alters people's perspectives and actions (*11*).

## THE PSYCHOMETRIC PARADIGM

One broad strategy for studying perceived risk is to develop a taxonomy for hazards that can be used to understand and predict responses to their risks. A taxonomic scheme might explain, for example, people's extreme aversion to some hazards, their indifference to others, and the discrepancies between these reactions and opinions of experts. The most common approach to this goal has employed the psychometric paradigm (*12, 13*), which uses psychophysical scaling and multivariate analysis techniques to produce quantitative representations or "cognitive maps" of risk attitudes and perceptions. Within the psychometric paradigm, people make quantitative judgments about the current and desired riskiness of diverse hazards and the desired level of regulation of each. These judgments are then related to judgments about other properties, such as (i) the hazard's status on characteristics that have been hypothesized to account for risk perceptions and attitudes (for example, voluntariness, dread, knowledge, controllability), (ii) the benefits that each hazard provides to society, (iii) the number of deaths caused by the hazard in an average year, and (iv) the number of deaths caused by the hazard in a disastrous year.

In the rest of this article, I shall briefly review some of the results obtained from psychometric studies of risk perception and out-line some implications of these results for risk communication and risk management.

## REVEALED AND EXPRESSED PREFERENCES

The original impetus for the pychometric paradigm came from the pioneering effort of Starr (*14*) to develop a method for weighing technological risks against benefits in order to answer the fundamental question, "How safe is safe enough?" His "revealed preference" approach assumed that, by trial and error, society has arrived at an "essentially optimum" balance between the risks and benefits associated with any activity. One may therefore use historical or current risk and benefit data to reveal patterns of "acceptable" risk-benefit trade-offs. Examining such data for several industries and activities, Starr concluded that (i) acceptability of risk from an activity is roughly proportional to the third power of the benefits for that activity, and (ii) the public will accept risks from voluntary activities (such as skiing) that are roughly 1000 times as great as it would tolerate from involuntary hazards (such as food preservatives) that provide the same level of benefits.

The merits and deficiencies of Starr's approach have been debated at length (*15*). They will not be elaborated here, except to note that concern about the validity of the many assumptions inherent in the revealed preferences approach stimulated Fischhoff *et al.* (*12*) to conduct an analogous psychometric analysis of questionnaire data, resulting in "expressed preferences." In recent years, numerous other studies of expressed preferences have been carried out within the psychometric paradigm (*16–24*).

These studies have shown that perceived risk is quantifiable and predictable. Psychometric techniques seem well suited for identifying similarities and differences among groups with regard to risk perceptions and attitudes (Table 1). They have also shown that the concept "risk" means different things to different people. When experts judge risk, their responses correlate highly with technical estimates of annual fatalities. Lay people can assess annual fatalities if they are asked to (and produce estimates somewhat like the technical estimates). However, their judgments of "risk" are related more to other hazard characteristics (for example, catastrophic potential, threat to future generations) and, as a result, tend to differ from their own (and experts') estimates of annual fatalities.

Another consistent result from psychometric studies of expressed preferences is that people tend to view current risk levels as unacceptably high for most activities. The gap between perceived and desired risk levels suggests that people are not satisfied with the way that market and other regulatory mechanisms have balanced risks and benefits. Across the domain of hazards, there seems to be little systematic relationship between perceptions of current risks and benefits. However, studies of expressed preferences do seem to support Starr's argument that people are willing

**TABLE 1** Ordering of perceived risk for 30 activities and technologies (22). The ordering is based on the geometric mean risk ratings within each group. Rank 1 represents the most risky activity or technology.

| Activity or Technology | League of Women Voters | College Students | Active Club Members | Experts |
|---|---|---|---|---|
| Nuclear power | 1 | 1 | 8 | 20 |
| Motor vehicles | 2 | 5 | 3 | 1 |
| Handguns | 3 | 2 | 1 | 4 |
| Smoking | 4 | 3 | 4 | 2 |
| Motorcycles | 5 | 6 | 2 | 6 |
| Alcoholic beverages | 6 | 7 | 5 | 3 |
| General (private) aviation | 7 | 15 | 11 | 12 |
| Police work | 8 | 8 | 7 | 17 |
| Pesticides | 9 | 4 | 15 | 8 |
| Surgery | 10 | 11 | 9 | 5 |
| Fire fighting | 11 | 10 | 6 | 18 |
| Large construction | 12 | 14 | 13 | 13 |
| Hunting | 13 | 18 | 10 | 23 |
| Spray cans | 14 | 13 | 23 | 26 |
| Mountain climbing | 15 | 22 | 12 | 29 |
| Bicycles | 16 | 24 | 14 | 15 |
| Commercial aviation | 17 | 16 | 18 | 16 |
| Electric power (non-nuclear) | 18 | 19 | 19 | 9 |
| Swimming | 19 | 30 | 17 | 10 |
| Contraceptives | 20 | 9 | 22 | 11 |
| Skiing | 21 | 25 | 16 | 30 |
| X-rays | 22 | 17 | 24 | 7 |
| High school and college football | 23 | 26 | 21 | 27 |
| Railroads | 24 | 23 | 29 | 19 |
| Food preservatives | 25 | 12 | 28 | 14 |
| Food coloring | 26 | 20 | 30 | 21 |
| Power mowers | 27 | 28 | 25 | 28 |
| Prescription antibiotics | 28 | 21 | 26 | 24 |
| Home appliances | 29 | 27 | 27 | 22 |
| Vaccinations | 30 | 29 | 29 | 25 |

to tolerate higher risks from activities seen as highly beneficial. But, whereas Starr concluded that voluntariness of exposure was the key mediator of risk acceptance, expressed preference studies have shown that other (perceived) characteristics such as familiarity, control, catastrophic potential, equity, and level of knowledge also seem to influence the relation between perceived risk, perceived benefit, and risk acceptance (12, 22).

Various models have been advanced to represent the relation between perceptions, behavior, and these qualitative characteristics of hazards. As we shall see, the picture that emerges from this work is both orderly and complex.

## FACTOR-ANALYTIC REPRESENTATIONS

Many of the qualitative risk characteristics are correlated with each other, across a wide range of hazards. For example, hazards judged to be "voluntary" tend also to be judged as "controllable"; hazards whose adverse effects are delayed tend to be seen as posing risks that are not well known, and so on. Investigation of these relations by means of factor analysis has shown that the broader domain of characteristics can be condensed to a small set of higher order characteristics or factors.

The factor space presented in Fig. 1 has been replicated across groups of lay people and experts judging large and diverse sets of hazards. Factor 1, labeled "dread risk," is defined at its high (righthand) end by perceived lack of control, dread, catastrophic potential, fatal consequences, and the inequitable distribution of risks and benefits. Nuclear weapons and nuclear power score highest on the characteristics that make up this factor. Factor 2, labeled "unknown risk," is defined at its high end by hazards judged to be unobservable, unknown, new, and delayed in their manifestation of harm. Chemical technologies score particularly high on this factor. A

third factor, reflecting the number of people exposed to the risk, has been obtained in several studies. Making the set of hazards more or less specific (for example, partitioning nuclear power into radioactive waste, uranium mining, and nuclear reactor accidents) has had little effect on the factor structure or its relation to risk perceptions (25).

Research has shown that lay people's risk perceptions and attitudes are closely related to the position of a hazard within this type of factor space. Most important is the horizontal factor "dread risk." The higher a hazard's score on this factor (the further to the right it appears in the space), the higher its perceived risk, the more people want to see its current risks reduced, and the more they want to see strict regulation employed to achieve the desired reduction in risk (Fig. 2). In contrast, experts' perceptions of risk are not closely related to any of the various risk characteristics or factors derived from these characteristics (25). Instead, as noted earlier, experts appear to see riskiness as synonymous with expected annual mortality (26). As a result, conflicts over "risk" may result from experts and lay people having different definitions of the concept.

The representation shown in Fig. 1, while robust and informative, is by no means a universal cognitive mapping of the domain of hazards. Other psychometric methods (such as multidimensional scaling analysis of hazard similarity judgments), applied to quite different sets of hazards, produce different spatial models (13, 18). The utility of these models for understanding and predicting behavior remains to be determined.

## ACCIDENTS AS SIGNALS

Risk analysis typically model the impacts of an unfortunate event (such as an accident, a discovery of pollution, sabotage, product tampering) in terms of direct harm to victims—deaths, injuries, and damages. The impacts of such events, however, sometimes

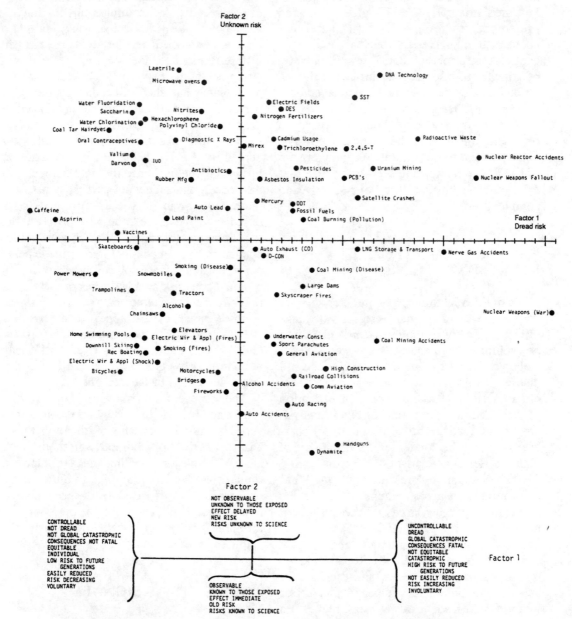

**FIGURE 1.** Location of 81 hazards on factors 1 and 2 derived from the relationships among 18 risk characteristics. Each factor is made up of a combination of characteristics, as indicated by the lower diagram (25).

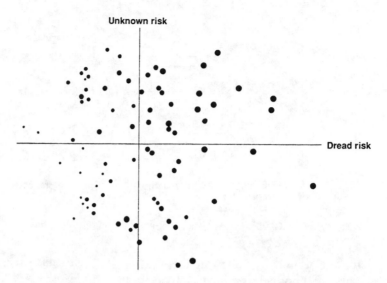

Unknown risk

Dread risk

**FIGURE 2.** Attitudes toward regulation of the hazards in Fig. 1. The larger the point, the greater the desire for strict regulation to reduce risk (25).

extend far beyond these direct harms and may include significant indirect costs (both monetary and nonmonetary) to the responsible government agency or private company that far exceed direct costs. In some cases, all companies in an industry are affected, regardless of which company was responsible for the mishap. In extreme cases, the indirect costs of a mishap may extend past industry boundaries, affecting companies, industries, and agencies whose business is minimally related to the initial event. Thus, an unfortunate event can be thought of as analogous to a stone dropped in a pond. The ripples spread outward, encompassing first the directly affected victims, then the responsible company or agency, and, in the extreme, reaching other companies, agencies, and industries.

Some events make only small ripples; others make larger ones. The challenge is to discover characteristics associated with an event and the way that it is managed that can predict the breadth and seriousness of those impacts (Fig. 3). Early theories equated the magnitude of impact to the number of people killed or injured, or to the amount of property damaged. However, the accident at the Three Mile Island (TMI) nuclear reactor in 1979 provides a dramatic demonstration that factors

besides injury, death, and property damage impose serious costs. Despite the fact that not a single person died, and few if any latent cancer fatalities are expected, no other accident in our history has produced such costly societal impacts. The accident at TMI devastated the utility that owned and operated the plant. It also imposed enormous costs (27) on the nuclear industry and on society, through stricter regulation (resulting in increased construction and operating costs), reduced operation of reactors worldwide, greater public opposition to nuclear power, and reliance on more expensive energy sources. It may even have led to a more hostile view of other complex technologies, such as chemical manufacturing and genetic engineering. The point is that traditional economic and risk analyses tend to neglect these higher order impacts, hence they greatly underestimate the costs associated with certain kinds of events.

Although the TMI accident is extreme, it is by no means unique. Other recent events resulting in enormous higher order impacts include the chemical manufacturing accident at Bhopal, India, the pollution of Love Canal, New York, and Times Beach, Missouri, the disastrous launch of the space shuttle Challenger, and the meltdown of the nuclear re-

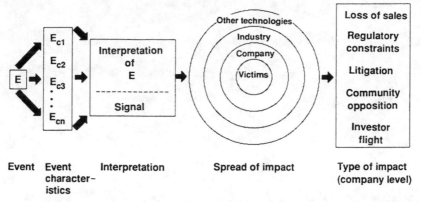

| Event | Event character- istics | Interpretation | Spread of impact | Type of impact (company level) |

**FIGURE 3.** *A model of impact for unfortunate events.*

actor at Chernobyl. Following these extreme events are a myriad of mishaps varying in the breadth and size of their impacts.

An important concept that has emerged from psychometric research is that the seriousness and higher order impacts of an unfortunate event are determined, in part, by what that event signals or portends (28). The informativeness or "signal potential" of an event, and thus its potential social impact, appears to be systematically related to the characteristics of the hazard and the location of the event within the factor space described earlier (Fig. 4). An accident that takes many lives may produce relatively little social disturbance (beyond that experienced by the victims' families and friends) if it occurs as part of a familiar and well-understood system (such as a train wreck). However, a small accident in an unfamiliar system (or one perceived as poorly understood), such as a nuclear reactor or a recombinant DNA laboratory, may have immense social consequences if it is perceived as a harbinger of further and possibly catastrophic mishaps.

The concept of accidents as signals was eloquently expressed in an editorial addressing the tragic accident at Bhopal (29).

> What truly grips us in these accounts is not so much the numbers as the spectacle of suddenly vanishing competence, of men utterly routed by technology, of fail-safe systems failing with a logic as inexorable as it was once—indeed, right

up until that very moment—unforeseeable. And the spectacle haunts us because it seems to carry allegorical import, like the whispery omen of a hovering future.

One implication of the signal concept is that effort and expense beyond that indicated by a cost-benefit analysis might be warranted to reduce the possibility of "high-signal accidents." Unfortunate events involving hazards in the upper right quadrant of Fig. 1 appear particularly likely to have the potential to produce large ripples. As a result, risk analyses involving these hazards need to be made sensitive to these possible higher order impacts. Doing so would likely bring greater protection to potential victims as well as to companies and industries.

## ANALYSIS OF SINGLE HAZARD DOMAINS

Psychometric analyses have also been applied to judgments of diverse hazard scenarios within a single technological domain, such as railroad transport (30) or automobiles (31). Kraus (30) had people evaluate the riskiness of 49 railroad hazard scenarios that varied with respect to type of train, type of cargo, location of the accident, and the nature and cause of the accident (for example, a high-speed train carrying passengers through a mountain tun-

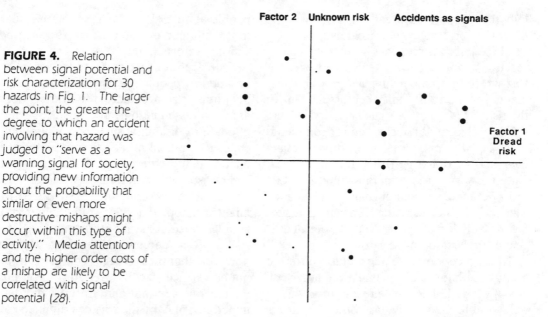

Factor 1
Dread
risk

**FIGURE 4.**  *Relation between signal potential and risk characterization for 30 hazards in Fig. 1.  The larger the point, the greater the degree to which an accident involving that hazard was judged to "serve as a warning signal for society, providing new information about the probability that similar or even more destructive mishaps might occur within this type of activity."  Media attention and the higher order costs of a mishap are likely to be correlated with signal potential (28).*

nel derails due to a mechanical system failure).  The results showed that these railroad hazards were highly differentiated, much like the hazards in Fig. 1.  The highest signal potential (and thus the highest potential for large ripple effects) was associated with accidents involving trains carrying hazardous chemicals.

A study by Slovic, MacGregor, and Kraus (*31*) examined perceptions of risk and signal value for 40 structural defects in automobiles.  Multivariate analysis of these defects, rated in terms of various characteristics of risk, produced a two-factor space.  As in earlier studies with diverse hazards, the position of a defect in this space predicted judgments of riskiness and signal value quite well.  One defect stood out much as nuclear hazards do in Fig. 1.  It was a fuel tank rupture upon impact, creating the possibility of fire and burn injuries.  This, of course, is similar to the notorious design problem that plagued Ford Pinto and that Ford allegedly declined to correct because a cost-benefit analysis indicated that the correction costs greatly exceeded the expected benefits from increased safety (*32*).  Had Ford done a psychometric study, the analysis might have highlighted this particular

defect as one whose seriousness and higher order costs (lawsuits, damaged company reputation) were likely to be greatly underestimated by cost-benefit analysis.

## FORECASTING PUBLIC ACCEPTANCE

Results from studies of the perception of risk have been used to explain and forecast acceptance and opposition for specific technologies (*33*).  Nuclear power has been a frequent topic of such analyses because of the dramatic opposition it has engendered in the face of experts' assurances of its safety.  Research shows that people judge the benefits from nuclear power to be quite small and the risks to be unacceptably great.  Nuclear power risks occupy extreme positions in psychometric factor spaces, reflecting people's views that these risks are unknown, dread, uncontrollable, inequitable, catastrophic, and likely to affect future generations (Fig. 1).  Opponents of nuclear power recognize that few people have died thus far as a result of this technology.  However, long before Chernobyl, they expressed great concern over the potential for catastrophic accidents.

These public perceptions have evoked harsh reactions from experts. One noted psychiatrist wrote that "the irrational fear of nuclear plants is based on a mistaken assessment of the risks" (34, p. 8). A nuclear physicist and leading advocate of nuclear power contended that ". . . the public has been driven insane over fear of radiation [from nuclear power]. I use the word 'insane' purposefully since one of its definitions is loss of contact with reality. The public's understanding of radiation dangers has virtually lost all contact with the actual dangers as understood by scientists" (35, p. 31).

Risk perception research paints a different picture, demonstrating that people's deep anxieties are linked to the reality of extensive unfavorable media coverage and to a strong association between nuclear power and the proliferation and use of nuclear weapons. Attempts to "educate" or reassure the public and bring their perceptions in line with those of industry experts appear unlikely to succeed because the low probability of serious reactor accidents makes empirical demonstrations of safety difficult to achieve. Because nuclear risks are perceived as unknown and potentially catastrophic, even small accidents will be highly publicized and may produce large ripple effects (Fig. 4).

Psychometric research may be able to forecast the response to technologies that have yet to arouse strong and persistent public opposition. For example, DNA technologies seem to evoke several of the perceptions that make nuclear power so hard to manage. In the aftermath of an accident, this technology could face some of the same problems and opposition now confronting the nuclear industry.

## PLACING RISKS IN PERSPECTIVE

A consequence of the public's concerns and its opposition to risky technologies has been an increase in attempts to inform and educate people about risk. Risk perception research has a number of implications for such educational efforts (36).

One frequently advocated approach to broadening people's perspectives is to present quantitative risk estimates for a variety of hazards, expressed in some unidimensional index of death or disability, such as risk per hour of exposure, annual probability of death, or reduction in life expectancy. Even though such comparisons have no logically necessary implications for acceptability of risk (15), one might still hope that they would help improve people's intuitions about the magnitude of risks. Risk perception research suggests, however, that these sorts of comparisons may not be very satisfactory even for this purpose. People's perceptions and attitudes are determined not only by the sort of unidimensional statistics used in such tables but also by the variety of quantitative and qualitative characteristics reflected in Fig. 1. To many people, statements such as, "the annual risk from living near a nuclear power plant is equivalent to the risk of riding an extra 3 miles in an automobile," give inadequate consideration to the important differences in the nature of the risks from these two technologies.

In short, "riskiness" means more to people than "expected number of fatalities." Attempts to characterize, compare, and regulate risks must be sensitive to this broader conception of risk. Fischhoff, Watson, and Hope (37) have made a start in this direction by demonstrating how one might construct a more comprehensive measure of risk. They show that variations in the scope of one's definition of risk can greatly change the assessment of risk from various energy technologies.

Whereas psychometric research implies that risk debates are not merely about risk statistics, some sociological and anthropological research implies that some of these debates may not even be about risk (5, 6). Risk concerns may provide a rationale for actions taken on other grounds or they may be a surrogate for other social or ideological concerns. When this is the case, communication about risk is simply irrelevant to the discussion. Hidden agendas need to be brought to the surface for discussion (38).

Perhaps the most important message from this research is that there is wisdom as well as error in public attitudes and perceptions. Lay people sometimes lack certain information about hazards. However, their basic conceptualization of risk is much richer than that of the experts and reflects legitimate concerns that are typically omitted from expert risk assessments. As a result, risk communication and risk management efforts are destined to fail unless they are structured as a two-way process. Each side, expert and public, has something valid to contribute. Each side must respect the insights and intelligence of the other.

## REFERENCES AND NOTES

1. For a comprehensive bibliography on risk assessment, see V. Covello and M. Abernathy, in *Technological Risk Assessment*, P. F. Ricci, L. A. Sagan, C. G. Whipple, Eds. (Nijhoff, The Hague, 1984), pp. 283–363.
2. "Risk in a complex society," report of a public opinion poll conducted by L. Harris for the Marsh and McClennan Company, New York (1980).
3. A. Wildavsky, *Am. Sci.* **67**, 32 (1979).
4. I. Burton, R. W. Kates, G. F. White, *The Environment as Hazard* (Oxford Univ. Press, Oxford, 1978).
5. J. F. Short, Jr., *Am. Sociol. Rev.* **49**, 711 (1984).
6. M. Douglas and A. Wildavsky, *Risk and Culture* (Univ. of California Press, Berkeley, 1982).
7. W. Edwards, *Annu. Rev. Psychol.* **12**, 473 (1961).
8. D. Kahneman, P. Slovic, A. Tversky, Eds. *Judgment Under Uncertainty: Heuristics and Biases* (Cambridge Univ. Press, New York, 1982).
9. M. Henrion and B. Fischhoff, *Am. J. Phys.*, in press.
10. R. Nisbett and L. Ross, *Human Inference: Strategies and Shortcomings of Social Judgment* (Prentice-Hall, Englewood Cliffs, NJ, 1980).
11. A. Tversky and D. Kahneman, *Science* **211**, 453 (1981).
12. B. Fischhoff *et al.*, *Policy Sci.* **8**, 127 (1978).
13. P. Slovic, B. Fischhoff, S. Lichtenstein, *Acta Psychol.* **56**, 183 (1984).
14. C. Starr, *Science* **165**, 1232 (1969).
15. B. Fischhoff, S. Lichtenstein, P. Slovic, S. L. Derby, R. L. Keeney, *Acceptable Risk* (Cambridge Univ. Press, New York, 1981).
16. G. T. Gardner *et al.*, *J. Soc. Psychol.* **116**, 179 (1982).
17. D. R. DeLuca, J. A. J. Stolwijk, W. Horowitz, in *Risk Evaluation and Management*, V. T. Covello, J. Menkes, J. L. Mumpower, Eds. (Plenum, New York, 1986), pp. 25–67.
18. E. J. Johnson and A. Tversky, *J. Exp. Psych. Gen.* **113**, 55 (1984).
19. M. K. Lindell and T. C. Earle, *Risk Anal.* **3**, 245 (1983).
20. H. J. Otway and M. Fishbein, *The Determinants of Attitude Formation: An Application to Nuclear Power* (RM-76-80 Technical Report, International Institute for Applied Systems Analysis, Laxenburg, Austria, 1976).
21. O. Renn and E. Swaton, *Env. Int.* **10**,557 (1984).
22. P. Slovic, B. Fischhoff, S. Lichtenstein, in *Societal Risk Assessment: How Safe is Safe Enough?*, R. Schwing and W. A. Albers, Jr., Eds. (Plenum, New York, 1980), pp. 181–216.
23. C. A. J. Vlek and P. J. Stallen, *Organ. Behav. Hum. Perf.* **28**, 235 (1981).
24. D. von Winterfeldt, R. S. John, K. Borcherding, *Risk Anal.* **1**, 277 (1981).
25. P. Slovic, B. Fischhoff, S. Lichtenstein, in *Perilous Progress: Managing the Hazards of Technology*, R. W. Kates, C. Hohenemser, J. X. Kasperson, Eds. (Westview, Boulder, CO, 1985), pp. 91–125.
26. P. Slovic, B. Fischhoff, S. Lichtenstein, *Environment* **21** (no. 3), 14 (1979).
27. Estimated at $500 billion [see *Electr. Power Res. Inst. J.* **5** (no. 5), 24 (1980)].
28. P. Slovic, S. Lichtenstein, B. Fischhoff, *Manage Sci.* **30**, 464 (1984).
29. The Talk of the Town, *New Yorker* **60** (no. 53), 29 (1985).
30. N. Kraus, thesis, University of Pittsburgh (1985).
31. P. Slovic, D. MacGregor, N. Kraus, *Accident Anal. Prev.*, in press.
32. *Grimshaw* vs. *Ford Motor Co.*, Superior Court, No. 19776, Orange County, CA, 6 February, 1978.
33. P. Slovic, B. Fischhoff, S. Lichtenstein in *Advances in Environmental Psychology*, A. Baum and J. E. Singer, Eds. (Erlbaum, Hillsdale, NJ, 1981), vol. 3, pp. 157–169.
34. R. L. Dupont, *Bus. Week*, 7 September 1981, pp. 8–9.
35. B. L. Cohen, *Before It's Too Late: A Scientist's Case for Nuclear Energy* (Plenum, New York, 1983).
36. P. Slovic, *Risk Anal.* **6**, 403 (1986).
37. B. Fischhoff, S. Watson, C. Hope, *Policy Sci.* **17**, 123 (1984).
38. W. Edwards and D. von Winterfeldt, *Risk Anal*, in press.
39. The text of this article draws heavily upon the author's joint work with B. Fischhoff and S. Lichtenstein. Support for the writing of the article was provided by NSF grant SES-8517411 to Decision Research.

# THEORIES OF RISK PERCEPTION:
## WHO FEARS WHAT AND WHY?

*Aaron Wildavsky* ✦ *Karl Dake*

In social science rival theories seeking to answer the same questions rarely confront one another. Indeed, a variety of perspectives has been employed in research on public perception of risk, but alternative formulations remain largely untested. Missing most of all is a focused comparison of rival hypotheses.

One could hardly find many subjects that are better known or considered more important to more people nowadays than the controversies over harm to the natural environment and the human body attributed to modern technology, whether this be from chemical carcinogens or nuclear power or noxious products introduced by industry into the land, sea, or air, or into water or food supplies. Thus we ask: Why are products and practices once thought to be safe (or safe enough) perceived increasingly as dangerous? Who (what sort of people) views technology as largely benign, and who as mostly dangerous? To what degree are different people equally worried about the same dangers, or to what extent do some perceive certain risks as great that

others think of as small? And how do concerns across different kinds of risk—war, social deviance, economic troubles as well as technology—vary for given individuals? Only by comparisons across types of danger can we learn whether individuals have a general tendency to be risk averse or risk taking, or whether their perceptions of danger depend upon the meaning they give to objects of potential concern. The test we shall put to each theory of risk perception is its ability to predict and explain what kinds of people will perceive which potential hazards to be how dangerous.

The most widely held theory of risk perception we call *the knowledge theory:* the often implicit notion that people perceive technologies (and other things) to be dangerous because they *know* them to be dangerous. In a critical review of *Risk and Culture*,[1] by Mary Douglas and Aaron Wildavsky, for instance, John Holdren's belief that perceivers are merely registering the actual extent of danger to themselves—the of-course-people-are-worried-they-have-lots-to-worry-about thesis—comes out clearly:

A much simpler description might suffice: people worry most about the risks that seem most directly to threaten their well being at the mo-

Reprinted by permission of *Daedalus*, Journal of the American Academy of Arts and Sciences, from the issue entitled, "Risk," Volume 119(4), Fall 1990, pp. 41–60.

ment; environmental concerns predominate only where and when people imagine the risks of violence and economic ruin to be under control. . . . What is wrong, after all, with the simple idea—paralleling Maslow's stages of wants—that worries about more subtle and complex threats will materialize if, and only if, thte most direct and obvious threats are taken care of?[2]

If Holdren is correct, perception of danger should accord with what individuals know about the risk in question. But do risk perceptions and knowledge coincide?

Another commonly held cause of risk perception follows from *personality theory*. In conversations we frequently hear personality referred to in such a way that individuals seem to be without discrimination in their risk-aversion or risk-taking propensities: some individuals love risk taking so they take many risks, while others are risk averse and seek to avoid as many risks as they can. We will test this common, if extreme, view. We will also examine a more moderate theory of personality: that stable individual differences among persons are systematically correlated with their perceptions of danger. Leaving aside the extraordinary Oblimov-like characters staying in bed all their lives, or Evel Knievels breaking bones on too-daring feats, this version of personality theory suggests that individuals are so constituted as to take or reject risks in an enduring manner.[3] But do traditionally assessed attributes of personality such as intrapsychic dynamics and interpersonal traits relate to risk perceptions and preferences in predictable ways?

The third set of explanations for public perceptions of danger follow two versions of *economic theory*. In one, the rich are more willing to take risks stemming from technology because they benefit more and are somehow shielded from adverse consequences. The poor presumably feel just the opposite. In "post-materialist" theory, the rationale is reversed, however: precisely because living standards have improved, the new rich are less interested in what they have (affluence) and what got them there (capitalism), than in what they think they used to have (closer social relations), and what they would like to have (better health).[4] Is it true, however, that the newly affluent aspire to post-materialist values, such as interpersonal harmony, and hence fear environmental pollution and chemical contamination?

Other explanations for public reactions to potential hazards are based on *political theory*. These accounts view the controversies over risk as struggles over interests, such as holding office or party advantage. The view of politics as clashing interests connects conflicts to different positions in society. The hope for explanatory power in such approaches to risk perception is thus placed on social and demographic characteristics such as gender, age, social class, liberal-conservative ratings, and/or adherence to political parties.[5]

Viewing individuals as the active organizers of their own perceptions, *cultural theorists* have proposed that individuals choose what to fear (and how much to fear it), in order to support their way of life.[6] In this perspective, selective attention to risk, and preferences among different types of risk taking (or avoiding), correspond to *cultural biases*—that is, to worldviews or ideologies entailing deeply held values and beliefs defending different patterns of social relations. *Social relations* are defined in cultural theory as a small number of distinctive patterns of interpersonal relationships—hierarchical, egalitarian, or individualist.[7] No causal priority is given to cultural biases or social relations; they are always found together interacting in a mutually reinforcing manner. Thus there are no relationships without cultural biases to justify them, and no biases without relations to uphold them.

Socially viable combinations of cultural biases and social relations are referred to in cultural theory as *ways of life* or as *political cultures*. More specifically, then, hierarchical, egalitarian, and individualist forms of social relations, together with the cultural biases that justify them, and are hypothesized to engender distinctive representations of what constitutes a hazard and what does not. Among all possible risks, those selected for worry or dismissal are functional in the sense that they strengthen one of these ways of life and weaken the others. This sort of explanation is at once more political (there is a political purpose to all of this perceiving—defending a way of life and attacking others) and less obvious (what has risk perception to do with ways of life?).

Since cultural biases are forms of ideology, there should be high correlations between certain biases and corresponding ideologies (e.g., egalitarianism and political liberalism). When we vary the kinds of possible dangers to which people react, however, we should see that the left-right distinction captures the cultural bias of egalitarianism but fails to distinguish between hierarchy and individualism. Hence we expect the three cultural biases to predict a broad spectrum of risk perceptions better than political ideology does.

According to cultural theory, adherents of hierarchy perceive acts of social deviance to be dangerous because such behavior may disrupt their preferred (superior/subordinate) form of social relations. By contrast, advocates of greater equality of conditions abhor the role differentiation characteristic of hierarchy because ranked stations signify inequality. Egalitarians reject the prescriptions associated with hierarchy (i.e., who is allowed to do what and with whom), and thus show much less concern about social deviance.

Individualist cultures support self-regulation, including the freedom to bid and bargain. The labyrinth of normative constraints and controls on behavior that are valued in hierarchies are perceived as threats to the autonomy of the individualist, who prefers to negotiate for himself. Social deviance is a threat to individualist culture only when it limits freedom, or when it is disruptive of market relationships. Our expectation is that individualists should take a stance between hier-

archists, to whom social deviance is a major risk, and egalitarians, to whom it is a minor risk at most.

Egalitarians claim that nature is "fragile" in order to justify sharing the earth's limited resources and to discomfort individualists, whose life of bidding and bargaining would be impossible if they had to worry too much about disturbing nature. On the contrary, individualists claim that nature is "cornucopian," so that if people are released from artificial constraints (like excessive environmental regulations) there will be no limits to the abundance for all, thereby more than compensating for any damage they do. Hierarchists have something in common with individualists: they approve of technological processes and products, provided their experts have given the appropriate safety certifications and the applicable rules and regulations are followed. In hierarchical culture, nature is "perverse or tolerant"; good will come if you follow their rules and experts, bad if you don't.

People who hold an egalitarian bias (who value strong equality in the sense of diminishing distinctions among people such as wealth, race, gender, authority, etc.) would perceive the dangers associated with technology to be great, and its attendant benefits to be small. They believe that an inegalitarian society is likely to insult the environment just as it exploits poor people. Those who endorse egalitarianism would also rate the risks of social deviance to be relatively low. What right has an unconscionably inegalitarian system to make demands or to set standards? The perceived risks of war among egalitarians would be low to moderate: they are likely to mistrust the military (a prototypical hierarchy); they also believe that the threat of war abroad is exaggerated by the establishment coalition of hierarchy and individualism in order to justify an inegalitarian system at home.

Cultural theory's predictions for the individualist bias are just the opposite: its adherents perceive the dangers of technology as minimal, in part because they trust that their institutions can control or compensate for the severity of untoward events. These same predictions hold for the cultural bias of hierarchy. How then do we distinguish between the worldviews of hierarchists and individualists? By varying the object of concern. A study of technological risks alone would leave these two cultures hopelessly confounded. Both are technologically optimistic, individualists because they see technology as a vehicle for unlimited individual enterprise—to them risk is opportunity—and hierarchists because they believe that technology endorsed by their experts is bound to improve the quality of life.

Due to the emphasis placed on obedience to authority within hierarchy, its supporters scorn deviant behavior. In contrast, individualists, who prefer to substitute self-regulation for authority, are much more willing to permit behavior that is the product of agreement. And yet, here too, a distinction must be made. If the object of attention is personal behavior, such as sex between consenting adults, individualists will be against allowing government to intervene. But if the subject is crime or violence against established institutions, they will be more disposed to support a governmental crackdown. In other words, if "order" signifies support for the stability and legitimacy necessary for market relationships, individualists will support government action toward that end.

Economic troubles represent a different kind of risk than those of technology or social deviance, since almost everyone has a reason to worry about them—egalitarians because lower living standards are especially harmful to the poorest people, and adherents of hierarchy because it weakens the system they wish to defend. We would expect individualists to fear economic failure more than others, however, because the marketplace is the institution most central to their life of negotiated contracts.

To test these rival theories—knowledge,

personality, economic, political, and cultural—we drew upon the risk-perception data archives established by Kenneth H. Craik, David Buss, and Karl Dake at the University of California's Institute of Personality Assessment and Research.[8] We used the *pro-risk* index of their *Societal Risk Policy* instrument to gauge the extent of an individual's endorsement of risk taking versus risk aversion in regard to technology. This pro-risk index assesses whether risk taking and risk management are viewed as opportunities for advancement, or rather as invitations to catastrophe at the societal level.

We assessed perceptions of risk associated with *technology and the environment, war, social deviance*, and *economic troubles* by using variables chosen from a list of 36 "concerns that people have about society today." Following procedures similar to those used in the most important pioneering study of risk perceptions, we selected average ratings of 25 technologies on *risk* and *benefit* for use.[9] These indices enable us to compare public responses to different kinds of dangers. Now we turn to the factors that have been used to explain such responses.

*Knowledge.* One measure of knowledge we have used is the individual's self-report of how much he or she knows about specific technologies. Another measure is self-report of educational level. Self-ratings are the simplest and best way to address some psychological phenomena (who knows better than the individual how much dread a perceived hazard evokes for him?), while in regard to other phenomena they are notoriously poor. To avoid the potential pitfalls of relying only on self-reported knowledge, we developed a measure of perceptual accuracy based on differences between public and expert judgments of annual fatalities associated with 8 technologies: contraceptives, nuclear power, diagnostic X rays, bicycles, lawn mowers, motor vehicles, home appliances, and commercial aviation.[10]

*Personality.* In order to explore the correlations among personality characteristics and risk perceptions, we have drawn upon a broad set of traditional personality measures, including the *Adjective Check List* and the *California Psychological Inventory*.[11]

*Political Orientation.* To evaluate predictions of risk perceptions based on political variables, we utilize measures of political party membership and liberal-conservative ideology (both self-rated and calculated on the basis of 20 policy issue stances).[12]

*Cultural Biases.* To test the relations among perceptions of danger and the worldviews justifying *hierarchy, individualism*, and *egalitarianism*, we developed new measures to assess individual endorsement of three cultural biases.

Our hierarchy index embodies support for patriotism ("I'm for my country, right or wrong"), law and order ("The police should have the right to listen in on private telephone conversations when investigating crime"), and strict ethical standards ("I think I am stricter about right and wrong than most people"). It also expresses concern about the lack of discipline in today's youth and supports the notion that centralization is "one of the things that makes this country great."[13]

Our index for the cultural bias of individualism expresses support for continued economic growth as the key to quality of life, and private profit as the main motive for hard work. It espouses the view that democracy depends fundamentally on the existence of the free market, and argues that "the welfare state tends to destroy individual initiative." The individualism scale also indicates support for less government regulation of business, and endorses private wealth as just rewards for economic endeavor: "If a man has the vision and ability to acquire property, he ought to be allowed to enjoy it himself."[14]

Our measure of egalitarianism is based on survey items written to assess attitudes toward equality of conditions. The egalitarianism scale centers on political solutions to inequality: "Much of the conflict in this world could be eliminated if we had more equal distribution of resources among nations," "I support federal efforts to eliminate poverty," and "I support a tax shift so that the burden falls more heavily on corporations and persons with large incomes." The egalitarianism index also covers perceived abuses by the other political cultures: "Misuse of scientific and expert knowledge is a very serious problem. . ." and "The human goals of sharing and brotherhood are being hindered by current big institutions. . . ."[15]

There are many theories that might account for the perceptions of risks, from those based on knowledge, personality, or economics, to those based on politics or culture. Our task is to discriminate among these rival theories by comparing their power to predict who fears what and why.

## CULTURAL BIASES BEST PREDICT RISK-PERCEPTION FINDINGS

If it were true that the more people know about technological risk, or about technology in general, the more they worry about it, it should follow that risk perception goes along with such knowledge. Using the measure of self-rated knowledge about technologies, and self-rated education, we see quite the opposite.

Our findings show that those who rate their self-knowledge of technologies highly also tend to perceive greater average benefits associated with technologies than those who are less confident about their knowledge.[16] Those who report higher levels of education tend to perceive less threat from the risks of war. Otherwise, self-rated knowledge and education bear only weak (that is, statistically insignificant) relations to preferences for societal risk taking or to perceived risks associated with technology and the environment, social deviance, and economic troubles.

The more an individual's annual fatality estimates correspond to expert estimates, in addition, the more likely that person is to rate other risks as small—at least compared with those who are less accurate. While on the whole those who are more in accord with expert mortality estimates perceive less risk, they are also less optimistic regarding the benefits of technology. *Overall, the conclusion is compelling that self-rated knowledge and perceptual accuracy have a minimal relationship with risk perception.*

With regard to personality, we find that those who feel our society should definitely take technological risks can be described as patient, forbearing, conciliatory, and orderly (i.e., the pro-risk measure is positively correlated with the personality traits "need for order" and "deference").[17] Advocates of societal risk taking tend not to be aggressive, or autonomous, or exhibitionistic, but are more likely to be cautious and shy and to seek stability rather than change. *This pattern is suggestive of a technologically pro-risk personality, which emerges as that of an obedient and dutiful citizen, deferential to authority.* Such a personality structure fits extremely well with the political culture of hierarchy.

By contrast, those citizens who perceive greater risk in regard to technology and the environment tend to turn up positive on exhibitionism, autonomy, and need for change, but negative on need for order, deference, and endurance (i.e., just the opposite of those who score as favoring societal risk taking). *This technologically risk-averse pattern of personality traits also holds for those who endorse egalitarianism.*

Those who endorse egalitarianism are also more likely to be personally risk taking, but societally risk averse, while those who favor hierarchy tend to be personally risk averse, but societally pro-risk with respect to technology and the environment. Thus, *we find no evidence for a personality structure that is*

*risk taking or risk averse across the board.* Risk taking and risk aversion are not all of a piece, but depend on how people feel about the object of attention. Cultural theory would predict, for example, that hierarchists would be risk averse when it comes to taking risks with the body politic.

Relative to conservatives, those who rate themselves as liberals tend to be technologically risk averse at the societal level, are more likely to rate the risks of technology and the environment as very great, and are comparatively unconcerned about the risks of social deviance. As the self-rating of liberal increases, the average ratings for the risks of the 25 specific technologies increases, and the average ratings of their benefits decreases.

Political party membership is less predictive of risk perceptions and preferences than left-right ideology, especially on the Democratic side (undoubtedly because Democrats are the more heterogeneous party). When we ask what it is about thinking of oneself as a liberal or a conservative that makes such a big difference compared with thinking of oneself as a Democrat or a Republican, the findings are informative. Whether by self-rating or policy designation, *liberals have strong tendencies to endorse egalitarianism ($r = 0.52$ and $r = 0.50$), and to reject hierarchy ($r = -0.55$ and $r = -0.51$) and individualism ($r = -0.37$ and $r = -0.31$).* Likewise, membership in the Democratic party is correlated with egalitarianism ($r = 0.30$), but is not predictive of agreement or disagreement with the hierarchical or individualist point of view. *Republicans have a penchant toward individualist ($r = 0.31$) and hierarchical biases ($r = 0.40$), and an equally strong proclivity for rejecting egalitarianism ($r = -0.45$).* These correlations among political party membership, left-right ideology, and cultural biases are huge by the standards of survey research.

How does cultural theory compare with other approaches to perceived risk? *Cultural biases provide predictions of risk perceptions and risk-taking preferences that are more pow-*

erful than measures of knowledge and personality and at least as predictive as political orientation. We find that egalitarianism is strongly related to the perception of technological and environmental risks as grave problems for our society ($r = 0.51$), and hence to strong risk aversion in this domain ($r = -0.42$). Egalitarianism is also related positively to the average perceived risks, and negatively to the average perceived benefits, of 25 technologies. One could hardly paint a worse picture of technology—little benefit, much risk, and the risks not worth taking.

Individualist and hierarchist biases, in contrast, are positively related to a preference for technological risk-taking ($r = 0.32$ and $r = 0.43$) and to average ratings of technological benefits ($r = 0.34$ and $r = 0.37$). Here the image is more sanguine: the benefits are great, and the risks small, so society should press on with risk taking to get more of the good that progress brings with it.

## DISCUSSION

We have shown that whether measured by cultural biases or by political orientation, perceptions of technology are predictable given the worldview of the perceiver. But one should not conclude that the establishment cultures of individualism and hierarchy always favor risk taking, or that egalitarians are always risk averse. *Perception of danger is selective; it varies with the object of attention.* For we find that compared with advocates of egalitarianism, those in greater agreement with individualism perceive greater risk in respect to war ($r = 0.15$ versus $r = 0.40$ respectively). Likewise, it is the hierarchical bias that is most highly correlated with perceived threat of social deviance ($r = 0.35$ compared with $r = 0.15$ for egalitarianism). Nor is it always the adherents of establishment cultures versus those of egalitarianism. As predicted by cultural theory, *it is not that devotees of individualism and hierarchy perceive no dangers in general,*

but that they disagree with those who favor egalitarianism about how dangers should be ranked. Just as technological and environmental risks are most worrisome to egalitarians, social deviance is deemed most dangerous to hierarchists, and the threat of war (which disrupts markets and subjects people to severe controls) is most feared by individualists.

It is obvious that culture neither causes nor influences demographic characteristics such as gender or age (though it may influence their social meanings). Thus we do not argue that the weak correlations we find between cultural biases and personal attributes like income or social class reveal the influence of political culture on those variables. Whether we look at knowledge, personality, political orientation, or demographic variables, however, we find that cultural theory provides the best predictions of a broad range of perceived risks and an interpretive framework in which these findings cohere.

The importance of using a wide range of risks in studying how people perceive potential dangers should now be apparent. Employing only dangers from technology, while better than nothing, is far less powerful than considering a panoply of dangers from the threat of war to social deviance to economic collapse. Broadening the spectrum of related questions to be considered allows for more discriminating tests of rival theories. With perceived dangers from technology as the only issue, moreover, one cannot tell whether the level of concern registered by an individual comes from aversion to or acceptance of risk in general, or is evoked differentially by various risks. By observing whether there is a variegated pattern of risk perception (now we know there is) and by ascertaining who rates each kind of risk in which way, we may study *patterns of risk perception.* Fitting these patterns to alternative explanations, we believe, is a superior test of competing theories.

Comparing rival theories, not just a single explanation, has similar advantages. Making the rival theories confront each other reduces the temptation to claim easy victories. It is not enough to show respectable correlations; it is also necessary to do better than the alternatives.

Viewed in this light, the cultural theory's greater power than alternative explanations is manifest in its ability to *generate* broader and finer predictions of who is likely to fear, not to fear, or fear less, different kinds of dangers. Having derived from cultural theory a number of explanations approximated in our findings, the next question is what this tells us about risk perception.

Our findings show that it is not knowledge of a technology that leads people to worry about its dangers. In the current sample, the difference between public and technical estimates of annual fatalities ranges up to several orders of magnitude in size. The enormous variation in these public perceptions is not accounted for by knowledge, leaving considerable room for other explanations. Indeed, if people have little knowledge about technologies and their risks, then public fears can hardly coincide with how dangerous various technologies have proved to be.

Wait a minute! Everyone knows that nuclear radiation and AIDS can kill. We agree. When these subjects become politicized, however, disagreement develops along the fault lines of policy differences, seizing upon whatever cracks of uncertainty now exist: What are the health consequences of prolonged exposure to low levels of radiation? Is there such a thing as an amount of radiation so small that exposure causes no harm? Can AIDS be passed along by social contact? Should sufferers from AIDS be quarantined? Should their sexual contacts be traced and informed?

Our findings on personality raise the question of why there are such interesting sets of correspondence among traditionally assessed traits and cultural biases. Part of the difficulty in interpreting these findings is that personality entails such a wide set of characteristics—from intrapsychic to interpersonal relations—that virtually no aspect of individual life is left out. Were there theories connecting particular aspects of personality to patterns of risk perception, interpretation would be easier, for then we could test these hypotheses. One possibility is that personal orientations may guide individuals to make commitments consistent with specific political cultures, while at the same time, cultures may select from among individuals those that support their way of life. Since there are no such theories, however, we are left to explore among personality characteristics to see what fits.

Assuming both personality and political culture are operative, which is more powerful in predicting perceptions of danger and preferences for risk taking? Clearly, the closer one gets to asking questions about policy preferences, the better one's predictions of the selective perceptions of danger should be. Since our measures of cultural biases are closer to public policy than traditional measures of personality are, we should expect our measures (other things being equal) to predict better. And they do. But if that were all there were to prediction—proximity of the explanation to the explained—then we would expect assessments of political ideology to predict risk preferences far better than cultural bias does. As we have shown, however, public policy stances and self-rated political orientation do not do as well as cultural biases in predicting risk preferences and perceptions—even though they are the most proximal to risk policy of all the variables we test.

How then do cultural biases, which are so remote from the evidence regarding risks, guide people in choosing what to fear? A detailed answer is presented in *Risk and Culture*. Here, we can say only that hierarchists favor technological risk taking because they see this as supporting the institutions that they rely on to make good their promises, to wit: technology can promote a stronger society and a safer future provided that their rules (and stratified social relations) are maintained. Individualists also deem technology to be good. They hold that following market principles (and individually negotiated social relations) will allow technological innovation to triumph, conferring creative human value on otherwise inert resources. They also believe that the enormous benefits of technological innovation will convey their premise that unfettered bidding and bargaining leaves people better off. If they believed that free market institutions are intrinsically ruinous to nature, individualists could no longer defend a life of minimum restraints. By the same token, egalitarians are opposed to taking technological risks because they see them as supporting the inegalitarian markets and coercive hierarchies to which they are opposed.

By this time readers are right to wonder, in view of the assertions we are making, whether other surveys support our claims. A recent one to come to our attention is supportive in many ways. Its subject is the irradiation of food as a preservative process, widely considered safe by scientists, but a topic of considerable worry to concerned consumers. The participants were 195 adult women chosen from Pennsylvania women's groups of various kinds—religious, civic, professional, social, and political. The respondents were given a questionnaire to fill out, then were shown different kinds of information about food irradiation, then filled out another questionnaire, and finally were engaged in group discussion. The authors, Richard Bord and Robert O'Conner, find, as we do, that knowledge (based on the information given to participants) is inversely related to fear of a technology: "Having accurate knowledge about the food irradiation process translates into greater acceptance." They add significantly that:

> whether respondents received a technical or nontechnical communication about the food irradiation process and whether they received a detailed discussion of the major arguments for and against food irradiation had no discernible effect on their judgments.

It is not knowledge per se, but confidence in institutions and the credibility of information that is at issue:

> Trust in business and industry in general, the food irradiation industry specifically, government regulators, and science as a provider of valid and useful knowledge is the major predictor of whether the respondent indicates she will or will not try irradiated food. . . . Learning that others have used food irradiation safely and of its approval by prestigious professional organizations enhanced its acceptability. . . . People who oppose big government and big business express greater fear of radiation.
>
> [One of the main topics of group discussion] was the respondents' view that complex technology bears a burden of too much uncertainty, too much greed on the part of its sponsors, and too little effective governmental control. The point was frequently made that even if the scientific-technical plan was flawless the people executing the plan and managing the technology would inevitably create serious problems.[18]

It is not only that "the facts" cannot by themselves convince doubters, but that behind one set of facts are always others relating to whether business and government can be trusted.

If there are any people to whom knowledge about hazards should make the most difference, it is those who are professionally employed in the analysis and management of risk. Yet a survey of risk professionals drawn from government, industry, environmental groups, and universities shows something dramatically different. Thomas Dietz and Robert Rycroft find that self-reported ideology:

> appears to have the strongest links to environmental attitudes and values of risk professionals. . . . For example, on the question of whether we are seeing only the tip of the iceberg with regard to technological risk, 88.5 percent of [the] very liberal . . . agreed . . . as did 74 percent of [the] liberals. Only 25 percent of [the] very conservative and 36.4 percent of [the] conservative respondents agreed.[19]

The more perceptions of contested subjects are studied, we believe, the more they will reveal the strong influence of cultural biases. In this respect, Paul Sabatier and S. Hunter's recent study of causal perceptions in belief systems is especially useful because, like the present analysis, it focuses on perceptual biases from more than one cultural direction:

> Environmentalists perceived water clarity to be getting worse, while those in favor of economic growth and property rights simply refused to believe the wealth of documented, and widely diffused, scientific evidence developed by one of the world's leading limnologists demonstrating statistically significant declines in water clarity over the previous 10–15 years. This suggests that in high-conflict situations, perceptions on even relatively straightforward technical issues can be heavily influenced by elites' normative presuppositions.[20]

This position reaffirming the importance of worldviews is bolstered by the "risk and benefit perceptions, acceptability judgments, and self-reported actions toward nuclear power" spoken of by Gerald Gardner and his coauthors. Their respondents were taken from environmental groups, blue-collar workers, college students, businesspeople, and technologists (scientists and engineers employed by a utility company). While education, sex, gender, religion, and other sociodemographic variables were not related to protests or other personal actions taken on nuclear power, Gardner et al. found that liberal-conservative ideology was predictive: "The most important correlate of reported action and 'acceptability' . . . appeared to represent a 'liberal/public interest group vs. a conservative/private enterprise' dimension."[21]

The power of the ideological explanation is strengthened further by Stanley Rothman and S. Robert Lichter, who analyzed questionnaires filled out by a sample of 1,203 congressional staff, civil servants, television and print journalists, lawyers, officials of public interest groups, moviemakers, military officers, energy and nuclear power experts. The results vary widely by group membership, with 98.7 percent of nuclear energy experts thinking nuclear power plants are safe, compared with only 6.4 percent of public interest offi-

cials, and 30.6 percent of journalists on television networks. Their major finding is that compared with a variety of demographic, social, and economic variables, political ideology was by far the most powerful predictor. "We hypothesize," they conclude, "that nuclear energy is a surrogate issue for more fundamental criticism of U.S. institutions."[22] This restates the thesis of *Risk and Culture* for nuclear technology.

Whenever other studies present comparable findings, they reveal that the most powerful factor for predicting risk perceptions is trust in institutions or ideology, which is largely about which institutions can be trusted. *Such findings show that, however conceptualized— whether as political ideology or cultural biases— worldviews best account for patterns of risk perceptions.*

In summary, the great struggles over the perceived dangers of technology in our time are essentially about trust and distrust of societal institutions, that is, about cultural conflict. Once we vary the object of concern, we do indeed discover that egalitarians (who fear social deviance less than hierarchists and individualists) fear technology a great deal— seeing in it, or so the cultural theory claims, the corporate greed they believe leads to inequality. Individualists, who believe in competition, and who are exceedingly loathe to place restraints on what they consider to be mutually profitable relationships, deem technology to be good. In contrast, hierarchists, who fear disorder and erosion of status differences, are more worried about social deviance and less worried about technological dangers than egalitarians.

We have shown that other surveys, with different assumptions, methods, and sample populations find, as we do, that risk perceptions and preferences are predictable given individual differences in cultural biases. It is the congruence of our analysis with others' that gives us the most confidence in our findings. We would have preferred to ask more subtle and differentiated questions about knowledge; but the other surveys we cite do that, and they also show the importance of cultural biases. Above all, we would have preferred more elaborate statistical analysis than small samples permit.[23]

Knowing what sorts of perceptions come from which kinds of people may allow for practical applications of cultural theory in a variety of policy contexts. Risk communication programs, for instance, might profitably focus on the underlying causes of risk perception—such as confidence (or lack of trust) in institutions, or the credibility of hazard information—rather than only on "the facts" regarding possible harms. Since cultural theory generates clues to the propensities of those with various worldviews to underestimate or overestimate specific kinds of risk, in addition, it can be used to tailor educational programs—say cigarette and alcohol warnings—to the plural rationalities represented in the general public.

It has been two decades since Chauncy Starr's seminal essay "Social Benefit versus Technological Risk" asked how much our society is willing to pay for safety.[24] Since then, a lively and spirited research community has grown up around the issues of technological risks.[25] We hope to have pointed the study of risk perception in the right direction by: (1) expanding the scope of the questions asked to include patterns of risk perception (not only technological hazard, but also war, social deviance, economic decline, etc.); and by (2) comparing rival explanations of public fears. As predicted by cultural theory, we find that individuals perceive a variety of risks in a manner that supports their way of life.

*Acknowledgments* The risk-perception data archives used in this paper were established in 1981 and 1982 by Kenneth H. Craik, with David Buss and Karl Dake, under National Science Foundation Grant PRA-8020017 to the University of California's Institute of Personality Assessment and Research. See David Buss, Kenneth H. Craik, and Karl Dake, "Perceptions of Decision Procedures for Managing and Regulating Hazards," in F. Homberger, *Safety Evaluation and Regulation* (New York: Karger, 1985), 199–208; "Contemporary Worldviews and Perception of the Technological System," in Vincent T. Covello, Joshua Menkes, and Jeryl Mumpower, *Risk Evaluation and Management* (New York: Plenum, 1986), 93–130. The writing of this paper was supported in part by the National Institutes of Health under Biomedical Research Support Grant 89–34 to the Survey Research Center. Acknowledgment is gratefully made to Dr. Kenneth H. Craik for his permission to use these data, and to Dr. Percy Tannenbaum for the support of the Survey Research Center.

### REFERENCES AND NOTES

[1] Mary Douglas and Aaron Wildavsky, *Risk and Culture: An Essay on the Selection of Technological and Environmental Dangers* (Berkeley: University of California Press, 1982).

[2] John Holdren, "The Risk Assessors," *Bulletin of the Atomic Scientists*, 39 (1983): quotation 36.

[3] K. R. MacCrimmon and D. A. Wehrung, *Taking Risks: The Management of Uncertainty* (New York: Free Press, 1986); R. G. Mitchell Jr., *Mountain Experience: The Psychology and Sociology of Adventure* (Chicago: University of Chicago Press, 1983).

[4] Ronald Inglehart, *The Silent Revolution: Changing Values and Political Styles among Western Publics* (Princeton: Princeton University Press, 1977).

[5] Stephen Cotgrove, *Catastrophe or Cornucopia: The Environment, Politics and the Future* (Chichester, England: John Wiley & Sons, 1982); Dorothy Nelkin and Michael Pollack, *The Atom Besieged: Antinuclear Movements in France and Germany* (Cambridge: MIT Press, 1981).

[6] Mary Douglas, "Cultural Bias," Occasional Paper 35 (London: Royal Anthropological Institute, 1978), republished in *In the Active Voice* (London: Routledge & Kegan Paul, 1982), 183–254; "Passive Voice Theories in Religious Sociology," *Review of Religious Research* 21 (1979): 51–56; *Essays in the Sociology of Perception* (London: Routledge & Kegan Paul, 1982). See also Douglas and Wildavsky, *Risk and Culture*; Michael Thompson, Richard Ellis, and Aaron Wildavsky, *Cultural Theory, or, Why All that is Permanent is Bias* (Boulder: Westview Press, 1990).

[7] Cultural theory delineates two additional cultures: fatalists and hermits; it also makes finer distinctions regarding nature, technology, and risk perception than are discussed here. Were it possible, we would prefer to measure cultural biases in their social context. Instead,

we have taken the approach suggested by the survey data at our disposal. We assess cultural biases as worldviews.

[8] Intensive assessments of 300 ordinary citizens were conducted, including measures of perceptions of technologies, preferences for societal decision approaches and societal risk policy, confidence in institutions, sociotechnological and political orientations, personal values, environmental dispositions, self-descriptions personal background, and more.

Two public samples were drawn from cities in the East Bay area of the San Francisco region: Richmond, Oakland, Piedmont, and Alameda. Stratified samples were selected on the basis of an analysis of social trends that provided detailed information regarding the median demographic characteristics of each postal zip code in the sample region. Participants were recruited via telephone directory sampling, letter of invitation, and telephone follow-up. Most, but not all, of the current findings are based on analysis of sample 2 (which had 134 participants), leaving sample 1 (which had 166 participants) available for replication of this study.

[9] Participants rated how risky, and how beneficial, they judged each of 25 technologies to be: refrigerators, photocopy machines, contraceptives, suspension bridges, nuclear power, electronic games, diagnostic X rays, nuclear weapons, computers, vaccinations, water fluoridation, rooftop solar collectors, lasers, tranquilizers, Polaroid photographs, fossil electric power, motor vehicles, movie special effects, pesticides, opiates, food preservatives, open-heart surgery, commercial aviation, genetic engineering, and windmills. See Baruch Fischhoff, Paul Slovic, Sarah Lichtenstein, Stephen Read, and Barbara Coombs, "How Safe is Safe Enough? A Psychometric Study of Attitudes toward Technological Risks and Benefits," *Policy Sciences* 9 (1978): 127–52.

[10] The measure of perceptual accuracy was motivated by Paul Slovic, Baruch Fischhoff, and Sarah Lichtenstein in "Facts and Fears: Understanding Perceived Risk," in R. Schwing and W. Albers, Jr., *Societal Risk Assessment: How Safe is Safe Enough?* (New York: Plenum, 1980), 181–216.

[11] Harrison Gough, *California Psychological Inventory Administrator's Guide* (Palo Alto: Consulting Psychologists Press, 1987); Harrison Gough and Alfred Heilbrun, *The Adjective Check List Manual* (Palo Alto: Consulting Psychologists Press, 1983).

[12] The measure of liberalism-conservatism based on policy preferences follows Edmond Costantini and Kenneth H. Craik, "Personality and Politicians: California Party Leaders, 1960–1976," *Journal of Personality and Social Psychology* 38 (1980): 641–61.

[13] Quotations, in order, are from Leonard Furguson, "The Isolation and Measurement of Nationalism," *Journal of Social Psychology* 16 (1942): 224, Hans Eysenck, *Sex and Personality* (London: Open Books, 1976), 153; Gough; David Buss, Kenneth H. Craik, and Karl Dake, *The IPAR Risk Perception Data Archives: Assessment II Instruments*, unpublished document, Institute of Personality Assessment and Research (Berkeley: University of California, 1982).

[14] Quotations, in order, are from Eysenck, 155, and from Furguson, 224.

[15] Quotations concerning a tax shift and the elimination of poverty are from Costantini and Craik; the balance are from Buss, Craik, and Dake, 1982.

[16] We report correlations throughout this essay, not means or mean differences. For sample 1 (134 participants), a correlation must be greater than 0.15 or less than −0.15 to be statistically significant. Nothing about average scores or group comparisons is implied.

[17] Gordon Allport, *Pattern and Growth in Personality* (New York: Henry Holt and Company, 1961); see also *Personality: A Psychological Interpretation* (New York: Henry Holt and Company, 1937).

[18] Richard Bord and Robert O'Conner, "Risk Communication, Knowledge, and Attitudes: Explaining Reactions to a Technology Perceived as Risky," (manuscript submitted for publication), quotations 14, 11–12, 14–15. Authors are at Pennsylvania State University.

[19] Thomas Dietz and Robert Rycroft, *The Risk Professionals* (New York: Russell Sage Foundation, 1987), quotation 47.

[20] Paul Sabatier and S. Hunter, "The Incorporation of Causal Perceptions into Models of Elite Belief Systems," *Western Political Quarterly* 42 (1989): quotation 253.

[21] Gerald Gardner, Adrian Tiemann, Leroy Gould, Donald Deluca, Leonard Doob, and Jan Stolwijk, "Risk and Benefit Perceptions, Acceptability Judgments, and Self-reported Actions toward Nuclear Power," *Journal of Social Psychology* 116 (1982): 116, quotations 194–95.

[22] Stanley Rothman and S. Robert Lichter, "Elite Ideology and Risk Perception in Nuclear Energy Policy," *American Political Science Review* 81 (1987): 81, quotation 395.

[23] Our findings call for multivariate statistical analysis of the interactions between cultural biases and the other classes of predictors. We are fully aware of the difficulties of regression or path analysis on small samples, so efforts are under way on larger samples. We are also sensitive to the fact that correlations do not necessarily imply causation. Cultural theory makes causal attributions, however, and the correlations we do find are consistent with its predictions.

[24] Chauncy Starr, "Social Benefit versus Technological Risk: What is Our Society Willing to Pay for Safety?" *Science* 165 (1969): 1232–38.

[25] National Research Council, Committee on Risk Perception and Communication, *Improving Risk Communication* (Washington, D.C.: National Academy Press, 1989).

# 14

# HUMAN ADJUSTMENT TO THE RISK
# OF ENVIRONMENTAL EXTREMES

*Dennis S. Mileti*

Relations of human aggregates with their natural environment have long been subject for investigation. Early efforts (Park, 1936) cast human collectives as adaptive units that respond to the natural world. Others (cf. Hawley, 1950; Alihan, 1938) emphasized examination of the mechanisms through which adjustment occurs and specification of physical-social relationships (Duncan, 1964). Labeled as human ecology, investigations seek the determinants of behavior in the natural environment, and the processes that facilitate human adjustment to the physical world through social organization (Duncan and Schnore, 1959).

An attempt to synthesize knowledge in this area by Micklin (1973) has identified four mechanisms of human adjustment to the natural environment. These are: (1) engineering mechanisms which include technological inventions and their application, (2) symbolic mechanisms which include culture and its constituent norms and roles, (3) regulatory mechanisms that define public policy and social control, and (4) distributional mechanisms which specify the movement of people, activities and resources. Most efforts to specify human interface with the natural environment which emanate from the human ecological perspective investigate human adjustment to the environmentally routine including, for example, factors such as local geography, food production, and rainfall.

Extremes in routine natural processes, when they impact a human collective, can cause disaster. Extremes in physical systems become disasters when the social systems they impact have only partially taken such extremes into account when adjusting to the physical world. Human aggregates typically emphasize adjustment to physical systems, through mechanisms such as those proposed by Micklin, on the basis of the probable routine of nature, rather than its sometimes equally predictable, albeit less frequent, extremes. The environmentally routine in a hydrological-water-system, for example, is the presence of a river, lake, average annual rainfall and other factors to which the human aggregate adjusts and on which it often depends. When extremes in this same physical system manifest themselves, for example, as flood, tsunami or drought, they are disaster. Environmental extremes are commonplace. Although they

Reprinted from *Sociology and Social Research* Volume 64 (3) (1980), pp. 327–47. Used with permission of the University of Southern California.

are of lesser probability in the short-term than is the environmental routine, they are a certainty over the long-term.

A good deal of the attention devoted to the study of human response to environmental extremes has been independent of the rich heritage of work in human ecology and has been cast under the label "disaster research." There have been several attempts to synthesize knowledge in this area. Efforts have been advanced by Quarantelli and Dynes (1977), Mileti, *et al.*, (1975), White and Haas (1975), Barton (1970), and Fritz (1968, 1961). Research into disasters began with works by Prince (1920), Kutak (1938) and others, and emphasized human adjustment or response to disaster only as it occurred or in its aftermath. Explanations have been traditionally placed in theories of collective behavior (cf. Barton, 1970), social organization and disorganization (cf. Dynes, 1970).

A parallel current in the area includes work advanced by White (1945), White, *et al.*, (1958), Burton and Kates (1964) and others, and more recently by Hutton and Mileti (1979), Burton, *et al.*, (1978), Kates (1978), Kunreuther (1978, 1974), White (1975, 1974, 1973), Slovic, *et al.*, (1974), and others. Pioneering work (White, 1945) investigated why certain adjustments to the risk of environmental extremes seemed preferred over others, and then despite investments in these adjustments, why loss was increasing (White, *et al.*, 1958). Subsequent investigations provided the theoretical basis on which further work was formed. This alternative current of work on disasters has emphasized specification and explanation of those social organizational mechanisms whereby human aggregates and individuals adopt and implement policies to mitigate the risk imposed by the possible occurrence of some future environmental extreme. This segment of disaster research, or hazards research as it has become known, is akin to human ecology in that the latter seeks explanations for human adjustment to the environmentally routine, while the former seeks explanations for human adjustment to the risk of environmental

extremes. Hazards research is also different from human ecology in that it has traditionally emphasized the importance of adjustment performed by individuals, although the adjustment of larger social units such as organizations (Haas and Mileti, 1976) and societies (Sorensen and White, 1980; White, 1974) have also been studied. Human ecology has emphasized adjustment of large collectives, although the adjustment of smaller social units such as individuals (Firey, 1947) has also received attention. Hazards research is different from disaster research in that the latter seeks explanations for response to disaster impact (Quarantelli and Dynes, 1977; Mileti, *et al.*, 1975; Barton, 1970; Fritz, 1968), while the former seeks explanations for adjustment to the risk of future disaster prevalent in everyday life (Burton, *et al.*, 1978; White and Haas, 1975).

In the several decades following the initial formulation of preliminary ideas in hazards research, good headway has been made in specifying the policies available for mitigating the risk of environmental extremes, and the causes of their adoption and implementation. It is the purpose of this paper to bring together the fruits of this work in an attempt to present the typology of policy adjustments to environmental extremes which has emerged, and the emerging theory of how and why human adjustment occurs to prepare for and lessen the risk imposed by future extremes in the natural environment. Emphasis is not placed on how people and social systems respond to disaster; rather attention has been focused on risk mitigation and preparedness for future environmental extremes.

## CONSEQUENCES OF RISK TO ENVIRONMENTAL EXTREMES

The inventory of risk-mitigating adjustments to environmental extremes, and the attempt to codify the emerging theory specifying the causes of their effectuation, is well prefaced by a short discussion of the consequences that environmental extremes can impose on a human collective. Were human adjustment to mitigate the risk of extreme environmental systems totally successful, disasters would be avoided. To the extent that adjustment is only partially successful, death and loss imposed by natural disaster is lessened. Given the rare probabilities of occurrence of some types of extremes, for example, an earthquake large enough to devastate the bulk of Los Angeles, total adjustment by a social system is impracticable, and, despite the potential losses to be incurred, not even wise from a cost-benefit perspective. Disasters will continue to occur despite attempts to mitigate risk because adjustments can not mitigate all risk. Some types of adjustment are more effective at risk-mitigation than others. As well, only partial levels of risk are worth mitigating; only limited application of risk-mitigating policies for environmental extremes is justifiably pursued; only within limits can society equitably prepare for disaster. For example, the monumental earthquake that could level all of Los Angeles has less than a one-in-a-million chance of occurring. Society is not willing to relocate Los Angeles nor impose the costs of building totally earthquake resistent structures to avoid loss from such an event because that event may not occur for another 5,000 years. Adjustment to the risk imposed by environmental extremes can always be exceeded by a physical system and result in disaster. Discovery of basic adjustment mechanisms, variables, causal links and process which define human adjustment to environmental extremes, if put to effective use by social engineers and drafters of policy, can at best reduce the risk imposed by environmental extremes to acceptable levels.

A variety of costs and benefits are produced when disaster occurs. Costs can be measured in different ways (Haas and White, 1975) including deaths, injuries, loss of property, dollar loss in disruption of economic systems, social disruption and psychic damage. In aggregate or average terms, loss from even large disasters appears small; average annual

losses are misleading because of the low probability of the occurrence of an environmental extreme in any one year. In disaster, some die, are injured, or are dislocated, many experience damage, and most are affected in other small ways (Bowden and Kates, 1974). Costs are unevenly distributed throughout the larger society (Cochran, 1975) and typically borne in the short run. Long-term effects of disaster have not been detected (Rossi, 1979).

The consequences of disaster are not all negative. Three kinds of benefits have been discerned (Sorensen and White, 1980). There are benefits of human occupancy of hazardous areas, direct benefits from a disaster, and benefits of human response. There are decided benefits from continuing to occupy hazardous areas; continued economic productivity from occupancy of the San Francisco peninsula, for example, far outweighs the costs of some future earthquake disaster. Floods may irrigate parched agricultural lands, and accordingly may be welcomed by some persons as a direct benefit rather than as a disaster. Human response to disaster typically enhances social cohesion and abates prior conflict patterns (Mileti, *et al.*, 1975; Fritz, 1961), facilitates adjustment to lessen future risk (Burton *et al.*, 1978; Haas *et al.*, 1977), and yields other direct benefits. Benefits and costs of disaster are difficult to measure with accuracy or consistency. Each risk-mitigating policy adopted also has associated costs and benefits. It is difficult to ascribe positive values to human adjustment to environmental extremes because each action taken to reduce risk has both costs and benefits all of which are rarely taken into account when adjustment occurs.

## A TYPOLOGY OF HUMAN ADJUSTMENTS OF ENVIRONMENTAL EXTREMES

In its effort to cope with the future possibility of environmental extremes, society and its constituent elements can take a variety of actions to reduce risk and uncertainty. These actions comprise the range of adjustment types to natural hazards. At any given time, a subset of these alternatives is implemented. Some adjustments are widely practiced, others remain figments of theoretical conjecture. The development, adoption and implementation of policies with regard to the risk of natural environmental extremes are functional in terms of the promise each holds to reduce risk and increase emergency preparedness (White, 1975).

Risk reduction is the consequence of adjustment policies which speed up, intensity or initiate activities which lower the potential for loss from future environmental extremes (Mileti, *et al.*, 1980). Adjustments which reduce risk are varied and include building codes, land use controls, deurbanization or sprawl, and others which all serve to lessen what is at risk or its concentration. Increased emergency preparedness, or the capacity of a social system to respond to disaster, refers to adjustment policies which reduce or redistribute the cost and anguish of recovery after a disaster. Adjustments which enhance emergency preparedness include evacuation plans, insurance and warning systems. Both risk reduction and emergency preparedness serve to reduce loss through human adjustment to the risk imposed by possible future disaster. The former does so by reducing what is at risk or enhancing its resilience; the latter does so by enhancing the ability of the social system to respond to disaster. Adjustments which enhance preparedness and reduce risk do not yield direct benefits until a low probability environmental extreme occurs; however, their associated costs begin to be incurred as soon as they are effected.

The theoretical range of human adjustments to environmental extremes is summarized in Table 1. This classification, based on a model advanced by Burton, *et al.*, (1978), suggests that man can purposely adjust to the risk of environmental extremes by a change in location or resource use, actions to reduce

loss, or the redistribution of loss. Other adjustments are incidental, but act in a manner which reduce losses. Finally, man may make changes which are unwittingly adaptive with respect to the risk of environmental extremes. Other typologies of adjustments to environmental extremes have been advanced (cf. Mileti, *et al.*, 1975; White and Haas, 1975); however, the elements of Table 1 capture the range of adjustments identified to date. This typology is similar to the typology advanced by Micklin (1973) of human adjustment to the environmentally routine. Both propose elements of engineering and technology, culture, regulation and policy, and techniques for redistribution as the key mechanisms whereby social systems adjust to routine and extreme natural environments.

Purposeful adjustments include actions to change land use, reduce or redistribute loss. The key means of changing land use is through management and regulation. For example, areas over earthquake fault lines slated for development can be zoned for low risk use such as parks or single-story low density housing; new development in areas that flood can be limited. Change in location involves leaving a hazardous area for one of lesser risk. Such can be the decision of individuals, or of a whole community resulting in the total abandonment of hazardous areas as occurred after the 1972 flood disaster in Rapid City. Although there are a few examples of policies in effect to enhance the use of land use management as a purposeful adjustment to reduce risk from environmental extremes, for example, the Alquist-Priolo Special Studies Act for the earthquake hazard in California and the National Flood Insurance Program (Hutton and Mileti, 1979), little application of this powerful risk-mitigating adjustment has been employed relative to the benefits it promises.

Loss reduction through risk-mitigation can also be achieved by modifying the environmental extreme itself. A hazard can be reduced by improving site selections for construction. It can also be reduced by

**TABLE 1**  Human Adjustments to the Risk of Environmental Extremes*

| Purposeful Adjustments | | |
| --- | --- | --- |
| Choose/Change | Reduce Loss | Redistribution of Loss |
| Change locations<br>• Abandonment[1] | Modify event<br>• Weather modification[1] | Share losses<br>• Insurance[4]<br>• Disaster relief[4]<br>• Charity[4] |
| Change use<br>• Land use planning[3] | Affect cause<br>• Hazard specific[1] | Bear loss<br>• Create reserve funds for<br>  anticipated loss[4] |
| | Prevent losses<br>• Warning systems[1]<br>• Building codes[3]<br>• Engineering works[1]<br>• Evacuation[4] | |

| Incidental Adjustments | | |
| --- | --- | --- |
| Redistribution of Loss | Reduce Loss | Choose Use/Location |
| • Savings[4] | • Fire codes[3]<br>• Transportation<br>  improvement[1]<br>• Fire fighting<br>  improvement[1] | • Land use regulation;<br>  non-hazard specific[3] |

| Unwitting Adaptations | |
| --- | --- |
| Biological | Cultural |
| • Undetermined | • Deurbanization (sprawl)[2]<br>• Shifts in aesthetic<br>  preferences[2]<br>• Shifts in family<br>  structure[2]<br>• Reduction in wealth[2] |

*Where 1 = engineering and technology, 2 = culture, 3 = regulation and policy,
4 = redistribution; adapted from Mileti, Hutton and Sorensen (1980).

diminishing associated secondary hazards, such as the landslide and fire hazards which accompany the earthquake hazard. These secondary hazards can be decreased through the application of, for example, slope stabilization or earthquake resistant firefighting systems.

A variety of purposeful adjustments can also be taken to prevent losses. With the development of techniques for hazard warning systems (Mileti, 1975), event-specific warnings inform the public of impending disaster. Preparedness measures in response to warnings include gearing up to provide emergency public services, medical assistance, or special planning for firefighting, evacuating dangerous areas, and mobilization of emergency organizations. Properly enforced building code regulations can make structures more resistant to major destruction by disaster. Selective or total evacuation of risk zones after or prior to disaster can result in preventing loss from the actual event as well as its induced effects.

It is impossible to abandon all hazard-prone areas permanently, or to reduce potential loss from environmental extremes to zero. Residual losses are, therefore, absorbed by society through both cost-sharing or cost-bearing. Losses can be redistributed in a variety of ways such as by insurance, tax deductions and programs of disaster relief.

Incidental adjustments are less obvious and more difficult to identify. A commonly cited example is that the building-code specifications for wind resistance in Boston also contribute to the seismic safety of that city; however, the extent and deaths from earthquakes are unknown. The most important incidental adjustments may come in the form of response to man induced problems such as fire or urban transportation, which become vital elements in reacting to a disaster and its consequences.

Finally, it can be speculated that man, over a longer time span, has unwittingly adapted to hazards through cultural or even biological changes. To date, no biological adaptations have been identified or even hypothesized. It is also still uncertain whether long-term cultural changes have enabled societies to adapt better to hazards. However, many changes in cultural patterns have resulted in new and different human response to hazards. A prime example is urban sprawl which has spread the population-at-risk over larger land areas thereby minimizing the population living in the most hazard-prone areas, for example, areas immediately along the bank of a river likely to flood.

Adjustments, be they purposeful, incidental or unwitting adaptations, differ in their ability to abate future risk. Some work well to negate the risk of small disasters; dams, for example, eliminate the risk of small floods, but actually may serve to escalate the risk of massive flood catastrophe by facilitating human occupance of the area beneath them. A subsequent environmental extreme large enough to not be contained by the flood protection work may be greater than if that work were not constructed in the first place (White and Haas, 1975). Other adjustments, for example, warning systems, may lack the necessary components to enable them to be ef-fective (Mileti, 1975) when put into use and, as a consequence, not reduce risk at all. Adoption of policy to enhance adjustment and actual subsequent adjustment is no guarantee that effective risk-mitigation has occurred.

An important aspect in understanding human adjustment to environmental extremes is the linkage and interaction between adjustments. Few if any adjustments are totally effective in mitigating risk when used alone (White and Haas, 1975). Effectiveness is often dependent on different adjustment types being applied in concert. Unfortunately, no current technique exists to define what would be the optimal mix of adjustments for a particular location (Sorensen and White, 1980). Hazard adjustment linkages are best viewed in terms of how adoption of one adjustment affects the adoption of others (White and Haas, 1975: 63). Sorensen (1977) has provided an inventory of possible linkages. First, one adjustment may cause the adoption of a second; for example, communities with engineering works typically become dependent on federal relief programs (Sorensen and White, 1980). Second, an independent factor may cause the adoption of one or several adjustments; for example, the National Flood Insurance Program enhances adoption of both insurance and land use controls (Hutton and Mileti, 1979; Kunreuther, 1978). As well, certain independent factors may inhibit the adoption of adjustments; for example, the prevailing cultural value of individualism in the United States negatively affects adoption of hazard-related land use controls (Hutton and Mileti, 1979); there are other examples (cf. Burton, et al., 1978; Sorensen, 1977). Finally, Sorensen and White (1980) propose that adjustments can interact randomly.

The inventory of human adjustments to the risk imposed by possible future environmental extremes includes engineering, symbolic, regulatory and distributional means to adjust to natural hazard risk. Although the specific terms used is reference to these mechanisms in the area of hazards research may well be different from those used in human ecology, the adjustment to routine environments studied by human ecologists and the adjustment to the risk of extreme environments investigated by hazard researchers are readily explained with identical theoretical constructs.

## CAUSES OF ADOPTION AND IMPLEMENTATION OF RISK MITIGATING POLICIES

A variety of research efforts have sought to specify the determinants of human adjustment to the risk of environmental extremes. Findings cross a range of units of analysis including individuals (Kunreuther, 1978; Kates, 1970; White, 1964; White, et al., 1958), organizations (Mileti, et al., 1980; Anderson, 1970; Haas, 1970), communities (Hutton and Mileti, 1979; Miller, et al., 1974; Murton and Shimabukro, 1974), and whole societies (White, 1974; Anderson, 1968). Research has been based on a variety of theoretical perspectives including social and complex theories of organization (Hutton and Mileti, 1979; Drabek, 1969; Lindbolm and Braybrooke, 1963), symbolic interactionism, decision making theory and especially the bounded rationality model of man (Kunreuther, 1978; Slovic, et al., 1974; Simon, 1956), and combinations and integrations of different perspectives (Hutton and Mileti, 1979; White and Haas, 1975; White, 1974). Contributions have come from most social sciences including geography (Burton et al., 1978; Kates, 1962), social psychology (Slovic, et al., 1974; Sims and Bauman, 1972), economics (Cochrane, 1975; Dacy and Kunreuther, 1969), and sociology (Hutton and Mileti, 1979; Quarantelli and Dynes, 1977; Anderson, 1970; Haas, 1970). Most forms of design and method have been used. Despite divergence in orientation, method, theoretical approach, adjustment type being studied, unit of analysis, and type of environmental extreme providing research focus, research has produced relatively consistent conclusions. Strong evidence exists concerning, first, the basic social process whereby adjustment is made to environmental extremes and, second, what are the key variables central to determining the outcome of that process.

## THE GENERAL RISK— MITIGATING POLICY ADOPTION PROCESS

The process that best describes how social units adopt policy to mitigate risks of environmental extremes has been defined in a simple ideal-type proposed by Slovic, et al., (1974). The process is comprised of four steps in which social units, be they individual, society or intermediate, first, assess the probabilities of a natural environmental extreme; second, review the alternative adjustment policies available to mitigate risk; third, evaluate the impacts of these alternative adjustment strategies in reference to both risk abatement and consequences for other aspects of social life; and fourth, choose none, one or more adjustment policies.

Rarely found to function this rationally in the real world, the process is often altered by other factors. The model is routinely altered by less than reasonable and accurate appraisals of risk, inadequate knowledge about the effectiveness of a particular adjustment policy, bias in the processing of information, self-serving decisions and other factors. For most social units, the adoption of adjustment policies is typically determined by other causal factors. In the absence of a less than perfect world which would enable the Slovic et al., adoption model to proceed, other factors have strong influence in shaping human adjustment to environmental extremes.

## THE EMERGING THEORY OF RISK MITIGATING ADJUSTMENT

Natural processes in physical systems on rare occasion, but often with specifiable probabilities, yield extremes. Typical labels for extremes include hurricane, flood, tornado, earthquake, drought, volcanic eruption and others. Extremes in physical systems occur independently of social systems, and are of little concern until they impact a social system and result in disaster. Indeed, environmental extremes which do not impact a human system, or which impact a human system which has adjusted effectively, can be a natural resource. It is only when an extreme in a physical system interacts with a social system that both positive and negative effects occur.

As illustrated in Figure 1, social systems and human actors are capable of continual adjustment to changes in the physical world (Burton, *et al.*, 1978). Human use systems and varied forms of social organization, for example, patterns in land use practices or building codes, act in concert with physical systems, for example, the geographical extent of a floodplain and the probability of an extreme in water volume, to define level of risk.

Risk from an environmental extreme is the chance that a physical system will exceed some "normal" level and cause damage to people, social and economic systems, and artifacts. Objective risk is determined, for floods for example, by three factors: geophysical characteristics, involving the hydrological system imposed on a particular landscape geomorphology; the use people and their varied forms of organization make of the area which could flood; and the set of actions or adjustments the community and people have taken to avoid flood risks. The first two of these factors determine exposure to risk; the latter determines how much of that risk has been reduced. All three determine net risk exposure of the human aggregate to the environmental extreme.

Level of risk affects the degree of risk which is perceived by social actors; both determine, along with other factors, the risk-mitigating adjustments made by social units. The char-

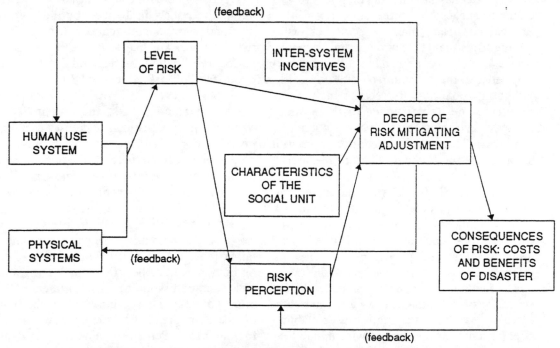

**FIGURE 1** Key concepts in the emerging theory of risk mitigating adjustment

acter of policies adopted and implemented define the extent to which risk is mitigated and consequently, the costs and benefits experienced by the human community when an environmental extreme does occur. It is the level of risk-mitigating adjustment which defines future disaster in that such adjustments alters aspects of human use of the physical world and the level of destruction experienced by a future environmental extreme. Disaster itself alters perception of risk and typically serves to enhance adoption of risk-mitigating adjustments after-the-fact.

In sum, risk from disaster is a consequence of the character of both physical and social systems. Disaster is averted or lessened by adoption and implementation of policies to mitigate risk and negative effects of environmental extremes. The system is dynamic as changes in human use patterns and activities, as well as in the physical environment, alter levels of risk which can require renewed or continued adjustment.

Research has mounted a good deal of evidence to suggest that three categories of theoretical constructs play the key roles in affecting adjustment to environmental extremes. These are: (1) risk perception, (2) such characteristics of the social unit considering a risk-mitigating policy as its social structure, its resources and its perception of the costs of policy effectuation, and (3) inter-system incentives. These three categories serve well to integrate the findings of recent research.

*Perceived Risk.* The adoption and implementation of policies to mitigate the risk of environmental extremes, by individuals, organizations, communities and other social units is influenced by the perceptions and interpretations people have and make about risk. Defined as cognition or belief in the seriousness of the threat of an environmental extreme, as well as the subjective probability of experiencing a damaging environmental extreme (Kunreuther, 1978; Slovic, *et al.*, 1974), a variety of factors have been found to con-

tribute to the formation of risk perception. These include the ability of a social unit to estimate risk (Burton, *et al.*, 1978; Slovic, *et al.*, 1974; White, 1974; Hewitt and Burton, 1971), perceived causes of environmental extremes (Burton, *et al.*, 1978), experience (Hutton and Mileti, 1979; Burton, *et al.*, 1978; Mileti, 1975; White and Haas, 1975; Kates, 1970; Anderson, 1969; Burton and Kates, 1964; Barton, 1962; Fritz, 1961; and others), propensity of people to deny risk (Mileti, *et al.*, 1980; Kunreuther, 1978; White and Haas, 1975; Mileti, 1974; Kates, 1970; Burton and Kates, 1964; and others), unit of analysis (Mileti, *et al.*, 1980; Kunreuther, 1978; National Academy of Sciences, 1978; Burton and Kates, 1964). Risk perception, through its mediating effects on image of damage and perceived benefits of effecting a risk-mitigating policy (Mileti, *et al.*, 1980; Kunreuther, 1978; Cochran, 1975; White and Haas, 1975), enhances human adjustment to environmental extremes.

Decisions to adopt and implement a risk-mitigating policy, influenced by the perceptions and interpretations that people have about risk, typically are made on the basis of imperfect information, biases, and difficulties in understanding risk concepts (Burton, *et al.*, 1978; Slovic, *et al.*, 1974). Despite faults in human cognition of risk, the probability of risk-mitigating adjustment increases as a positive function of risk perception through the mediating effect that perceived risk has on the variables of image of damage and perceived benefits of such adjustment (Hutton and Mileti, 1979). Image of damage is what social units think will happen to themselves, possessions and community were an environmental extreme to occur; it has a positive effect on both perceived benefits of risk-mitigating policy and on risk-mitigating adjustment. The more potential damage imputed on the basis of risk, the more likely a social unit will adjust to that risk. Perceived benefits, positively affect the probability of risk-mitigating adjustment to the extent that anticipated benefits are worth the costs of policy implementation (Hutton and Mileti, 1979; Kunreuther, 1978). Established relationships between these causes and consequences of risk perception, as they direct the effectuation of adjustment to environmental extremes, are illustrated in Figure 2.

Aspects of perceived risk that have been examined are diverse. Tversky and Kehneman (1974), for example, illustrated that people are poor probabilistic thinkers, yet are able to estimate the frequency probability of some environmental extremes better than others (Hewitt and Burton, 1971). People over-estimate small frequencies and under-estimate large frequencies of major environmental extremes. At the same time (White, 1964), people rarely take scientific factors into account in estimating risk and defining its perception (Burton and Kates, 1964). Burton, et al., (1978) have shown that the perceived cause of disaster—God, nature, technology,

societal choices—affect perceived risk and the choice of adjustment policy solutions selected. For example, those who define the cause of an earthquake as the will of God would rarely elect for a social solution to earthquake risk. Such solutions typically flow from perceptions of cause in human decisions to occupy hazardous areas. Actual experience with disaster greatly enhances perceived risk (Burton, et al., 1978; Mileti, et al., 1975; White and Haas, 1975, and others), but its effect is reduced as time passes. Policies designed to adjust to environmental extremes are typically more likely installed in the aftermath of disaster (Mileti, 1975; White and Haas, 1975; Danzig, et al., 1958) when risk perception are high than in the years prior to or long after disaster.

Perceived risk has also been found to differ by virtue of the unit of analysis under study (Mileti, et al., 1975; White and Haas, 1975). There is total certainty at the national level,

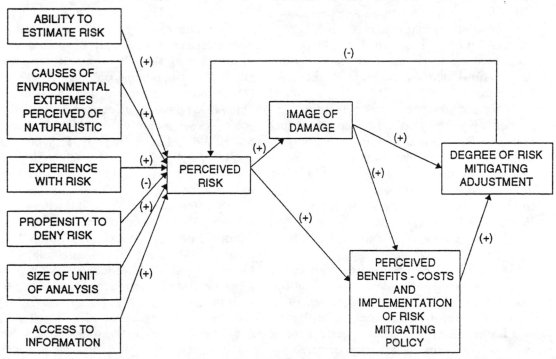

**FIGURE 2** Causes and consequences of risk perception

for example, that a disastrous flood will occur every year; Federal flood risk perception is therefore high. The probability that a flood will occur in a specific local community is low in a relative sense. Local community risk perception to this hazard is consistently lower than Federal risk perception; as well, risk perception of individuals in that same community is even lower. Decisions to adopt risk mitigating policies are typically more commonplace as the unit of analysis is increased because objective risk is increased.

Another aspect of perceived risk is the level to which already in-place adjustment policy yield a false sense of security; for example, people may think there is no longer a flood problem due to the presence of a flood protection work in their community. In such cases risk perception is typically low despite high levels of risk from a great flood catastrophe which exceeds the level of protection provided by the engineering solution (White and Haas, 1975). Mileti, et al., (1980) have also illustrated this in reference to the earthquake hazard. They found that business organizations and corporations that hold earthquake insurance and thereby have reduced risk through the distributional mechanism, refrain from further adjustment and adoption of other risk-mitigating policies.

Other determinants of risk perceptions are propensity to deny risk and access to information. Residents of hazardous areas typically deny the risk imposed by environmental extremes by discounting the possibility that anything truly serious will ever happen to themselves or possessions (Kunreuther, 1978; White and Haas, 1975; Mileti, et al., 1975; Burton and Kates, 1964, and others). Risk perception has also been shown to be upgraded to more accurate levels as the access of the unit to scientific information about the character of risk increases (Kunreuther, 1978; National Academy of Sciences, 1978; Burton and Kates, 1964). In general, certain social units and decision makers have more access

to information about environmental extremes than others. Some organizations, for example, have the resources to employ staff whose job it is to get, process and refine information of interest or concern to the organization. Others lack the perceived need for such employees or the resources to hire them. Social units which are fortunate enough to have access to good information gain more accurate perceptions of risk than others which lack that information.

The research findings accumulated on the causes of risk perception and its consequences for risk-mitigating adjustment are readily summed in two straight-forward phrases. First, social units adjust to the risk imposed by environmental extremes to the extent they think there is reason to adjust, and that the costs and bother to adjust are worth benefits that could be gained. Second, outcomes of policy decisions are made on the basis of available information, which is often biased and incomplete, as well as on perceptions of risk that are often inaccurate.

*Characteristics of Social Units.* Two key characteristics of social units play major roles in determining the adoption and implementation of risk-mitigating policies. These are: (1) the perceived costs of policy implementation based on social values and interest group goals (Sorensen and White, 1980; Hutton and Mileti, 1979; Kunreuther, 1978; White and Haas, 1975; Stallings, 1971; Clifford, 1956; and others), (2) the capacity of the social unit to implement the policy under consideration, largely determined by aspects of structure including social, power and political differentiation (Hutton and Mileti, 1979; Mileti, 1975; Dynes and Wenger, 1971; Quarantelli, 1965; Moore, et al., 1964; Sjoberg, 1962; Fritz, 1961) and resources (Sorensen and White, 1980; Hutton and Mileti, 1979; Burton, et al., 1978; Kennedy, 1970; Sjoberg, 1962; Fritz, 1961, and others). The relationships of perceived costs, capacity to implement, and their antecedent factors to the degree of risk-mitigating adjustment is illustrated in Figure 3.

Social values and the goals of special interest groups impact adjustment to environmental extremes because they determine the costs of such adjustment (Hutton and Mileti, 1979). Although the specific values held by members of any social unit contemplating adjustment are the most germane (cf. Sorensen and White, 1980; Kunreuther, 1978; Stallings, 1971; Clifford, 1956; and others), values at the societal level may also be predictive. For example, the general American values for self-reliance, science and technology, and change all serve to enhance community adoption of the land use controls policy contained in the National Flood Insurance Program (Hutton and Mileti, 1979), whereas values for achieved status, upward mobility, individualism and a general perception of abundance in a free enterprise system all serve to work against adoption of that particular risk-mitigating policy.

Any members of a social system could take a stand on an issue surrounding risk-mitigating policy for an environmental extreme. Land use controls, for example, restrict the uses to which land can be put, often alter the value of land, and foster opposition from owners of land who perceive themselves as losers (Hutton and Mileti, 1979; Haas and White, 1975). As well, the goals of environmentalists are readily served by adoption of these controls. Relatively few members of a social system are strongly affected by a risk-mitigating policy in terms of personal goals not associated with the risk imposed by an environmental ex-

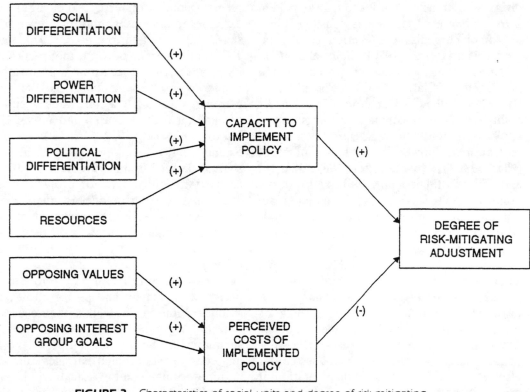

**FIGURE 3**  Characteristics of social units and degree of risk-mitigating adjustment

treme. However, such goals typically provide strong motivation to express interests and steer the outcome of a policy decision (Hutton and Mileti, 1979; White and Haas, 1975) through the basis they form for the definition of perceived secondary adjustment impacts.

Research on public policies of varied forms and types has established that characteristics of jurisdictions are related to community policy adoption. This generalization holds true with respect to policies to mitigate risk from environmental extremes adopted by communities, states and societies (cf. Hutton and Mileti, 1979; Dynes and Wenger, 1971; Quarantelli, 1965; Moore, et al., 1963; and others).

Elements of structure most frequently assessed in generic policy adoption research have been the differentiation of social, power and political systems. Typical measures of social differentiation include ethnic and economic diversity (Lauman, 1966; Walton, 1970; Aiken, 1969; and others), as well as density of linkages among organizations (Warren, et al., 1974; Perrucci and Pilisuk, 1970), and level of membership heterogeneity in voluntary association membership (Clark, 1968b). The differentiation of individuals or groups in terms of ability to influence others is power structure and has frequently been measured as the concentration of power in the hands of elites (Clark, 1968a, 1968b; Walton, 1967; Banfield and Wilson, 1963; Dahl, 1961; and others). Political structure or the differentiation of actors in terms of normative authority to influence adjustment decisions, has been taken to mean level of centralization of authority (Bingham, 1976; Lineberry and Fowler, 1967; and others).

Recent work in the arena of human adjustments to environmental extremes suggests that these three typical determinants of policy adoption and implementation—differentiation of social, power and political systems— also have an effect on the implementation of risk-mitigating policies (Dynes and Wenger, 1971; Fritz, 1961; Moore, et al., 1963; and others). This suggests that social units differ in their ability to implement a policy with respect to an environmental extreme once it has been formally adopted. Capacity to implement has been defined by Hutton and Mileti (1979) as the ability of a social unit to splice adjustment work into ongoing work, and is high if implementation work can be performed without much change from the pre-adjustment status quo (Quarantelli, 1965). Resources and differentiation enhance the capacity of a social unit to implement a policy and as such escalate the probability of risk-mitigating adjustment.

The findings accumulated detail three types of resources each of which has a positive effect on the capacity of the social unit to implement an adjustment policy and its preceding adoption. First, areas in which intensive use of resources like capital and labor is in effect, such as urban centers, are more likely to implement policy with respect to environmental extremes than areas without high level of resource intensity (Burton, et al., 1978; Kates, 1977; and others). Second, the level of capital available to a social unit is positively related to policy implementation; cost of implementing is less likely a constraint when level of wealth is high (Sorensen and White, 1980; Hutton and Mileti, 1979; Burton, et al., 1978; Kennedy, 1970; and others). Finally, resources defined in non-monetary terms such as the availability, level of professionalism, and technical expertise of personnel enhance the capacity of a social system to implement adjustment policy (Hutton and Mileti, 1979; Dynes and Wenger, 1971; Fritz, 1961; and others).

The evidence gathered on the roles of capacity to implement and perceived negative secondary impacts, as well as their causes, in determining adjustment outcome is easily summarized. Social units adjust to the risk imposed by environmental extremes if the work required to maintain the adjustment is easily incorporated into ongoing work, and if opposition to adjustment does not exceed intolerable levels.

*Inter-system incentives.* The third major construct to affect human adjustment to the risk of environmental extremes is inter-system incentives. Social units, be they individuals, groups, organizations, states or others, do not adjust to environmental extremes independent of the rest of the social world. Indeed, little adjustment would occur by social units which perceive low levels or risk were policy and regulation not implemented at larger levels of human aggregation, which perceive greater risk, to enhance the adjustment of these smaller social units. Dahl and Lindbolm (1953) have provided a typology of government policy which aptly defines the character of policies to enhance adjustment to environmental extremes. This three-class typology is comprised of: (1) information, (2) incentives, which is economic power to alter the incentive structure through payments or penalties, and (3) command or regulation to compel conformance. Burton, *et al.*, (1978) discuss policy guides to hazard adjustment using a related typology. They distinguish between four types of purposeful guides: preempt, mandate, influence and facilitate.

Policy designed at larger levels of aggregation comprised of information dissemination, incentives, regulations, or combinations of these, enhances adjustment of lower-level social units to environmental extremes. However, the effect of macro-level policy on micro-level adjustment to environmental extremes has been studied only to a limited extent because few policies of this character have been coined. Illustrative studies by Hutton and Mileti (1979) and Kunreuther (1978) have focused on different types of risk-mitigating adjustment within the National Flood Insurance Program.

## SYNTHESIS AND CRITIQUE

The risk of extremes in the physical systems which comprise the natural world and to which human populations are subject is the result of interaction between physical and human use systems. A variety of policies can be installed and adjustments accomplished to mitigate or reduce risk from future disaster by either increasing emergency preparedness or reducing human vulnerability. These adjustments cut across technological, cultural, redistributional and regulatory mechanisms for adjustment, and include those that are purposeful—including actions to change use and reduce or accept losses—those that are incidental, and unwitting adaptations. Purposeful adjustments are made because of several key factors. Generally, social units adjust to mitigate the risk imposed by environmental extremes: (1) if they think there is reason to adjust, (2) the costs and bother to adjust are seen as worth the benefits that could be gained, (3) on the basis of available information which is often biased and incomplete, (4) on the basis of risk perceptions which are typically inaccurate, (5) if the work required to maintain an adjustment does not require much change from the pre-adjustment status quo of life, (6) if opposition to adjustment which typically arises on the basis of alternative goals is not great, and (7) if larger level units of human aggregation provide adjustment incentives.

Several gaps in knowledge exist at this time which detract from theoretical closure. First, little is known about the true character of interaction between adjustments. Most researchers are ready to suspect that the adoption of one adjustment affects the adoption of others. However, specification of how this occurs is lacking. Second, the array of independent, mediating and dependent variables brought together in this attempt to synthesize knowledge have been gathered from a diverse set of research efforts which have typically emphasized only one or several of these variables and relationships. No attempt has ever been made to test the relative power of alternative suggested causes through the use of multivariate statistical techniques appraising large portions of the emerging theory. Finally, what emphasis there has been on theory development has been squarely laid on the adoptions of adjustment policies rather than their actual implementation. The act of adopting a risk-mitigating policy need not alter risk if that policy is not well implemented. For example, a city may adopt risk-mitigating building codes but not implement then comprehensively, or it may adopt elaborate plans for emergency preparedness but have few persons read them. Policy adoption must, therefore, be recast as a mediating variable between its causes and implementation before we will know if the human aggregate has actually, or if it just thinks it has, adjusted. Until these theoretical questions are answered and gaps in knowledge filled, practical application of assembled knowledge may be counterproductive, of no utility or helpful. The problem is simply that we do not know which is the case.

## REFERENCES

Aiken, Michael, 1969. "Community Power and Community Mobilization." *Annals of the American Academy of Political and Social Science* 385 (3).

Ailhan, Milla A., 1938. *Social Behavior*. New York: Columbia University Press.

Anderson, Jon, 1968. "Cultural Adaptation to Threatened Disaster." *Human Organization* (27): 298–307.

Anderson, William A., 1970. "Tsunanic Warning in Crescent City, California and Hilo, Hawaii." In Committee on the Alaska Earthquake of the Natural Research Council (Ed.) *The Great Alaska Earthquake of 1964*. Washington, D.C.: National Academy of Sciences.

Anderson, William A., 1969. "Disaster Warning and Communication Processes in Two Communities." *The Journal of Communication* (19/2): 92–104.

Banfield, E. C. and J. A. Wilson, 1963. *City Politics*. Cambridge, Washington: Harvard University Press.

Barton, Allan, 1970. *Communities in Disaster*. New York: Doubleday.

Bingham, Richard D., 1976. *The Adoption of Innovation by Local Government*. Lexington, Massachusetts: D.C. Health and Co.

Bowden, M., and R. Kates, 1974. "The Coming San Francisco Earthquake: After the Disaster," Pp. 62–81 in H. Cochrane, *et al.*, *Social Science Perspectives on the Coming San Francisco Earthquake*. Boulder: University of Colorado, Institute of Behavioral Science, Natural Hazards Paper 25.

Burton, Ian, 1962. *Types of Agricultural Occupance of Flood Plains in the United States*. Chicago: University of Chicago, Department of Geography, Research Paper 75.

Burton, Ian, and Robert Kates, 1964. "The Perception of Natural Hazards in Resources Management." *Natural Resources Journal* (3): 412–41.

Burton, Ian, Robert Kates, and Gilbert White, 1978. *The Environment as Hazard*. New York: Oxford University Press.

Clark, Terry N., 1968a. *Community Structure and Decision-Making: Comparative Analyses*. San Francisco: Chandler.

Clark, Terry N., 1968b. "Community Structure, Decision-Making, Budget Expenditures, and Urban Renewal in 51 American Communities." *American Sociological Review* 33.

Clifford, Roy, 1956. *The Rio Grande Flood: A Comparative Study of Border Communities*. Washington, D.C.: National Academy of Sciences, National Research Council.

Cochrane, Harold, 1975. *Natural Hazards: Their Distributional Impacts*. Boulder: University of Colorado, Institute of Behavioral Science, Monograph 14.

Craik, K., 1970. "The Environmental Disposition of Environmental Decision-Makers." *Annals of the American Academy of Political and Sociological Science* (May): 87–94.

Cyert, R., and J. March, 1963. *A Behavioral Theory of the Firm*. Englewood Cliffs, New Jersey: Prentice-Hall.

Dacey, D., and H. Kunreuther, 1969. *The Economics of Natural Disaster*. New York: The Free Press.

Dahl, R. A., 1961. *Who Governs? Democracy and Power in an American City*. New Haven: Yale University Press.

Dahl, R. A. and C. E. Lindblom, 1953. *Politics, Economics, Welfare*. New York: Harper & Row.

Danzig, Elliott, Paul Thayer, and Lila Galanter, 1958. *The Effects of a Threatening Rumor on a Disaster-Stricken Community*. Washington, D.C.: National Academy of Sciences, National Research Council.

Drabek, Thomas, 1969. "Social Processes in Disaster: Family Evacuation." *Social Problems* 16 (3): 336–49.

Duncan, Otis D., 1964. "Social Organization and the Ecosystem." Pp. 37–82 in Robert E. L. Faris (ed.) *Handbook of Modern Sociology*. Chicago: Rand McNally.

Duncan, Otis, D., and Leo F. Schnore, 1959. "Cultural, Behavioral, and Ecological Perspectives for the Study of Social Organization." *American Journal of Sociology* 65 (September): 132–46.

Dynes, Russell, 1970. *Organized Behavior in Disaster*. Lexington, Massachusetts: D.C. Health and Co.

Dynes, Russell, and Dennis Wenger, 1971. "Factors in the Community Perception of Water Resources Problem." *Water Resources* Bulletin 7(4): 644–51.

Firey, Walter, 1947. *Land Use in Central Boston*. Cambridge, Mass.: Harvard University Press.

Fritz, Charles E., 1968. "Disasters." Pp. 202–207. *In International Encyclopedia of the Social Sciences.* New York: Macmillan.

Fritz, Charles E., 1961. "Disaster." Pp. 651–94 in Merton and Nisbet (Eds.). *Contemporary Social Problems.* New York: Harcourt.

Haas, J. Eugene, 1970. "Lessons for Coping with Disaster." Pp. 39–51 in Committee on the Alaska Earthquake of the National Research Council (Ed.) *The Great Alaska Earthquake of 1964.* Washington, D.C.: National Academy of Sciences.

Haas, J. Eugene, and Dennis S. Mileti, 1976. "Socioeconomic Impact of Earthquake Prediction on Government, Business and Community," *California Geology* 30 (7): 147–57.

Haas, J., R. Kates, and M. Bowden (eds.), 1977. *Reconstruction Following Disaster.* Cambridge: MIT Press.

Hawley, Amos H., 1950. *Human Ecology: A Theory of Community Structure.* New York: The Ronald Press.

Hewitt, K., and Ian Burton, 1971. *The Hazardousness of a Place.* Toronto: University of Toronto Press.

Hutton, Janice, and Dennis Mileti, 1979. *Analysis of Adoption and Implementation of Community Land Use Regulations for Floodplains.* San Francisco: Woodward-Clyde.

Hutton, Janice, and Dennis Mileti, 1978. "Social aspects of earthquake." Paper presented at the second International Conference on Microzonation. San Francisco.

Kates, Robert, 1978. *Risk Assessment of Environment Hazard.* New York: John Wiley and Sons.

Kates, Robert, 1970. "Human Adjustment to Earthquake Hazard." Pp. 7–31 in Committee on the Alaska Earthquakes of the National Research Council (eds.) *The Great Alaska Earthquake of 1964.* Washington, D.C.: National Academy of Sciences.

Kennedy, Will C., 1970. "Police Departments: Organization and Tasks in Disaster." *American Behavioral Scientist* 13 (3): 354–62.

Kunreuther, Howard, 1978. *Disaster Insurance Protection: Public Policy Lessons.* New York: John Wiley and Sons.

Kunreuther, Howard, 1976. "Limited Knowledge and Insurance Protection." *Public Policy* (24): 227–61.

Kunreuther, Howard, 1974. "Economic Analysis of Natural Hazards: An Ordered Choice Approach." Pp. 206–14 in G. White (Ed.) *Natural Hazards: Local, National, Global.* New York: Oxford University Press.

Kutak, Robert L., 1938. "The Sociology of Crises." *Social Forces* 17 (2): 66–72.

Laumann, E. O., 1966. *Prestige and Association in an Urban Community: An Analysis of an Urban Stratification System.* Indianapolis, Indiana: Bobbs-Merrill.

Lindbolm, Charles E., and David Braybrooke, 1963. *A Strategy of Decision.* New York: Free Press.

Lineberry, R. L. and E. P. Fowler, 1967. "Reformism and Public Policies in American Cities." *American Political Science Review* 61.

Micklin, Michael, 1973. *Population, Environment and Social Organization.* Hinsdale, Illinois: Dryden Press.

Mileti, Dennis S., 1975. *Natural Hazard Warning Systems in the United States.* Boulder: Institute of Behavioral Science, Monograph 13.

Mileti, Dennis, Janice Hutton and John Sorensen, 1980. *Earthquake Prediction Response and Options for Public Policy.* Boulder: University of Colorado, Institute of Behavioral Science (forthcoming).

Mileti, D., T. Drabek, and J. Haas, 1975. *Human Systems in Extreme Environments: A Sociological Perspective.* Boulder: University of Colorado, Institute of Behavioral Science, Monograph 2.

Miller, D. J., W. Brinkmann, and R. Barry, 1974. "Windstorms: A Case Study of Wind Hazard for Boulder, Colorado." Pp. 80–86 in Gilbert White (ed.) *Natural Hazards: Local, National, Global.* New York: Oxford University Press.

Moore, Harry E. *et al.*, 1963. *Before the Wind: A Study of Response to Hurricane Carla.* Washington, D.C.: National Academy of Sciences, National Research Council.

Murton, Brian, and Shinzo Shimabukro, 1974. "Human Adjustment to Volcanic Hazard in Puna District, Hawaii," Pp. 151–59 in Gilbert White (Ed.) *Natural Hazards: Local, National, Global.* New York: Oxford University Press.

National Academy of Sciences, 1978. *A Program of Studies on the Socioeconomic Effects of Earthquake Predictions.* Washington, D.C.: National Academy of Sciences, National Research Council.

Park, Robert E., 1936. "Human Ecology." *American Journal of Sociology* 42 (July): 1–15.

Perrucci, R. and M. Pilisuk, 1970. "Leaders and Ruling Elites: The Interorganizational Bases of Community Power." *American Sociological Review* 35.

Prince, Samuel H., 1920. *Catastrophe and Social Change.* New York, Columbia University, Unpublished Ph.D. Dissertation.

Quarantelli, E. L., 1965. "Mass Behavior and Government Breakdown in Major Disaster." *The Police Yearbook* (21): 105–12.

Quarantelli, E. L., and Russell R. Dynes, 1977. "Response to Social Crisis and Disaster." *Annual Review of Sociology* 3: 23–49.

Rossi, Peter, *et al.*, 1978. "Are There Long Term Effects of American Natural Disasters?" *Mass Emergencies* (3): 117–32.

Simon, H. A., 1956. "Rational Choice and the Structure of the Environment." *Psychological Review* (63): 129–38.

Sims, J., and D. Bauman, 1972. "The Tornado Threat: Coping Styles of the North and South." *Science* (176): 1386–92.

Sjoberg, Gideon, 1962. "Disasters and Social Change." Pp. in George Baker and Dwight Chapman (eds.) *Man and Society in Disaster.* New York: Basic Books.

Slovic, Paul, Howard Kunreuther, and Gilbert White, 1974. "Decision Processes, Rationality, and Adjustment to Natural Hazards." Pp. 187–204 in G. White (Ed.) *Natural Hazards: Local, National, Global*. New York: Oxford University Press.

Smith, Thomas B., 1973. "The Policy Implementation Process." *Policy Sciences* (4): 197–209.

Sorensen, John, 1977. *Interaction of Adjustments to Natural Hazard*. Boulder, University of Colorado, Unpublished Ph.D. Dissertation.

Sorensen, John, and Gilbert F. White, 1980. "Natural Hazards: A Cross-Cultural Perspective." In I. Altman, A. Papaport and J. Wohwill (eds.) *Human Behavior and the Environment*. New York: Planum Press.

Stallings, Robert A., 1971. *A Comparative Study of Community as Crisis Management Systems*. Columbus: The Ohio State University, Unpublished Ph.D. Dissertation.

Tversky, A., and D. Kahnerman, 1974. "Judgment under Uncertainty: Heuristics and Biases." *Science* (185): 1124–31.

Walton, J., 1970. "A Systematic Survey of Community Power Research." In M. Aiken and P. E. Mott (eds.) *The Structure of Community Power*. New York: Random House.

Walton, J., 1967. "Vertical Axis and Community Power." *Southwestern Social Science Quarterly* 48.

Warren, R. L. *et al.*, 1974. *The Structure of Urban Reform*. Lexington, Massachusetts: D.C. Health.

White, Anne, 1974. "Global Summary of Human Response to Natural Hazards: Tropical Cyclones." Pp. 255–64 in Gilbert White (ed.) *Natural Hazards: Local, National, Global*. New York: Oxford University Press.

White, Gilbert F. (ed.), 1974. *Natural Hazards: Local, National, Global*. New York: Oxford University Press.

White, Gilbert F., 1973. "Natural Hazards Research." Pp. 193–216 in R. Chorley (ed.) *Directions in Geography*. London: Methuen.

White, Gilbert F., 1964. *Choice of Adjustments to Floods*. Chicago: University of Chicago, Department of Geography, Research Paper 93.

White, Gilbert F., 1945. *Human Adjustment to Floods*. Chicago: University of Chicago, Department of Geography, Research Paper 29.

White, Gilbert F. *et al.*, 1958. *Changes in Urban Occupancy of Flood Plains in the United States*. Chicago: University of Chicago, Department of Geography, Research Paper 57.

White, Gilbert, F., and J. Eugene Haas, 1975. *Assessment of Research on Natural Hazards*. Cambridge: MIT Press.

# 15

# PUBLIC RESPONSE TO EARTHQUAKE
# HAZARD INFORMATION

*Risa I. Palm*

Consumer protection legislation has been based on two assumptions about human behavior: first, that individuals are risk-averse, and second, that decisions are generally rational, given only limited knowledge of alternatives and their consequences. The first assumption implies that when given a choice, individuals will prefer less exposure to major losses even if this choice also means less opportunity for large gains.[1] Such risk-averse behavior produces a utility function with a concave form, and is the basis for economic theory predicting the conditions under which insurance is purchased: the individual trades a small, fixed loss (the premium) against protection from larger losses.

The second assumption implies that the course of decision-making can be manipulated, to some extent at least, by constricting or expanding the amount of information available to individuals.[2] The combination of these ideas provides the framework for such consumer protection legislation as the mandatory inclosure of warnings of side effects associated with particular medications, the publication of the sur-

geon general's warning on packages of cigarettes, and other forms of mandated disclosure.

There have been several forms of consumer protection legislation as related to residential purchases. For example, federally insured financial institutions are now required to make a complete disclosure of the full costs to be incurred by the borrower when a home mortgage application is filed. Similarly, several laws require that *environmental* information be provided to buyers: the federal government requires that lenders notify prospective buyers that property is located within a flood hazard area as defined by the Federal Insurance Administrator when communities are part of the federally-subsidized flood insurance program; the Department of Housing and Urban Development requires that real estate agents in Boulder and Jefferson counties (Colorado) inform buyers within ten miles of the Rocky Flats Plant that there is an emergency response plan which would go into effect in the event of accidental release of radioactive materials from the plant; and the Santa Clara County California Board of Supervisors requires sellers of property partly or wholly within flood, landslide, and fault-rupture zones to provide a statement of geologic risk to prospective homebuyers.[3] The legislation con-

Reprinted from *Annals of the Association of American Geographers* Volume 71 (3) (1981), pp. 389–99. Used with permission of the Association of American Geographers.

sidered in this paper is one of this genera: the California Alquist-Priolo Special Studies Zones Act which requires disclosure to prospective buyers of the location of a property within one-eighth mile of a fault trace. It is argued that the failure of this legislation to change the behavior of homebuyers is the result not only of incomplete compliance with the law, but also and more importantly the incorrect premise linking information provision with "rational" behavior.

## THE ALQUIST-PRIOLO SPECIAL STUDIES ZONES ACT

The California state legislature has several times been spurred to action on seismic legislation by major damaging earthquakes.[4] The most recent impetus was the earthquake in San Fernando-Sylmar in February, 1971. The legislative response to this earthquake was the passage, eleven months later, of the Alquist-Priolo Geologic Hazards Zones Act which directed the state geologist to delineate by the end of 1973 "appropriate wide special studies zones to encompass all potentially and recently active traces of the San Andreas, Calaveras, Hayward, and San Jacinto Faults," as well as any other faults which were "a potential hazard to structures from surface faulting or fault creep."[5] These zones were to be one-quarter mile in width or less. City or county approval was required for all new real estate development or structures for human occupancy in the original legislation, although this language was later modified to exempt single-family frame dwellings not a part of large developments.

In 1975 the act was amended; the names of the zones were changed from "geologic hazard zones" to "special studies zones" and the new act provided for disclosure to purchasers that property was within the special studies zones.[6] The disclosure amendment stated that "a person who is acting as an agent for a seller of real property which is located within a delineated special studies zone, or the seller if he is acting without an agent, shall disclose to any prospective purchaser the fact that the property is located within a delineated special studies zones."[7] Enforcements included the threat of revocation of license by the State Department of Real Estate, as well as a legal precedent proscribing misrepresentations and requiring "the fullest disclosure of all material facts concerning the transaction."[8] Specific directions, such as the type of materials to be used in the disclosure or when in the sales process disclosure was to take place, were not specified in the law.

The legislation was met with a great deal of apprehension. Real estate agents were uncertain as to how mortgage lenders would respond to the zonation, and therefore its impact on sales. These apprehensions were soon laid to rest, however, when it was learned that mortgage lenders took almost no account of special studies zones location.

A standardization of disclosure practice soon took place. This standardization was assisted by the publication in 1977 of a manual on special studies zone disclosure by the California Association of Realtors, the development of a contract addendum to the deposit receipt which was made available to California Realtors, and the production of colored maps by several local boards of Realtors.[9] As far as the regulatory department and the USGS were concerned, the disclosure legislation seemed to be "working."[10] But the important issue, of whether homebuyers were being "informed" by the disclosure, and whether this information was being translated into behavior which would mitigate their exposure to risk, had not been determined.

## THEORIES OF INFORMATION PROVISION AND BEHAVIOR CHANGE

The assumption that the provision of information may result in behavior change is based on the postulate that because most behavior is risk-averse, the individual should respond to seismic risk information either by attempting to avoid location within the area or lessening the potential hazard by mitigation measures. These measures might include the purchase of earthquake insurance, structural changes in the house, or even self-insurance in the form of bargaining for a lower price which is in effect "traded" for the willingness to assume the risk of major structural damage. There have been many experimental tests directed at the notion of risk aversion, with a surprising number of results which suggest that individuals may prefer risks rather than avoid them.[11] A number of suggestions have been made to account for this seeming preference for risk assumption. Three are particularly applicable to earthquake hazard response: the existence of a probability threshold, the difficulty of analyzing individual aspects of a complex decision, and the role of the information agent.

The existence of a probability threshold has been suggested by several authors studying the purchase of hazards insurance.[12] This concept refers to the observation that if the probability of a hazard is very low, it seems to be treated as if it were zero. One suggested explanation for this phenomenon is the fact that people face a multitude of hazards and problems in their everyday lives, and give their attention to those which recur most frequently. There have been few time-specific predictions of seismic risk for localized areas within California, but it is likely that even where such probabilities are estimated and widely publicized, they fall below a critical threshold at which they would be taken into account.

In any complex decision, it is very difficult to analyze the impacts of a single element. To the researcher, a portion of that decision, such as the refusal to respond to seismic risk warnings, may seem as if it were risk-taking behavior. However, when the complex decision is viewed as an entirety, the conjunction of all of the related decisions may fit the risk avoidance or utility maximization model.[13] In short, what is observed and labeled as risk-taking behavior for a portion of the decision, would actually be perceived by the decision-maker as risk-averse behavior given all of the elements which made up the final decision. It is difficult to analyze portions of the home purchase decision, particularly that portion of it dealing with environmental hazards, apart from the other constraints and utilities of the household, a fact which must constantly be kept in mind when labeling the migration behavior as risk-averse or risk-taking in nature.

The third factor which may account for the observation of risk-taking behavior is the role of the information agent. A great deal of research has been addressed to the general topic of the impact of the change agent on behavior modification. In the field of earthquake and flood insurance adoption, some research has found that it was not so much the objective nature of the risk as the institutional structure of the insurance industry which affected the likelihood that insurance would be adopted. Specifically, the commission structure for the sale of insurance had as great an effect on the purchase of insurance as any determination of utility functions by homeowners.[14]

The information agent may also affect the way in which a message is received by the techniques used to communicate the information and also the agent's personal and role characteristics. The timing and materials used in the presentation of new information will greatly affect the extent to which they are understood and heeded, as well as the presence or absence of behavior-relevant information or practical suggestions for responses incorporated into the message. In addition, the information will have more impact if the information agent has high credibility, a function of one's trustworthiness and ability to provide knowledge on the subject at hand.[15]

Previous research also suggests that new information has greater impact if the information is perceived as important and instrumental to the attainment of the goals of the recipient and if there is institutional support for the suggested behavior change.[16] If information concerning special studies zones is seen as an important aspect in the drive to acquire safe and secure housing, it will be attended to and included within the decision-making process. Similarly, if other societal institutions, such as the mass media, mortgage lenders, or employers reinforce the notion that special studies zones are significant residential considerations, any information about their existence will seem more important to the homebuyers.

Finally, individual and cultural factors have an effect on the impact of any new information. Individuals may vary in the extent to which they can comprehend the significance of the disclosure, and their awareness of alternative responses. In addition, "external attitudes" or beliefs, not directly related to the decision at hand, may affect the response to information.[17] For example, if one generally believes that there is little an individual can or should do to prevent injury or damage from "an act of God," then no amount of information will produce mitigation behavior. These external attitudes are very difficult to assess without lengthy observations or field interviews, but may affect the process of response to new information.

Although the research reviewed here suggests that the relationship between information provision and behavior change is extremely complex, legislation has been adopted as if there were a simple and straightforward connection between new information and behavior change. Assuming for a moment that the complexities can be ignored, one might expect that disclosure legislation would have an immediate and measurable effect. In this case, knowledge of the risks associated with living in a particular area would result in responses including the bargaining for a lower sales price, the avoidance of house purchase within the special studies zone, or the adoption of mitigation measures in the new resi-

dence. Any such change in behavior should be noted in a reduction in average house prices, since demand would be weakened and the cost of mitigation measures should enter in the estimation of the cost of living in the surface fault rupture environment.

## ORGANIZATION
## OF THE STUDY

Two California housing submarkets were selected for intensive surveys of recent homebuyers, associated real estate salespersons, and housing market behavior. These areas, Berkeley and central Contra Costa County, were chosen to minimize the social and economic contrasts of the study areas while varying the physical appearance and geologic characteristics of the areas, as well as the organization of the board of Realtors (Figs. 1 and 2). Although both are East Bay communities suburban to San Francisco, they function as separate housing submarkets. Both are generally inhabited by white, upper-middle class households in single-family, detached dwellings. They differ in that Berkeley is located on the Hayward fault and has more visible damage from fault creep to the retaining walls, houses and curbs, while central Contra Costa County is located primarily on the Calaveras fault, and visible signs of fault creep are difficult to find.

Evidence concerning the impacts of mandated disclosure was garnered from three sources: a survey of recent homebuyers, a survey of real estate agents active in special studies zones sales, and a study of house price trends in neighborhoods within and adjacent to the zones. The survey of homebuyers was a set of three surveys conducted throughout 1979. The first was a telephone survey of 207 homebuyers who had purchased houses within the special studies zones within the six months prior to being interviewed. The purpose of this survey was to ascertain whether buyers recalled a disclosure, the impact which disclo-

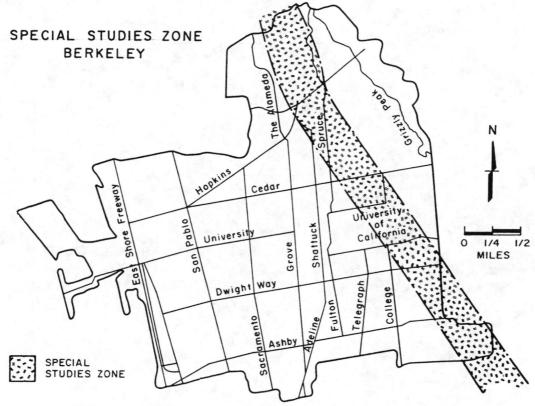

SPECIAL STUDIES ZONE
BERKELEY

**FIGURE 1.** The special studies zones in Berkeley along the Hayward fault.

sure might have had on search and purchase behavior, and general attitudes towards earthquake hazards and special studies zones. The second survey was directed at 100 households that had purchased homes near but not actually within the special studies zones. It was assumed that these homebuyers might have different attitudes toward earthquake risks, since they might include those who had received a disclosure while looking at a house within the special studies zone and had responded by moving to a similar area outside the zone. The third survey was a mail re-survey of the 96 homebuyers within the zones who had responded that they were aware that their homes were within special studies zones, and seemed to understand the significance of this fact. Its purpose was to ascertain whether disclosure had spurred these buyers to take

mitigation measures other than avoiding the area.

The survey of real estate agents was directed at only those salespersons who were identified by recent special studies zones homebuyers as having "helped" them with their home purchase. This screening limited the sample to licensees actively selling real estate, and to those who should be most familiar with the special studies zones disclosure procedures.[18] The survey attempted to discover the methods used for disclosure, the extent to which the real estate agents understood what it was they were disclosing, and the general response of real estate agents to the legislation.[19]

The third portion of the empirical work was a hedonic price analysis based on data describing over 7,000 sales as collected by the

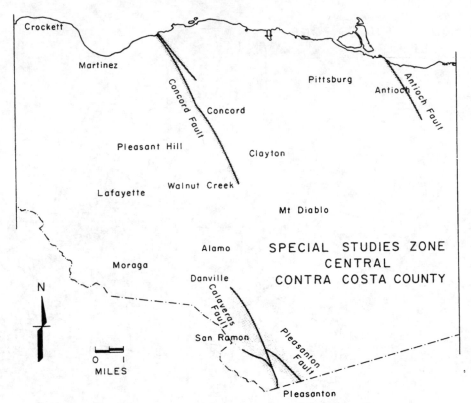

**FIGURE 2.** The special studies zones in central Contra Costa County along the Calaveras fault.

Society of Real Estate Appraisers. The independent effects of location within a special studies zone on prices were tested for both before disclosure had been mandated and after it was safely in place.

## RESPONSES TO DISCLOSURE

Disclosure legislation had little measurable impact on buyer or market behavior. First, surveys of recent homebuyers both within the special studies zones and in nearby similar neighborhoods outside the zones showed that distance from an active earthquake fault has little or no influence on search behavior or the final purchase decision. When recent buyers in both study areas and in both types

of neighborhoods were asked to indicate whether each of fifteen factors was "very important," "somewhat important," "not important," or "did not consider" with respect to the decision to buy the home, the two factors consistently evaluated as most important were investment potential and price; those evaluated as least important were location out of a floodplain and distance from an active earthquake fault. When asked more directly, "Did the location of earthquake hazard zones make any difference in your decision to buy this particular home?" only 18.7 percent of the within zone residents and 20.8 percent of the residents outside the zones said "yes." The location of a fault trace was not important in the purchase decision, even to those to whom a disclosure had been made.

This response by buyers was corroborated by the experience of real estate agents. When asked if they had ever had a client decide not to buy a home after being informed that it was in a special studies zone, only twelve agents said they had had such a refusal. Of these, only four could recall more than one client balking at the special studies zone designation.

Most homebuyers, whether located within or outside the special studies zones, were not aware that such districts existed. Even more significantly there was no difference in the likelihood that individuals would know of the existence of the zones by neighborhood; those living within the zones usually had never heard the term "special studies zones," nor were they aware their house was in any type of designated seismic hazard district. Although the survey could not gather direct evidence that disclosures were not being made, responses clearly indicate that the disclosure is not remembered, even only six months after it is made. It should be noted that although real estate agents are charged by law to make a disclosure, not even all of the real estate agents could associate the term "special studies zone" with seismic risks or earthquake faults; a full 16 percent of the real estate agents interviewed defined special studies zones as flood hazard areas or even as areas in which transportation planning surveys were slated.

There was a difference in the interpretation of the special studies zones by those who lived inside as opposed to outside. Those who lived within the zones said that they were no more susceptible to damage than those who lived elsewhere in the Bay Area (65 percent), but those who lived outside disagreed (62 percent said such residents were more susceptible to losses). Neither group, however, felt that location in a special studies zone would impair either the ability to sell the house or its selling price.

Objective evidence for this belief was reflected in the analysis of house prices. Data on sales prices, housing, and neighborhood characteristics for 1972 and 1977 in the two study areas were analyzed. In addition, a third study area, southern Alameda County, was added to further generalize the impacts of zonation on house prices. Hedonic price indices were calculated for square footage of dwelling space, age of the house, quality of the house, condition of the house, size of the lot, the presence of a swimming pool, fireplace, or "view lot," the economic status of the area, and housing stock composition. Location with respect to the special studies zone was coded as a dummy variable: property was classified as within the zone, adjacent to the zone (within one mile), or outside the zone (beyond one mile).[20]

The research hypothesis stated that in 1972 the special studies zone was unrelated to house price (the coefficient was zero), but in 1977 was negatively related. In addition, in 1977 location near the special studies zone should have a positive regression coefficient because of a build-up of demand for housing near but not actually in the zones, and location outside the zones should continue to have no effect on house prices. The results of a set of single-step ordinary least squares equations for the three study areas are complex (Table 1). For central Contra Costa County, the results are nearly as hypothesized, suggesting that even the very few people who were aware of, and concerned with, proximity to an active fault trace may have been a sufficient force in the marketplace to weaken prices within the zones. However, house prices in the other two areas did not perform as hypothesized. In southern Alameda County, special studies zones locations had no independent effect on house prices, and in Berkeley, location in the zones had a *positive* impact on house prices. The Berkeley results should not be interpreted as reflecting a preference for the special studies zones, but rather show the continued demand for houses in the lower portions of the Berkeley hills. Because of the great variation in the effects of zonation on house prices, it is probable that it is not the zonation disclosure

**TABLE 1.**  Effects of Location in Special Studies Zones on House Prices

| | 1972 Beta for Price Impact in Dollars (significance) | 1977 Beta for Price Impact in Dollars (significance) | Hypothesized Effect | Observed Effect |
|---|---|---|---|---|
| **Southern Alameda County** | | | | |
| Inzone | −741 (.166) | −243 (.807) | negative | none |
| Adjacent | 807 (.030) | −1,062 (.101) | positive | none |
| Outside | −422 (.234) | 1,121 (.078) | none | positive at .10 |
| R² = | .74 | .75 | | |
| **Berkeley** | | | | |
| Inzone | 2,617 (.000) | 9,618 (.092) | negative | positive at .10 |
| Adjacent | 1,162 (.061) | 9,118 (.092) | positive | positive at .10 |
| Outside | −3,121 (.000) | −1,315 (.004) | none | negative at .01 |
| R² = | .84 | .74 | | |
| **Central Contra Costa County** | | | | |
| Inzone | −912 (.307) | −4,182 (.000) | negative | negative at .01 |
| Adjacent | −473 (.620) | 1,500 (.048) | positive | positive at .05 |
| Outside | −623 (.377) | 1,705 (.007) | none | positive at .01 |
| R² = | .55 | .67 | | |

but rather some correlated neighborhood characteristics omitted from the equation that have affected prices. On the basis of the price equations for two of the three study areas, one may conclude that disclosure has not had a negative impact on house price levels.

Finally, disclosure has resulted in little measurable increase in mitigation measures. Because the Bay Area respondents were higher in income and education than the general population, and because they had recently received information that their homes were located within one-eighth mile of an active fault trace, they should have been more likely than the general Los Angeles population to adopt mitigation measures. This expectation was not borne out by the survey findings (Table 2). A comparison of the 96 Berkeley and Contra Costa special studies zones residents with 1,450 randomly sampled Los Angeles County households showed that only a minority of the residents of either area has taken any of the mitigation measures, the only exceptions being the possession of a working flashlight, battery radio, and first aid kit.[21] Residents of the special studies zones were more likely than the Los Angeles respondents to purchase earthquake insurance, to structurally reinforce their homes, and to replace cupboard latches, but most other measures were more frequently adopted by Los Angeles residents. A possible explanation for the greater frequency of adoption by the general Los Angeles population of measures such as emergency procedures at residence and family plans for reunion after an earthquake may be the combination of a fairly recent earthquake experience in 1971 and the widespread dis-

TABLE 2. *Mitigation Measures*

| | Have Done Primarily Because of Earthquake Threat | | Total Percentage Who Have Done | |
|---|---|---|---|---|
| | Bay Area | L.A. | Bay Area | L.A. |
| Inquired about earthquake insurance | 41.4%* | 23.1% | 41.4* | 23.1 |
| Bought earthquake insurance | 24.1* | 12.8 | 24.1* | 12.8 |
| Instruct children what to do in an earthquake | 20.0 | 47.6 | 22.2 | 50.4 |
| Emergency procedures at residence | 15.6 | 26.1 | 25.4 | 34.1 |
| Family plans for reunion after earthquake | 14.0 | 19.9 | 16.0 | 22.1 |
| Replace cupboard latches | 13.8* | 4.5 | 22.4* | 10.2 |
| Have a working battery radio | 8.6 | 11.1 | 53.4 | 54.6 |
| Structurally reinforce home | 8.6* | 4.7 | 13.8* | 11.1 |
| Have a working flashlight | 6.9 | 10.8 | 86.2* | 71.5 |
| Rearrange cupboard contents | 5.2 | 9.7 | 12.1 | 16.3 |
| Contacted neighbors for information | 3.4 | 9.8 | 15.5 | 19.5 |
| Have first aid kit | 3.4 | 6.0 | 68.9 | 50.1 |
| Store food | 1.7 | 8.0 | 20.7 | 26.8 |
| Store water | 1.7 | 8.0 | 5.1 | 17.1 |
| Set up neighborhood responsibility plans | 1.7 | 4.0 | 12.0 | 12.2 |

*Bay Area respondents exceed Los Angeles respondents.

cussion surrounding the so-called Palmdale bulge as a precursor of major movement along the southern portion of the San Andreas fault running through Los Angeles. But whatever the explanation, one must conclude that special studies zones disclosure, even when understood and remembered, was not associated with increased adoption of mitigation measures.

## DISCUSSION

The empirical study of homebuyers within the special studies zones, homebuyers in nearby areas outside the zones, real estate agents, and housing market behavior demonstrates that mandated disclosure has had little effect on buyer behavior or market performance. Although we have no direct information on the probabilities assigned to earthquakes by recent homebuyers and the existence of a probability threshold, it is apparent that seismic risks are given a very low priority in the home purchase decision. In addition, when some of the many factors considered within the pur-

chase decision are considered individually, it is apparent that respondents give little weight to earthquake hazards or other environmental disamenities in their priority system. Instead, the primary motivation for a given home purchase decision is the minimization of the price paid for a dwelling unit of given characteristics and the maximization of its potential resale value. This motivation is very clear from the priority assigned to price and resale value by buyers in both study areas, both within and outside the zones. The house is treated as an economic investment, rather than a locus for family activity for a ten- to twenty-year period. Given this motivation, buyers may not hesitate to buy a home in the special studies zone as long as it is believed to have good potential resale value, and will not adopt costly hazard mitigation measures which cannot be recouped in the subsequent sales price. Information provided by the mandated disclosure is simply treated as irrelevant to the decision at hand.[22]

The role of the information agent also interferes with the translation of disclosure into measurable behavior. To be effective, the real estate salesperson should have a generally high credibility to the buyers; yet it is well documented that buyers do not place their trust in real estate agents as a source of information about all aspects of the house purchase process, a wariness which may be partly attributed to the uncertainty concerning which party the real estate agent is representing in the sales process.[23] A related problem is the fact that real estate agents might provide misinformation concerning the special studies zones or may reinforce wishful thinking on the part of the buyer that the zones are not meaningful. Misinformation may be attributed to the simple lack of understanding of the meaning of special studies zones. The reinforcement of wishful thinking may come about from the desire of the real estate agent to close the sale, and the sincere belief on the part of the agent that the zones are not particularly important or meaningful, an attitude held by two-thirds of the real estate agents surveyed.

The nature of the zones themselves has also proved confusing. The maps provided by the office of the state geologist were not at a scale at which the location of individual properties could be located. This problem of accurate portrayal of individual parcels has become so severe that boards of realtors have urged their members not to make assessments of questionable parcels, but rather to recommend seeking the advice of a consulting geologist. Even more serious is the fact that the zones themselves, while they do include the known fault traces, do not encompass areas most susceptible to earthquake-associated damage. For example, although the 1906 earthquake in San Francisco is recalled as particularly severe and damaging, no special studies zone runs through the city. The zones do not include areas susceptible to liquifaction, shaking, or ground failure, since these are related to bedrock conditions as well as simple proximity to the fault. Although the special studies zones were relatively easy to define by legislation, they do not necessarily portray the areas of greatest seismic hazard.

Finally, there are problems with the disclosure legislation itself. Because the method of disclosure was not specified within the law, it has been possible to develop methods that minimize the impact of disclosure on buyers. The three standard formats used are the information in the Multiple Listing Service pages (used by thirty percent of the real estate agents interviewed), a map of the area with special studies zones drawn in (used by seventy percent), and a contract addendum (used by ninety-one percent of the respondents). The Multiple Listing Service form presents little information to the buyer. In Berkeley, disclosure on this form is simply a typed line stating "in Alquist-Priolo zone" or "in Alquist-Priolo district." To the uninitiated buyer, such a statement might mean anything. In Contra Costa County, the form includes a line stating "special studies zone" and a box marked "yes" or "no." This disclosure tells the buyer nothing about the *meaning* of the zones. The map, used particularly in Contra Costa County at the time of the survey, is a detailed street map of the region with the "hundred year floodplain" in blue, the special studies zone in yellow, and areas of combined hazard in green. Terms are not defined on the map, and the districts can be used by the real estate agents to demonstrate to the prospective buyers that many other properties share the same characteristics, and that therefore such a zonation cannot be very important since so many houses are at risk. The third disclosure method is the signing of a contract addendum. This addendum, until recently, stated that "the property is or may be situated in a Special Studies Zone." No definition of the special studies zone is presented, although the form does note that construction for human occupancy on the property may be subject to the findings of a geologic report unless such buildings are single-family, wood-frame dwellings or were in existence prior to May 4, 1975. The words, "seismic," "earthquake," or "fault" are nowhere mentioned in the contract addendum. Although the prescription of disclosure methods within the law would not guarantee that the information would be heeded, it is possible that the memorability of the disclosure could be increased. At present, real estate agents are disclosing at the least sensitive time, and using methods which convey a minimum of information, with the result that a majority of those who presumably received a disclosure are unable to remember it.

## CONCLUSIONS

Mandated provision of information by real estate agents has had little impact on homebuyers. Among the factors diminishing the effectiveness of this method of informing the general public about seismic hazards are the possible poor credibility of the real estate agent as a source of environmental information, the possible role conflict of the real estate agent in both selling property and protecting prospective buyers from environmental hazards, the differential degree of understanding by real estate agents of the location and meaning of the special studies zones, and the belief held by real estate agents that the zones are unimportant and therefore not worthy of attention. Equally important reasons for the nonresponse of buyers are the emphasis homebuyers place on the house as a financial investment overwhelming most other considerations, and the inadequacies of the zones themselves in outlining areas susceptible to greatest damage in the event of a major earthquake. An additional factor that may have lessened buyer response is the widespread belief that there are few real alternatives: that "all California is earthquake country," and that in an economic situation in which there is an excess of demand over supply buyers have little choice but to purchase any home they can afford whenever and wherever it becomes available.

The linkage between consumer protection legislation and behavioral response is based on the notion that improved information results in more "rational behavior." However, when what is "rational" to the individual differs from the assessment of the same situation by the policymaker, the predicted or intended response may not be achieved. In the case of provision of information about special studies zones, flaws within the legislation were compounded because the hazard homebuyers were being warned against generally lies outside the realm of concerns, constraints, and boundaries within which location decisions are made. In such a situation, mandated disclosure was doomed to failure.

A recurring theme in this study has been the attempt by a legislature to protect individuals from environmental hazards. In order to improve their performance, the legislature should realistically deal with the full range of hazards, both natural and human-made, facing state residents. If in such an assessment it is decided that earthquake hazards are more important than some others, and if the legislature truly wishes to take effective action to mitigate these hazards, it must deal with a more comprehensive definition of earthquake risks, and reconsider legislation which would better inform residents both about the hazards and feasible mitigation strategies. It is important that lawmakers recognize that the mere provision of environmental information to homebuyers who are constrained by other aspects of the purchase process is insufficient as a hazard mitigation or consumer-protection measure: mitigation can come only through forceful and direct measures such as zoning, land use regulation, or the compensation of current residents in exchange for the condemnation of dwellings in particularly hazardous environments.

*Acknowledgments* The research reported here was supported by the National Science Foundation under grant no. PFR78-04775. Any opinions, findings, and conclusions or recommendations expressed herein are those of the author and do not necessarily reflect the views of the National Science Foundation. I am grateful for the work of David Kuntz, Rene DuFort, Vickie Kendrick, and Gladys Bloedow, and for the helpful comments of Gilbert F. White, David Greenland, and Stuart W. Cook.

## NOTES

[1] The notion of risk aversion and its relationship to expected utility models was presented in M. Friedman and L. J. Savage. "The Utility Analysis of Choices Involving Risk," *Journal of Political Economy*, Vol. 56 (1948), pp. 279–304; W. Edwards, "The Prediction of Decisions Among Bets," *Journal of Experimental Psychology*, Vol. 50 (1955), pp. 201–14; and F. Mosteller and P. Nogee, "An Experimental Measurement of Utility," *Journal of Political Economy*, Vol. 59 (1951), pp. 371–404. In these papers, decision-making under conditions of uncertainty is examined, and choices are predicted using either the expected utility model or its variant, the subjective expected utility model.

[2] This postulate is derived from studies of manipulative communication or persuasion which evaluates the impact of the form of the message and the order in which information is presented on attitude change. Influential reviews of this literature include C. I. Hovland, I. L. Janis, and H. H. Kelley, *Communication and Persuasion* (New Haven, Connecticut: Yale University Press, 1953); and M. Sherif and C. I. Hovland, *Social Judgment: Assimilation and Contrast Effects on Communication and Attitude Change* (New Haven, Connecticut: Yale University Press, 1961). A study specifically investigating the ways in which real estate agents manipulate information to achieve desired sales patterns is T. R. Smith and F. Mertz, "An Analysis of the Effects of Information Revision on the Outcome of Housing-Market Search, with Special Reference to the Influence of Realty Agents," *Environment and Planning A*, Vol. 12 (1980), pp. 155–74.

[3] For a discussion of the impacts of the federally subsidized insurance program as it has affected the management of floodplains, see D. R. Anderson, "The National Flood Insurance Program—Problems and Potential," *Journal of Risk and Insurance*, Vol. 41 (1974), pp. 579–99; and R. Platt, "The National Flood Insurance Program: Some Midstream Perspectives," *Journal of the American Institute of Planners*, Vol. 42 (1976), pp. 303–13.

[4] A summary of the pre-1975 legislation is presented in Joint Committee on Seismic Safety, *Meeting the Earthquake Challenge: Final Report to the Legislature of the State of California* (Sacramento: Joint Committee on Seismic Safety, 1974).

[5] California Public Resources Code, Section 2621.1.

[6] Given the strength of the real estate lobby in California, it might have been expected that this amendment would have generated public controversy; instead, the act passed virtually unopposed in the legislature after a few amendments were modified in the assembly and unnoticed in the general press. Part of the reason for the absence of controversy and the acquiescence to disclosure on the part of the California Association of Realtors was the package of amendments of which the disclosure provision was a part. Along with disclosure, several changes were added which were favorable to real estate developers and agents including the exemption of new single-family frame dwellings not part of large developments from geologic reports, the exemption of mobile homes and condominium conversions from reports, and the exemption of alterations and additions to structures not exceeding fifty percent of the value of the structure. The California Association of Realtors would have preferred that disclosure be the responsibility of the seller, but they agreed to the language given the rest of the package. The name change was favored not only by real estate interests, but also by state geologists. It was felt that the term "geologic hazards zones" might be misinterpreted, since the zones delimited in the legislation did not include many of the areas susceptible to geologic hazards. The scientists preferred a term implying that the identified areas were targeted for future special study, and therefore the change of terminology was seen as an improvement in precision. Proponents of the disclosure provision viewed the final package of amendments as a compromise in which they had traded the exemption of single-family dwellings for the disclosure provision.

[7] California Public Resources Code, Section 2621.9.

[8] 98 Ca. Reptr. 242, 20 C.A. 3d 785.

[9] California Association of Realtors, *Disclosure of Geologic Hazards* (Los Angeles: California Association of Realtors, 1977).

[10] W. J. Kockelman, "Examples of the Use of Earth-science Information by Decision-makers in the San Francisco Bay Region, California," *USGS Open File Report No. 80–124*, 1980.

[11] Experimental findings running counter to strictly risk aversion are presented in A. Tversky, "Elimination by Aspects: A Theory of Choice," *Psychological Review*, Vol. 79 (1972), pp. 281–99; S. Lichtenstein and P. Slovic, "Reversals of Preference Between Bids and Choices in Gambling Decisions," *Journal of Experimental Psychology*, Vol. 89 (1971), pp. 46–55; H. R. Lindman, "Inconsistent Preferences Among Gambles," *Journal of Experimental Psychology*, Vol. 89 (1971), pp. 390–97; P. Slovic, "Choice Between Equally Valued Alternatives," *Journal of Experimental Psychology: Human Perception and Performance*, Vol. 1 (1975), pp. 280–87; and D. M. Grether and C. R. Plott, "Economic Theory of Choice and the Preference Reversal Phenomenon," *American Economic Review*, Vol. 69 (1979), pp. 623–38. The implications of these findings for expected hazards response are outlined in Kunreuther et al., op. cit., footnote 1; and P. Slovic, H. Kunreuther, and G. F. White, "Decision Processes, Rationality and Adjustment to Natural Hazards," in G. F. White, ed., *Natural Hazards: Local, National and Global* (New York: Oxford University Press, 1974).

[12] Kunreuther et al., op. cit., footnote 1; and Slovic et al., op. cit., footnote 1.

[13] B. P. Pashigian, L. L. Schkade, and G. Menefee, "The Selection of an Optimal Deductible for a Given

Insurance Policy," *Journal of Business*, Vol. 39 (1966), pp. 35–44.

[14] Kunreuther et al., op. cit., footnote 1.

[15] W. J. McGuire, "The Nature of Attitudes and Attitude Change," in G. Lindzey and E. Aronson, eds., *The Handbook of Social Psychology*, 2nd ed., vol. 3 (Reading, Massachusetts: Addison-Wesley, 1969), pp. 136–314; and P. Zimbardo and E. B. Ebesen, *Influencing Attitudes and Changing Behavior* (Reading, Massachusetts: Addison-Wesley, 1970).

[16] R. H. Weigel and L. S. Newman, "Increasing Attitude-behavior Correspondence by Broadening the Scope of the Behavioral Measure," *Journal of Personality and Social Psychology*, Vol. 33 (1976), pp. 793–802.

[17] Weigel and Newman, op. cit., footnote 16.

[18] Only about fifty percent of those persons holding licenses in California actively sell real estate according to the President of the California Association of Realtors, Clark Wallace (personal communication, 1979).

[19] A more complete version of the survey results is available from the author.

[20] The use of hedonic price indexes to assess the impacts of individual environmental variables has been widespread in economics. This literature has focused on the impacts of air quality on land values, as summarized in A. M. Freeman, "Hedonic Prices, Property Values and Measuring Environmental Benefits: A Survey of the Issues," *Scandinavian Journal of Economics*, Vol. 81 (1979), pp. 154–73. A hedonic price study of the impacts of special studies zones on land values is D. S. Brookshire and W. D. Schulze, *Methods Development for Valuing Hazards Information* (Laramie, Wyoming: University of Wyoming, Institute for Policy Research, October 1980). In this study of Los Angeles County, special studies zones were found to have had a negative impact on sales price, reducing home value by an average of $6,140 in 1978–79.

[21] R. Turner, J. Nigg, D. Pas, and B. Young, *Earthquake Threat: The Human Response in Southern California* (Los Angeles: University of California, Institute for Social Science Research, 1979).

[22] J. C. Hershey and P. Schoemaker, "Risk Taking and Problem Context in the Domain of Losses: An Expected Utility Analysis," *Journal of Risk and Insurance*, Vol. 47 (1980), pp. 111–32.

[23] D. Hempel, *The Role of the Real Estate Broker in the Home Buying Process* (Storrs, Connecticut: Department of Marketing, University of Connecticut, Center for Real Estate and Urban Economic Studies, 1969).

# 16

# EVACUATION DECISION-MAKING
# IN NATURAL DISASTERS

*Ronald W. Perry*

When seeking to manage the consequences of natural disaster, evacuation
is an important tool in the hands of authorities. In particular evacuation which is instituted *before* disaster impact can result in the
preservation of life, reduction of personal injuries, and the protection of property. Thus,
when there is sufficient forewarning—and
evacuation is an appropriate coping strategy—
pre-impact evacuation of a threatened population is an effective means of mitigating the
negative consequences of natural disaster.

As a means of adjusting to natural hazards,
evacuation is a process with a very long history. As early as the fifth century B.C., the
Greek historian Herodotus described Egyptian evacuations in the face of the seasonal
flooding of the Nile River. Indeed, evacuation is a concept which pervades journalistic,
popular and professional literature on disaster. In spite of its apparent ubiquity, however, very little attention has been devoted to
examining variables which are important in
individuals' decisions to evacuate in response
to a disaster warning.

This paper will review empirical studies of

warning response, particularly focusing upon
pre-impact evacuation of threatened populations, and summarize the available findings.
The summary may be seen as a conceptual
framework of inter-related hypotheses describing the relationships among variables
which *past research suggests* are important in
individuals' decisions to evacuate. This paper should not be seen as an attempt to develop a formal theory. Instead, it represents
an effort to order the empirical literature by
organizing existing findings into a general conceptual framework. The remainder of the
paper is structured around three primary tasks:
(1) development of a theoretical perspective
of evacuation behavior; (2) a review of empirical literature; and (3) assembling findings
into a tentative framework.

## EVACUATION
## IN THE CONTEXT
## OF WARNING RESPONSE

Much of the early social scientific research on
evacuation was conducted within the framework of man-made rather than natural hazards. Following World War II, a number of
studies were released which focused on Ger-

Reprinted from *Mass Emergencies* Volume 4 (1979),
pp. 25–38. Used with the permission of the author.

man (United States Strategic Bombing Survey, 1947a), Japanese (United States Strategic Bombing Survey, 1947b), and English (Titmuss, 1950) efforts to remove people from threatened cities and at the same time maintain the production and flow of military support goods. This literature however, tends to focus upon either the social psychology of stressful life events, or problems in the logistics of administration in evacuated areas (cf. Titmuss, 1950: 177–180). Although some research focused upon psychological factors in motivating families to evacuate and remain away from their homes (Janis, 1951), most studies were reports of the relative success of failure of various evacuation efforts and not specific attempts to integrate findings or develop principles for understanding the process of evacuation.

Although several studies conducted in the past decade have focused upon problems of evacuation (Drabek and Boggs, 1968; Drabek, 1969; Drabek and Stephenson, 1971; Mileti and Beck, 1975), there remains a paucity of analytic models which identify important variables and specify patterns of expected relationships. While research on evacuation in natural hazards has largely examined preimpact attempts to remove threatened families, the literature is fairly small and widely scattered. Furthermore, some of the most comprehensive research was conducted as long ago as the middle 1950's (cf. Lammers, 1955, on the Holland floods). In this section, a theoretical context for evacuation will be developed as an initial step toward laying the basis for a framework describing the determinants of evacuation of natural hazards. The focus here is upon describing factors in *voluntary evacuation* which takes place prior to disaster impact.

Historically, students of natural hazards have treated evacuation as one possible protective measure which may be taken in response to a hazard warning message. Hence in the literature of disaster research, the study of evacuation is usually subsumed under the general rubric of warning systems and individuals' adaptive or protective responses. Most of these studies have used a social systems' view of warning processes (Chapman, 1962; Janis, 1962; Williams, 1964; McLuckie, 1970; Mileti, 1974). Such models typically emphasize macroprocesses operating on an aggregate level; concern is with detection–dissemination subsystems, response subsystems and the like. For example, Harry B. Williams (1964) basic warning response model is composed of parameters which are structural in nature. This essentially four-phase model represents the warning process in terms of:

1. Sources of information about the nature of the threat;
2. The official decision to issue a warning;
3. The channels through which the warning is communicated to the public; and
4. The response of the public.

Systems models have proved quite useful for research in which the primary unit of analytic interest is the community (cf. Dynes et al., 1972). Taken alone however, system models of community warning processes are less useful in representing factors that shape individual or family behavior.

Consequently, investigators interested in the individual as a primary unit of analysis have usually supplemented the social systems' framework with some social-psychological model. Such models have included psychoanalytic and psychodynamic framework (Tyhurst, 1951; Menninger, 1952; Leopold and Dillon, 1963), reference group or role theory approaches (Anderson, 1968; Form and Loomis, 1956), symbolic interactionism (Drabek and Boggs, 1968; Drabek, 1969), and behavioral principles (Mileti, 1974:35–46). The theoretical problem which must be given careful consideration when choosing a supplementary framework centers upon the difficulty some models have in allowing for the fact that processes important in warning response decisions proceed simultaneously at two levels of abstraction. For example, we know that the individual's assessment of personal risk and the reality of the disaster threat are important factors in his decision to evacuate. But it is also known that two additional factors related to community preparedness, the content of the warning message and number of times it was received, contribute to assessments of risks and the formation of warning belief. Thus, we must be concerned *both* with aspects of the individual and attributes of the community system.

Psychoanalytic theories are not easily adapted to the consideration of factors *external* to the individual in explaining warning response (cf. Lifton and Olsen, 1976) and Mileti (1974:163–166) reports empirical difficulties in adapting behavioral principles to studies of warning responses. Descriptive research using a symbolic interactionist perspective appears promising, but as yet no systematic attempt has been made to explicitly formulate an analytic model based upon this approach. Recently some attention has been devoted to integrating one or more of the social-psychological perspectives with a systems model, thereby producing a flexible analytic paradigm which either implicitly or explicitly acknowledges the operation of processes at several levels of abstraction (cf. Drabek, 1969:336–337; Drabek and Stephenson, 1971:199–202; Gillespie and Perry, 1976:303–305; Perry et al., 1974:115–119).

Gillespie and Perry (1976), in an attempt to refine Barton's (1970) systems model of disaster behavior, have argued that by integrating the premises of an emergent norm approach to collective action (cf. Turner, 1964:389–392; Turner and Killian, 1972:21–25) with a systems perspective, one gains the flexibility necessary to adequately characterize processes and interrelationships which prevail at both individual and community levels of abstraction. The adoption of an integrated systems-emergent norm approach also provides a framework for the *temporal ordering of factors* in personal reaction to warnings and helps to isolate important theoretical dimensions.

The emergent norm approach has been developed primarily by Ralph Turner and Lewis Killian (1972) as an alternative to contagion and convergent perspectives on collective behavior. Emergent norm perspectives focus upon the development of situational norms and expectations which arise as a function of some crisis or change in the social or physical environment which renders traditional norms inappropriate (cf. Fritz, 1957:8, Form and Nosow, 1958:14–28; Barton, 1963:20–22; Anderson, 1969:92–93; Anderson, 1968:298–307). Drabek (1968:143–144) has succinctly summarized the emergent norm orientation to disaster behavior:

> Societies are composed of individuals interacting in accordance with an immense multitude of norms, i.e., ideas about how individuals *ought* to behave . . . Our position is that activities of individuals . . . are guided by a normative structure in disaster just as in any other situation . . . In disaster, these actions . . . are largely governed by *emergent* rather than established norms, but norms nevertheless.

Thus, human behavior in disaster can be conceptualized as nontraditional behavior in response to a changing or changed social and physical environment. The emergent norm perspective is concerned with the *process* beginning with changes in the stimulus environment which requires a new (or different) "definition of the situation" and ending with some change in the individual's behavior which is responsive to this different definition of the situation. With respect to the study of warning response, warning itself serves as the event which signals that a pending change in the environment could render traditional (established) norms inoperative. Thus, a redefinition of the situation is required, and situational norms are developed to cope with the changed social and physical environment. The human response of interest is the act of evacuation—and the cluster of situational norms which encourage individuals to undertake that act or some other appropriate personal protective measure.

The adoption of an emergent norm perspective highlights important processes typically ignored when systems theory is used alone:

namely, those processes that operate *between* the issuance of a call to evacuate and the public response. Furthermore, a careful review of existing theoretical and empirical literature permits the isolation of specific processes that occur as part of post-warning redefinition of this situation. From the standpoint of an emergent norm approach, whether or not the desired response (e.g., evacuation) occurs is dependent upon the outcome of this process of redefining the situation. The major processes involved in post-warning attempts to restructure the normative environment are illustrated in Fig. 1. First, there is an initial milling process which focuses upon *confirming* the warning message, gathering further information and establishing a warning belief. Assuming that a warning belief is established, further milling centers upon the problem of assessing personal risk—the determination of one's *proximity* to the impact area and the individual's perception of the *certainty*, and probable *severity* of disaster impact. Finally, if personal risk appears high, necessitating some adaptive response, individuals assess the logistics of making such a response. Only *after* these processes have operated are we able to suggest that any protective response (evacuation or otherwise) will be undertaken.

It should be emphasized here that concern is with the application of certain tenets of emergent norm thinking to the problem of warning response in disaster. It is *not* the goal of this paper to elaborate an emergent norm "theory" of human behavior and derive from it propositions which explain evacuation decision-making. While such an undertaking would be commendable and in keeping with the highest standards of the philosophy of social science (cf. Homans, 1967; Blalock, 1969; Wallace, 1971), it is both beyond the scope of the present paper and unnecessary for the task at hand. Indeed, to utilize emergent norm imagery (or any other theoretical perspective for that matter) to structure a problem which is subsequently addressed in terms of the existing empirical literature certainly does not depend upon a formalized elab-

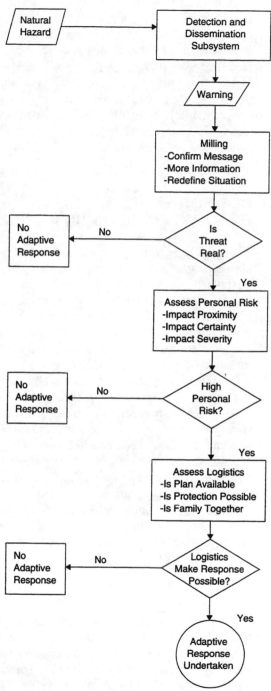

**FIGURE 1.** Flow diagram of systems-emergent norm issues in individual response to natural hazard warning (arrows represent temporal paths).

oration of the theory. Finally, it is acknowledged that, as an approach to the study of collective behavior, emergent norm theory (like any other theory) has strong and weak points (see Tierney, 1977 for a thoughtful critique). Note, however, that emergent norm theory does unequivocally provide what is sought for this paper; namely, a set of ordering assumptions upon which a general framework for viewing evacuation behavior may be based (cf. Weller and Quarantelli, 1973:674).

## DETERMINANTS OF EVACUATION BEHAVIOR

The emergent norm perspective, as outlined above, suggests three variables of major importance in evacuation: (1) the presence of an *adaptive plan*, (2) the individual's *definition of threat as real* (i.e., the development of a warning belief) and (3) the level of *perceived risk*. In addition to these variables which are included on theoretical grounds, there is considerable consensus in the empirical literature regarding the importance of three more variables: the *family context* in which the warning is received, the patterns of *kin relationships* in which the family is enmeshed and the level of *community involvement*. Using these six variables as a starting point, the following sections review pertinent research findings to identify other variables which are antecedent to these and assess the interrelationships among all of the variables, as well as the impact of each upon the target response—evacuation.

### Adaptive Plan

Studies of evacuation often overlook the fact that in order to effectively clear an area, residents must either have prior knowledge of some standing evacuation plan or be informed of such a plan by officials. Hamilton et al. (1955:120) interviewed a disaster victim who reported:

> We couldn't decide where to go or what to do. So we grabbed our children and stuff and were

just starting to move outside. Where if it had just been ourselves, we might have taken out. But we didn't want to risk it with the children.

This family received a warning to evacuate, but had no plan which identified safe routes for exit or appropriate destinations. Thus, even if one wants to evacuate, the absence of a plan for so doing is sufficient to hinder any adaptive response.

The problems of families *not* evacuating (or evacuating to an even more dangerous location) when evacuation routes and destinations are not well known or well published has been widely documented with a variety of types of natural hazard (cf. Windham et al., 1977:15). Although not universally implemented, most manuals for community predisaster preparation encourage the planning and dissemination of evacuation routes (cf. Leonard, 1973; Healy, 1969). Therefore, the possession of adaptive plan, at a very minimum, a route and safe destination, is positively related to the probability of evacuation.

Research indicates that individuals' knowledge of an adaptive plan depends largely upon two factors: warning content and prior experience. Williams (1964:91–93) has pointed out that the most effective warning messages are clear, specific, consistent and, if possible, contain instructions for an appropriate protective response. The warning message itself, whether delivered via official channels or through kin and friendship networks, is one source of an adaptive plan. It has also been reported that the *experience* of warning and/or evacuation provides the individual with a potential adaptive plan for future hazards, simply by replicating his past behavior (Bernert and Ikle, 1952:133–135).

### Perception of Threat as Real

For *any* adaptive response to be defined as necessary, the individual must perceive the threat described in the hazard warning to be real. Hence, unless the warning is believed to be valid, no action (protective measure) is likely to be undertaken (cf. Janis, 1962:59–66; Williams, 1964:94; Anderson, 1968:299–304; Mileti, 1975:21–23; Janis and Mann, 1977). The importance of establishing a situational definition of real threat is underscored by the existence of a large empirical literature detailing the ways in which people go about *confirming* hazard warnings (Drabek, 1969:344; Drabek and Stephenson, 1971:195). In general, the empirical literature indicates that the greater the perceived threat the greater the probability of evacuation.

Four factors important in producing a warning belief or enhancing the individual's perception of threat may be derived from existing research findings: warning content, prior experience, the number of warnings received and the warning source. Warning content and prior experience with disaster are important in that each affords information to the individual upon which an assessment of threat is made (Lachman et al., 1961:1409). A warning that contains precise information about the hazard and its probable consequences is more likely to create the "reflexive fear" which contributes to the individual's belief that the warning is to be taken seriously (Janis, 1962:59–66; Levanthal et al., 1967:314). Anderson (1969:100) has also pointed out that individuals who have recently experienced a natural hazard are more sensitive to warnings and are more likely to attempt some adaptive response. Indeed, a study of people who left their homes in response to what turned out to be a false warning reports that "few of the evacuees complained about being misled by the false alarm: the vast majority said that they would evacuate again under the same circumstances" (Janis, 1962:85). Assuming that a warning is confirmed, it can be argued that previous experience with disaster (even with false alarms) enhances warning belief and the probability of an adaptive response. A recent study of evacuation in response to hurricanes on the Gulf coast would seem to contradict this contention by reporting that a large proportion of people who failed to evacuate

were long-time residents of the area who presumably had previously experienced the problem (Windham et al., 1977). Two qualifiers must be considered in the interpretation of this finding. First, the data are *not* unequivocal; two communities were studied and in one of them, the proportion of people with prior experience is nearly the same among both "stayers" and "leavers." Secondly, hurricanes on the Gulf coast are a recurrent disaster threat, with possibly *several impacts* in a single season. Although empirical studies are not available, it is possible to speculate that under such conditions prior experience becomes a "constant" with *no* relationship to evacuation. This is another possible interpretation of the Windham et al. (1977:29) study. A more cautious interpretation of the data, however, would suggest that recent experience with a valid warning contributes positively to the development of a warning belief (i.e., perception of real threat).

There is evidence that the number of warnings an individual receives increases the chance of an adaptive response. Drabek (1969:340–341) has indicated that a frequent response to an initial warning is disbelief; although subsequent warnings may provide no new information, it is argued that repetition increases warning belief (Mileti and Beck, 1976). Numerous studies have shown that individuals rarely evacuate after hearing a single warning and that additional warnings enhance the probability of evacuation (Mileti, 1974; Fritz and Marks, 1954:41; Fritz and Mathewson, 1957:51–53).

Finally, research indicates that the more credible the source from which one receives a warning, the more likely one is to believe that the threat is real. Drabek and Boggs (1968:445), in a study of flood warning response, report that "families were warned through three distinct processes: (1) authorities, (2) peers, (3) mass media . . . those warned by an authority were less likely to be skeptable of the warning." The warning confirmation literature generally supports this contention, suggesting that people more often seek other sources for confirmation when they receive warnings from peers or media than when the source is an authority or kinsman.

## Personal Risk

In assessing personal risk, concern centers upon the individual's perception of the probability that impact of the hazard agent will result in great damage or destruction to his person or property (cf. Fritz and Marks, 1954:29–31; Diggory, 1956). Withey (1962:104–107), in developing a series of stages of reaction to threat, emphasizes the importance of the individual's perception of "the probability of the impending event occurring [and] the severity, to the individual, of such a development." Put slightly differently, personal risk may be thought of in terms of individual's conception of his proximity to impact and the certainty and severity of the impact. Studies have shown that the perception of personal risk has a direct affect on the nature of the individual's response to a warning (Menninger, 1952:129; Williams and Fritz, 1957:46–50; Withey, 1964:86).

Glass (1970:64–67) in discussing "contingency responses" to threat points out that unless a person is convinced that impact is certain and that he is within the danger area, there is general reluctance to cooperate with emergency plans. Menninger (1952), in reporting evacuation problems during the summer of 1951's Kansas floods found that "an amazing number of people refused to believe that the flood would hit them . . . [and] . . . would not move themselves or their belongings out of their houses." Similar findings have been reported by Tyhurst (1951:764, and Danzig et al., 1958:51–53). Thus, it can be argued that a direct relationship exists between personal risk and the probability of evacuation.

Warning content and prior experience have been shown to be closely related to the individuals' level of perceived personal risk. It is fairly routine that warning messages of an impending natural hazard contain information on where, when, and the probable force of disaster impact (Mogil and Groper, 1977; Moore et al., 1963:31–33; Williams, 1957; 15–19). Furthermore, the individual's prior experience, as well as his reading of environmental cues, contributes to the perception of personal risk (Drabek and Boggs, 1968:445–447; Windham et al., 1977:49).

## Family Context

As Killian (1952) has indicated, the study of human behavior in disaster must take into account the network of community and family roles in which the individual is immersed. Furthermore, research to date suggests that in the event of conflicting responsibilities among various roles, "the majority of persons involved in such dilemmas resolve them in favor of loyalty to the family . . ." (Killian, 1952:311). In particular, we know that families faced with disaster seek to protect members (Quarantelli, 1960) and generally perform as units when undertaking any protective behavior. Studies of evacuation have revealed that "when they did evacuate, families left as units . . . these data provide additional support for the hypothesis that families move as units and remain together, even at the cost of overriding dissenting opinions" (Drabek and Boggs, 1968:446). Support for the contention that families evacuate as a group may also be drawn from studies of the bombing of London during World War II (Titmuss, 1950:172; Bernert and Ikle, 1952:133–135). The primary consequence of these findings is to introduce an additional constraint on evacuation: unless the family is together or missing members are safety accounted for, evacuation will not occur. Thus, family context is positively related to the probability of evacuation.

Ethnicity and age indirectly effect family context. Family structure as well as kin relations, vary among ethnic groups. Staples (1976:123), for example, reports that ". . . the Black kinship network is more extensive and cohesive than kinship bonds among the White population . . . a larger proportion of Black families take relatives into their households." As indicated above, the structure of the household is important in defining family context. Age of the head of household is important for family context in two ways. In general, age correlates with the life-cycle position of the family, and especially among minorities, with the generational depth of the extended family household (Lansing and Kish, 1957:512). Furthermore, the presence of aged persons in a household (or a household composed of aged people), has been cited by disaster researchers as a factor which complicates family evacuation by increasing the complexity of family role responsibilities. Thus, Ellemers (1955:421) points out that "old and sick people were unable to leave their homes at the time of the flood warning." Hill and Hansen (1962:186) have also observed that the "extended family household is poorly organized to meet threats and hardships, for its very young and very old members are often ill-equipped to meet such sudden challenges." In terms of our knowledge of family tendencies to adapt to disaster as a unit, it becomes clear that such "deficiencies" of individual family members reflect upon the performance of the group.

## Kin Relationships

Kin relationships are here conceived in terms of people's interaction and exchange patterns with their kinsmen. Several studies of communities in disaster have pointed out that very close kin relationships promote post-disaster recovery success among victims (Drabek et al., 1975:486; Bolin, 1976:268; Drabek and Key, 1976:90). Although less often studied, kin relationships also play an important role in the warning dissemination process and, consequently, in the promotion of successful adaptation to disaster warning (Clifford, 1956:113–124). In particular, one's interaction patterns with kin have an impact upon the content, source and number of warnings an individual receives.

Drabek and Stephenson (1971:199) report that "extended family relationships were crucial as warning messages and confirmation sources . . . telephone conversation with relatives during the warning period were usually a key factor." Official warning messages broadcast via mass media are sometimes vague, often not heard by all the potential victims, and are usually confirmed via some other source (Mileti and Beck, 1975:30; Drabek, 1969:341; Bates et al., 1963:11–13). Thus, kinsmen can supply both additional information (i.e., warning content), and serve as confirmation sources for hazard warning. It has also been found that people who hear disaster warnings relay the information to kinsmen who reside within the probable impact area (Drabek and Boggs, 1968:445–447). This has the immediate effect of *increasing* the number of warning messages received by potential victims. Furthermore, during such contacts help from kin which may promote a successful adaptive response to the warning may be offered or solicited. In studying flood evacuations, Drabek (1969:345) found many victims who evacuated in response to a relative's invitation to "come over and spend the evening just in case this thing might be serious."

Studies have shown that the nature and frequency of relationships with primary kin are very much affected by ethnicity, and age (cf. Litwak and Szelenyi, 1969:465; Sussman and Burchinal, 1968; 1962). In general, Anglo-American elderly are relatively more socially isolated (Bennett, 1973:179), and exhibit greater variation in income and wealth accumulation than minority elderly (Terrell, 1971:363). Aged Blacks, in contrast, tend to be more uniformly poor but are also more actively involved in kin network (cf. Jackson, 1971; Kent, 1971; Babchuk and Ballweg, 1971). Jackson (1972:271) in a study of southern Black grandparents concludes that "these findings help to debunk the myth of . . . the disintegrating or ephemeral kinship ties between aged and aging Blacks . . . they indicate that many Black grandparents serve as a point of anchorage for grandchildren and provide kinds of support for them unavailable from their own parents." Perry and Perry (1959:45–49) have also commented on the greater cohesion among relatives in Black as opposed to White communities facing disaster. Elderly American Indians, perhaps even more than Blacks, tend to fall into low income brackets and be deeply involved with the extended family (Taylor and Peach, 1974:154).

Finally, it should be pointed out that age and ethnicity have a direct impact on warning response, in addition to the indirect relationship through kinship described above. McLuckie (1970:38) indicates that "different classes or ethnic groups have varying conceptions of what constitutes adaptation to a threat, or credibility of community organizations which might be involved in issuing warning messages." Hence, what might be perceived as appropriate protective behavior by Blacks or Indians may not at all be related to the adaptive behavior promoted by some official agency in its warning message. It is also known that aged people in general respond less enthusiastically to disaster warning and calls for evacuation (Friedsam, 1962:155–157). Studies of English evacuation during World War II indicate that even as the bombs were falling some aged citizens still refused to be evacuated (Titmuss, 1950:451; Ikle, 1951:135).

## Community Involvement

Community involvement refers to the individual's patterns of interaction with friends (neighbors), and his participation in voluntary associations and other community organizations. Barton (1970:63–124) has pointed out that the extent of people's integration into the community affects the content, source, and a number of warnings received in much the same way as kin relationships. The greater one's social contacts, the more likely one is to receive more information regarding a potential hazard. It is generally agreed that when both kin and community contacts are available, kin relationships are more important in evacuation decision making (Drabek and Boggs, 1968). The reason for including community involvement, however, is that when kin bonds are weak or absent, ties to the community can serve a similar function as far as a model of evacuation behavior is concerned.

As with kin relationships, community involvement varies with age and ethnicity. Blacks, in particular, are cited as relatively more involved in voluntary and formal organizations than Whites or other minority groups (Babchuck and Thompson, 1962; Oram, 1966; Renzi, 1968).Tomeh (1973:99) points out that, with respect to membership in voluntary associations, "studies . . . show higher participation rates for Blacks at all social class levels, especially lower class." Previous research which did not control for social class erroneously reported Black participation rates lower than White rates (Wright and Hyman, 1966:32).

Although it is generally argued that social isolation—shrinking friendship networks and decreased affiliations with organizations (Watson and Maxwell, 1977:59–66)—characterizes most aged people, it has recently been acknowledged that the variance along this dimension is greater than previously believed" (Cottrell, 1974:49–57). Broadly defined, however, community involvement tends to decline with increasing age of the head of household (Harry, 1970).

## SUMMARY OF THE DETERMINANTS OF EVACUATION

Based upon the literature review presented above, one can begin to sort out and order research findings, and summarize the data which bear upon the important dimensions derived from a systems-emergent norm framework for warning response. The most important function of this review is that it provides information on empirical findings which allow one to construct (inductively) images of causal order and to infer possible relationships among variables. As Blalock (1969:8) points out such inductive reasoning combined with information deduced from theoretical frameworks, represents the initial step in the development of causal theory.

Based upon information from the review of the empirical literature, one can define a series of hypotheses which identify factors important in evacuation and specify the interrelationships among these factors. The identification of factors (variables) to be interrelated is based upon theoretical considerations—that is, important issues derived from an emergent norm-system perspective—and existing empirical evidence—that is, research-based data regarding the relationship between each factor and the target adaptive response of evacuation. Each numbered statement characterizes the hypothesized relationship of a major variable to evacuation; sub-statements specify the relationship between a major variable and variables antecedent to it.

1. The more precise the individual's adaptive plan, the higher the probability of evacuation.
   a. Recent prior experience with disaster impact increases the chance that an individual will have a precise adaptive plan.
   b. The more detailed the warning message, the more likely it is to provide an adaptive plan.

2. The greater the individual's perception of real threat (warning belief), the greater the probability of evacuation.
   a. The more recent the individual's prior experience with disaster, the more likely he is to perceive the threat as real.
   b. The more precise the warning message the greater the chance that the individual will perceive the threat as real.
   c. As the number of warnings received increases, so does the degree to which the threat is perceived as real.
   d. Receipt of a warning from a credible source (or confirmation by a credible source) increases the degree to which the threat is perceived as real.

3. The higher the level of perceived personal risk, the greater the probability of evacuation.
   a. Recent prior experience with disaster is positively related to the level of perceived personal risk.
   b. To the extent that the warning message specifies the location of impact, level of perceived personal risk for persons within the impact area will increase.
   c. To the extent that the warning message specifies that disaster impact will be severe, persons within the projected impact area will experience increased levels of perceived risk.

4. To the extent that family (household) members are together or accounted for, the probability of evacuation is increased.

5. The closer one's relationship to extended kinsmen, the more likely one is to evacuate.
   a. The closer one's relationship to extended kinsmen, the more likely one is to receive additional warning information through these contacts.
   b. The closer one's relationship to extended kinsmen, the greater the number of potential credible sources for warning information.
   c. The closer one's relationship to extended kinsmen, the more warnings the individual will receive from these contacts.

6. The greater one's participation in the community, the more likely he is to evacuate.
   a. The greater one's participation in the community, the more likely one is to receive additional warning information from these contacts.
   b. The greater one's participation in the community, the greater the number of potential credible sources for warning information.
   c. The greater one's participation in the community, the more warnings the individual will receive from these contacts.

7. Families headed by aged persons, or extended family households containing aged, are less likely to evacuate in response to hazard warnings.
   a. As age increases, the frequency of contacts with kinsmen decreases.
   b. As age increases, the level of community participation decreases.
   c. In general, with other factors constant, as age increases, the number of warnings received decreases.

8. Cultural factors (race/ethnicity) influence the extent to which a family is likely to evacuate.
   a. Cultural factors (race/ethnicity) play an important role in defining appropriate patterns of interaction with kinsmen.

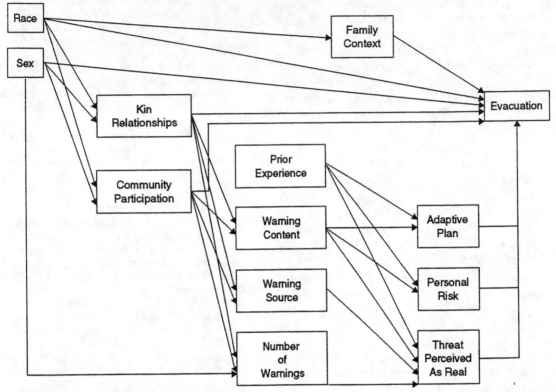

**FIGURE 2.** Diagram of variables important in individual decisions to evacuate (arrows represent relationships, either direct or indirect).

b. Cultural factors (race/ethnicity) play an important role in defining the ways in which the family relates to the community.

c. Cultural factors (race/ethnicity) play an important role in definition of family role responsibilities.

The above statements summarize empirical findings from studies of individual's response to warnings as they related to the problem of evacuation. The assumed causal order implied in the propositions reflects four basic theoretical assumptions resulting from the adoption of an emergent norm perspective. First, it is assumed that the warning itself serves as a stimulus to individuals which requires (or suggests) that an environmental change is pending and some restructuring of the situation is necessary. Second, it is acknowledged

that social restructuring takes the form of several "milling processes" wherein assessments are made of the extent to which the threat is real and the level of personal risk involved. Third, this restructuring involves the collection of information which may be drawn from the warning message, the source of the warning, and one's prior experiences under similar circumstances as well as kin or community contacts. Finally, it is acknowledged that certain characteristics of the individual (e.g., family role responsibilities, kin relationships, community development, age, race/ethnicity) must be taken into account as factors which affect either directly or indirectly the outcome of the posited milling process (this outcome being evacuation).

The relationships specified above are

graphed in Fig. 2. This conceptual framework represents a sketch of the interrelationships among factors believed to be important in individual decisions to evacuate in the face of natural disaster. The value of this framework lies in the fact that it is an explicit attempt to integrate existing research findings along explicit theoretical lines. As such, the hypotheses highlight important variables, conditions and relationships which should be given careful scrutiny when addressing the problem of understanding people's decision-making processes relative to evacuation warnings.

*Acknowledgments* This study was supported by the National Science Foundation, Grant number ENV 77-23697. The opinions expressed are those of the author and do not necessarily represent the National Science Foundation. The author wishes to thank William A. Anderson, Thomas Drabek, Michael K. Lindell, and Michael T. Wood for helpful comments on earlier versions of this paper.

## REFERENCES

Anderson, Jon (1968). Cultural adaptation to threatened disaster. *Human Organization* 27 (Winter): 298–307.

Anderson, William A. (1969). Disaster warning and communication in two communities. *Journal of Communication* 19 (June): 92–104.

Babchuk, Nicholas and Ballweg, John (1971). Primary extended kin relations of Negro couples. *Sociological Quarterly* 12 (Winter): 69–77.

Babchuk, Nicholas and Thomson, Ralph (1962). The voluntary association of Negroes. *American Sociological Review* 27 (October): 647–55.

Baker, George and Chapman, Dwight (1962). *Man and Society in Disaster*. New York: Basic Books.

Barton, Allen (1963). *Social Organization Under Stress*. Washington, DC: National Academy of Sciences—National Research Council.

Barton, Allen (1970). *Communities in Disaster*. New York: Anchor Books.

Bates, Frederick L. et al. (1963). *The Social and Psychological Consequences of Natural Disaster*. Washington, DC: National Academy of Sciences.

Bennett, Ruth (1973). Living conditions and everyday needs of the elderly with particular reference to social isolation. *International Journal of Aging and Human Development* 4 (Summer): 179–198.

Bernert, E.H. and Ikle, F.C. (1952). Evacuation and the cohesion of urban groups. *American Journal of Sociology* 58 (September): 133–38.

Blalock, Hubert M. (1969). *Theory Construction: From Verbal to Mathematical Models*. Englewood Cliffs, NJ: Prentice-Hall.

Bolin, Robert (1976). Family recovery from natural disaster. *Mass Emergencies* 1(4): 267–77.

Clifford, Roy A. (1956). *The Rio Grande Flood*. Washington, DC: National Academy of Sciences—National Research Council.

Cottrell, Fred (1974). *Aging and the Aged*. Dubuque: William C. Brown Company.

Danzig, Elliott, Thayer, Paul and Galanter, L.R. (1958). *The Effects of a Threatening Rumor on a Disaster-Stricken Community*. Washington, DC: National Academy of Sciences—National Research Council.

Diggory, James (1956). Some consequences of proximity to a disaster threat. *Sociometry* 19 (March): 47–53.

Drabek, Thomas (1968). *Disaster in Isle 13*. Columbus, OH: Disaster Research Center, Ohio State University.

Drabek, Thomas (1969). Social processes in disaster: Family evacuation. *Social Problems* 16 (Winter): 336–47.

Drabek, Thomas and Boggs, Keith (1968). Families in disaster: Reactions and relatives. *Journal of Marriage and the Family* 30 (August): 443–51.

Drabek, Thomas and Key, W. (1976). The impact of disaster on primary group linkages. *Mass Emergencies* 1 (2): 89–105.

Drabek, Thomas, Key, W., Erickson, P. and Crowe, J. (1975). The impact of disaster on kin relationships. *Journal of Marriage and the Family* 37 (August): 481–94.

Drabek, Thomas and Stephenson, John (1971). When disaster strikes. *Journal of Applied Social Psychology* (2): 187–203.

Dynes, Russel, Quarantelli, Enrico and Kreps, Gary (1972). *A Perspective on Disaster Planning*. Columbus, OH: Disaster Research Center, Ohio State University.

Ellemers, J.E. (1955). General conclusions. In: Instituut voor Sociaal Onderzoek van het Nederlandse Volk, *Studies in Holland Flood Disaster*, Washington, DC: National Academy of Science—National Research Council.

Form, William and Loomis, Charles (1956). The persistence and emergence of social and cultural systems in disaster. *American Sociological Review* 21 (April): 180–85.

Form, William and Nosow, Sigmund (1958). *Community in Disaster*, New York: Harper and Brown Brothers.

Friedsam, H.J. (1962). Older persons in disaster, in Baker and Chapman (eds.) *Man and Society in Disaster*. New York: Basic Books.

Fritz, Charles (1957). Disaster. In R. Merton and R.

Nisbet, *Contemporary Social Problems*. New York: Harcourt.

Fritz, Charles and Marks, Eli (1954). The NORC studies of human behavior in disaster. *Journal of Social Issues* 10: 26–41.

Fritz, Charles and Mathewson, J.H. (1957). *Convergence Behavior in Disaster: A Problem in Social Control*. Washington, DC: National Academy of Sciences—National Research Council.

Gillespie, David and Perry, Ronald W. (1976). An integrated systems and emergent norm approach to mass emergencies. *Mass Emergencies* 1 (4): 303–12.

Glass, Albert (1970). The psychological aspects of emergency situations, in Abram (ed.) *Psychological Aspects of Stress*. Springfield, IL: Charles C. Thomas.

Hamilton, R., Taylor, R.M. and Rice, G. (1955). *A Social Psychological Interpretation of the Udall, Kansas Tornado*. University of Wichita Press, Kansas.

Harry, Joseph (1970). Family localism and social participation. *American Journal of Sociology* 75 (March): 821–27.

Healy, Richard J. (1969). *Emergency and Disaster Planning*. John Wiley and Sons, New York.

Hill, Rueben and Hansen, Donald (1962). Families in disaster, in Baker and Chapman (eds.) *Man and Society in Disaster*. New York: Basic Books.

Homans, George (1967). *The Nature of Social Science*. Englewood Cliffs, NJ: Prentice-Hall.

Ikle, Fred C. (1951). The effect of war destruction upon the ecology of cities. *Social Forces* 29 (May): 383–91.

Idle, Fred and Kincaid, Harry (1956). *Social Aspects of Wartime Evacuation of American Cities*. Washington, DC: National Academy of Sciences—National Research Council.

Jackson, J.J. (1971). Negro aged: Toward needed research in social gerontology. *The Gerontologist* 11 (Spring): 52–57.

Jackson, J.J. (1972). Aged Blacks: A potpourri in the direction of the reduction of inequalities. *Phylon* 43 (Winter): 260–80.

Janis, Irving L. (1951). *Air War and Emotional Stress: Psychological Studies of Bombing and Civilian Defense*. New York: McGraw-Hill.

Janis, Irving L. (1962). Psychological effects of warnings, in Baker and Chapman (eds.) *Man and Society in Disaster*. New York: Basic Books.

Janis, Irving and Mann, Leon (1977). Emergency decision making. *Journal of Human Stress* 3 (June): 35–45.

Kent, Donald (1971). The Negro aged. *The Gerontologist* 11 (Spring): 48–51.

Killian, Lewis (1952). The significance of multiple-group membership in disaster. *American Journal of Sociology* 57 (January): 309–14.

Lachman, Roy, Tatsuoka, Maurice and Bonk, William (1961). Human behavior during the Tsunami of May 1960. *Science* 133 (May 5): 1405–09.

Lammers, C.J. (1955). Survey of evacuation problems. In: Instituut voor Sociaal Onderzoek van het Nederlandse Volk, *Studies in Holland Flood Disaster*. Washington, DC: National Academy of Science—National Research Council.

Lansing, John and Kish, Leslie (1957). Family life cycle as an independent variable. *American Sociological Review* 22 (August): 512–19.

Leonard, V.A. (1973). *Policy Pre-Disaster Preparation*. Springfield, IL: Charles C. Thomas.

Leopold, R.L. and Dillon, Harold (1963). Psycho-Anatomy of Disaster. *American Journal of Psychiatry* 119 (April): 913–21.

Leventhal, Howard, Watts, J.C. and Pagano, F. (1967). Effects of fear and instructions on how to cope with danger. *Journal of Personality and Social Psychology* 6 (3): 313–21.

Lifton, Robert and Olson, Eric (1976). The human meaning of total disaster. *Psychiatry* 39 (February): 1–18.

Litwak, Eugene and Szelenyi, Ivan (1969). Primary group structures and their functions: Kin, neighbors and friends. *American Sociological Review* 34 (August): 465–81.

McLuckie, Benjamin (1970). *The Warning System in Natural Disaster Situations: A Selective Analysis*. Columbus, OH: Disaster Research Center, Ohio State University.

Menninger, W.C. (1952). Psychological reactions in an emergency. *American Journal of Psychiatry* 109 (August): 128–30.

Mileti, Dennis S. (1974). *A normative causal model analysis of disaster warning response*. Boulder: University of Colorado. Doctoral dissertation.

Mileti, Dennis S. (1975). *Natural Hazard Warning Systems in the United States: A Research Assessment*. Boulder: Institute for Behavioral Science, University of Colorado.

Mileti, Dennis and Beck, E.M. (1976). Communication in crisis. *Communication Research* 2: 24–49.

Mogil, H. Michael and Groper, Herbert S. (1977). NWS's severe local storm warning and disaster preparedness programs. *Bulletin of the American Meteorological Society* 58 (4): 318–19.

Moore, Harry et al. (1963). *Before the Wind: A Study of the Response to Hurricane Carla*. Washington, DC: National Academy of Sciences—National Research Council.

Oram, Anthony (1966). A reappraisal of the social and political participation of Negros. *American Journal of Sociology* 72 (July): 32–46.

Perry, H.S. and Perry, Stewart (1959). *The Schoolhouse Disasters: Family and Community as Determinants of the Child's Response to Disaster*. Washington, DC: National Academy of Sciences—National Research Council.

Perry, Ronald, Gillespie, David and Mileti, Dennis (1974). System stress and the persistence of emergent organizations. *Sociological Inquiry* 42 (December): 113–21.

Quarantelli, E.L. (1960). A note on the protective function of the family in disaster. *Marriage and Family Living* 22 (August): 263–64.

Renzi, Marrio (1968). Negroes and voluntary associations. *Research Reports in the Social Sciences* 2 (Spring): 63–71.

Staples, Robert (1976). *Introduction to Black Sociology.* New York: McGraw-Hill.

Taylor, Benjamin and Peach, Nelson (1974). Social and economic characteristics of elderly Indians in Phoenix, Arizona. *Journal of Economics and Business* 26 (2): 151–55.

Terrell, H.S. (1971). Wealth accumulation of Black and White families. *Journal of Finance* 26: 363–79.

Tierney, Kathleen (1977). Emergent norm theory as theory. Paper read at the meetings of the North Central Sociological Association.

Titmuss, Richard (1950). *Problems of Social Policy.* London: HM Stationery Office.

Tomeh, Aida K. (1973). Formal voluntary organizations: Participation, correlates and inter-relations. *Sociological Inquiry* 43 (3): 89–122.

Turner, Ralph (1964). Collective behavior, in Faris (ed.) *Handbook of Modern Sociology.* Chicago, IL: Rand McNally.

Turner, Ralph and Killian, Lewis (1972). *Collective Behavior.* Englewood Cliffs, NJ: Prentice-Hall.

Tyhurst, J.S. (1951). Individual reactions to community disaster. *American Journal of Psychiatry* 107 (April): 764–69.

U.S. Strategic Bombing Survey (1947a). *The Effects of Strategic Bombing on German Morale.* Vols. I and II. Washington, DC: U.S. Government Printing Office.

U.S. Strategic Bombing Survey (1947b). *The Effects of Strategic Bombing on Japanese Morale.* Vols. I and II. Washington, DC: U.S. Government Printing Office.

Wallace, Walter (1971). *The Logic of Science in Sociology.* Chicago: Aldine.

Watson, Wilbur and Maxwell, Robert (1977). *Human Aging and Dying: A Study in Social Gerontology.* New York: St. Martin's Press.

Weller, Jack and Quarantelli, E.L. (1973). Neglected characteristics of collective behavior. *American Journal of Sociology* 79: 665–85.

Williams, Harry B. (1957). Some functions of communication in crisis behavior. *Human Organization* 16 (Summer): 15–19.

Williams, Harry B. (1964). Human factors in warning and response systems, in Grosser (ed.) *The Threat of Impending Disaster.* Cambridge, MA: MIT Press.

Williams, Harry B. and Fritz, Charles (1957). The human being in disaster: A research perspective. *The Annals* 309 (January): 42–51.

Windham, G.O. et al. (1977). *Reactions to Storm Threat During Hurricane Eloise.* Mississippi State University, State College.

Withey, Stephen (1962). Reaction to Uncertain Threat, in Baker and Chapman (eds.) *Man and Society in Disaster.* New York: Basic Books.

Withey, Stephen (1964). Sequential accommodations to threat, in Grosser (ed.) *The Threat of Impending Disaster.* Cambridge: MIT Press.

Wright, Charles R. and Hyman, Herbert (1966). Who belongs to voluntary associations, in Glaser, William and Sills, David (eds.) *The Government of Associations.* Totowa, NJ: Bedminister Press.

# EVACUATION BEHAVIOR IN RESPONSE TO NUCLEAR POWER PLANT ACCIDENTS

*Donald J. Zeigler* ❖ *James H. Johnson, Jr.*

The accident at the Three Mile Island (TMI) nuclear generating station in March 1979 presented social scientists with their first opportunity to study actual evacuation behavior during a radiological emergency and to compare it with the known parameters of evacuation behavior in response to natural and other technological disasters. Within one month of the TMI accident we served on a team of geographers which surveyed south central Pennsylvania residents to determine how they responded to a limited evacuation advisory [3, 41]. More recently we participated in the design of a social survey of Long Island, NY residents who were asked how they were likely to behave in the event of a general emergency at the Shoreham Nuclear Power Station in Suffolk County, 60 mi (96 km) east of New York City. The purpose of the Shoreham survey was to provide a behavioral data base on which to build a workable emergency response plan that would protect the public from exposure to ionizing radiation during an accident at the plant. Our purpose in the present study is to ascertain whether the results

of the Shoreham survey support our hypotheses (1) that human spatial behaviors during nuclear emergencies are fundamentally different from behaviors during other emergencies and (2) that these differences must be taken into consideration in planning for possible nuclear power plant accidents. The TMI experience is used as a basis for comparison.

Evacuation planning at TMI took place during the accident, not before. Prior to 1979, a major reactor accident with off-site consequences was assumed to be a highly unlikely event [27]. Emergency plans were required only for the site itself and the surrounding 2–3 mile low population zone. After the TMI accident the Nuclear Regulatory Commission (NRC) issued new regulations requiring that off-site emergency plans be approved by the Federal Emergency Management Agency (FEMA) and the NRC before a plant such as Shoreham may be licensed [28, 29]. These upgraded regulations purportedly build upon the lessons learned from TMI to insure that "adequate protective measures can and will be undertaken in the event of [another] radiological emergency" [28, p. 55402]. Evacuation planning is now required for a generic 10-mi (16 km) plume exposure pathway emergency planning zone (EPZ) around all nuclear

Reprinted from *Professional Geographer* Volume 36 (2) (1984), pp. 207–15. Used with the permission of the Association of American Geographers.

power plant sites. No evacuation plans are required for the more distant 10–50 mi (16–81 km) area known as the ingestion exposure emergency planning zone because "the probability of large doses [of radiation] drops off substantially at about 10 miles from the reactor" [27, p. 1–37]. The decision to limit evacuation planning to the 10-mi (16 km) EPZ ignores the recommendations of studies which analyzed the evacuation behavior of south central Pennsylvania residents and which identified a set of behavioral responses different in many respects from other experiences with evacuation [2, 3, 4, 7, 12, 13, 41].

## CONCEPTUAL OVERVIEW

Knowledge of evacuation behavior, whether in response to natural or technological threats, is "crucial for the successful design and implementation of community emergency plans" [33]. What is known about evacuation behavior comes primarily from studies of natural disasters [8, 25, 31, 32, 33] and, more recently, non-nuclear technological ones [24, 35]. Only in response to the accident at TMI was it possible to amass an empirical data base on evacuation behavior during a nuclear emergency [2, 3, 7, 12, 41]. While there are always situational contingencies influencing the success or failure of mass evacuations, spatial behaviors during non-nuclear emergencies tend to manifest some general patterns: individuals and families seem to evacuate only when confronted with direct sensory evidence or explicit warning messages convincing enough to persuade them that they are in the hazard zone of a life- or limb-threatening disaster agent. Most often, the problem which materializes during mass evacuations is one of trying to convince those who see no reason to evacuate that they need to leave their homes for reasons of personal safety. When evacuees do leave, they flee as family units, travel not much farther than the edge of the hazard zone, and take refuge with family and friends rather than in public shelters.

Evidence from the nuclear technological disaster at TMI suggests several dimensions of evacuation behavior that may be unique to nuclear accidents. In contrast to the model outlined above, response to the Pennsylvania Governor's limited evacuation advisory reflected a decided over-response among local residents, not the under-response anticipated from previous evacuation studies. That is, what was intended to be a limited evacuation turned into one of mammoth proportions. Zeigler, Brunn, and Johnson identified this process as the evacuation shadow phenomenon: "the tendency of an official evacuation advisory to cause departure from a much larger area than was originally intended" [41, p. 7]. The process itself has been termed spontaneous evacuation [4]. We believe the evacuation shadow phenomenon characterizes nuclear emergencies because of the dread with which people view radiation hazards [5, 17, 21, 22, 26, 30, 37, 38], probably due in large part to the delayed effects, including cancer and transgenerational injury, and to the catastrophic potential of nuclear releases [18].

During the TMI emergency, pregnant women and pre-school children within 5 mi (8 km) of the threatening reactor were advised to evacuate. Sheltering (staying indoors) was recommended for all others within 10 mi (16 km) of the plant. If this advisory had been followed, only about 3,000 pre-school children and 444 pregnant women would have left the area within 5 mi of the plant [14]. Ninety-five percent of the population aged 0–5 did evacuate as did 90 percent of the pregnant women. Altogether approximately 144,000 people within a 15 mi (23 km) radius of the plant decided to evacuate [12]. Our study reported that 9 percent of those surveyed in three communities 25 mi (40 km) distant from the plant also evacuated, indicating that the evacuation shadow extended at least five times the distance to which the advisory applied [41]. In the words of Lindell and Perry, "even if one assumes that of all families with any pregnant women or pre-school children evacuated as a unit, and that therefore as many as 10,000 persons evacuated 'appropriately,' this is still an over-response of more than an order of magnitude" [23, p. 423]. Thus, even though the Governor's evacuation advisory was geographically limited and issued with the proviso that "an excess of caution is best" [36], an extensive evacuation shadow was cast over a six-county area. Another dimension of evacuation behavior also differed from the norm established for other disasters. Evacuees from TMI fled unprecedentedly long distances to their temporary destinations, a behavior which again seems to reflect extreme fear of the disaster agent. These dramatically different behaviors have been largely ignored in formulating revised procedures for radiological emergency response planning. Even behaviors which conform to expectations, including the tendency for evacuees to use self-selected rather than designated evacuation routes and to choose the homes of relatives and friends over public shelters, argue for emergency response plans built on a behavioral as well as logistical base

## THE LONG ISLAND EVACUATION SURVEY

The Long Island Lighting Company's Shoreham Nuclear Power Station is located in the town of Brookhaven on the north shore of Long Island and within the suburban expansion zone of New York City. The island comprises Nassau County to the west and Suffolk County to the east. Within 10 mi (16 km) of the Shoreham site reside an estimated 57,000 people; between 10–50 mi (16–80 km) reside slightly over 5 million. Only three other nuclear power plant sites have a larger population living within fifty miles. Compounding the problem of high population densities, particularly to the west of the plant, is the elongated configuration of Long Island which is

only 14 mi (20 km) wide at the Shoreham location. In addition, bridges leading off the island are limited to the extreme western end in New York City.

In 1982, Suffolk County sponsored a sample survey of Nassau and Suffolk County households to help determine what role the local public sector should play in managing the off-site consequences of an accident. Since location with respect to the reactor site is one of the most important variables in the decision to evacuate, the population of telephone subscribers in the two-county area was stratified on the basis of distance and direction from the Shoreham plant. Four zones were delineated and three-digit telephone exchanges were identified within each zone. The last four digits of each telephone number were then generated at random. In all, 2,595 telephone interviews were completed [39]. About one-third of the eligible respondents contacted refused to participate. Nevertheless, the characteristics of the sample households closely approximated the population parameters for Long Island. The survey instrument itself was designed by an interdisciplinary team of social scientists including the authors of this paper. The scenario method was chosen in part because it has been successfully used in other hazard studies [6, 9, 10, 16, 40] and in part because the President's Commission on the Accident at Three Mile Island recommended that emergency plans be designed on the basis of alternative disaster scenarios for any given plant [34].

## THE EVACUATION SHADOW PHENOMENON ON LONG ISLAND

The Shoreham evacuation questionnaire asked respondents to consider three increasingly serious protective action advisories. Each advisory was preceded by the statement: "Suppose that you and your family were at home and there was an accident at the Shoreham

Nuclear Power Plant." In scenario I, all people who lived within 5 miles of the plant were advised to stay indoors. In scenario II, all pregnant women and pre-school children living within 5 miles of the plant were advised to evacuate and everyone else within 10 miles was advised to remain indoors. In scenario III, everyone living within 10 miles of the plant was advised to evacuate (the most extreme evacuation advisory for which plans would be in place).

All survey respondents in both Suffolk and Nassau counties were asked how they would react to each scenario. Their responses are summarized in Table 1 which presents data for the total two-county population and for five subzones. These figures provide estimates of the magnitude and geographic extent of spontaneous evacuation. In scenario I, no one was advised to evacuate but 25 percent of all households said they would leave. In

scenario II, modeled after the TMI advisory, 34 percent of the total population indicated their intentions to evacuate. In scenario III, the most severe, only 3.6 percent of Long Island's population should have indicated their intentions to evacuate. Instead, half of the total population were projected to leave their homes in search of safer quarters. If conditions actually warranted the advisory in the first scenario, 215,000 families could be expected to evacuate. In the second scenario (which should have precipitated the flight of only 2,700 families) 289,000 families indicated they would leave; and in the third scenario the number was 430,000 families.

A comparison of the intended evacuation rates of the total population and the population within 5 mi (8 km) of the Shoreham plant suggests a decided distance-decay phenomenon. In fact, distance from the reactor has been found to be the single most important

**TABLE 1** Behavioral Responses to Selected Protective Action Advisories (figures in percentages)

| | Evacuate | Shelter | Business as Usual | Do Not Know |
|---|---|---|---|---|
| Scenario I | | | | |
| Total | 25 | 42 | 30 | 3 |
| 0–5 miles | 40 | 52 | 4 | 5 |
| 6–10 miles | 40 | 49 | 8 | 3 |
| Eastern Suffolk | 22 | 50 | 22 | 5 |
| Western Suffolk | 34 | 43 | 20 | 2 |
| Nassau | 18 | 39 | 40 | 3 |
| Scenario II | | | | |
| Total | 34 | 41 | 23 | 3 |
| 0–5 miles | 57 | 38 | 2 | 3 |
| 6–10 miles | 52 | 39 | 5 | 3 |
| Eastern Suffolk | 30 | 45 | 19 | 5 |
| Western Suffolk | 44 | 42 | 13 | 1 |
| Nassau | 25 | 41 | 31 | 3 |
| Scenario III | | | | |
| Total | 50 | 31 | 16 | 3 |
| 0–5 miles | 78 | 18 | 1 | 3 |
| 6–10 miles | 78 | 17 | 2 | 3 |
| Eastern Suffolk | 46 | 36 | 14 | 4 |
| Western Suffolk | 63 | 29 | 7 | 2 |
| Nassau | 39 | 34 | 25 | 3 |

Source: Social Data Analysis [39]. Scenario I: all people within 5 miles of the plant advised to stay indoors; Scenario II: all pregnant women and pre-school children within 5 miles of plant advised to evacuate and everyone else within 10 miles to stay indoors; Scenario III: everyone within 10 miles advised to evacuate.

determinant of evacuation rates [19]. The distance-decay curves presented in Figure 1 show that in all three scenarios families closer to the plant are more likely to evacuate than those farther away. Regardless of scenario, there appears to be very little distance decay within 10 mi (16 km) of the reactor site. In response to the advisory in scenario I, approximately 40 percent of the population in both the 5 mile and 6–10 mile zones are likely evacuees. Beyond the 10-mile zone, the distance-decay effect is more apparent. All three curves exhibit a fairly constant rate of decline until about 25 or 30 miles from the accident site, whereupon they level off or exhibit more irregular variations in response to increasing distance. While the proportion of households intending to evacuate declines as a function of distance, the absolute number of evacuees increases dramatically with each additional mile. Another significant characteristic of the distance-decay curve for scenario III is the indication that not everyone within 10 miles of the plant would follow the instructions to evacuate. In fact, almost one-fifth of the population of the plume exposure EPZ indicated they would remain behind even if advised to leave.

For scenario III there is also likely to be a directional bias in the pattern of evacuation from a nuclear accident at Shoreham (Figure 2). Residents west of the plant are more likely to evacuate than their counterparts to the east. Beyond the 10-mile radius of the site, the distance-decay effect is much more pronounced to the east than to the west. The population of eastern Suffolk County is not nearly as likely to choose evacuation as a response to the scenarios as the population of western Suffolk. The gap which separates their comparative evacuation rates widens with each increasingly serious scenario. Those east of the plant are likely to find themselves in a quandary. Although downwind from the plant, they cannot evacuate to the west without passing through or near the hazard zone, thus increasing their risk of exposure to radiation. At the same time, they would probably be reluctant to evacuate eastward for fear of being trapped in a cul de sac should conditions at the plant deteriorate and threaten an even larger area. Moreover, the sparser population to the east diminishes the likelihood of evacuees being able to find a satisfactory place to stay, which would, in itself, discourage evacuation.

These findings suggest that a high degree of spontaneous evacuation is likely to occur on Long Island, just as it did around TMI, during a radiological emergency. Depending

**DISTANCE DECAY EVACUATION CURVES**

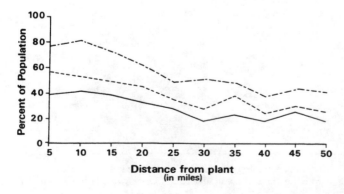

**FIGURE 1.** Distance Decay Evacuation Curves on Long Island. (*Source: Johnson and Zeigler* [20]).

**EVACUATION BY DISTANCE AND DIRECTION FROM THE PLANT**
Scenario 3

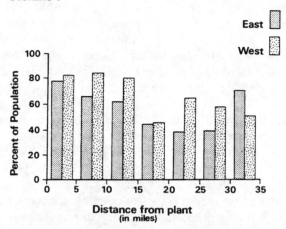

**FIGURE 2.** Rates of Intended Evacuation by Distance and Direction from the Shoreham Nuclear Power Plant. Scenario III: everyone within 10 miles of the plant advised to evacuate. (*Source: Johnson and Zeigler* [20]).

on the scenario, 25 to 50 percent of the Long Island population is likely to evacuate in response to very limited protective action advisories. The evacuation shadow is likely to affect the entire two-county area. Even though the Shoreham evacuation survey asked only for an indication of behavioral intentions, we believe that these intentions would be acted upon in an actual emergency for the following reasons. First, the intended spatial behaviors at Shoreham closely parallel the actual spatial behaviors in response to the TMI accident. Within the 5 mile zone, for instance, the TMI population registry indicated that 64 percent of the population evacuated at some time during the crisis [14]. The intended evacuation rate within 5 miles of the Shoreham plant in response to scenario II was 57 percent. Within 15 miles of TMI, an estimated 39 percent of the population evacuated; the comparable figure for Long Island was 53 percent. Second, the theory of reasoned action, developed extensively in the writings of Ajzen and Fishbein [1], suggests that the most powerful predictor of actual behavior is intended behavior. That is, unless something changes their intentions before they have a chance to act, Long Island residents who say they would evacuate are likely to do just that. Third, the longstanding controversy over the Shoreham plant, com-

pounded by the recency of the TMI accident, helped assure that survey respondents had given the issue of emergency response some previous thought.

## THE SEARCH FOR EVACUATION QUARTERS

Not only do people evacuate from an unprecedentedly large area when threatened by a radiological emergency, but they also flee longer distances than in other types of disasters. TMI area evacuees traveled distances greater than had been recorded in any previous study, according to Hans and Sell's analysis [15] which found the longest median evacuation distance on record to be 80 mi (128 km) in response to a Gulf Coast hurricane. Zeigler, Brunn, and Johnson [41] found the median distance at TMI to be 85 mi (137 km) and Flynn [12] found that half of the population put at least 100 mi (161 km) between themselves and the threatening reactor. The mean distance reported by Barnes et al. was 112 mi (180 km) [2]. These spatial behaviors dramatically illustrate the difference between perceptions of nuclear emergencies and perceptions of other emergencies. The Mississauga train derailment in 1979, for instance, forced the evacuation of an estimated quarter of a million people, but the vast majority remained within the Toronto-Hamilton corridor, putting only a few miles between themselves and the spreading chlorine gas cloud [24]. Similarly, natural disasters such as floods generally send evacuees not much farther than the projected high-water line.

Like the population of the TMI area, the people of Long Island indicated that they would evacuate greater distances than should be necessary to protect themselves. Respondents to the Shoreham survey were asked where they would go if they were to evacuate and how far away the evacuation destination was from home. The results are presented in Table 2. Several observations from these data have important planning implications. First, responses are very similar from one scenario to another. Despite the fact that no one was ordered to evacuate in scenario I and everyone within 10 miles of the plant was ordered to evacuate in scenario III, the space-searching behavior of evacuees remained relatively

unchanged. This finding suggests that once families feel threatened by a radiological emergency, the seriousness of the emergency does not seem to influence decisions about where to seek shelter. Second, almost two-thirds of all evacuees would seek out quarters which were more than 50 mi (80 km) from home. A directional bias would characterize the resultant traffic flow, since only about one in seven evacuees would intend to remain in Suffolk or Nassau Counties. At least seven out of ten would attempt to leave Long Island for New York City or a destination beyond. Third, about 60 percent of all families intending to evacuate would stay in someone else's home, with friends or relatives. Even hotels and motels were preferable to officially designated public shelters which would attract only about 3 percent of the evacuees. The tendency to avoid mass accommodations seems to be a universal finding in disaster studies and conforms with what we know about non-nuclear evacuation behavior [25, 31, 32].

**TABLE 2**  Selected Evacuation Behaviors (figures in percentages)

| | Evacuation Destinations | | |
|---|---|---|---|
| | Scenario I | Scenario II | Scenario III |
| Suffolk County | 4 | 6 | 5 |
| Nassau County | 8 | 9 | 9 |
| New York City | 24 | 19 | 22 |
| Some Other Place | 49 | 51 | 51 |
| Don't Know | 14 | 16 | 13 |
| | Distance From Home | | |
| | Scenario I | Scenario II | Scenario III |
| Less than 50 Miles | 36 | 34 | 33 |
| More than 50 Miles | 62 | 63 | 63 |
| Don't Know | 2 | 3 | 3 |
| | Evacuation Quarters | | |
| | Scenario I | Scenario II | Scenario III |
| Somebody Else's Home | 59 | 58 | 61 |
| Hotels or Motels | 19 | 19 | 18 |
| Public Shelters | 6 | 6 | 8 |
| Somewhere Else | 6 | 6 | 6 |
| Don't Know | 10 | 11 | 8 |

Source: Social Data Analysis [39]. Scenario I: all people within 5 miles advised to stay indoors; Scenario II: all pregnant women and pre-school children within 5 miles of plant advised to evacuate and everyone else within 10 miles to stay indoors; Scenario III: everyone within 10 miles advised to evacuate.

## SOME IMPLICATIONS FOR EMERGENCY PLANNING

The distance-decay curves presented in Figure 1 suggests that evacuation planning, including the calculation of evacuation time estimates, needs to encompass, if not the entirety of Long Island, at least an area extending to 25–30 mi (40–48 km) from the plant, the point at which the distance-decay curves level off. A 10-mile limit on evacuation planning, with no attention to what may occur beyond that zone, would establish a procedure for handling a maximum of 31,000 families. If the results of the Shoreham survey are accepted as an indication of actual behaviors, about fourteen times that many families would be on the road. Because not all families in the 10-mile zone would voluntarily evacuate even if advised to do so, only about one out of eighteen evacuating households would originate from within the zone for which evacuation plans had been put in place. These evacuation-resistent families, which would number about 22 percent of the population within 10 miles of the plant under scenario III, present another problem for emergency planning personnel. Just as plans must be made to discourage spontaneous evacuation beyond the designated hazard zone, they must also address the problem of how to motivate the residual population within that zone to take action.

The evacuees beyond the 10-mile zone are likely to make it more difficult for those closer to the plant to get away from the threatening reactor as quickly as possible. According to FEMA [11] the time from the initiating event in the reactor to the start of an atmospheric release could be as short as one-half hour or as long as one day. It would take only one-half to two hours for the release to travel 5 miles from the plant, and from one to four hours for the release to travel 10 miles, that is, to the edge of the plume exposure EPZ. Such a time frame underlines the necessity for being able to completely clear the hazard zone as quickly as possible. Yet, road congestion beyond the planning zone, cross-town traffic generated in an effort to assemble families, and the movement of emergency personnel into the area could all retard the speedy exodus of those closest to the plant. In short, the evacuation shadow phenomenon could significantly lengthen the time it would take to evacuate the EPZ. In addition, 60–76 percent of the evacuees originating east of the Shoreham plant indicated their intentions to head for shelter in Nassau County, New York City, or beyond. This finding suggests there will be a sizable flux of traffic through the plume exposure EPZ, adding yet another variable to the evacuation time equation.

In general, both the experience at TMI and the results of the Shoreham survey indicate that people have their own ideas about how to behave during a nuclear accident and cannot be counted on to adhere to protective action advisories issued by public officials. This finding suggests two strategies for increasing the effectiveness of radiological emergency response plans. First, in the short run, efforts should be made to determine how people would intend to react in a radiological emergency so that response plans can capitalize on these behaviors. Second, in the long run, efforts should be made to teach people the importance of following directions. These efforts should be directed at both the under-reactors close to nuclear plant sites and at over-reactors father away. These are the groups that either endanger society by congesting travel arteries or themselves by remaining in place. Such an educational campaign will not be easy to conduct, particularly in light of the emotional nature of nuclear issues and the long-standing controversy over nuclear risk [38].

## SUMMARY AND CONCLUSION

The intended behavior of Long Island residents in response to a general emergency at Shoreham is likely to parallel the actual behavior of TMI area residents during the 1979 mishap. The behavioral response to nuclear accidents appears to be quite different from responses to other emergencies, particularly in terms of the spatial dimensions of the evacuation process. Spontaneous evacuation and its geographic manifestation, the evacuation shadow phenomenon, seem to place nuclear power plant accidents in a class by themselves. We are unable to cite any other class of accidents or disasters which has precipitated such extreme evacuation behavior, particularly in its geographic dimensions. Unlike pre-impact evacuations in response to natural disasters, during which it is often difficult to get people to move at all, evacuations in response to nuclear power plant accidents are likely to be characterized by an extreme over-response to limited protective action advisories. Both in terms of the magnitude of the evacuation and the geographic extent of the evacuation shadow, the results of the Shoreham evacuation survey seem to support our hypothesis that human behaviors during nuclear emergencies cannot be predicted on the basis of what we know about other emergencies. Another finding which supports our hypothesis is the distance people feel compelled to put between themselves and the threatening reactor. At TMI the median distance people traveled was 85–100 miles. In the Shoreham survey, more than six out of ten respondents said they would select an evacuation destination more than 50 miles from home. Only in terms of their disdain for public evacuation shelters do nuclear evacuees seem to fit the model developed for nonnuclear emergencies.

Our results suggest several implications for planning and several directions for future research. First, the extent of the evacuation shadow suggests that limiting evacuation planning to the 10-mile plume exposure EPZ would be under-planning for a nuclear accident because so few of the evacuees would actually originate in that zone. Second, because resistence to evacuation among those closest to nuclear plant sites does not seem to be totally eliminated, strategies need to be devised for encouraging total compliance with evacuation advisories as well as for discouraging spon-

taneous evacuation farther from plant sites. Third, the evacuation shadow phenomenon needs to be taken into consideration in calculating evacuation times for the removal of people closest to nuclear plant sites. The development of realistic models on which to base such calculations merits additional research. These models should take into consideration people's likely behaviors and be adaptable to local conditions, both physical and social, as they vary from site to site. Fourth, social surveys should become a required part of emergency response planning in order to assure that, within the context of local circumstances, planning capitalizes on the natural behavioral inclination of a potentially impacted population. And fifth, additional work is needed on the formulation of a comprehensive hazard theory which accounts for cross-hazard differences and provides explanations for the varying ways people perceive hazards.

### REFERENCES

1. Ajzen, Icek and Martin Fishbein. *Understanding Attitudes and Predicting Social Behavior.* Englewood Cliffs, NJ: Prentice-Hall, 1980.

2. Barnes, K., J. Brosius, S. Cutter, and J. K. Mitchell. *Responses of Impacted Population to the Three Mile Island Nuclear Reactor Accident: An Initial Assessment.* Discussion Paper No. 13. New Brunswick, NJ: Department of Geography, Rutgers University, 1979.

3. Brunn, S. D., J. H. Johnson, Jr., and D. J. Zeigler. *Final Report on a Social Survey of Three Mile Island Area Residents.* East Lansing, MI: Department of Geography, Michigan State University, 1979.

4. Chenault, W. W., G. D. Hilbert and S. D. Reichlin. *Evacuation Planning in the TMI Accident.* Washington: Federal Emergency Management Agency, 1979.

5. Cook, Earl. "The Role of History in the Acceptance of Nuclear Power." *Social Science Quarterly* 63 (1982), 3–15.

6. Crow, Kelley and Justin Friberg. "The Scenario Method: Its Application to Natural Hazard Research." Paper presented at the Kentucky Academy of Science, Lexington, KY, 1980.

7. Cutter, Susan and Kent Barnes. "Evacuation Behavior and Three Mile Island." *Disasters*, 6 (1982), 116–24.

8. Drabek, Thomas. "Social Processes in Disaster: Family Evacuation." *Social Problems*, 16 (1969), 336–47.

9. Ericksen, Neil J. *A Scenario Approach to Assessing*

*Natural Hazards*. Ph.D. dissertation, Department of Geography, University of Toronto, 1974.

10. ———. "A Tale of Two Cities: Flood History and the Prophetic Past of Rapid City, South Dakota." *Economic Geography*, 51 (1975), 305–20.

11. Federal Emergency Management Agency. *Dynamic Evacuation Analyses*. Washington: 1981.

12. Flynn, Cynthia B. *Three Mile Island Telephone Survey: Preliminary Report on Procedures and Findings*. Prepared for U.S. Nuclear Regulatory Commission (NUREG/CR-1093). Tempe, AZ: Mountain West Research, Inc., 1979.

13. ———. "Reactions of Local Residents to the Accident at Three Mile Island." In *Accident at Three Mile Island: The Human Dimensions*, pp. 49–63. Edited by D. L. Sills et al. Boulder, CO: Westview Press, 1982.

14. Goldhaber, Marilyn K. and James E. Lehman. "Crisis Evacuation During the Three Mile Island Nuclear Accident: The TMI Population Registry." Paper presented at the 1982 Annual Meeting of the American Public Health Association, Montreal, Quebec, updated 1983.

15. Hans, Joseph M., Jr. and Thomas C. Sell. *Evacuation Risks—An Evaluation*. Las Vegas, NV: U.S. Environmental Protection Agency, 1974.

16. Harris, Stephen L. "When Mt. Shasta Erupts." In his *Fire and Ice*, pp. 269–79. Seattle, WA: Pacific Search Press, 1976.

17. Hohenemser, Christoph, Roger Kasperson and Robert Kates. "The Distrust of Nuclear Power." *Science* 196 (April 1, 1977), 25–34.

18. ———. Robert W. Kates, and Paul Slovic. "The Nature of Technological Hazard." *Science* 220 (22 April 1983), 378–84.

19. Johnson, James H., Jr., and Donald J. Zeigler. "Distinguishing Human Responses to Radiological Emergencies." *Economic Geography*, 59 (1983), 386–402.

20. ——— and ———. "A Spatial Analysis of Evacuation Intentions at the Shoreham Nuclear Power Station." In *Nuclear Power: Assessing and Managing Hazardous Technology*, pp. 279–301. Edited by M. J. Pasqualetti and K. D. Pijawka. Boulder, CO: Westview Press, 1984.

21. Kuklinski, J. H., D. S. Metlay, and W. D. Kay. "Citizens Knowledge and Choices on the Complex Issue of Nuclear Energy." *American Journal of Political Science*, 25 (1982), 615–42.

22. Lifton, R. J. "Nuclear Energy and the Wisdom of the Body." *Bulletin of the Atomic Scientists*, 32 (1976), 16–20.

23. Lindell, Michael K., and Ronald W. Perry. "Protective Action Recommendations: How Would the Public Respond?" *Transactions of the American Nuclear Society*, 41 (1982), 423–24.

24. Liverman, Diana and J. P. Wilson. "The Mississauga Train Derailment and Evacuation, 10–16 November 1979." *Canadian Geographer*, 25 (1981): 365–75.

25. Moore, Harry E. et al. *Before the Wind*. Washington: National Academy of Sciences, 1963.

26. Nowotny, H. "Social Groups—Fear of Nuclear Power and Economic Interest." *Transactions of the American Nuclear Society*, 24 (1976), 99–100.

27. Nuclear Regulatory Commission. *Reactor Safety Study: An Assessment of Accident Risks in U.S. Commercial Nuclear Power Plants*. Washington: 1975 (NUREG 75/104 [WASH-1400]).

28. ———. Emergency Planning; Final Regulations." *Federal Register*, 45 (August 19, 1980).

29. ——— and Federal Emergency Management Agency. *Criteria for Preparation and Evaluation of Radiological Emergency Response Plans and Preparedness in Support of Nuclear Power Plants*. Washington: 1980 (NUREG 0654).

30. Otway, H. J. "Risk Assessment and the Social Response to Nuclear Power." *Journal of the British Nuclear Energy Society*, 16 (1977), 327–33.

31. Perry, Ronald W. "Evacuation Decision-Making in Natural Disasters." *Mass Emergencies*, 4 (March 1979), 25–38.

32. ———. "Incentives for Evacuation in Natural Disaster." *Journal of the American Planning Association*, 45 (1979), 440–47.

33. Perry, Ronald W., Michael K. Lindell, and Marjorie R. Greene. *Evacuation Planning in Emergency Management*. Lexington, MA: Lexington Books, 1981.

34. President's Commission on the Accident at Three Mile Island. *The Need for Change: The Legacy of TMI*. Washington: U.S. Government Printing Office, 1979.

35. Quarantelli, E. L. *Evacuation Behavior: Case Study of the Taft, Louisiana Chemical Tank Explosion Incident*, Columbus, OH: Disaster Research Center, 1983.

36. Rogovin, M., and G. T. Frampton, Jr. *Three Mile Island: A Report to the Commissioners and the Public*, Vol. 1. Washington: Nuclear Regulatory Commission, 1980.

37. Slovic, P., B. Fischhoff, and S. Lichtenstein. "Rating the Risks." *Environment*, 21 (April 1979), 14–39.

38. ———. "Psychological Aspects of Risk Perception." In *Accident at Three Mile Island: The Human Dimensions*, pp. 11–20. Edited by D. L. Sills et al. Boulder, CO: Westview Press, 1982.

39. Social Data Analysts, Inc. *Attitudes Toward Evacuation: Reactions of Long Island's Residents to a Possible Accident at the Shoreham Nuclear Power Plant*. Setauket, NY: 1982.

40. White, Gilbert F. and J. Eugene Haas. *Assessment of Research on Natural Health Hazards*. Cambridge, MA: MIT Press, 1975.

41. Zeigler, D. J., S. D. Brunn and J. H. Johnson, Jr. "Evacuation from a Nuclear Technological Disaster." *Geographical Review*, 71 (1981), 1–16.

# PART IV

# IMPROVING MANAGEMENT

We are constantly asked to make choices about risks and hazards. Every time we get into our automobiles or go to the beach, we are intuitively balancing the risks of injury with the benefits that we perceive by engaging in these activities. But how does society weigh risks and hazards against benefits to figure out what is acceptable? How do risk and hazards managers compare such diverse risks and hazards as nuclear power plant failures, pesticide residues on foods, and floods, and then develop policies to manage them?

One management technique that has generated widespread enthusiasm during the last decade is risk assessment. Risk assessment is a tool for comparing different types of risk by using quantitative or probability estimates on the likelihood of injury or death. Quantitative risk assessment (QRA) has both proponents and opponents, as the articles in this section demonstrate. Some say they are meaningless, purporting to be "scientific" when in fact they are educated guesses based on incomplete or uncertain data. Others claim they are extremely useful in illustrating the potential increase in human health risks from selected activities. Still others claim they have little to do with general environmental conditions, since their sole focus is on human health, not environmental quality. Regardless of one's position, risk assessments are widely used in hazards management as a tool for identifying risks and for developing hazard reduction policies.

Wilson and Crouch (Reading 18) provide an excellent overview of the methods risk managers use to conduct risk assessments: historical evidence, event trees, and analogy (used almost exclusively in determining carcinogenic risks). Yet, even they concede that it is still difficult to compare all types of risks, especially in management situations, because there are often different meanings on the magnitude of risk. Which is riskier—the small probability of a large nuclear power plant accident, or the cumulative risks of burning fossil fuels? This is the unanswered question.

To improve the technique, Freudenburg (Reading 19) argues for a broader framework for risk assessments, an expansion of the quantifiable elements to include "people factors." He contends that risk assessment is really a social process because so much of the science is inexact. In suggesting a broader based view of risk assessment in hazard management, Freudenburg offers a number of suggestions for improvements in the technique. Since the social context within which the risk assessment occurs is often just as important as the risk assessment itself, both are needed to help improve management. The public is not always as misinformed as experts think. Rather, they have entered other contextual information into their risk assessments and thus are often acting in a very prudent and responsible manner. To improve the usefulness of risk assessments in the management process, these factors must be considered.

Weinberg (Reading 20) echoes the theme of risk assessment as being more art than science. His article explains why the inexactness of science (which he calls limits) poses difficult problems for regulators or managers. Uncertainty regarding the likely impacts, or the loss of predictive ability when focusing on rare or singular events pose the greatest headaches for regulators. Accidents (nuclear, chemical, or otherwise) are particularly troublesome, as are the more chronic hazards (Weinberg calls these low-level insults) such as toxic contamination. To reduce uncertainty, the dominant management strategy usually involves some form of technological fix such as the implementation of redundant safety systems. Another management approach is the *de minimis principle*. Here, very low levels of risk are simply ignored because there is no such thing as zero risk. For example, if the risk is naturally occurring, such as radiation, then we need not worry about the additional amount of radiation put into the environment from human-made exposures unless they exceed the natural or background levels. Weinberg's message is clear: We must understand the limits of science. When we disregard them, bad management policies are the usual result.

In a slightly different vein, Lave (Reading 21) illustrates how risk assessments can be used in a regulatory environment of competing priorities. For example, he questions why hazards managers spend so much time and energy on regulating relatively low human health risks (chemical exposures) instead of those producing the greatest number of fatalities (smoking and obesity). In defense of risk assessments, Lave concedes their usefulness in focusing the public debate on crucial issues as well as scientific attention on future informational needs. But, he adds, they do not answer what should or should not be regulated.

Most management systems include the following components: identification and assessment of the risk and hazard, legislation, regulation, inspection, enforcement, and emergency response. The goal of risk and hazards assessment is to minimize surprises for the hazards manager, especially scientific uncertainty. In addition to incomplete knowledge on impacts, there are other constraints on these systems, as Kates (Reading 22) illustrates. Geographic scale and transboundary movements of risks or hazards are two such examples. What happens when the risks or hazards are produced in one place, yet their effects transcend the local environment, such as the Chernobyl accident? Another is the focus on prevention (before the risk or hazard appears) versus response (after the event). Depending on the approach, very different management systems are required. Finally, issues of equity regarding who bears the burdens of risks and hazards and who benefits from their management are crucial considerations. In tracing the evolution of hazards management institutions during the last fifteen years, Kates describes their limitations—uncertain scientific knowledge, bureaucracy, and moral choice. All of these factors ultimately govern the successes and failures in managing risks and hazards.

# DISCUSSION POINTS

1. What is risk assessment? How do you conduct risk assessments? What types of data or informational needs are required?

2. What are the advantages and disadvantages in using risk assessments to manage risks and hazards?

3. Describe some of the constraints to effectively managing hazards and risks, especially those in a regulatory context.

4. How might you go about determining which of the following pairs of risk or hazards were the most serious, requiring some form of institutional response to mitigate their impacts on society?
   a. Nuclear power plants, coal-burning electrical plants
   b. Acid rain, global climate change
   c. Drought, floods
   d. Pesticides, food additives
   e. Earthquakes, volcanoes

# ADDITIONAL READING

Comfort, L. K. (Ed.), 1988. *Managing Disaster: Strategies and Policy Perspectives*. Durham, NC: Duke University Press.

Crouch, A. E. C., and R. Wilson, 1982. *Risk/Benefit Analysis*. Cambridge, MA: Ballinger.

Dietz, T. R. and R. W. Rycroft, 1987. *The Risk Professionals*. New York: Russell Sage Foundation.

Fischhoff, B., S. Lichtenstein, P. Slovic, S. L. Derby, and R. L. Keeney, 1981. *Acceptable Risk*. New York: Cambridge University Press.

Kahneman, D., P. A. Slovic, and A. Tversky (Eds.), 1982. *Judgment under Uncertainty: Heuristics and Biases*. New York: Cambridge University Press.

Kates, R. W., 1978. *Risk Assessment of Environmental Hazard*. New York: John Wiley & Sons, SCOPE 8.

Kates, R. W., C. Hohenemser, and J. X. Kasperson, 1985. *Perilous Progress: Managing the Hazards of Technology*. Boulder: Westview Press.

Krimsky, S. and A. Plough, 1988. *Environmental Hazards: Communicating Risks as a Social Process*. Dover, MA: Auburn House.

Lave, L. (Ed.), 1982. *Quantitative Risk Assessment in Regulation*. Washington, D.C.: Brookings Institution.

Lowrance, W. W., 1976. *Of Acceptable Risk: Science and the Determination of Safety*. Los Altos, CA: William Kaufman.

National Research Council, 1983. *Risk Assessment in the Federal Government: Managing the Process*. Washington, D.C.: National Academy Press.

Schwing, R., and W. Albers Jr. (Eds.), 1980. *Societal Risk Assessment: How Safe Is Safe Enough?*. New York: Plenum.

Schrader-Frechette, K. S., 1991. *Risk and Rationality*. Berkeley: University of California Press.

Whyte, A. V. and I. Burton, 1980. *Environmental Risk Assessment*. New York: John Wiley & Sons, SCOPE 15.

# RISK ASSESSMENT AND COMPARISONS:
# AN INTRODUCTION

*Richard Wilson* ❖ *E. A. C. Crouch*

Every day we take risks and avoid others. It starts as soon as we wake up. One of us lives in an old house that had old wiring. Each time he turned on the light, there was a small risk of electrocution. Every year about 200 people are electrocuted in the United States in accidents involving home wiring or appliances, representing a risk of death of about $10^{-6}$ per year, or $7 \times 10^{-5}$ per lifetime. To reduce this risk, he got the wiring replaced. When we walk downstairs, we recall that 7000 people die each year in falls in U.S. homes. But most are over 65, so we pay little attention to this risk since both of us are younger than that.

How should we go to work? Walking is probably safer than using a bicycle, but would take five times as long and provide less healthful exercise. A car or, better, public transport would be both safer and faster. Expediency wins out, and the car comes out of the garage. Fortunately, the choice nowadays is not between horse or canoe—both of which are much more dangerous. The day has just begun, and already we are aware of several risks, and have made decisions about them.

Most of us act semi-automatically to minimize our risks. We also expect society to minimize the risks suffered by its members, subject to overriding moral, economic, or other constraints. In some cases these constraints will dominate, in others there will be trade-offs between the values assigned to risks and the constraints. Risk assessments, except in the simplest of circumstances, are not designed for making judgments, but to illuminate them (1). To effectively illuminate, and then to minimize, risks requires knowing what they are and how big they are. This knowledge usually is gained through experience, and the essence of risk assessment is the application of this knowledge of past mistakes (and deliberate actions) in an attempt to prevent new mistakes in new situations.

The results of risk assessments will necessarily be in the form of an estimate of probabilities for various events, usually injurious. The goal in performing a risk assessment is to obtain such estimates, although we consider the major value in performing a risk assessment is the exercise itself, in which (ideally) all aspects of some action are explored. The results, goals, and values of performing the risk assessment must be sharply contrasted with the cultural values assigned to the results. Such cultural values will presumably be factors influencing societal decisions and may differ even for risk estimates that are identical in probability.

## RISK AND UNCERTAINTY

The concept of risk and the notion of uncertainty are closely related. We may say that the lifetime risk of cancer is 25%, meaning that approximately 25% of all people develop cancer in their lifetimes. Once an individual develops cancer, we can no longer talk about the risk of cancer, for it is a certainty. Similarly if a man lies dying after a car accident, the risk of his dying of cancer drops to near zero. Thus estimates of risks, insofar as they are expressions of uncertainty, will change as knowledge improves.

Different uncertainties appear in risk estimation in different ways (2). There is clearly a risk that an individual will be killed by a car if that person walks blindfolded across a crowded street. One part of this risk is stochastic; it depends on whether the individual steps off the curb at the precise moment that a car arrives. Another part of the risk might be systematic; it will depend on the nature of the fenders and other features of the car. Similarly, if two people are both heavy cigarette smokers, one may die of cancer and the other not; we cannot tell in advance. However there is a systematic difference in this respect between being, for instance, a heavy smoker and a gluttonous eater of peanut butter, which contains aflatoxin. Although aflatoxin is known to cause cancer (quite likely even in humans), the risk of cancer from eating peanut butter is much lower than that from smoking cigarettes. Exactly how much lower is uncertain, but it is possible to make estimates of how much lower and also to make estimates of how uncertain we are about the difference.

Some estimates of uncertainties are subjective, with differences of opinion arising because there is a disagreement among those assessing the risks. Suppose one wishes to assess the risk (to humans) of some new chemical being introduced into the environment, or of a new technology. Without any further information, all we can say about any measure of the risk is that it lies between zero and unity. Extreme opinions might be voiced; one person might say that we should initially assume a risk of unity, because we do not know that the chemical or technology is safe; another might take the opposite extreme, and argue that we should initially assume that there is zero risk, because nothing has been proven dangerous. Here and elsewhere, we argue that it is the task of the risk assessor to use whatever information is available to obtain a number between zero and one for a risk estimate, with as much precision as possible, together with an estimate of the imprecision. In this context, the statement "I do not know" can be viewed only as procrastination and not responsive to the request for a risk estimate (although this should not be read as condemning procrastination in all circumstances).

The second extreme mentioned, the assumption of zero risk, can arise because people and government agencies have a propensity to ignore anything that is not a proven hazard. We argue that this attitude is inconsistent if the objective is to improve the public health, may also lead to economic inefficiencies, and often leads to unnecessary contention between experts who disagree strongly. Fortunately, if risk assessors have been diligent in searching out hazards to assess, few hazards posing large risks will be missed in this way, so that there may be minor direct danger to human health from a continuation of the attitude.

## RISK ESTIMATION BASED ON HISTORICAL DATA

The way in which risks are perceived is strongly correlated with the way in which they are calculated. Risks based on historical data are particularly easy to understand and are often perceived reliably. It is therefore easy to illustrate a risk calculated from historical data to understand some characteristics of risk estimation. There are plenty of data on automobile accidents (although never enough to make risk assessors happy). One thing that these data can tell us is the frequency of such accidents in the past and their trend through time. To make predictions, however, we must use a model. The simplest model is that there will be as many accidents next year as last, to within a statistical error of the square root of the number. A slightly more complicated, but perhaps more accurate, model might be to fit a mathematical function to numbers from previous years and to argue that next year's accidents will follow the trend given by this function. A possibly better and possibly more accurate model still might use all available information that might influence accident trends. For example, an oil embargo with a concomitant rise in oil price and reduction in automobile travel would be likely to reduce the risk of accident. In any event, it becomes clear that it is impossible to calculate any risk without a model of some sort, even the simple one that tomorrow will be like today.

## RISKS OF NEW TECHNOLOGIES

We can only use the historical approach to estimating risks when the hazard (for example, technology, chemical, or simply some action) has been present for some time and the risk is large enough to be directly measured (although when it is not large enough to be measured, an upper limit may be calculated, if one assumes some sort of model). If there is no historical database for the hazard (a new power plant or industrial facility, for instance), one approach is to consider it in separate parts, calculating the risks from each part and adding them together to estimate a risk for the whole. For example, all possible chains of events from

an initiator to a final accident are followed in an "event tree," with the probabilities of each event in the tree being estimated from historical data in different situations.

A particularly well-known example is the calculation of the probability of a severe accident at a nuclear power plant (3). That this procedure has at least a partial validity is due to the fact that the design of nuclear power plants proceeded in approximately this factorable way; attempts were made to imagine all major accident possibilities, "maximum credible accidents" or "design basis accidents," and then to add an independent device to prevent this accident from having severe consequences. To the extent that the added safety device is independent, the failure probability is independent, and the small overall accident probability is the product of individual failure probabilities which are larger.

## RISKS BY ANALOGY: CARCINOGENIC RISKS

Some carcinogenic risks may be estimated from historical data. But this is complicated by the time delay between the insult and the final cancer, one reason why causality is hard to prove if the risk is small. This is the difficult field of epidemiology.

Although some of the largest cancer risks have been identified through the use of epidemiology (4), preventive public health suggests that we endeavor to estimate risks even where no historical data exist and the risk is small. This is often done by analogy with the cancer risks to animals, usually rodents, which are deliberately exposed to large enough quantities of pollutant so that an effect is observed. To use these data to estimate the risk at low doses in people involves (to oversimplify matters) two difficult steps: the comparison of carcinogenic potency in animal and man (5–7) and the extrapolation from a high dose to a low dose. Because both steps require a certain amount of theory, they are controversial. Indeed, there are those who regard the uncertainty as so great that they prefer not to provide numerical estimates of risk (8, 9), although they may order materials in carcinogenic potency. The difference between this and providing a numerical estimate is important, but is one of presentation rather than substance.

If there are no animal data, or if in an animal experiment there is no statistically significant effect, it does not necessarily mean that there is no risk. If the experimenters have been diligent, the risk is probably small, although never zero, even though that may be the best estimate. Various attempts are made to use data even less direct than the animal bioassays to estimate risks in such cases. These include simple analogies based on chemical similarity (10), and comparison with outcomes other than cancer—for example, mutagenesis (11) and acute toxicity (12, 13). Not surprisingly, these more indirect procedures arouse even more controversy than the animal bioassays.

There have been few attempts to perform risk assessments for biological end points other than cancer. However, it is known that the pollutants in cigarette smoke cause at least as many deaths through heart problems as by cancer (14), and we should not be surprised if other carcinogens were to produce chronic effects other than cancer. For now, the cancer risk assessment has to act as surrogate for these other risks also.

## RISK VALUE VERSUS CERTAINTY OF INFORMATION

After risks of a number of situations have been assessed, we often want to order them in order to decide which should command our attention. It is not always the order of increasing risk that is used for such purposes. There have been proposals to order potential carcinogens on other factors (8, 15), such as the certainty of information.

Vinyl chloride gas has been found to cause angiosarcomas both in people and in rats. Since an angiosarcoma is a rare tumor, the risk ratio (the ratio of the observed number of cancers in those exposed to the number expected by chance) is of order 100 or more in some cases. If an angiosarcoma is seen in a vinyl chloride worker, the attribution to vinyl chloride exposure is almost certain. On the other hand, the number of persons who have been heavily exposed to vinyl chloride is small, so that only about 125 angiosarcomas have been seen among vinyl chloride workers worldwide in the last 20 years. Now that exposures in the workplace have been greatly reduced, no angiosarcomas attributable to recent occupational exposure have been seen. We do not know the dose-response relation, but it is generally believed that the response falls at least linearly as the exposure is reduced, so that no more than one cancer is expected in several years.

We can compare this with the possible cancer incidence that was predicted by the Food and Drug Administration (FDA) in 1977 from use of saccharin (16). This was based on experiments with rats, leading to an additional uncertainty. More people ate saccharin than were exposed to vinyl chloride, and nearly 500 cancers per year were estimated for the United States alone. For vinyl chloride we therefore have the situation that the individual risk is now low, yet there is considerable certainty that there is a risk. For saccharin the risk is higher, but there is more uncertainty about the value of the risk. Some persons, in some situations, may demand that more attention be given to the risk from vinyl chloride than to the risk from saccharin; for other persons or situations the reverse may be the case.

## COMPARISON OF RISKS

The purpose of risk assessment is to be useful in making decisions about the hazards causing risks, and so it is important to gain some perspective about the meaning of the magnitude of the risk. Comparisons can be useful. We are not born with an instinctive feeling for what a risk of one in a million per lifetime means, although we do learn that some risks are small and others large. It is particularly helpful to compare risks that are calculated in a similar way. For example, the risk of traveling by automobile can be compared to that of traveling by horse with the use of historical data.

Another common procedure is to compare exposures only. Table 1 shows a list of radiation exposures in typical situations (17). The dose-response relation for radiations with similar energy deposition per unit track length will be similar, although there may be some correction required for dose-rate effects, so that ordering by exposure should be similar to ordering by risk. In estimating the number of lethal cancers on a linear hypothesis, we have here assumed approximately 8000 man-rems per cancer (at low doses), in itself uncertain by 30% or more.

An example of comparison of risks that are similarly calculated is the comparison of risks of various chlorinated hydrocarbons in drinking water. The risks to humans are estimated from carcinogen bioassays in rodents (rats and mice). Since these are similar materials, we might expect that the dose-response relationships have the same shape. Chloroform, which is produced by interaction of chlorine with organic matter during the chlorination of surface waters to kill bacteria, produces cancer in animals 20 times as readily as does trichloroethylene, an industrial solvent that is occasionally found in well waters as a result of accidental pollution. Although neither is known to cause cancer in people, we might expect that chloroform would do so about 20 times as readily.

Table 2 shows a variety of risks calculated in various ways and our estimate of the uncertainty. They are deliberately jumbled to provoke thought by juxtaposition. [Risk estimates quoted by the Environmental Protection Agency (EPA) for carcinogens tend to be greater than those shown in Table 2 by a factor approximately equal to the uncertainty factor—this is not accidental (5, 18).]

## CONTRASTING RISKS

Objections have been raised to risk comparisons on the ground that they are misleading. This would be true if all risks of the same numerical magnitude were treated in the same way. But they are not. In some cases it is

**TABLE 1** Comparison of Several Common Radiation Risks.

| Action | Dose (mrem/year) | Cancers If All U.S. Population Exposed (assuming linearity) |
|---|---|---|
| Medical x-rays | 40 | 1100 |
| Radon gas (1.5 pCi/liter, equivalent dose)* | 500 | 13,500 |
| Potassium in own body | 30 | 1000 |
| Cosmic radiation at sea level | 40 | 1100 |
| Cosmic radiation at Denver | 65 | 1800 |
| Dose to average resident near Chernobyl first year | 5000 | Not relevant |
| One transcontinental round trip by air | 5 | 135 |
| Average within 20 miles of nuclear plant | 0.02 | >1 |

*The radon exposure is to the lungs and cannot be directly compared to whole body external exposure. The comparison here is on the basis of the same magnitude of risk. The uncertainty of the radon number is at least a factor of 3.

**TABLE 2**  Some Commonplace Risks (Mean Values with Uncertainty).

| Action | Annual Risk | Uncertainty |
|---|---|---|
| Motor vehicle accident (total) | $2.4 \times 10^{-4}$ | 10% |
| Motor vehicle accident (pedestrian only) | $4.2 \times 10^{-5}$ | 10% |
| Home accidents | $1.1 \times 10^{-4}$ | 5% |
| Electrocution | $5.3 \times 10^{-6}$ | 5% |
| Air pollution, eastern United States | $2 \times 10^{-4}$ | Factor of 20 downward only |
| Cigarette smoking, one pack per day | $3.6 \times 10^{-3}$ | Factor of 3 |
| Sea-level background radiation (except radon) | $2 \times 10^{-5}$ | Factor of 3 |
| All cancers | $2.8 \times 10^{-3}$ | 10% |
| Four tablespoons peanut butter per day | $8 \times 10^{-6}$ | Factor of 3 |
| Drinking water with EPA limit of chloroform | $6 \times 10^{-7}$ | Factor of 10 |
| Drinking water with EPA limit of trichloroethylene | $2 \times 10^{-9}$ | Factor of 10 |
| Alcohol, light drinker | $2 \times 10^{-5}$ | Factor of 10 |
| Police killed in line of duty (total) | $2.2 \times 10^{-4}$ | 20% |
| Police killed in line of duty (by felons) | $1.3 \times 10^{-4}$ | 10% |
| Frequent flying professor | $5 \times 10^{-5}$ | 50% |
| Mountaineering (mountaineers) | $6 \times 10^{-4}$ | 50% |

useful to contrast risks to indicate the different ways in which they are treated in society. In Table 3 we give an example by comparing and contrasting the carcinogenic effects of aflatoxin B1 and dioxin, both among the carcinogenic chemicals known. The difference in treatment of these two materials is perhaps a reflection of different values assigned to various aspects of the problems caused by their presence.

Aflatoxin and dioxin have similar toxicities and carcinogenic potency (perhaps within a factor of 10, although both measures for both chemicals vary substantially with species tested). The certainty of information for aflatoxin is great. There is less information about carcinogenicity of dioxin. Dioxin may be a promoter and pose a minuscule risk at low doses, whereas aflatoxin is almost certainly an initiator also. Nonetheless such standards as

**TABLE 3**  Comparison of Two Very Toxic Chemicals, Aflatoxin B1 (22) and Dioxin (23)

| Measure | Aflatoxin B1 | Dioxin |
|---|---|---|
| Acute toxicity | High | Equal |
| Carcinogenic potency to people [(kg · day)/mg] | ~500 | Unknown |
| Carcinogenic potency to rats [(kg · day)/mg] | ~5000 | ~5000 |
| Mutagenic | Yes | No |
| Certainty of information on human carcinogenicity | High | Low |
| Activity (initiator or promoter) | Initiator | Promoter (?) |
| Possibility of threshold dose response | Low | High |
| Source | Natural | Artificial |
| Common knowledge | Little known | Agent Orange |
| FDA action level in peanuts (ppb) | 20 | |
| CDC level of concern in soil (ppb) | | 1 |

CDC, Centers for Disease Control

there are appear to be more stringent for dioxin, possibly because dioxin is an artificial chemical and possibly because it was a trace component of a chemical mixture (Agent Orange) that was used in warfare.

The small risk of a large accident in a nuclear power plant can also be contrasted with the more numerous small accidents or events that occur every day in the mining, transport, and burning of coal. One feature that is brought out clearly here is that we do not always compare the risk averaged over time, but worry more about risks that are sharply peaked in time.

## EXPRESSION OF RISKS

Just as a comparison of risks is an aid in understanding them, so is a careful selection of the methods of expression. It is hard to comprehend the statistical (stochastic) nature of risk. There are ways to mitigate this difficulty in comprehension. We are almost all used to one such statistical concept—the expectation of life. When we talk about the expectation of life being 79 years (for a nonsmoking male in the United States) we all know that some die young and that many live to be over 80. Thus the expression of a risk as the reduction of life expectancy caused by the risky action conveys some of the statistical concept essential to its understanding. One particular calculation of this type can be used as an anchor for many people, because it is easy to remember. The reduction of life expectancy by smoking cigarettes can be calculated from the risk, one in 2 million, of smoking one cigarette, multiplied by the difference of the average life-span of a nonsmoker and a lung cancer victim. This turns out to be 5 minutes, or the time it takes to smoke the one cigarette.

It is important to realize that risks appear to be very different when expressed in different ways (19). One example of this can be seen if we consider the cancer risk to those persons exposed to radionuclides after the Chernobyl disaster. According to the Soviets (20), the 24,000 persons between 3 and 15 kilometers from the plant, but excluding the town of Pripyat, received and are expected to receive 1.05 million man-rems total integrated dose, or about 44 rems average. Even if we assume a linear dose-response relation, with 8000 man-rems per cancer, the risk may be expressed in different ways. Dividing 1.05 million man-rems by 8000 gives 131 cancers expected in the lifetimes of that population. This is larger than, and for some people more alarming than, the 31 people within the power plant itself who died within 60 days of acute radiation sickness combined with burns. Dividing the 131 again by the approximately 5000 cancer deaths expected from other causes, the accident caused "only" a 2.6% increase in cancer. This seems small compared to the 30% of cancers attributable to cigarette smoking. The difference is even more striking if we consider the 75 million people in Byelorussia and the Ukraine who received, and will receive, 29 million man-rems over their lifetimes. On the linear dose-response relation this leads to 3500 "extra cancers," surely a large number for one accident. But dividing by the 15 million cancers expected in this population leads to an "insignificant" increase of 0.0047%. Of course, none of the methods of expressing the risk can be considered "right" in an absolute sense. Indeed, it is our belief that a full understanding of the risk involves expressing it in as many different ways as possible.

## COST OF REDUCING A RISK

Another interesting and instructive way of comparing risks is by comparing the amount people have paid in the past to reduce them. It might be thought that people would try to adjust their activities until the amount spent is roughly the same. Cohen (21) has shown that the amounts spent vary by a factor of more than a million. He shows that it would be possible even for an American to save lives in Indonesia by aiding in immunization at $100 per life saved. Society is willing to spend more on environmental protection to prevent cancer (over $1 million per life) than on cures (about $50,000 per life with the high value of $200,000 for kidney dialysis raising some objections). This ratio is in rough accord with the maxim "an ounce of protection is better than a pound of cure." People are willing to spend still more on radiation protection at nuclear power plants and on waste disposal. Economists and others often argue that efficiency depends on adjusting society until the amounts spent to save lives in different situations are equalized. It seems to us that society does not work that way. People are aware of the order of magnitude of these differences, and approve of them. Nonetheless, we believe that providing this information to a decision-maker is essential for an informed decision.

## REFERENCES AND NOTES

1. L. B. Lave, *Science* **236**, 291 (1987).
2. R. Wilson, E. A. C. Crouch, L. Zeise, in *Risk Quantitation and Regulatory Policy* (Cold Spring Harbor Laboratory Press, Cold Spring Harbor, NY, 1985), Banbury Report, vol. 19, pp. 133–47.
3. N. C. Rasmussen *et al.*, "Reactor safety study—an assessment of accident risks in U.S. commercial nuclear power plants" (WASH 1400, NUREG 75/014, U.S. Nuclear Regulatory Commission, Washington, DC, 1975). See also D. Okrent, *Science* **236**, 296 (1987).
4. R. Doll and R. Peto, *J. Natl. Cancer Inst.* **66**, 1191 (1984).
5. E. L. Anderson *et al.*, *Risk Anal.* **3**, 277 (1983).
6. E. A. C. Crouch and R. Wilson, *J. Taxicol. Environ. Health* **5**, 1095 (1979).
7. E. J. Calabrese, *Principles of Animal Extrapolation* (Wiley, New York, 1983).
8. R. Peto, in *Assessment of Risk from Low-Level Exposure to Radiation and Chemicals*, A. D. Woodhead, C. J. Shellabarger, V. Pond, A. Hollaender, Eds. (Plenum, New York, 1985), pp. 3–6.
9. B. N. Ames, R. Magaw, L. S. Gold, *Science* **236**, 271 (1987).
10. "Control of trihalomethanes in drinking water," proposed rule, *Fed. Regist.* **43**, 5756 (1968). See also the advanced notice [*ibid.* **41**, 28991 (1976)] and the final rule [*ibid.* **44**, 68624 (1979)].
11. M. Meselson and K. Russell, in *Origins of Human Cancer*, H. H. Hiatt, J. D. Watson, J. A. Winsten, Eds. (Cold Spring Harbor Laboratory, Cold Spring Harbor, NY 1977), p. 1473.
12. S. Parodi, M. Taningher, P. Boero, L. Santi, *Mutat. Res.* **93**, 1 (1982).
13. L. Zeise, R. Wilson, E. A. C. Crouch, *Risk Anal.* **4**, 187 (1984).
14. *Smoking and Health, a Report of the Surgeon General* (PHS79-50066, Public Health Service, Washington, DC, 1979).
15. R. A. Squire, *Science* **214**, 877 (1981).
16. "Saccharin and its salts," proposed rule and hearing, *Fed. Regist.* **42**, 19996 (1977).
17. R. Wilson and W. J. Jones, *Energy, Ecology and the Environment* (Academic Press, New York, 1974), table 9-6. Other entries may be readily calculated from data in the reports of the United Nations scientific committee on the effects of atomic radiation ["Sources and effects of ionizing radiation" (United Nations, New York, 1977)] and the report of the Committee on the Biological Effects of Ionizing Radiations ["The effects on populations of exposure to low levels of ionizing radiations" (National Academy Press, Washington, DC, 1980)].
18. M. Russell and M. Gruber, *Science* **236**, 286 (1987).
19. A. Tversky and D. Kahneman, *ibid.* **211**, 453 (1981). See also P. Slovic, *ibid.* **236**, 280 (1987).
20. U.S.S.R. State Committee for the Utilization of Atomic Energy, "The accident at the Chernobyl Nuclear Power Plant and its consequences," working document for the Post Accident Review Meeting, 25–29 August 1986, International Atomic Energy Agency, Vienna.
21. B. L. Cohen, *Health Phys.* **38**, 33 (1980).
22. H. R. Roberts, "The regulatory outlook for nut products," paper presented at the Annual Convention of the Peanut Butter Manufacturers and Nut Salters Association, West Palm Beach, FL, November 1977.
23. R. D. Kimbrough, H. Falk, P. Stehr, G. Fries, *J. Taxicol. Environ. Health* **14**, 47 (1984).
24. Our work on risk assessment has been supported by donations from Clairol, Inc., the Dow Chemical Company, the Cabot Corporation, the General Electric Foundation, and the Monsanto Corporation.

# 19

# PERCEIVED RISK, REAL RISK: SOCIAL SCIENCE AND THE ART OF PROBABILISTIC RISK ASSESSMENT

*William R. Freudenburg*

"Why doesn't anybody believe us anymore? This question was recently asked by a Ph.D.-level biologist, a man who has risen over the years to a position of considerable authority in a federal resource management agency. The specific context was the public insistence that his agency stop promoting a "risky technology," even though the evidence convinced him the risk was low. The broader problem is that he is not the only scientist asking such questions lately.

As the average life-span increases, the public perceives many of the risks around them to have become more severe (*1*). As scientists, we often assume that public perceptions are simply at variance with the real risks. The scientific method, however, calls for us to test our assumptions before accepting them uncritically, and there is "no law of nature that requires us to abandon the scientific method merely because questions of human behavior are involved" (*2*, p. 27).

Instead, this is a problem that calls for input from the social and behavioral sciences, which offer at least three contributions to the burgeoning field of risk assessment. The first and most obvious is in providing tools—and increasingly, a set of relevant findings—to help clarify the differences between the scientific community and the general public in the assessment of technological risks. Such a contribution is generally seen to lie in the area of risk communication, risk perception, and risk management. A second but less-understood contribution, however, comes from social scientists' input to risk assessments themselves—to actual calculations of the probabilities and consequences of undesired outcomes. Third, social sciences offer insights into the processes by which risk assessments are carried out. In each of these three areas, the cost of ignoring social scientists' contributions can be an unnecessary bias.

In this article I discuss three potential sources of error—in calculating risk consequences, in calculating risk probabilities, and in paying insufficient attention to the person-intensive nature of performing the assessments themselves. I conclude with a discussion of the rationality of public risk perceptions.

## RISK CONSEQUENCES

The potential social and economic consequences of technological failures tend to be reasonably easy to comprehend and to be in-

Reprinted from *Science* Volume 242 (1988), pp. 44–49. Copyright 1988 by the AAAS. Used with permission of the author and the AAAS.

creasingly recognized in the technical community. Three categories of consequences, however, deserve greater attention than they have received to date.

## Impacts of Serious Accidents

Easiest to consider are the consequences of rare but genuinely serious accidents—including accidents less extreme than those at Bhopal or Chernobyl. An examination of the Goiânia event in Brazil (3) is instructive. Two men entered an abandoned clinic in search of scrap metal; they found a small capsule and pried it open, releasing approximately 100 grams of cesium-137. This led to 121 cases of skin contact with the material and four deaths, with another three to five expected in the next 5 years. The death toll was not wildly out of line with "any other industrial accident," but just the labor costs of decontamination exceeded $20 million (U.S.) by December 1987. The broader economic and social costs were far greater. Within 2 weeks of the time the event was announced, the wholesale value of agricultural products from the entire state of Goiás fell by 50%; demand for manufactured goods (including textiles, clothing, and other finished products) was also affected—even though the study was unable to find "even a published suggestion" that the agricultural products or manufactured goods could have been contaminated (3). Severe impacts were also felt through treatment and research costs, declining property values, and a decline in the tourist trade. More than 100,000 residents lined up at monitoring stations to be checked for radioactive contamination, and more than 8,000 residents requested and received certification that they were not contaminated (4).

## Uncertainty Costs

In the second category are the less obvious costs of dealing with uncertainty and risk, even when "nothing goes wrong." When people buy automobile, health, home, or fire insurance, they incur real costs even if the insurance is never "needed." Insurance companies keep the premiums even if the house does not burn down, the automobile is not involved in an accident, the insured person is not hospitalized, and so on. Insurance premiums provide examples of costs created by the possibility that something may go wrong, not by the actual occurrence of the event itself (5). Similarly, real costs are incurred when communities invest in emergency-preparedness training or the preparation of evacuation plans, when societal strains are created by inequitable distributions of technological risks, or even when individuals "invest" in the psychic costs of worrying about potential disasters, whether such disasters actually occur or not (6).

## "Signal" Incidents

The third category includes a special class of the cases in between: not cases where best hopes or worst fears are realized, but cases when problems indicate that a technology may not be fully under control. Slovic refers to such events as having "signal" value—as signifying or sending a signal to the broader public that there may be reasons for concern (7)—and Kasperson et al. note that such events have the potential to lead to the "social amplification of risk" (8).

The explosion of the space shuttle Challenger sent a "signal" not only to the public and the policy system, but to the scientific and technical community as well. In the context of major airline accidents, the quantitative death toll from the Challenger was small, yet the accident led to an expensive, lengthy reanalysis of the nation's space program. A clearer example is the accident at Three Mile Island (TMI), which was found by official investigations to have released very little radioactivity, although it did lead to significant mental health consequences for nearby populations (9). More broadly, TMI sent a signal that nuclear power plants were less safe than the public had previously been led to believe (*10*), and its consequences were little short of disastrous for the nuclear power industry.

## RISK PROBABILITIES

Less widely recognized is the need for social science input in calculating risk probabilities. Three sets of factors illustrate the types of systematic biases that can be created by the failure to pay adequate attention to human behaviors.

### Human Error and Human Factors

When the President's commission on the accident at TMI began its investigation, its members expected to focus on problems of "the technology," meaning the nuclear hardware, rather than the ways in which the hardware was managed and operated. Ultimately, however, the commission found "the fundamental problems [were] people-related problems and not equipment problems" (*11*, p. 8). Included were several human errors that helped to turn a malfunctioning valve into the most expensive accident in the history of domestic nuclear power.

"Human error" is a value-laden term that has often been used to describe situations that actually involve mismatches between people and machinery, including those at TMI (*12*). With this caveat in mind, however, "human error" has been implicated in accidents that range from the Chernobyl disaster to the more prosaic problems of transporting hazardous materials, where more than 50% of the risk may be due to "driver error" and other human factors (*13*). Such figures make it clear that attention to "hardware" of vehicles, sign designs, road materials, and so forth, although valuable, may address only the minority of the causes of transportation accidents or other technological risks.

## Organizational Error and "Organizational Factors"

The behaviors that need to be explained, moreover, include not just those of individuals, but those of organizations. Empirical research on organizational and institutional factors often leads to conclusions that appear counterintuitive to persons trained in non-social science fields (*14*). The people problems identified at TMI were not limited to the operators in the plant at the time of the accident, but included "problems with the 'system' that manufactures, operates, and regulates nuclear power plants. There are structural problems in the various organizations, there are deficiencies in various processes, and there is a lack of communication among key individuals and groups" (*11*, p. 8; *15*). Even after TMI, some utilities have operated nuclear facilities well, but others have been less successful. The Peach Bottom nuclear power plant, 60 kilometers downstream from TMI, was extensively criticized both by the Nuclear Regulatory Commission (NRC) and the nuclear industry's own Institute of Nuclear Power Operations, which called the plant "an embarrassment to the industry and to the nation," calling attention to "the very culture of the company and how it was managed" (*16*, p. 6; *17*). Problems of operators sleeping on the job were eventually reported not by the utility, but by engineers from the reactor manufacturer, leading the NRC to close down the plant in March 1987, "the first time a nuclear power plant had been closed by the agency for a nonmechanical reason" (*16*, p. 6).

In another example that received broad attention, the immediate cause of the loss of the Challenger may have been the failure of an O-ring, but the official investigation into the accident also found problems in the broader management and organization of the National Aeronautics and Space Administration (NASA). Investigators recommended sweeping reforms of NASA's management structure, flight operations, and safety and risk analysis procedures—in addition to changes in the O-ring itself (*18*).

More broadly, it is no surprise to persons who study complex organizations, although it sometimes surprises or frustrates many of us in the technical community, that behaviors of janitors, operational personnel, and others are often different in empirical reality than in organizational policy. Such departures from official expectations can cause the actual risks to be significantly different from risk estimates that are based on the assumption that official policies will simply be followed (*19*).

## External Social Factors

Receiving still less attention are behaviors outside of the responsible organizations altogether. The obvious examples are threats of terrorism and sabotage. Although risks from terrorism appear to be relatively low in the United States, at least to date, they may be nonnegligible, particularly for technologies that are highly controversial. Disruptions might be more likely in cases where people feel they have not been accorded an adequate voice through traditional or acceptable channels, where they believe they are responding to a higher form of morality, or where they are interested in altering public perceptions of a technology's safety. Probabilities might be further increased for acts of sabotage that carry relatively low risks of capture but have high signal value—as in ambushing nuclear waste trucks on a little-patrolled section of an interstate highway or contaminating city water supplies with genetically engineered organisms stolen from experimental farms. For the longer term, there is a need for systematic research on terrorism, as well as other external factors such as political and economic pressures, to allow decisions to be guided by more than conjecture. For the near term, however, prudence suggests that external human factors cannot be safely ignored. For certain types of technologies, the probability of disruption might ultimately prove to be

anything but negligible—perhaps closer to the range of several percent per year than to a one-in-a-million chance.

## "HUMAN ERROR" IN ESTIMATION TECHNIQUES

Intriguingly, the group that generally receives the least attention of all is the one that may have the greatest influence on the assessments—those of us who do the calculations. A growing body of evidence, however, suggests that scientists may be subject to some of the same foibles that affect the general public, and to a few more besides (*20*). Three sets of problems, in particular, are worthy of mention here.

## Calibration and Overconfidence

Like other human beings, scientists may fail to foresee all factors that can introduce errors into estimates. Examples are provided by difficulties in seeing all the ways in which components of a system are interrelated (*21*), by failing to foresee interactions among individually minor problems (*22*), by temptations to overlook the aspects of technological systems that are "nontechnical" or outside a given field of expertise (*23*), by insufficient sensitivity to the fragility of assumptions or to the problems of small sample sizes (*24*), or even by conscious decisions to simplify analysis by excluding low-probability events from consideration.

Perhaps partly because of these problems, even scientists may have excessive confidence in estimates, as illustrated by the well-developed science of physics and the value of a quantity as fundamental as the speed of light. An instructive compilation found 27 published surveys of the speed of light between 1875 and 1958 that included formal estimates of uncertainty. The measurements differed from the official 1984 value by magnitudes that would be expected to occur less than 0.005 of the time, by chance alone, when the original estimators' own calculations of the uncertainties in their estimates were used (25). The straightforward conclusion is that "the respective investigators' uncertainties . . . must [have been] significantly underestimated" (25, p. 793). Although the absolute magnitude of the errors declined significantly over time, there was no significant improvement in the degree to which the remaining uncertainty was estimated. The 1984 estimate of the speed of light (which has since been used to calibrate the length of a meter, rather than vice versa) falls entirely outside the range of standard error (1.48 × "probable error") for all the recommended values for the true velocity of light that were reported between 1930 and 1970 (25, figure 2).

Other examples can be reported for scientists ranging from engineers to physicians. In one study a group of internationally known geotechnical engineers were asked for their 50% confidence bands on the height of an embankment that would cause a clay foundation to fail; when an actual embankment was built, not one of the expert's bands was broad enough to enclose the true failure height (26). In another study, physicians reviewed medical histories and examined more than 1500 patients with coughs; of the group they diagnosed as having more than an 80% chance of having pneumonia, less than 20% actually did (27). Other studies of the ability to assess probabilities accurately, the problem of calibration, have found that calibration errors are unaffected by differences in intelligence or expertise (28) but may be increased by the importance of the task (29). Most prosaically, faculty members routinely underestimate the time required for faculty meetings, and I underestimated by 50% the time that would be needed to finish this article, despite having enough experience in writing such papers to be presumed to "know better" (30).

Overall, one would expect that only about 2% of the estimates made with a confidence level of 98% would prove to be surprises, but nonspecialist assessors may have a "surprise index" on the order of 20 to 40% (31), and even technical specialists can exhibit overconfidence. Indeed, it may be more instructive to turn to the relatively rare examples of experts who have been found not to exhibit high degrees of overconfidence. I am aware of only two such groups: weather forecasters (32) and the group of experts who publish forecast prices for horse bets (33). Both sets of experts receive enough feedback to calibrate the accuracy of their estimates, and both are subject to considerable scrutiny from lay experts if they fail to recognize calibration errors themselves.

## Statistical Vulnerability of Low-Probability Estimates

In one important area, moreover, those of us with training in probability theory may be subject to a potential bias that is rarely found among the general public. Nearly all of us have had the frustration of attempting to explain to the lay public, or even to students, that events with one-in-a-million probabilities are not impossible; a low estimated probability is not necessarily called into question if an unlikely event does in fact occur. Of all the events that are expected to occur only once every thousand years or so, some can be expected to occur each year, and a tiny proportion may even occur more than once per year. Yet our understanding of these principles may sometimes cause our hypotheses about extremely low probabilities to become effectively nonfalsifiable.

The familiar statistical problem of type I and type II errors—of rejecting hypotheses that are ultimately found to be true, on the one hand, or failing to reject those that are actually false, on the other—takes on new complexity in cases of incidents that are expected to occur once in a million reactor years, for example, but that actually occur twice in a single year (34). If empirical operating experience is limited, there is little scientific basis for deciding whether probability estimates are too low or too high. If we stick with our estimates, we avoid discarding them on the basis of what may prove to be isolated experiences, but in doing so, we make a de facto decision to trust reasoning that may be incorrect. Although many areas of risk assessment provide enough experience to correct such errors, events that are truly rare, or technologies that are still new or untried, may provide too little information to permit the needed corrections.

The problem is exacerbated by the statistical power of the hidden flaw. Low-probability estimates are especially vulnerable to the inaccuracies created when calculations fail to foresee even a small number of problems. Contrary to "common sense" expectations, the failure to recognize a problem in one portion of a probabilistic analysis is often not offset by an exaggerated conservatism in another portion of the analysis.

Consider a technology estimated to have a one-in-a-million chance of failing. For simplicity's sake, assume that risk assessors had succeeded in identifying all potential risk factors but two—one of which made the technology safer than the official estimate, and the other of which made it less safe. Imagine that the technology would still have a one-in-a-million level of risk during 80% of its operational life, but that 10% of the time, the real risk would be 1 in 1000, and 10% of the time the risk would be one in a billion. Then the true risk of the technology would be $(0.1 \times 10^{-9} + 0.8 \times 10^{-6} + 0.1 \times 10^{-3})$, that is, 10% times $10^{-9}$ (one in a billion), plus 80% times one in a million, plus 10% times 1 in 1,000, respectively—for an overall probability of 0.0001008001, or slightly more than 1 in 10,000. Rather than being offset by the presence of the unexpected safety factor, the unexpected problem dominates the ultimate probability. Indeed, even if the risk assessment were to have been so conservative in other respects that the "real" risks were to be no higher than one in a trillion except for the 10% of the operating experience where the 1 in 1,000 estimate would hold, the overall probability would still be higher than 1 in 10,000.

## Monetary and Political Pressures

So far in this article I have limited the discussion to cases in which no deliberate bias or distortion has been introduced into the risk assessment process. The biases or failings that have been noted are those of good-faith scientists and practitioners who are not subject to political, economic, or other pressures that might create biases of their own. The empirical world, however, is not always so tidy; scientists are sometimes subjected to distinctly unscientific pressures, and scientific results are not immune to being used by persons who may not fully share a scientific commitment to the fair and balanced reporting of evidence (35).

One obvious source of pressure would be the need to control costs, particularly in a competitive environment. Heimer (36) examined the willingness of technically trained industrial workers to take risks, even though most people tend to be quite reluctant to take risks for the sake of monetary gain (37). Drawing from analyses of accidents involving North Sea oil exploration (38), Heimer suggests the workers take risks "to avoid costs rather than to make gains. . . . When they are deciding whether to take a risk or not, offshore oil workers typically are not considering whether they will get some bonus for taking the risk, but instead whether they will be fired if they refuse" (36, p. 503).

Incentives to minimize or understate risks can also be created by political motivations, such as the desire to avoid public embarrassment, and by high levels of commitment to organization goals. It is easy to see the potential for abuse in totalitarian countries, or to be skeptical of the claim about Chernobyl and nearby reactors in the February 1986 issue of *Soviet Life* that "the odds of a meltdown are one in 10,000 years" (*39*, p. 4A). Yet our own system may also be vulnerable to such pressures—a point that can be illustrated by the Challenger incident.

As made clear in part through independent assessments and news reports (*40, 41*), NASA's official estimate of space shuttle risks, 1 in 100,000, is spectacularly at variance from the empirical record. Historical rates of failure have been on the order of 1 in 25 to 1 in 50, depending on the calculations used. Yet NASA "pressured one consultant to produce a more optimistic estimate of booster safety and disregarded even more pessimistic predictions contained in two subsequent studies (*41*, p. 1). A 1981 report calculated the historic failure rate of solid-fuel rockets to be 1 in every 57 firings, but concluded shuttle booster risks would be in the range of 1 in 1,000 to 1 in 10,000; a second study in 1983 concluded that the chance of a booster blowup was 1 in 70, and a third, "commissioned by the Air Force in 1984 to resolve the discrepancies between the first two, suggested the booster failure rate would be about 1 in 210" (*41*, p. 1). Finally, officials of NASA's Marshall Space Flight Center published their own risk estimate in February 1985, predicting a failure rate of 1 in 100,000, a rate of safety approximately 2,000 times better than actual experience (*42*). With the Challenger accident occurring on the 25th shuttle launch, and with two boosters per launch, the failure rate ultimately proved to be almost identical to the historical track record.

## IMPLICATIONS: REASSESSING "RATIONALITY"

Despite flaws in risk assessments, many scientists worry about the problems that could be created by allowing a greater public role in technological decisions (*43*). Scientific workshops, for example, focus on topics such as "risk assessment and the 'misinformed' public." To some extent, of course, excluding the public from decisions is a luxury that a democracy does not offer. Some observers, moreover, have argued we would not want to indulge in this luxury even if it were available; the role of the scientist is to serve society, not to run it (*44, 45*). But how can these beliefs be reconciled with the supposedly "well-known fact" that members of the general public "are poor decision makers" (*46*, p. 2)?

I suggest three ways. The first is to keep in mind that "science often fluctuates between hubris and humility" (*47*, p. 24). It is important to remember in the policy realm what we often tell students in the classroom—that science is not infallible and that scientists do err from time to time. The need to avoid hubris, or even simple overconfidence, may be particularly high when our probability estimates are particularly low.

The second suggestion is to realize that the social science literature on decision-making in the general public is less clear-cut than is sometimes assumed. Berkeley and Humphreys (*48*) found that the *Science* article on decision-making by Tversky and Kahneman (*49*) had been cited 227 times between 1975 and 1980, with roughly one-fifth of the citations in sources outside the field of psychology—100% of which used the citation to support the unqualified claim that people are poor decision-makers. Yet the original article was somewhat less sweeping in its claims, and the broader literature provides a more mixed picture. Another survey found 84 articles in which the key words decision-making, judgment, and problem-solving were used (and a comparison to an explicit normative model was provided); 47 reported poor performance in decision-making, and 37 reported good performance (*50*). Other fields of the social and behavioral sciences tend to report that people's decision-making processes are more rational than they may at first appear (*51*). In short, just as scientists' estimates may need to be treated with something less than reverence, the views of the public may need to be treated with something better than contempt.

The third suggestion is to reconsider the notion of "irrationality" even in cases when the public does fail to understand the scientific details of a technological decision. This suggestion requires that we go beyond the notion of the public as irrational, but also beyond the notion of the public as economically rational—making selfish risk-benefit comparisons on the basis of whatever information and values may apply—and consider instead the possibility of prudence.

## The Public as Prudent

It is instructive to turn from the large number of people who make up the general public to the small number who make up boards of trustees and boards of directors—the people who are expected to exercise prudence in directing society's largest and most influential organizations. In general, trustees and directors need to keep an eye on the big picture, not the little details; their job is to establish policy, not to pursue the particulars. A university's board of trustees is expected to look after the welfare of the overall organization, not the dealings between deans and departmental chairs, and certainly not the daily performance of tasks in research laboratories.

Given that the specialist is expected to look after specific details, there is a potential for friction between the trustee and the technician, even if each is doing his or her job in a way that is basically appropriate; at a minimum, the job of one is to ask questions that are of little concern to the other. At what point, however, would a prudent trustee decide not to defer to specialists who have been hired to look after the technical details? I asked this question to a small group of high-level policy-makers; they identified two sets of factors (52). The first has to do with characteristics of the technical specialists or experts, and the second with characteristics of the broader situation.

There are four problematic characteristics of specialists, the first of which occurs when an expert might have a vested interest in outcomes; for example, the suspicions of trustees tend to be raised if one department claims to need more money than all of the other departments combined. Concerns are also raised if the expert's recommendations or activities have implications for other parts of the organization, or if the expert fails to recognize constraints imposed by the larger picture. A third reason for concern is created if a given expert has been wide of the mark or has caused problems and difficulties in the past. The fourth arises when another expert warns the policy-maker that something seems to be seriously amiss. In all four cases, the gut-level credibility of the expert in question is likely to play a key role in guiding the policy-maker's response.

The second set of factors includes three types of situations that call for particular scrutiny. First are situations that incorporate a large element of the unknown: Activities that are familiar or draw on a large body of experience are not nearly as worrisome as those that get into areas in which the organization has less of an experience base, or in which the ratio of knowledge to guesswork is lower. Second are situations when the potential consequences of a mishap, specifically including an "unexpected" mishap, might be especially severe. It is unwise to put all of the organization's eggs into any one basket, for example, even if it is a good-looking basket. Third are situations when potential problems, if experienced, would be difficult or impossible to correct.

A more exhaustive search could develop a longer list, but the seven interrelated points in Table 1 should be sufficient. To return to technological controversies, everything on this list could be applied to examples such as nuclear power and to the group that in many ways is the ultimate board of trustees—the public. The public has little knowledge of technical details of the industry or the nuclear fuel cycle, just as trustees or directors tend

**TABLE 1** Typical "Warning Signs" That Would Cause a Prudent Board of Trustees to Question the Recommendations of Their Technical Experts.

---

*Characteristics of specialists*

Specialists have direct interest in outcomes
Specialists' past recommendations were wrong
Specialists' activities and recommendations have
  broader implications
Other experts indicate there may be reason for worry

*Characteristics of situations*

Those that contain a large element of the unknown
Those in which potential consequences of mistakes
  could be especially severe
Those in which errors have potential to be irreversible

---

not to have detailed familiarity with given technical processes. What the public does have, however, is a set of all seven of the kinds of warning signals that would draw the attention of a "real" board of directors (53).

Although the key proponents of nuclear power tend to be scientifically trained, they stand to benefit from its implementation, if only through research and employment opportunities. Early efforts to promote the industry may have been too enthusiastic in describing the benefits or downplaying its problems. Rather than producing electricity "too cheap to meter" (54), nuclear power plants completed in recent years have suffered pervasive problems with cost overruns (55), and attention to management and waste control at some facilities has been less vigorous than it might have been (16, 17). A widespread commitment to nuclear power has been criticized as having serious implications, ranging from disposing of nuclear wastes to the threat of increased "police state" characteristics (56), and difficulties in identifying sites for nuclear waste disposal have lent some credence to the argument (57). Some respected scientists argue that nuclear technologies are unsafe or insufficiently understood, and TMI and Chernobyl have given the criticisms an increased credibility.

A similar picture emerges when we consider the characteristics of the broader situation. Nuclear power is still widely seen as involving a large element of the unknown. Events of the recent past have sent signals not only that the technology is less well understood than people might hope, but that the consequences of accidents could be extremely severe. The notion of "irreversible" implications takes on additional meaning when the public hears that the official planning horizon for a high-level nuclear waste repository is 10,000 years—a period roughly double the age of written records.

Many of these points, of course, are associated with counterarguments. Cost overruns resulted from a variety of factors, some of which were largely outside the control of the industry. The early promotional literature was intended to get people to think about the technology in terms of peacetime potential rather than mushroom-shaped clouds, and downplaying of the problems or uncertainties might have seemed necessary at the time. Many of the earlier problems have been recognized and remedied. We all underestimate problems that might crop up in implementing an idea, whether in the form of writing a new paper or installing a new kitchen sink.

All these objections are legitimate, but to a cetain extent, they all miss the point. The public is often castigated for "irrationality" in its reactions to nuclear or other controversial technologies, and most people do know as little about the technical details as we might expect directors to know about a given technology in an organization that is served by literally thousands of different technologies. Even so, the public may be playing something like the role that "ought" to be played—particularly in managing controversial technologies that require decisions about values as well as about technical details (58).

## CONCLUSION

It is tempting to assume that risk management can be improved by settling scientific facts before worrying about any social implications

(59), or to assume that scientists identify "real" risks, with additional public concerns being due to misinformation or irrationality. Such assumptions may cause few problems when the stakes are low, consensus is high, experience is vast, and decisions do not impose burdens on one group for the benefit of another. The assumptions are clearly problematic, however, for controversies that involve high stakes, low consensus, new technologies, and unequal distributions of burdens and benefits. These kinds of technological controversies are often precisely those for which the perceived-versus-real argument is pushed with the greatest passion.

In studies that examine the ways in which citizens and scientists assess risks, investigators have found that citizens often reach ill-advised conclusions—and that scientists do as well. Although citizen judgments often incorporate misinformation, they can also reflect a deeper kind of prudence than is commonly realized.

Scientists' errors appear to be most problematic in two areas—those involving human and social factors, rather than physical and biological ones, and those requiring guesswork or judgment in the face of limited or nonexistent evidence. Monetary or political pressures can create additional problems and distortions.

There are no easy solutions to the problems of political, monetary, or personal biases, although such problems cannot be safely ignored. Nor are there easy solutions to the nonavailability of necessary data—although experts' resultant guesswork can be significantly worse than is often realized. Data gaps on human and social influences, however, are often correctable, but only if the needed research is performed.

In short, the social sciences should be asked to provide not just an improved understanding of public perceptions, but also significantly improved quantitative estimates of the probabilities as well as the consequences of important risks. This will require, however, that all of us explicitly begin to integrate human behaviors into our thinking about "technological" systems—and that we begin devoting approximately the same level of resources to understanding the human components of technological systems as to the hardware. Although often overlooked, human and social factors play vital roles in technological systems; real-world risks, far from being free of such inconvenient "people factors," are indeed often dominated by them.

## REFERENCES AND NOTES

1. L. Harris et al., *Risk in a Complex Society* (Marsh & McLennon, New York, 1980); V. T. Covello, J. Menkes, J. Mumpower, *Risk Evaluation and Management* (Plenum, New York, 1986); B. Fischhoff, P. Slovic, S. Lichtenstein, in *Analysis of Actual Versus Perceived Risks*, V. T. Covello, G. Flamm, J. Rodericks, R. Tardiff, Eds. (Plenum, New York, 1983), pp. 235–40.

2. W. R. Freudenburg and E. A. Rosa, Eds., *Public Reactions to Nuclear Power: Are There Critical Masses?* (Westview/AAAS, Boulder, CO, 1984).

3. J. S. Petterson, in *Waste Management '88*, R. G. Post and M. E. Wacks, Eds. (Univ. of Arizona Press, Tucson, in press).

4. This "was not simply a case of 'ignorant peasants' flopping around in confusion—or of pervasive cultural or information-based differences. For example, doctors and dentists trained in the U.S. routinely refused to treat patients without certificates . . . politically and economically well-placed individuals sought preferential treatment [certificates of noncontamination] to protect against stigma" (3).

5. If the possibility of a negative event is accompanied by uncertainty or "ambiguity" about the actual probability of the event, this creates further costs. A fuller discussion of insurance actuaries' pricing of risk premiums under situations of ambiguity is given by R. M. Hogarth and H. Kunreuther (*J. Risk Uncertainty*, in press.); H. J. Einhorn and R. M. Hogarth (*J. Bus.* **59**, S225, 1986).

6. J. F. Short, *Am. Soc. Rev.* **49**, 711 (1984); A. Baum, in *Cataclysms, Crises and Catastrophes: Psychology in Action*, G. R. VandenBos and B. K. Bryant, Eds. (American Psychological Association, Washington, DC, 1987), pp. 5–53.

7. P. Slovic, *Science* **236**, 280 (1987).

8. R. E. Kasperson, O. Renn, P. Slovic, *Risk Anal.* **8**, 177 (1988).

9. B. P. Dohrenwend et al., in *The Three-Mile Island Nuclear Accident: Lessons and Implications*, T. H. Moss and D. L. Sills, Eds. (New York Academy of Science, New York, 1981), pp. 159–74; D. L. Sills, C. P. Wolf, V. B. Shelanski, Eds., *Accident at Three-*

*Mile Island: The Human Dimensions* (Westview, Boulder, CO, 1982).

10. W. R. Freudenburg and R. K. Baxter, *Policy Stud. Rev.* **5**, 96 (1985); *Soc. Sci. Q.* **65**, 1129 (1984); E. A. Rosa and W. R. Freudenburg, in (2), pp. 3–37.

11. President's Commission on the Accident at Three Mile Island, *The Need for Change: The Legacy of TMI* (Government Printing Office, Washington, DC, 1979), p. 8.

12. C. B. Flynn, in (2), pp. 205–32; J. R. Egan, *Technol. Rev.* **85**, 23 (1982).

13. Office of Technology Assessment, *Transportation of Hazardous Materials* (OTA-SET-304, Government Printing Office, Washington, DC, 1986); B. E. Sabey and H. Taylor, in *Societal Risk Assessment: How Safe is Safe Enough?* R. C. Schwing and W. A. Albers, Eds. (Plenum, New York, 1980), pp. 43–65.

14. S. G. Hadden, *Risk Analysis, Institutions, and Public Policy* (Associated Faculty Press, Port Washington, NY, 1984); A. W. Lerner, *Admin. Soc.* **18**, 334 (1986). One paradoxical example is that the longer a technology operates without accidents, the harder it may be to sustain alertness and to continue operating safely. The problem may be worsened if redesigns cause operators to "depend on the equipment" instead of feeling sometimes-stressful levels of personal responsibility; G. I. Rochlin, T. R. LaPorte, K. H. Roberts, *Nav. War Coll. Rev.* **40**, 77 (1987). See also (36).

15. Nuclear Regulatory Commission Special Inquiry Group, *Three-Mile Island: A Report to the Commissioners and to the Public* (Government Printing Office, Washington, DC, 1980); Nuclear Regulatory Commission, *TMI-2 Lessons Learned Task Force Final Report* (NRC, Washington, DC, 1979).

16. M. L. Wald, "The Peach Bottom syndrome," *New York Times*, 27 March 1988, section 3, p. 1.

17. B. Dumaine, *Fortune* **114**, 40 (27 October 1986).

18. W. Rogers *et al.*, *Report to the President by Presidential Commission on the Space Shuttle Challenger Accident* (NASA, Washington, DC, 9 June 1987).

19. Larger organizations may be more prone to experience such departures from official policies, along with problems of top-level managers not being fully aware of problems, if only because of the number of communications links involved. In a hypothetical organization with an average correlation of 0.7 between what a given employee knows about a problem and what that person's supervisor knows, just two layers of bureaucracy would yield a correlation of less than 0.5 ($0.7 \times 0.7 = 0.49$), and seven layers would reduce the correlation to 0.082 ($0.7^7$). Overcoming this "organizational attenuation of information" would appear to require a concerted effort to highlight rather than to hide bad news. However, this is just the opposite of what often happens, particularly if an organization sees itself as facing a risk of failing to meet an important deadline, losing a key contract, or seeming not to be in control of a problem.

20. See the summary provided by B. B. Johnson and V. T. Covello [B. B. Johnson and V. T. Covello, Eds., *The Social and Cultural Construction of Risk: Essays on Risk Selection and Perception* (Reidel, Dordrecht, 1987)].

21. C. Perrow, *Normal Accidents: Living with High-Risk Technologies* (Basic, New York, 1984).

22. P. Slovic, B. Fischhoff, S. Lichtenstein, in (2), pp. 115–35.

23. J. P. Holdren, *Bull. At. Sci.* **32**, 20 (1976); J. P. Holdren, G. Morris, I. Mintzer, *Annu. Rev. Energy* **5**, 241 (1980).

24. B. Fischhoff, S. Lichtenstein, P. Slovic, S. L. Derby, R. L. Keeney, *Acceptable Risk* (Cambridge Univ. Press, New York, 1981), pp. 33–46.

25. M. Henrion and B. Fischhoff, *Am. J. Phys.* **54**, 791 (1986). In general, measurements of the physical constants are from experiments done in a single laboratory, providing a less satisfactory error estimate than from experiments involving different laboratories. The article notes highly similar problems of overconfidence in estimated errors for values such as $h$ (Planck's constant), $e$ (electron charge), $m_e$ (electron mass), and $N$ (Avogadro's number).

26. M. Hynes and E. VanMarcke, in *Mechanics in Engineering*, R. N. Darbey and N. C. Lind, Eds. (Univ. of Waterloo Press, Waterloo, Ontario, 1976), pp. 367–84.

27. J. J. J. Christensen-Szalanski and J. B. Bushyhead, *J. Exper. Psych.* **7**, 928 (1982). For related studies, see also A. A. De Smet, D. G. Fryback, J. R. Thornbury, *Am. J. Radiol.* **43**, 139 (1979); S. Lichtenstein, B. Fischhoff, L. D. Phillips, in *Judgment Under Uncertainty: Heuristics and Biases*, D. Kahneman, P. Slovic, A. Tversky, Eds. (Cambridge Univ. Press, New York, 1982), pp. 306–34.

28. S. Lichtenstein and B. Fischhoff, *Org. Behav. Human Perform.* **20**, 159 (1977).

29. J. E. Sieber, *J. Pers. Soc. Psychol.* **30**, 688 (1974).

30. Part of the problem is that these are often not statistical confidence intervals but statements of the expert's degree of belief in an estimate. "One's degree of belief in an outcome is based only on information selected as relevant from that which is available, while there is no way of knowing if even all the available knowledge is sufficient, let alone complete" [S. Rayner, in *Risk Analysis, Institutions, and Public Policy*, S. G. Hadden, Ed. (Associated Faculty Press, Port Washington, NY, 1984), p. 152]. Otway puts it more boldly: "Probability is no more than a degree of belief in an event or proposition whose truth has not been ascertained" [H. Otway, *Risk Anal.* **5**, 271 (1985)].

31. For example, B. Fischhoff, P. Slovic, S. Lichtenstein, *J. Exper. Psychol.* **3**, 552 (1977); see also (27).

32. A. H. Murphy and R. I. Winkler, *Nat. Weather Dig.* **2**, 2 (1977).

33. J. Dowie, *Economica* **43**, 139 (1976).

34. For example, E. Marshall, *Science* **220**, 280 (1983); see also (*27*).

35. A Schnaiberg, *The Environment: From Surplus to Scarcity* (Oxford Univ. Press, New York, 1980).

36. C. A. Heimer, *Annu. Rev. Sociol.* **14**, 491 (1988).

37. D. Kahneman and A. Tversky, *Sci. Am.* **246**, 160 (January 1982).

38. W. G. Carson, *The Other Price of Britain's Oil: Safety and Control in the North Sea* (Rutgers Univ. Press, New Brunswick, NJ, 1982).

39. Translation, in "Soviets praised industry's safety," *USA Today*, 29 April 1986, p. 4A. The article continued with the claim that "the plants have safe and reliable controls that are protected from any breakdowns."

40. E. Marshall, *Science* **232**, 1596 (1986).

41. M. Thompson, "NASA defined 5 years of warnings," *Wisconsin State Journal*, 1 June 1986, p. 1.

42. Marshall Space Flight Center, *Space Shuttle Data for Planetary Mission RTG Safety Analysis* (NASA, Marshall Space Flight Center, AL, 15 February 1985).

43. B. Cohen, *Risk Anal.* **5**, 1 (1985).

44. For example, O. Renn, in *Regulating Industrial Risks: Science, Hazards, and Public Protection*, H. Otway and M. Peltu, Eds. (Butterworths, London, 1985), pp. 111–27; A. M. Freeman III, in *People, Penguins, and Plastic Trees: Basic Issues in Environmental Ethics*, D. Van De Veer and C. Pierce, Eds. (Wadsworth, New York, 1986), pp. 218–27.

45. Particularly significant problems are introduced by the fact that many decisions "about technology" actually require value-based along with fact-based decisions, especially given that the values of scientists are often quite different from those of the general public. For illustrations, see S. M. Nealey and J. A. Hebert, in *Too Hot to Handle: Social and Policy Issues in the Management of Radioactive Wastes*, C. E. Walker, L. C. Gould, E. J. Woodhouse, Eds. (Yale Univ. Press, New Haven, 1983), pp. 94–111; R. E. Dunlap and M. E. Olsen, *Policy Stud. J.* **13**, 413 (1984); also (*20*) and (*24*).

46. L. L. Lopes, paper presented at *Symposium in Mass Communication* (Madison, WI, 19 November 1987).

47. R. Kates, *Nat. Res. Counc. News Rep.* **35**, 24 (7 July 1985).

48. D. Berkeley and P. Humphreys, *Acta Psychol.* **50**, 201 (1982).

49. A. Tversky and D. Kahneman, *Science* **185**, 1124 (1974).

50. J. J. J. Christensen-Szalanski and L. R. Beach, *Am. Psychol.* **39**, 75 (1984).

51. H. J. Einhorn and R. M. Hogarth [*Annu. Rev. Psychol.* **32**, 53 (1981)] note that the superiority of formal decision-making models over people's everyday rules of thumb may be less impressive in the "messiness" of the real world than in the artificial neatness of psychological laboratories. There is a clear possibility that "evolution is nature's way of doing cost/ benefit analysis" (*ibid.*, p. 54), and that everyday rules of thumb may have survived the tests of natural selection, although this is a possibility rather than a proven fact. The authors note the possibility that reports of "flawed" decision-making may be subject to half-empty–half-full interpretations on the part of experimenters, particularly given that research on "lower animals" often reports behaviors that are impressively "consistent with optimizing principles. . . . The danger of such pictures is that they are often painted to be interesting rather than complete" (*ibid.*, p. 55). See also (*45*).

52. Discussion drawn from W. R. Freudenburg [in *Waste Management '87*, R. G. Post, Ed. (Univ. of Arizona Press, Tucson, AZ, 1987), pp. 109–15].

53. In this context, it may be worth reexamining the frustration that many feel when members of the public ignore the technical considerations that scientists find important and focus instead on what seem to be impossibly broad or global questions. The "trustees" metaphor would suggest that these may be precisely the types of "overall policy" questions that should be answered by someone other than the scientist, except when the scientist is acting as a citizen-trustee.

54. L. L. Straus, remarks to National Association of Science Writers, New York City, 16 September 1954. Reprinted in *Background Info* (Atomic Industrial Forum, Washington, DC, 1987), and in D. Ford, *The Cult of the Atom: The Secret Papers of the Atomic Energy Commission* (Simon and Schuster, New York, 1982, p. 50.

55. Department of Energy, *1983 Survey of Nuclear Power Plant Construction Costs* [DOE/EIA-0439(83), Energy Information Administration, Washington, DC, 1983].

56. The classic statement is by A. B. Lovins, *Foreign Aff.* **55**, 65 (October 1976); see also the compilation by S. H. Murdock *et al.* [*Nuclear Waste: Socioeconomic Dimensions of Long-Term Storage* (Westview, Boulder, CO, 1983)]; C. E. Walker, L. C. Gould, E. J. Woodhouse, Eds., *Too Hot to Handle: Social and Policy Issues in the Management of Radioactive Wastes* (Yale University Press, New Haven, 1983).

57. After extended debate, Congress attempted to resolve the issue with the Nuclear Waste Policy Act in 1982, but lawmakers amended the act within 5 years, dropping a carefully developed process in favor of sticking "a pin in the map" at a potential site in Nevada [E. Marshall, *Science* **239**, 15 (1988)]. Like earlier "solutions," this one may or may not last. D. S. Zinberg, in (*2*), pp. 233–53.

58. I refer to "the" public in the interest of simplicity, but there are of course many publics, not all of which are equally likely to be castigated for irrationality. Thus it may be instructive that the segment of the public made up by the insurance industry, with a long history of being held accountable for pragmatic risk assessments, has become reluctant to provide liability insurance for nuclear installations and environmental hazards, in other words, for some of the same types of installations that inspire aversive

reactions among the publics at large. Public opinion polls do not suggest that the public is willing to shut down properly operating reactors at this time, but there is mounting evidence that new nuclear facilities, ranging from nuclear waste repositories to recently completed reactors, are increasingly seen as unacceptable. B. C. Farhar-Pilgrim and W. R. Freudenburg, in (2), pp. 183–203; see also (10).

59. National Academy of Science, *Risk Assessment in the Federal Government: Managing the Process* (National Academy Press, Washington, DC, 1983).
60. I thank D. Anderson, K. T. Erikson, A. M. Freeman III, R. E. Kasperson, H. Kunreuther, E. Nichols, R. A. Rappaport, P. Slovic, R. Stevenson, and several additional colleagues in the informal Risk Interest Group at the University of Wisconsin–Madison, for comments on this article.

# 20

# SCIENCE AND ITS LIMITS: THE REGULATOR'S DILEMMA[1]

*Alvin M. Weinberg*

In his essay "Risk, Science, and Democracy," William D. Ruckelshaus expresses very clearly what I call the regulator's dilemma. During the past 15 years, Ruckelshaus notes, there has been a shift in public emphasis from visible and demonstrable pollution problems, such as smog resulting from automobiles and raw sewage, to potential and largely invisible problems, such as the effects of low concentrations of toxic pollutants on human health. This shift is important for two reasons. First, it has changed the way that science is applied to practical questions of public health protection and environmental regulation. Second, it has raised difficult questions about managing chronic risks within the context of free and democratic institutions.[2]

When the environmental concern was patent and obvious—such as the problem of smog in Los Angeles—science could and did provide unequivocal answers. Smog, for example, comes from the gas emissions from burning liquid hydrocarbons, and the answer to the smog problem lies in controlling these emissions. The regulator's course was rather straightforward because the science upon which regulatory decisions are made was operating well within its power. However, when the environmental concern is subtle—for example, how much cancer is caused by an increase of 10 percent in mean background radiation—science is being asked a question that lies beyond its power; the question is trans-scientific. Yet the regulator, by law, is expected to regulate even though science can hardly help him; this is the regulator's dilemma.

Although my essay is subtitled The Regulator's Dilemma, many of the same issues arise in the adjudication of disputes over who is to blame and who is to be compensated for damage allegedly caused by rare events, such as nuclear accidents. The regulator's dilemma is also faced by the judge who is presiding over a tort case involving, for example, a claim for damages blamed on a toxic waste dump. Indeed, the regulator's dilemma could equally be called the toxic tort dilemma.

A lawsuit involving alleged injury from chemical pollutants is unlike the traditional liability case. If my car injures a pedestrian, I am liable to be sued. What is at issue, however, is not whether I have injured a pedestrian. Rather, it is whether I am at fault. On the other hand, if the lead from my car's exhaust is alleged to cause bodily harm, the issue is not whether my car emitted the lead but whether the lead actually caused the alleged harm. The two situations are quite different. In the first example the relation between cause and injury is not at issue. In the second it is the issue.

In this essay, therefore, I try to delineate more precisely those limits to science that give rise to the regulator's dilemma. I speculate on how these intrinsic limits to science seem to have catalyzed a profound attack on science by some sociologists and public-interest activists. In addition, I offer a few ideas that may help the harried regulators finesse these trans-scientific issues.

## II

Science deals with regularities in our experience; art deals with singularities. It is no wonder that science tends to lose its predictive or even explanatory power when the phenomena it deals with are singular, irreproducible, and one of a kind—in other words, rare. Although science can often analyze a rare event after the fact—for example, the extinction of dinosaurs during the Cretaceous-Tertiary period following the presumed collision of the earth and an asteroid—it has great difficulty predicting when such an uncommon event will occur.

I distinguish here between two sorts of rare events—accidents and low-level insults, whose potential to cause injury is unknown. Accidents are large-scale malfunctions whose etiology is not in doubt, but whose likelihood is very small. The partial nuclear reactor meltdown at Three Mile Island in 1979 and the release of toxic gas from a chemical plant at Bhopal, India, in 1984 are examples of accidents. The precursors to these specific events—for example, the condition of the auxiliary water feed system and other components at Three Mile Island—and the way in which the accidents unfolded are well

understood. Estimates of the likelihood of the particular sequence of malfunctions are less firmly grounded. As the number of individual accidents increases, prediction of their probability becomes more and more reliable. We can predict very well how many automobile fatalities will occur in 1986; we can hardly claim the same degree of reliability in predicting the number of serious reactor accidents in 1986.

Low-level insults are rare in a rather different sense. We know that about 100 rems of radiation will double the mutation rate in a large population of exposed mice. How many mutations will occur in a population of mice exposed to 100 millirems of radiation? In this case the mutations, if induced at all by such low levels of exposure, are so rare that to demonstrate an effect with 95 percent confidence would require the examination of many millions of mice. Although such an effort is not impossible in principle, it is in practice. Moreover, even if we could perform so heroic a mouse experiment, the extrapolation of these findings to humans would still be fraught with uncertainty. Thus, human injury or abuse from low-level exposure to radiation is a rare event whose frequency cannot be accurately predicted.

## III

When dealing with events of this sort, science resorts to the language of probability. Instead of saying that this accident will happen on that date, or that a particular person exposed to a low-level dose of radiation will suffer a particular fate, it tries to assign probabilities for such occurrences. Of course, where the number of instances is very large or the underlying mechanisms are fully understood, the probabilities are themselves perfectly reliable. In quantum mechanics there is no uncertainty as to the probability distribution of the phenomenon being described. In the class of phenomena considered here, however, even though the likelihood of an event happening or of a disease being caused by a specific exposure is given as a probability, the probability itself is very uncertain. One can think of a somewhat fuzzy demarcation between what I have called science and trans-science. The domain of science covers phenomena that are deterministic or whose probability of occurrence can itself be stated precisely; in contrast, trans-science covers those events whose probability of occurrence is itself highly uncertain.

Despite the difficulties, scientific mechanisms have been devised for estimating, however imperfectly, the probability of rare events. For accidents the technique is probabilistic risk assessment (PRA); for low-level insults various empirical and theoretical approaches are used.

Although probabilistic risk assessment had been used in the aerospace industry for a long time (for example, to predict the reliability of components), it first sprang into public prominence in 1975 with a reactor safety study directed by nuclear engineer Norman C. Rasmussen.[3] The Rasmussen study, sponsored by the Atomic Energy Commission (now known as the Nuclear Regulatory Commission), was designed to estimate the public risks involved in potential accidents at commercial nuclear reactors.

Probabilistic risk assessment, when applied to nuclear reactors, seeks to identify all sequences of subsystem failures that may lead to a failure of the overall system; it then tries to estimate the consequences of each subsystem failure so identified. The result is a probability distribution, $P(C)$; that is, the probability, $P$, per reactor year, of a consequence having magnitude $C$. Consequences include both material damage and health effects. Usually, the probability of accidents having large consequences is less than the probability of accidents having small consequences.

A probabilistic risk assessment for a reactor requires two separate estimates: first, an estimate of the probability of each accident sequence; second, an estimate of the consequences—particularly the damage to human health—caused by the uncontrolled radioactive effluents released in the accident. An accident sequence is a series of equipment or human malfunctions, such as a pump that fails to start, a valve that does not close, or an operator confusing an ON with an OFF signal. We have statistical data for many of these individual events; for example, enough valves have operated for enough years so that we can, at least in principle, make pretty good estimates of the probability of failure.

Uncertainties still remain, however, because we can never be certain that we have identified every relevant sequence. Proof of the adequacy of probabilistic risk assessment must therefore await the accumulation of operating experience. For example, the median probability of a core melt in a light water reactor, according to the 1975 Rasmussen study, was 1 in every 20,000 reactor-years; the core melt at Three Mile Island's number two reactor (TMI-2) occurred after only 700 reactor-years. The number two reactor, however, differed from the reactors Rasmussen studied, and in retrospect, one could rationalize most of the discrepancy between his estimate and the seemingly premature occurrence at TMI-2.

Since the core melt at Three Mile Island, the world's light water reactors have accumulated some 1,500 reactor-years of operation without a core melt. This performance places an upper limit on the a priori estimate of the core-melt probability. Thus, if this probability were as high as 1 in every 1,000 reactor years, the likelihood of surviving 1,500 reactor-years would not be more than 22 percent; put otherwise, we can say with 78 percent confidence that the core-melt probability is not as high as 1 in 1,000 reactor years. With 500 light water reactors on line in the world, should we survive until the year 2000 without another core melt, we could then say with 95 percent confidence that the core-melt probability is not higher than 1 in 3,000 reactor-years. In the absence of such experience, one is left with rather subjective judgments. Although Harold W. Lewis, in his critique of Rasmussen's 1975 study,[4] asserts that he could not place a bound on the uncertainty of probabilistic risk assessment, Rasmussen argued that his estimate of core-melt probability may be in error by about a factor of 10 either way— that is, the probability may be as high as 1 in 2,000 reactor-years or as low as 1 in 200,000 reactor-years.

As we see, after 3,000 reactor-years of operation without a core melt, we can say with about 78 percent confidence that Rasmussen's upper limit (1 in 2,000 reactor-years) is not too optimistic. Furthermore, if we survive to the year 2000 without a core melt, the confidence level with which we can make this assertion rises to 95 percent. Our confidence in probabilistic risk assessment can eventually be tested against actual, observable experience. Until this experience has been accumulated, however, we must concede that any probability we predict must be highly uncertain. To this degree our science is incapable of dealing with rare accidents, but time, so to speak, annihilates uncertainty in estimates of accident probability.

Unfortunately, time does not annihilate uncertainties over consequences as unequivocally as it does uncertainties over frequency of accidents. A large reactor or chemical plant accident can cause both immediate, acute health effects and delayed, chronic effects. If the exposure either to radiation or to methyl isocyanate is high enough, the effect on health is quite certain. For example, a single exposure of about 400 rems will cause about half of the people exposed to die. On the other hand, in a large accident many people will also be exposed to smaller doses—indeed, to doses so low that the resulting health effects are undetectable. At Bhopal many thousands of people were exposed to methyl isocyanate but they recovered. We cannot say positively whether or not they will suffer some chronic disability.

The very worst accident envisaged in the Rasmussen study, with a probability of 1 in 1 billion reactor-years, projected an estimated 3,300 early fatalities, 45,000 early illnesses, and 1,500 delayed cancers per year among 10 million exposed people. Almost all of the estimated delayed cancers are attributed to exposures of less than 1,000 millirems per year—a level at which we are very hard put to estimate the risk of inducing cancer. Similarly, the American Physical Society's critique of the Rasmussen study attributed an additional 10,000 deaths over 30 years among 10 million people exposed to cesium-135 distributed in a very large accident.[5] The average exposure in this case was assumed to be 250 millirems per year—again, a level at which our estimates of the health effects are extremely uncertain.

Has the nuclear community, particularly its regulators, figuratively shot itself in the foot by trying to estimate the number of delayed casualties as a result of these low-level exposures? In retrospect, I think the Rasmussen study would have been on more solid ground had it confined its estimates to those health effects resulting from exposures at higher levels, where science makes reliable estimates. For the lower exposures the consequences could have been stated simply as the number of man-rems (the number of people multiplied by the number of rems) of exposure of individuals whose total exposure did not exceed, say, 5,000 millirems, without trying to convert this man-rems number into numbers of latent cancers. Thus, health consequence would be reported in two categories: (1) for highly exposed individuals, the number of health effects; and (2) for slightly exposed individuals, the total man-rems or even the distribution of exposures accrued by the large number of individuals so exposed. Perhaps such a scheme could be adopted in reporting the results of future probabilistic risk assessments; at least it has the virtue of being more faithful than the present convention to the state of scientific knowledge.

## IV

In both of my examples of accidents (Bhopal and nuclear accidents), many people are exposed to low-level insult. The uncertainties inherent in estimating the effects of such low-level exposure are heaped on top of the uncertainties in estimating the probability of the accident that may lead to exposure in the first place.

Science has exerted great effort to ascertain the shape of the dose-response curve at low dose—but very little, if anything, can be said with certainty about the low-dose response. Thus, to quote the report of the National Research Council, *The Effects on Populations of Exposure to Low Levels of Ionizing Radiation: 1980* (also known as BEIR-III, for the committee that prepared it, the Committee on the Biological Effects of Ionizing Radiation), "The Committee does not know whether dose rates of gamma or x-rays of about 100 mrads/yr are detrimental to man. . . . It is unlikely that carcinogenic and teratogenic effects of doses of low-LET radiation administered at this dose rate will be demonstrable in the foreseeable future."[6] This prompted Philip Handler, then president of the National Academy of Sciences, to comment in his letter of transmittal to the Environmental Protection Agency, which had requested the study, "It is not unusual for scientists to disagree . . . (and) . . . the sparser and less reliable the data base, the more opportunity for disagreement. . . . The report has been delayed . . . to permit time . . . to display all of the valid opinions rather than distribute a report that might create the false impression of a clear consensus where none exists."[7]

This forthright admission that science can say little about low-level insults I find admirable. It represents an improvement over the unjustified assertion in the BEIR-II report of 1972 that 170 millirems per year over 30 years, if imposed on the entire U.S. population, would cause between 3,000 and 15,000 cancer deaths per year.[8] I do not quarrel with the estimated upper limit—which amounts to 1 cancer per 2,500 man-rems, but I regard placing the lower limit at 3,000 rather than at zero as unjustified. Moreover, I think it has caused great harm. The proper statement should have been that at 170 millirems per year, we estimate the upper limit for the number of cancers to be 15,000 per year; the lower limit may be zero.

Since the appearance of the BEIR reports, two other developments have added to the burden of those who must judge the carcinogenic hazard of low-level insults: an awareness and study of (1) natural carcinogens, and (2) ambiguous carcinogens.

### Natural Carcinogens

Is cancer environmental in the sense of being caused by technology's effluents, or is it a natural consequence of aging? In the past few years we have seen a remarkable shift in viewpoint; whereas 15 years ago most cancer experts would have accepted a primarily environmental etiology for cancer, today the view that natural carcinogens are far more important than are manmade ones has gained many converts. In his 1983 article in *Science*, biochemist Bruce N. Ames marshaled powerful evidence that many of our most common foods contain naturally occurring carcinogens.[9] Indeed, biochemist John R. Totter, former director of the Atomic Energy Commission's division of biology and medicine, has offered evidence for the oxygen radical theory of carcinogenesis: that we eventually get cancer because we metabolize oxygen and subsequently produce oxygen radicals that can play havoc with our DNA.[10] As such views of the etiology of cancer acquire scientific support, I think that the trans-scientific question, as to how much cancer is caused by a tiny chemical or physical insult will be recognized as irrelevant. One does not swat gnats when pursued by elephants.

## Ambiguous Carcinogens

To further complicate the cancer picture, there is evidence that some agents, such as dioxin, various dyes, and even moderate levels of radiation, seem to diminish the incidence of some cancers while simultaneously increasing the incidence of others. The lifespan of the animals exposed to these agents in laboratory tests on average exceeds that of animals not exposed.[11] A most striking example, given by biostatistician Joseph K. Haseman, is yellow dye number 14 given to leukemia-prone female rats. This dye completely suppresses leukemia, which is always fatal, but causes liver tumors, most of which are benign.

I mention these two findings—or perhaps they should be considered points of view—to stress my underlying point: that when we are concerned with low-level insult to human beings, we can say very little about the cancer dose-response curve. Saying that so many cancers will be caused by so much low-level exposure to so many people, a practice that terrifies many people, goes far beyond what science actually can say.

## V

Does the scientific community accept the notion that there are intrinsic limits to what it can say about rare events; that as events become rarer, the uncertainty in the probability of occurrence of a rare event is bound to grow? Perhaps a better way of framing this question is: Of what use can we put scientific tools of investigation of rare events, such as probabilistic risk assessment and large-scale animal experiments, if we concede that we can never get definitive answers?

I believe that probabilistic risk assessment with an uncertainty factor as high as 10 is often useful, especially if one uses the technique for comparing risks. For example, the 1,500 reactor-years already experienced since the Three Mile Island accident suggest that a reactor core-melt probability is likely to be less than 1 in 1,000 reactor-years and may well be as low as less than 1 in 10,000 reactor-years. This is to be compared with dam failures whose probability, based on many hundreds of thousands of dam-years (where time has annihilated uncertainty), is around 1 in 10,000 dam-years. Even with an uncertainty factor of 10, we can judge how safe reactors are compared to dams.

When one compares the relative intrinsic safety of two very similar devices—such as two water-moderated reactors—probabilistic risk assessment is on much more solid ground. Here one is not asking for absolute estimates of risk, but rather for estimates of relative safety. If reactors A and B differ in only a few details—say reactor A has two auxiliary water feed trains whereas B has only one—the ratio of core-melt probabilities should be much more reliable than their absolute values because the ratio requires an estimate of failure of a single subsystem, in this instance the extra auxiliary water feed on reactor A.

Not only can one say with reasonable assurance how much safer reactor A is than reactor B, but as a result of the detailed analysis one can identify the subsystems that contribute most to the estimated failure rate. Even if probabilistic risk assessment is inaccurate, it is very useful in unearthing deficiencies; one can hardly deny that a reactor in which deficiencies revealed by probabilistic risk assessment have been corrected is safer than one in which they have not been corrected, even if one is unwilling to say how much safer.

Somewhat the same considerations apply to low-level insult. An agent that does not shorten lifespan at high dose will not shorten lifespan at low dose. An agent that is a very powerful carcinogen at high dose is more likely to be a carcinogen at low dose than one that is a less powerful high-dose carcinogen. Thus, animal experiments surely are useful in deciding which agents to worry about and which not to worry about. Of course, the Ames test (which determines by a relatively simple procedure whether a substance is mutagenic) has made at least some preliminary screening of carcinogens more feasible because substances that cause mutations are considered to be potential carcinogens. The difficulty today seems to be not so much identifying agents that at high dose may be carcinogens as it is prohibiting exposures far below levels at which no effect can be, or perhaps ever will be, demonstrated. The regulator and the concerned citizen are inclined to approve the Delaney clause of the Federal Food, Drug, and Cosmetic Act, which prohibits the use of any food additive that has been shown to cause cancer in laboratory animals or humans. This clause, however, is of no help in resolving such issues as the relative risks of, say, cancer induction by nitrosamines (carcinogenic compounds that can be formed in the body from nitrites) and digestive disorders caused by meat untreated with nitrites.

The Delaney clause is the worst example of how a disregard of an intrinsic limit of science can lead to bad policy by overenthusiastic politicians. Harvard physicist Harvey Brooks has often pointed out that one can never prove the impossibility of an event that is not forbidden by a law of nature. Most will agree that a perpetual motion machine is impossible because it violates the laws of thermodynamics. That one molecule of a polychlorinated biphenyl (PCB) may cause a cancer in humans is a proposition that violates no law of nature: hence many, even within the scientific community, seem willing to believe that this possibility is something to worry about. It was this error that led to the Delaney clause.

When is an event so rare that the prediction of its occurrence forever lies outside the domain of science and therefore within the domain of trans-science? Clearly we cannot say, and perhaps as science progresses, this boundary between science and trans-science will recede toward events of lower frequency. At any stage, however, the boundary is fuzzy, and much scientific controversy boils over deciding where it lies. One need only read the violent exchange between Edward P. Radford

and Harald H. Rossi over the risk of cancer from low levels of radiation to recognize that where the facts are obscure, argument— even ad hominem argument—blossoms.[12] Indeed, Alice Whittemore in her "Facts and Values in Risk Analysis for Environmental Toxicants," has pointed out that facts and values are always intermingled at this "rare event" boundary between science and trans-science.[13] A scientist who believes that nuclear energy is evil because it inevitably leads to proliferation of nuclear weapons (which is a common basis for opposition to nuclear energy) is likely to judge the data on induction of leukemia from low-level exposures at Nagasaki differently than is a scientist whose whole career has been devoted to making nuclear power work. Cognitive dissonance is all but unavoidable when the data are ambiguous and the social and political stakes are high.

## VI

No one would dispute that judgments of scientific truth are much affected by the scientist's value system when the issues are at or close to the boundary between science and trans-science. On the other hand, as the matter under dispute approaches the domain of science, most would claim that the scientist's extrascientific values intrude less and less. Soviet scientists and U.S. scientists may disagree on the effectiveness of a ballistic missile defense, but they agree on the cross section of $U^{235}$ or the lifetime of the pi meson.

This all seems obvious, even trite. Yet in the past decade or so a school of sociology of knowledge has sprung up in Great Britain that claims that "scientific views are determined by social (external) conditions, rather than by the internal logic of scientific tradition and inherent characteristics of the phenomenal world,"[14] or that "all knowledge and knowledge claims are to be treated as being socially constructed: genesis, acceptance, and rejection of knowledge [is] sought in the domain of the Social World rather than . . . the Natural World."[15]

The attack here is not on science at the boundary with trans-science, in particular—the prediction of the frequency of rare events. At least the more extreme of the sociologists of knowledge claim that using traditional ways of establishing scientific truth—by appealing to nature in a disciplined manner—is not how science really works. Scientists are seen as competitors for prestige, pay, and power, and it is the interplay among these conflicting aspirations, not the working of some underlying scientific ethic, that defines scientific truth. To be sure, these attitudes toward science are not widely held by practicing scientists; however, they are taken seriously by many political activists who, though not in the mainstream of science, nevertheless exert important influence on other institutions—the press, the media, the courts—that ultimately influence public attitudes toward science and its technologies.

If one takes such a caricature of science seriously, how can one trust a scientific expert? If scientific truth, even at the core of science, is decided by negotiation between individuals in conflict because they hold different nonscientific beliefs, how can one say that this scientist's opinion is preferable to that one's? Furthermore, if the matter at issue moves across the boundary between science and trans-science, where all we can say with certainty is that uncertainties are very large, how much less able are we to distinguish between the expert and the charlatan, between the scientist who tries to adhere to the usual norms of scientific behavior and the scientist who suppresses facts that conflict with his political, social, or moral preconceptions?

One way to deal with these assaults on scientists and scientific truth would be to define a new branch of science, called regulatory science, in which the norms of scientific proof are less demanding than are the norms in ordinary science. I should think that a far more honest and straightforward way of dealing with

the intrinsic inability of science to predict the occurrence of rare events is to concede this limitation and not to ask of science or scientists more than they are capable of providing. Instead of asking science for answers to unanswerable questions, regulators should be content with less far-reaching answers. For example, where the ranges of uncertainty can be established, regulate on the basis of uncertainty; where the ranges of uncertainty are so wide as to be meaningless, recast the question so that regulation does not depend on answers to the unanswerable. Furthermore, because these same limits apply to litigation, the legal system should recognize, much more explicitly than it has, that science and scientists often have little to say, probably much less than some scientific activists would admit.

The expertise of scientific adversaries is often at the heart of litigation over personal injury alleged to be caused by subtle, low-level exposures. Each side presents witnesses whose scientific credentials it regards as impeccable. Because the issues themselves tend to be trans-scientific, one can hardly decide the validity of the assertions of either side's witnesses. Under the circumstances, I suppose, one is justified in regarding a scientific witness no differently than any other witness; his credibility is judged by his past record, behavior, and general demeanor, as well as the self-consistency of his testimony. Such, at least, was the way in which a federal district court judge, Patrick Kelley, settled *Johnston v. United States*, in which the issue was the claim that exposure to radiation from reworking old aircraft instrument dials had caused injury; Kelley impugned, on grounds no different from those one would invoke in an ordinary lawsuit, the competence if not the integrity of some of the plaintiff's scientific witnesses.

## VII

There are various ways to provide some assurance of safety despite uncertainty. Here I briefly describe two of these ways—which

I call the technological fix and de minimis—without claiming that these are the most important, let alone the only, ones.

## Technological Fix

Science cannot exactly predict the probability of a serious accident in a light water reactor or the likelihood that a radioactive waste canister in a depository will dissolve and release radioactivity to the environment. Can one design reactors or waste canisters for which the probability of such occurrences is zero—or at least, where the prevention of such mishaps relies on immutable laws of nature that can never fail rather than on the less than reliable intervention of electromechanical devices? Surprisingly, this approach to nuclear safety has come into prominence only in the past five years. Kåre Hannerz in Sweden and Herbert Reutler and Günter H. Lohnert in West Germany have proposed reactor systems whose safety does not depend on active interventions, but rather on passive, inherent characteristics.[16] Although one cannot say that the probability of mischance has been reduced to zero, there is little doubt that the probabilities are several, perhaps three, orders of magnitude lower than the probabilities of mischance for existing reactors. To the extent that such proposed reactors embody the principle of inherent safety, their adoption would avoid much of the dispute over reactor safety, the limits on nuclear accident liability contained in the Price-Anderson Act, repetition of the Three Mile Island accident, and so forth. In short, such a technological fix enables one largely to ignore the uncertainties in any prediction of core-melt probabilities.

The idea of incorporating inherent or passive safety into the design of chemical plants had been proposed by Trevor A. Kletz of the Loughborough University of Technology in 1974, shortly after the disaster at the Flixborough cyclohexane plant, which killed 28 people.[17] I suspect that one of the main consequences of the Bhopal disaster will be the incorporation of inherent safety features into new chemical plants; again, a way of finessing uncertainty in predicting failure probabilities.

## De minimis

A perfect technological fix, such as a totally safe reactor or a crash-proof car, is usually not available, at least at an affordable cost. Some low-level exposure to materials that are toxic at high levels is inevitable, even though we can never accurately establish the risk of such exposure. One way of dealing with this situation is to invoke the principle of de minimis. This principle, as Howard I. Adler and I suggested several years ago, argues that for insults that occur naturally and to which the biosphere has always been exposed and presumably to which it has adapted, one should not worry about any additional man-made exposure as long as the man-made exposure is small compared to the natural exposure.[18] The basic idea is that the natural level of a ubiquitous exposure (such as cosmic radiation), if it is deleterious, cannot have been very deleterious because in spite of its ubiquity, humans have survived. Moreover, we do not know and can never know what the residual effect of that natural exposure really is. An additional exposure that is small compared to natural background radiation should be acceptable; at the very least, its deleterious effect, if any, cannot be determined.

Adler and I suggested that for radiation whose natural background is well known, one may choose a de minimis level as the standard deviation of the natural background. This turns out to be around 20 percent of the mean background, around 20 millirems per year; this value has been used as the Environmental Protection Agency standard for exposure to the entire radiochemical fuel cycle.

Scientists know more about the natural incidence and biological effects of radiation than they do about any other agent. It would be natural, therefore, to use the standard established for radiation as a standard for other agents. This approach has been used by chemist T. Westermark of the Royal Institute of Technology in Stockholm. He has suggested that for naturally occurring carcinogens such as arsenic, chromium, and beryllium, one may choose a de minimis level to be, say, 10 percent of the natural background.[19]

Clearly, a de minimis level will always be somewhat arbitrary. Nevertheless, it seems to me that unless such a level is established, we shall forever be involved in fruitless arguments, the only beneficiaries of which will be the toxic tort lawyers. Could the principle of de minimis be applied in litigation in much the same way it may be applied to regulation—that is, if the exposure is below de minimis, then the blame is intrinsically unprovable and cannot be litigated? I would imagine that the legal de minimis may be set higher than the regulatory de minimis; for example, the legal de minimis for radiation could be the background (after all, the BEIR-III committee concedes there is no way of knowing whether or not such levels are deleterious). The regulatory de minimis could justifiably be lower, simply on grounds of erring on the side of safety.

One approach may be to concede that there is some level of exposure that is beyond demonstrable effect. This defines a trans-scientific threshold. A de minimis level could then be established at some fraction, say one-tenth, of this beyond-demonstrable-effect level. For example, if we take 100 millirems per year of radiation as the beyond-demonstrable-effect level for general somatic effects (damaging somatic cells as opposed to germline cells), which is the value according to the BEIR-III committee, a de minimis level could be set at 10 millirems per year. Of course, such a procedure would evoke much controversy as to what is the beyond-demonstrable-effect level or whether 10 is an ample safety factor. This example demonstrates, however, that at least in the case of low-level radiation, a scientific committee has been able to agree on a be-

yond-demonstrable-effect level. As for the safety factor of 10, this cannot be adjudicated on scientific grounds. The most one can say is that tradition often supports a safety factor of 10—for example, the old standard for public exposure (500 millirems per year) was set at one-tenth of the tolerance level for workers (5,000 millirems per year).

Can the principle of de minimis be applied to accidents? What I have in mind is the notion that accidents that are sufficiently rare may be regarded somehow in the same category as acts of God and be compensated accordingly. We already recognize that natural disasters should be compensated by the society as a whole. One can argue that an accident whose occurrence requires an exceedingly unlikely sequence of untoward events may also be regarded as an act of God. Thus, the Price-Anderson Act could be modified so that, quite explicitly, accidents whose consequences exceeded a certain level, and whose probability as estimated by probabilistic risk assessment would be less than, say, 1 in 1 billion per year, would be treated as acts of God. Compensation in excess of the amount stipulated in the revised act would be the responsibility of Congress. The cutoff for either compensation or for probabilities would be negotiable, and perhaps it would be revised every 10 years or so. One not entirely fanciful suggestion may be to set any probability of the order of 1 in 10 million to 1 in 100 million per year to be a de minimis cutoff, this being the frequency at which the earth may have been visited by the cometary asteroids that may have caused the extinction of species in past geologic eras.

## VIII

As in most such questions, identifying and characterizing the problem is easier than solving it. That the dilemma of the regulator and the toxic tort judge is rooted in science's inability to predict rare events cannot be denied.

Getting the regulator and the toxic tort judge off the horns of the dilemma is far from easy, and my two suggestions—the technological fix and de minimis—are offered tentatively and with diffidence.

Equally obvious is the intrinsic social dimension of the issue. In an open, litigious democracy such as ours, any regulation and any judicial decision can be appealed, and if the courts offer no redress, Congress, in principle, can do so. These legal mechanisms are ponderous, however. The result seems to me to be a gradual slowing of our technological-social engine as we become more and more enmeshed in fruitless argument over unresolvable questions.

Western society was debilitated once before by such fruitless tilting with Don Quixotian windmills. I refer of course to the devastating campaign against witches from the fourteenth century to the early seventeenth century. As ecologist William Clark has put it so vividly, society took it for granted during that period that death, disease, and crop failure could be caused by witches.[20] To avoid such catastrophes, one had to burn the witches responsible for them—and consequently some million innocent people were burned. Finally, in 1610, the Spanish inquisitor Alonzo Salazar y Frias realized there was no demonstrated connection between catastrophe and witches. Although he did not prohibit the burning of witches, he did prohibit use of torture to extract confessions. The burning of witches, and witch hunting generally, declined precipitously.

I have recounted this story many times by now. Yet it still seems to me to capture the essence of our dilemma: the connection between low-level insult and bodily harm is probably as difficult to prove as the connection between witches and failed crops. I regard it as an aberration that our society has allowed this issue to emerge as a serious social concern, which in the modern context is hardly less fatuous than were the witch hunts of the past. That dark phase in western society died

out only after several centuries. I hope our open, democratic society can regain its sense of proportion far sooner and can get on with managing the many real problems we always will face rather than waste its energies on essentially insoluble, and by comparison, intrinsically unimportant, problems.

## NOTES

[1] This article was adapted from a paper delivered at a June 3–4, 1985, National Academy of Engineering symposium on "Hazards: Technology and Fairness." A report on that symposium will be published in book form by the National Academy Press.

[2] William D. Ruckelshaus, "Risk, Science, and Democracy," *Issues in Science and Technology* I (Spring 1985): 19–38.

[3] U.S. Nuclear Regulatory Commission, *Reactor Safety Study: An Assessment of Accident Risk in U.S. Commercial Nuclear Plants* (WASH-1400, NUREG 75/014) (Washington, D.C., 1975).

[4] U.S. Nuclear Regulatory Commission, *Risk Assessment Review Group Report to the U.S. Nuclear Regulatory Commission* (NUREG/CR-0400) (Washington, D.C., September 1978), vi.

[5] "Report to the American Physical Society by the Study Group on Light Water Reactor Safety," *Reviews of Modern Physics* 47 (Supplement 1) (Summer 1975).

[6] National Research Council, *The Effects on Populations of Exposure to Low Levels of Ionizing Radiation: 1980* (BEIR-III), (Washington, D.C.: National Academy Press, 1980), 2.

[7] National Research Council, *The Effects on Populations of Exposure to Low Levels of Ionizing Radiation: 1980* (BEIR-III), iii.

[8] National Research Council, *The Effects on Populations of Exposure to Low Levels of Ionizing Radiation* (BEIR-II), (Washington, D.C.: National Academy Press, 1972), 2.

[9] Bruce N. Ames, "Dietary Carcinogens and Anticarcinogens," *Science* 221 (Sept. 23, 1983): 1249, 1256–64.

[10] John R. Totter, "Spontaneous Cancer and Its Possible Relationship to Oxygen Metabolism," *Proceedings of the National Academy of Sciences* 77 (April 1980): 1763–67.

[11] Alvin M. Weinberg and John B. Storer, "On 'Ambiguous' Carcinogens and Their Regulation," *Risk Analysis* 5 (June 1985): 151–55.

[12] National Research Council, *The Effects on Populations of Exposure to Low Levels of Ionizing Radiation: 1980* (BEIR-III), 287–321.

[13] Alice Whittemore, "Facts and Values in Risk Analysis for Environmental Toxicants," *Risk Analysis* 3 (March 1983): 23–33.

[14] John Ben-David, "Emergence of National Tradi-

tions in the Sociology of Science: The United States and Great Britain," *Sociological Inquiry* 48, nos. 3 and 4 (1978): 209.

[15] Trevor J. Pinch and Wiebe E. Bijker, "The Social Construction of Facts and Artefacts: Or How the Sociology of Science and the Sociology of Technology Might Benefit Each Other," *Social Studies of Science* 14 (1984): 401.

[16] Kåre Hannerz, *Towards Intrinsically Safe Light Water Reactors* (ORAU/IEA-83-2(M) Rev.) (Oak Ridge, Tenn.: Oak Ridge Associated Universities, Institute for Energy Analysis, June 1983); Herbert Reutler and Günter H. Lohnert, "The Modular High Temperature Reactor," *Nuclear Technology* 62 (July 1983): 22–30.

[17] Trevor A. Kletz, *Cheaper, Safer Plants or Wealth and Safety at Work: Notes on Inherently Safer and Simpler Plants* (Rugby, England: The Institution of Chemical Engineers, 1984).

[18] Howard I. Adler and Alvin M. Weinberg, "An Approach to Setting Radiation Standards," *Health Physics* 34 (June 1978): 719–20.

[19] T. Westermark, *Persistent Genotoxic Wastes: An Attempt at a Risk Assessment* (Stockholm: Royal Institute of Technology, 1980).

[20] William C. Clark, *Witches, Floods, and Wonder Drugs: Historical Perspectives on Risk Management* (RR-81-1) (Laxenburg, Austria: International Institute for Applied Systems Analysis, March 1981).

# RISK ASSESSMENT
# AND REGULATORY PRIORITIES

*Lester B. Lave*

A number of surveys taken over time, from the early 1960s through quite recently, show that the public supports a "clean environment."[1] One may say to the public, "Don't you understand that a clean environment is going to cost money? That your utility bill is going to go up?" The public answers, "Yes, we understand that, but we want a clean environment." One replies, "But wait a minute, we may have to shut down some plants and lose some jobs because of it." And the public says, "Yes, but we want a clean environment." Whatever way one asks the question, the public comes back and overwhelmingly answers, "We want a clean environment. Let's get on with it!"

Despite this apparent public support, however, it is clear that legislative goals in the environmental area have not yet been achieved. We continue to discharge pollutants into the waterways of the United States. Air quality levels do not yet "provide an ample margin of safety to protect the public health."[2] While levels of suspended particles and sulfur oxides in the ambient air have been reduced signif-

icantly, the ozone problem is worse.[3] Most sewage is treated, but disposal of toxic wastes has proven a more difficult problem than expected and the levels of carcinogens in water do not appear to have been reduced.[4] In general, while it is clear that there has been some improvement in the state of the environment, it is also clear that environmental regulation has not completely succeeded. The amount of improvement has probably tended to be overstated.

A great deal of scholarly literature investigating regulatory agencies has found them to be inefficient, if one measures the amount of pollution abatement achieved per dollar expended.[5] Political scientists point out that while a variety of purposes have been served by the creation of regulatory agencies, neither efficiency nor effectiveness were dominant or even important goals to Congress or to agencies historically.[6]

One way of illustrating the problems facing the Environmental Protection Agency is to look at the sheer number of chemicals in the environment. There are currently about 60,000 chemicals in common use.[7] A National Academy of Science panel surveying the extent of scientific knowledge available about these chemicals concluded that in rel-

Reprinted from *Columbia Journal of Environmental Law* Volume 14 (2) (1989), pp. 307–14. Used with permission of the Columbia Journal of Environmental Law.

ative terms, science knows absolutely nothing about the human health effects of nearly all of them.[8] There is reasonably complete evidence as to human toxicity for only about 1,000 of these 60,000 chemicals.[9] This lack of data does not indicate indifference on the part of agencies and companies. Rather, it illustrates that toxicological data is extremely expensive to obtain. In all likelihood, therefore, it is technically and economically impossible to get complete data on all 60,000 chemicals currently in use.

For most of the chemicals that we know much about, our knowledge is derived from inadvertent human exposure, that is, human beings have been our test animals.[10] Science has historically obtained its best and most reliable information by using human beings as our "guinea pigs." Unsuspecting workers exposed to asbestos in the workplace played a bigger role than did laboratory rats in discovering the link between asbestos and mesothelioma. Consequently, there has been little controversy about the asbestos data and its interpretation, because the data concerned human beings. Asbestos is not an anomaly in this respect. However, the obvious problem with this approach is that while using humans as test subjects for carcinogens gives much information, it is unethical. Using chemicals without prior testing gives rise to social disasters—such as those that occurred with asbestos and thalidomide—which are simply not acceptable.

Setting regulatory goals has proven difficult however. In the early 1970s, the Assistant Administrators of EPA would typically have breakfast together every Monday morning to map out their regulatory strategy. Those breakfasts were dominated by discussions of whatever "cancer scare" story had appeared in the *New York Times* or *Washington Post* over the weekend. Nearly every weekend, there was apparently at least one newspaper story identifying a "hot spot" area and the chemicals that might have increased the risk of cancer in that area. At the Monday break-

fast meeting attention was directed to that story and agency staff were assigned to spend time working on the problem. This syndrome came to be called the "Carcinogen of the Month." Staff people ran around helter-skelter seeking to address the latest crisis, with little sustained effort among them. While this approach showed EPA to be responsive to public concerns, it did not produce much in the way of "regulatory strategy."

Today we arguably have much more experience with environmental regulation, yet we still have difficulty setting rational policy goals. This is well illustrated by an interesting set of calculations done by Bernard Cohen, a physicist at the University of Pittsburgh. Table 1 shows Cohen's calculations of the reduction in life expectancy in the United States due to a range of causes.

Interestingly, our society spends a great deal of time and resources regulating things that are considered to be very low risk on Table 1, such as environmental chemicals,[11] and not very much time regulating things that are considered to be high risk on Table 1, such as cigarette smoking and obesity. According to Cohen's research, if we could lower the death rate directly attributable to heart disease, smoking or obesity by even a little bit, then we could increase longevity in the population by a considerable amount. Instead, we tend to spend our time worrying about things that have relatively minor effects on life expectancy.

To illustrate, consider the standard criterion used by the Food and Drug Administration: if a substance causes more than one cancer per million lifetimes among those exposed, then the Agency regards it as a "non-trivial risk."[13] If it causes less than one cancer per million lifetimes, then the FDA regards it as a "trivial risk."[14] A chemical with a one-in-one-million risk of cancer would reduce the life expectancy of the average exposed person by 1/100 of a day. Regulatory agencies such as the FDA often set as goals the elimination of risks that could reduce an exposed person's life expectancy by as little as one day.

**TABLE 1.[12]** Estimated Life Expectancy Reduction from Risks and Activities

| Activity or risk | Days LER |
|---|---|
| Heart disease | 2100 |
| Being unmarried | 2000 |
| Cigarette smoking | 1600 |
| Cancer | 980 |
| Being 30 lbs. overweight | 900 |
| Grade school dropout | 800 |
| Unskilled laborer | 700 |
| Stroke | 520 |
| Vietnam army duty | 400 |
| Mining or construction work (due to accidents only) | 300 |
| Motor vehicle accidents | 200 |
| Pneumonia, influenza | 130 |
| Homicide | 90 |
| Drowning | 40 |
| Poison + suffocation + asphyxiation | 37 |
| Energy production and use | 25 |
| Diet drinks | 2 |
| Hurricanes, tornadoes | 1 |
| Airline crashes | 1 |
| All-nuclear electricity | 0.04-2[a] |
| Harrisburg area residents (from TMI accident) | 0.001 |
| Radioactive waste burial ground leaks, risk to nearest neighbors | 0.0001 |
| Sky-Lab fall | 0.00000002 |

[a]The lower number is the estimate of government-sponsored scientists, and the higher number is the estimate of nuclear critics.

Risk assessment is a tool which is used in this environment of fluctuating and competing priorities. Therefore it is important to understand some of the assumptions behind risk analysis. Risk analysis requires a considerable amount of data. The first task is often to determine what population is at risk. While at first glance this may seem to be a trivial task, it turns out in practice to be an extraordinarily difficult problem. For example, suppose there is a contaminant in the groundwater and we want to assess the health risks it poses. Who is being exposed to it? All the people who get water from that site? But are there people who are pumping water out of their own wells? Is a municipal water treatment plant filtering out the contaminant? Surprisingly, simply figuring out who is exposed is difficult.

The need to determine levels of exposure compounds the difficulty. Some people may be exposed to the contaminant at a one part per million level; other people, at the one part per trillion level. We also need to know what the exposures are like over time, not just at one moment in time, so that we can figure out what the total dose looks like. This requires assumptions about the rate at which the contaminant will be seeping into the groundwater over the next fifty or seventy years.

To determine the likelihood that a contaminant will increase the risk that an exposed population will contract a given disease, we also need to know the dose-response relationship for each relevant disease. The fact that we know very little about dose-response relationships for chronic diseases other than cancer makes it difficult to estimate the increase in risk of those diseases.

When trying to figure out the dose/response relationship for a given disease, we use data either from studying laboratory animals or from epidemiological studies.[15] Each has its own problems. Data from laboratory animals requires extrapolation from mouse to man. We have all heard a bit about this, because when saccharin was an item of front page news, it was reported that in order to ingest the amount of saccharin that the laboratory rats in several studies consumed, a human being would probably have to drink extraordinarily large quantities of the soft drink Tab a day. The animal bioassays thus raised questions about whether one could extrapolate to humans and whether the very high doses received by the animals overwhelmed their immune system or otherwise led to an abnormal metabolic uptake.[16]

How can we extrapolate human effects from a mouse, which obviously is not identical to a human being? How can we extrapolate from a very high dose to a very low dose? We make assumptions in order to perform these extrapolations, because we do not know how to do anything better. For instance we reason that a mouse is a mammal and a human being is a mammal. Therefore, we say, anything that causes cancer in a mouse is likely to cause cancer in a human. The wisdom of that assumption is questionable. We also assume that when going from high to low doses, we can simply extrapolate proportionality, another questionable assumption. Again, we do not know what else to do.

Alternatively, we can rely on epidemiological studies, which are usually studies of workers who have been exposed to high levels of a chemical. We are able to detect disease in workers more easily because they are exposed to very high doses. To use this data for environmental or consumer regulation, we extrapolate from the very high doses of workers to the much lower doses of consumers. The usual problem in the epidemiological study is that there is typically little exposure data. There is generally a latency period between the time workers are exposed and the point at which a tumor or other manifestation arises. There are also other confounding factors to consider, such as multiple exposures.

In regulation and in litigation, whenever one is using a risk analysis, whether using data from a laboratory animal or from epidemiology, there are thus many uncertainties. That fact leads to a more basic question: how accurate are risk assessments? Are they accurate to plus or minus ten percent or are they off by a factor of ten million? The answer depends on the particular situation. For risk estimates associated with cigarette smoking, the uncertainty is probably within the ten percent range.[17] A risk estimate of saccharin is probably uncertain by about a factor of ten million.[18] Unfortunately, many risk assessments are more like saccharin than they are like cigarette smoking.[19]

Generally, risk assessment methodological procedures are designed to provide "reasonable upper-bound estimates"[20] of a given risk. Thus, by design, the point estimate of the risk derived from risk assessment is at the top of the actual risk distribution curve. The usual risk assessment procedures choose an estimate so high that it is unlikely to actually

occur. To illustrate, think of 100 similar chemicals. Rather than choosing an estimate representing the middle of the distribution, the estimate chosen by risk assessment is, for example, fifth from the top.

In virtually every case, the reasonable lower-bound estimate of a given risk is zero.[21] In other words, for chemicals that are carcinogenic in rodents, but for which there is no human evidence of carcinogenicity, it is a reasonable assumption that the chemical may not be a human carcinogen. Even when a chemical is known to be a human carcinogen, the relationship has only been shown to exist at much higher doses than those to which humans are usually exposed. A reasonable lower-bound assumption is that there is a threshold dose below which the risk to an exposed person is zero. It follows from this that in virtually every case, the reasonable range of uncertainty extends from zero to the point estimate derived from risk analysis. This range can be quite broad, as in the example of saccharin.

Recognition of this broad range of uncertainty leads to another question: how useful is risk assessment? I believe it is useful for several purposes. First, risk estimates help to focus public debate. They are an excellent tool to guide inquiry, because they show very quickly the crucial assumptions, and the need for additional data or interpretation. Risk analysis improves science by focusing on what we need to know in the future. However, it does not answer the most basic questions about what should and should not be regulated in the present. As we present this information it is important that we acknowledge the uncertainty explicitly.

These methodological obstacles aside, however, I believe that the most difficult and important step in risk management is setting policy goals. In the past, risk management was regarded as the province of "professionals." These professionals generally did not know and did not care to know public concerns. They believed that the public did not understand the nature of risk and thus had little to contribute. This led the system, through its risk managers, to address a number of concerns of little importance to the public and to ignore a number of concerns of great importance to the public. Consequently, we have spent far too much time in this country worrying about what scientists or professionals think ought to be done and far too little time worrying about what the public thinks ought to be done.

What risk assessment provides us is a systematic approach to analyzing complex problems. As long as we are trying to set policy in a scientific and systematic way, there is nothing better at our disposal than these admittedly imperfect risk assessments, no matter how uncertain they are. Scientific knowledge is not in a position, at this stage, to give confident answers as to how risky it is to drink water that has a certain contaminant in it at a certain dose. At least at the moment, and probably for any foreseeable future, uncertainty is ubiquitous and inevitable. Litigators will have much material for litigation. We will undoubtedly continue to be faced with difficult questions to resolve, and scientific experts alone will not be able to resolve them. But with risk assessment we can provide the best available health effects data and a systematic approach to estimating that risk so that more informed decisions are made by the public and its appointed decisionmakers. Risk assessment is a tool that should be used to present the evidence to the population.

As professionals, whether lawyers, scientists, managers or regulators, we thus need to concern ourselves a lot more with risk communication. Unfortunately, the previous attempts at risk communication have been terrible. When we are speaking to the public, risk communication must be a two-way street. We must seek to uncover the public's concerns, and make those concerns the primary source of regulatory priorities. What the public wants, in the end, is what should prevail. Risk assessment is not so much a means of convincing scientists or even judges. Rather it is a tool

that should be used to present the evidence to the jury, the population in general, so that more informed policy decisions can be made.

## NOTES

[1] See, e.g., G. Gallup, Forecast 2000: George Gallup, Jr. Predicts the Future of America (1984).

[2] Clean Air Act, 42 U.S.C. § 7412(b)(1)(B) (1982).

[3] Conservation Foundation, The State of the Environment: A View Toward the Nineties (1987).

[4] Id.; Toxic Substances, Health and the Environment (L. Lave & A. Upton eds. 1987)

[5] See, e.g., J. Mendeloff, The Dilemma of Toxic Substance Regulation (1988): Crandall, Gruenspecht, Keller & Lave, Regulating the Automobile (The Brookings Institution 1986).

[6] J. Q. Wilson, The Politics of Regulation (1980).

[7] National Research Council, Toxicity Testing—Strategies to Determine Needs and Priorities (1984).

[8] Id.

[9] Id.

[10] Lave, Ennever, Rosenkranz & Omenn, Information Value of the Rodent Bioassays, 336 Nature 631 (1988).

[11] Cohen's assertion that the risk associated with environmental chemicals is relatively small is a view shared by numerous others. See, e.g., Doll & Peto, The Causes of Cancer: Quantitative Estimates of Avoidable Risks of Cancer in the United States Today, 66 J. Nat'l. Cancer Inst. 1192 (1982) (estimating that only two to four percent of all cancers are associated with environmental chemicals).

[12] Nuclear Energy: A Sensible Alternative 322 (Ott & Spinrad eds. 1985).

[13] 42 Fed. Reg. 10,421 (Feb. 22, 1977). See also Hutt, The Basis and Purpose of Government Regulation of Adulteration and Misbranding of Food, 33 Food Drug Cosm. L. J. 505 (1978): Merrill, Regulating Carcinogens in Food: A Legislator's Guide to the Food Safety Provisions of the Federal Food, Drug and Cosmetic Act, 77 Mich. L. Rev. 171 (1978).

[14] Id.

[15] M. Pike, Epidemiology and Risk Assessment: Estimation of GI Cancer Risk from Asbestos in Drinking Water and Lung Cancer Risk from PAHs in Air, Banbury Report 19: Risk Quantification and Regulatory Policy 55 (Hoel, Merrill & Perera eds. 1985).

[16] L. Lave, The Strategy of Social Regulation: Decision Frameworks for Policy (The Brookings Institution 1981).

[17] L. Lave, Health and Safety Risk Analysis, 236 Science 291 (1987).

[18] Id.

[19] Banbury Report 19, supra note 15.

[20] E. Anderson, The Risk Analysis Process, in Carcinogenic Risk Assessment (C. Travis ed. 1988).

[21] U.S. National Cancer Institute, Cancer Control Objectives for the Nation 1985–2000 (1986).

# SUCCESS, STRAIN, AND SURPRISE[1]

*Robert W. Kates*

In the past decade and a half, citizens of most industrialized countries have become concerned about the hazards of technology, have created a new set of institutions and activities to control them, and have profoundly changed the ways in which technologies are designed, produced, and used. Over the next decade and a half, more subtle hazards will confront us, strains and contradictions will emerge in the new institutions, and we will still be surprised at the strange ways in which our technologies unintentionally injure us.

Fifteen years after Earth Day 1970, much progress has been made in the United States in controlling air and water pollution, somewhat more with the former than the latter.[2] At the same time, the hazards that we cope with today have changed markedly. There has been a shift in emphasis from visible problems of automobile smog and raw sewage to less-visible problems posed by low concentrations of toxic pollutants.[3] We are less concerned with the acute consequences of a hazardous technology such as the automobile (which are measured well by the National

Safety Council) than we are with the chronic consequences of a hazardous technology such as toxic chemicals (which are not measured well) because either we do not understand the causation or the effects are still latent. Our concerns have shifted in temporal scale as well. We are less worried about the daily recurrence of commonplace accidents than about confronting the frightening possibility of rare but catastrophic accidents. And in spatial scale we are shifting attention from the local to the regional and global—from local improvement in water or air quality, achieved in almost every industrialized country, to regional frustration in dealing with acid rain, stratospheric ozone depletion, and tropospheric ozone enrichment, and to global uncertainty about carbon dioxide, trace gas enrichment, and nuclear winter.

This shift from better-understood hazards to less-understood hazards has placed an enormous burden on science to identify hazards and to assess their risks. Scientists often fluctuate between humility and hubris, and scientific risk assessment has manifested these fluctuations as well. Currently, humility appears to be in ascendence as the limits to knowledge emerge and experts routinely contradict each other in the press or the court-

Reprinted with permission from *Issues in Science and Technology* Volume II (1), pp. 46–58. Copyright © 1985, by the National Academy of Sciences, Washington D.C.

room. But it should be equally recognized that while the media, the public, or the courts may demand more of science than it can give, some scientists have promised more than they can deliver. Scientists have implied that they know the significant ways in which technologies fail, that they are close to understanding the fundamental causes of cancer and arteriosclerosis, and that hazardous waste can be safely collected, transported, and stored. Thus, while some scientists would limit the burden on science, others continue to extend it, either from hubris, from a desire to reassure the public, or because they relish the challenge and opportunities for further research.

## II

As we shift our focus to more elusive hazards, the institutions of hazard management become more concrete. The institutionalization of government regulation is well known. New technological hazards that are wholly unregulated are rare. Indeed, recently perceived hazards, such as those of genetic engineering, come under regulation by at least five government agencies. Despite occasional resistance based on ideological reflex, the principle that hazards should be regulated is not seriously disputed. It is the wisdom, magnitude, social cost, and implementation of the specific regulations that fuel controversy.

The excessive focus on public regulation has obscured the institutionalization of hazard management that has taken place outside of government. For example, groups claiming to represent the public interest consistently monitor most domains of hazard. Often underfunded, such public-interest groups have nevertheless established legal and scientific competence that enables them to participate actively in public regulatory processes and to use the courts to guarantee implementation of congressional intent or to obtain compen-

sation for injury. As these groups become more professional, they increasingly join with their counterparts in industry in efforts to implement regulations and to provide alternatives to regulation through voluntary efforts, environmental mediation, and the like.

Another set of hazard management institutions are judicial ones arising from the remarkable innovation of contingency representation in class-action compensation and liability cases. This institutionalization of representation follows a long-term shift in both public attitude and legal precedent, from concern with acts of God to acts of persons and from private risk to public risk. When a resigned public once hesitated to sue a steamship company for injury from a boiler explosion, an indignant public now blames the government for causing floods. In a time when lawyers routinely advertise for clients, insurance and self-insurance reserves provided for hazard compensation are being seriously strained.

But the most remarkable institutional change—and one receiving the least formal study—has occurred within industry itself. For the last two years I have been part of a modest research effort seeking to learn more about how corporations manage hazards.[4] Our studies are limited; the corporate world appears reluctant to allow external inspection or to encourage internal self-examination. But even our limited studies suggest a remarkable set of changes over the last 15 years in corporate goals and in resources devoted to the management of hazardous technologies.

Since 1970, when General Motors established its public policy committee, U.S. industry has institutionalized the social responsibility movement, creating codes of ethics in many corporations and board of directors' committees in some of them, and, most important, establishing specific policies and operational guidelines in major corporations, such as Dow, Du Pont, and Monsanto, that produce hazardous products. Common elements in these operational guidelines include intensive efforts to screen for potential toxicity; attention to worker health and safety; concern that consumers are protected and that company products are used safely; protection of the environment by waste reduction, pollution control, effluent cleanup and safe waste disposal; and disclosure and dissemination of correct product information.

Industry has also committed resources. A recent Chemical Manufacturers Association survey in which about half the industry responded found that the mean firm surveyed had 84 health and environmental specialists, or about 2 percent of its U.S. employees, and spent almost 4 percent of annual sales on toxicity testing and environmental pollution control.[5] One industry leader that our group studied had a 125-employee headquarters health and safety unit, including an epidemiologist and a risk-assessment unit, a toxicology lab with 20 toxicologists, and 30 industrial hygienists assigned to plants—resources probably exceeding those of most state governments.

The resources of an industry leader may be misleading. Our limited studies have focused on large and prosperous corporations, and even within this set there are large variations in commitment and resources. Nonetheless, the major changes in approach and staffing since 1970 are widespread. Indeed, we have found that a corporate health and safety regulatory system exists that matches that of the public sector. This shadow government employs a variety of standards. Generally, these include the relatively few standards that are mandated by government, the many more industry consensus standards, and some internal corporate standards developed for new products and facilities or to maintain a higher standard of workplace health and safety. This standard-setting process is replete with many of the conflicting roles and motivations found in the public process, and in some corporations it is based on an extremely sophisticated risk-assessment process. As with the governmental process, it is easier to set standards than to enforce them. But a genuinely committed firm may be able to enforce its own standards more easily than the government can regulate that firm.

## III

Underlying the institutionalization of hazard management is a vital change in public attitudes. In the late 1960s three powerful movements—those concerned with the environment, consumerism, and, more recently, personal health—began to overlap and coalesce. The strongest of the three, the environmental, is founded on deeply held values and strong, persistent attitudes. Polls have consistently shown that concern with air and water pollution, support for strong government regulation, and sympathy for the environmental movement claim favor with two-thirds of the population despite liberal-conservative political affiliations and fluctuations.[6] The more recent and somewhat amorphous personal health movement can only reinforce these concerns about hazards.

Although techniques and institutions vary, a commitment to the management of technological hazards is deeply entrenched in most industrialized countries. The costs are substantial. Modern industrialized countries commonly devote between 1 percent and 2 percent of their gross national product (GNP) just to prevent and reduce pollution. Our study group calculated that in 1979 the social cost of coping with hazards associated with technology in the United States was equivalent to between 7 and 12 percent of GNP, with about half devoted to hazard management and the remainder incurred as damages to people, material, and the environment.[7] These expenditures reflect the stable commitment to environmental values forged over the last two decades and its implementation in major political and economic institutions. Coping with the emergent issues of the next 15 years begins on that base, and the maturation of nonfederal institutions offers new opportunities.

One goal of professional risk and hazard assessment is to minimize surprise. Notwithstanding the substantial resources now devoted to the management of technological hazards, one of the distinguishing features of the last 15 years is that surprise persists and, paradoxically, grows.

While the partial core meltdown at the Three Mile Island nuclear plant on March 28, 1979, was within the range of uncertainty postulated by some assessors of the risk of nuclear accidents, the public and most scientists were surprised by the event. Other examples of major surprises include acquired immune deficiency syndrome (AIDS), the Bhopal disaster, Legionnaires' disease, natural carcinogens, the nuclear winter scenario, suicide truck bombs, toxic shock syndrome, and poisoned Tylenol. Hazardous surprises seem to occur at a frequency of about twice a year worldwide.

Paradoxically, success in managing hazards fuels our surprise. The conquest of many common infectious diseases makes the outbreak of new infectious diseases surprising, not only because of the complexity of the pathogens involved but also because of the mix of social behavior and technologies involved in their transmission: conventions, homosexuality, blood-distribution networks, superabsorbent tampons, and air-conditioning systems. The remarkable record of purity in our food and drugs heightens the impact of a mass poisoner. Our vaunted military strength makes our vulnerability to the truck bomber seem astonishing.

But there are also genuine scientific surprises. Forty years after Hiroshima, nuclear winter, a major new consequence of nuclear war, is hypothesized.[8] Long after the identification of natural carcinogens such as aflatoxins in peanut butter, we are still surprised by their extent and potential toxicity.[9]

Surprising hazards are an inevitable outgrowth of technological change. One of the positive developments of the past 15 years has been the growing public understanding that all technology is hazardous and that some technologies are substantially more hazardous than others. I believe that the demand for totally safe technologies has diminished. Technological innovation will surely produce new hazards, and many of these will prove quite surprising despite the successful institutionalization of the processes of ensuring early hazard identification and product safety.

Identifying potentially hazardous technologies may become more difficult because of a troubling characteristic of the so-called high technologies. Since 1939 three major high technologies have dominated the innovation process: nuclear engineering, solid state electronics, and biotechnology. A characteristic of the development of these technologies has been the blurring of the roles of the basic scientist, the technologist, and the entrepreneur—a blurring hailed by many as leading to quicker innovation, application, and use. This pattern began with the atomic bomb, when Albert Einstein wrote to President Roosevelt about the military implications of atomic energy and a generation of physicists worked as technologists to make it a reality. It is most evident today in the development of biotechnology, in which leading scientists assume all three roles. What is troubling is that this blurring of roles denies to hazard management one of its strongest sources of early hazard identification—knowledgeable but independent basic scientists. Such scientists, knowledgeable about a technology but independent of its development or production, are society's best bulwark against technological surprise.

Finally, it appears that it is intrinsically more difficult to predict the hazardous nature of some technologies than others. Scientific theories of comparative degrees of hazard are just being developed. My own research group recognizes five distinguishing characteristics of extremely hazardous technologies: intentional design as biocides (chemical pesticides); the combination of latency and long potency in materials (asbestos); the potential for catastrophic effects (jumbo jets); the persistent, ubiquitous capacity to inflict harm (motor vehicles); and the as yet unquantified capacity to cause damage of global extent (acid rain).[10] But high hazardousness is not necessarily surprising. Rather, the surprise or unpredictability of some hazardous technologies may lie in the qualities Charles Perrow identifies as high hazard—the combination of technological complexity, the tight coupling of components so that the failure of one component starts a process that cannot be arrested, and catastrophic potential.[11] Or the surprise may reside in the complex dynamics of biological and technological systems which C.S. Holling and his colleagues have studied, and in which there is great potential for serious, unwanted, and hazardous surprises (for example, in the interaction between pesticides and pesticide-resistant insects).[12]

## IV

The capability and competence of the hazard-management system that has evolved in recent years is substantially limited. Without major breakthroughs in our fundamental understanding of the mechanisms of low-level effects of toxic chemicals and radiation, for example, there are clear limits to our ability to quantify these effects. Similarly limited is our ability to anticipate and prevent catastrophic accidents. And society as a whole cannot reduce *all* hazards or reduce *any* hazard to zero risk.

A second set of limits is institutional. Ten or more years after most regulatory agencies were created, considerable doubt remains whether they can carry out their legislative mandate to set standards and to force compliance with them. Federal air pollution standards exist for only a few of the hundreds of known airborne toxic substances. There are even greater doubts as to the judicial system's capacity to deter negligence and to compensate efficiently and justly for injury. In-surance companies and other risk-sharing institutions linked to that judicial system appear to be in crisis, overwhelmed by the legacies of old hazards such as asbestos, changing concepts of responsibility, and the potential for future Bhopal-like catastrophes. Finally, our exemplary tradition of voluntary action, public-interest initiative, and corporate good citizenship seems inadequate in the face of such overwhelming tasks as the cleanup of thousands of existing hazardous waste sites.

A third set of limits relates to moral choice— the perennial conflict between efficiency and equity. Hazards pose special and subtle problems. These include the separation in space and time of those who receive the benefits of technologies from those who experience the risks; the wide differences in hazard susceptibility between individuals, ages, sexes, and even ethnic groups; and the uncertainty as to both cause and responsibility when compensating for injuries inflicted by substances such as asbestos or Agent Orange.

These limits are hardly absolute; their boundaries still need to be explored. Some believe that the effectiveness of the hazard-management system, recently created in this country and still in flux, has yet to be thoroughly tested. Nonetheless, an interesting search for alternatives is already under way. It includes a search for ways to finesse the uncertainty imposed by the limits to our scientific knowledge, to diminish the catastrophic potential of technology, and to choose an agenda of hazards that pose the greatest threat to our society. It seeks alternatives to government regulation and litigation and new ways to link equity and efficiency in compensating victims of hazards. It is a search for technological and behavioral fixes.

## V

I served my apprenticeship in hazard science 25 years ago at the University of Chicago under the leadership of Gilbert F. White, whose work involved analyzing the failure of the major engineering works that had been used since 1936 to reduce riverine flood hazard in the United States.[13] As engineering projects they worked well, but as social engineering efforts they had a perverse effect. While they reduced the frequency and magnitude of flooding, they also encouraged the development of floodplains. Thus, there were fewer floods but greater damages. A behavioral fix was needed to complement the prevailing technological fix, and we sought to develop one with a broad program of management innovations in the form of scientific information, floodplain regulation, insurance, emergency evacuation, and incentives for floodproofing individual buildings. I started my career, therefore, skeptical of technological fixes and with a bias toward behavioral fixes.

Like the engineering projects that reduce the numbers of floods while amplifying potentially catastrophic floodplain development, there are similar perverse combinations of factors found in many hazard-management situations. For example, students taking driver training courses in high school have somewhat safer driving records than untrained ones. However, because the institution of such courses has been accompanied by a widespread lowering of the age of licensing, the number of young drivers, as well as their overall accident toll has increased.[14]

Confidence in single fixes—technological or behavioral—is usually misplaced. The projected diminution in risk or consequences from the single fix is often overestimated, partly because of the energetic advocacy of its proponents. More often, however, the single fix overlooks some process elsewhere in the chain of causation that either increases the releases, exposures, or consequences of the hazard or introduces a new chain of hazard. Thus, even the most successful of recent simple technological fixes, the childproof drug container that has substantially reduced child-related deaths, creates painful frustration for elderly arthritics and annoyance for all of us. On the other hand, well-managed hazards, exemplified by commercial aviation, employ a spectrum of fixes at every stage in the chain of hazard causation. They combine both behavioral and technological fixes—better crew training and better aircraft. In what follows, I suggest some alternative fixes for coping with technological hazards, but with the warning that although they could be useful, they are not universally applicable.

## VI

One desirable class of technological fixes—inherently safe processes—depends on immutable laws of nature rather than the intervention of humans or electromechanical devices.[15] Two such systems have been proposed for nuclear reactors, and suggestions for somewhat similar processes are available in chemical engineering. Inherently safe waste disposal is also possible where pretreatment, high-temperature incineration, or constant recycling reduce the toxicity of waste by many orders of magnitude. It now seems clear that efficient hazard reduction poses as great an engineering challenge as efficient product production.

I would also offer a second class of technological fixes—the inherently simple fix. I have cited one example, the childproof drug container. Related to it are the post-Tylenol sealed containers. Innovations in chain-saw safety following the epidemic of accidents that resulted from greater use of firewood during the recent energy crisis are another example. Inherently simple fixes require a well-understood hazard, some motivation to cope with it, and a bit of old-fashioned ingenuity.

Now that government has attempted to regulate almost every hazard, it is clear that there are real limits to regulation. These limits stem from ideological distaste for government regulation and, more pragmatically, from the inherent shortcomings of rulemaking processes and compliance efforts. These limits suggest a need for behavioral fixes—changes in human behavior—using the recently created resources of industry and public-interest groups in hazard prevention and reduction.

Theoretically, the hazard makers should be the best hazard managers. If they can be persuaded to do so, those who design and manufacture products are in the best position to identify potential hazards and to correct or control them. Much of the persuasion has been accomplished already. Controversy over the grim legacies of the past or the safety of existing products has obscured the real progress made in the design of new, less-hazardous products. Novel reporting requirements, such as the so-called squeal law provisions of the Toxic Substance Control Act (requiring manufacturers to report any knowledge of substantial risks), attempt to use industry's own considerable scientific resources. Many more voluntary and creative experiments using industry, public-interest groups, and the scientific community are needed.

Let me illustrate one such possible effort. Recently, the Board on Toxicology of the National Research Council (NRC) completed a shocking study on available toxicity data.[16] Based on a sample of some 53,500 distinct chemical entities, the board found that minimal toxicity information was available for only a third of the drugs and pesticides, a quarter of the cosmetics, and a fifth of the chemicals in commerce. In contrast to the virtues of the de minimis approach that proposes to ignore very low levels of hazard, our society seems to have adopted a de ignoramis approach that avoids knowing about many hazards.

Industry may well quibble with the standards of minimal knowledge that require new chemical compounds to be tested for toxicity by rodent studies that are expensive and time consuming. But the NRC study employed and, for the most part, was limited to publicly available data. Industry conducts an extraordinary amount of proprietary testing of new chemical products, the results of which may be withheld if the corporation does not develop the product further. Corporate executives are reluctant to disclose these data because screening tests are a significant business expense and they do not wish to reveal to competitors their search strategy for new chemicals. Given the cooperation of industry, it should be possible for an independent scientific body, such as the NRC, to review these test data confidentially and prepare a composite list of mutagenic chemicals to be issued annually, with the proprietary sources held in confidence.

The final form of the procedure is not important, but the principle is. As a society we need to use our collective hazard-management resources in ways that avoid the ponderousness of the regulatory system and the competitiveness of the marketplace.

Similarly, we need some new behavioral fixes in our procedures for compensating the victims of hazards. Compensation itself is a failure of hazard management, usually inadequate to match the loss, pain, and suffering incurred. Our current system is in crisis and extremely costly. Insurance premiums continue to rise; some malpractice insurance costs are prohibitive; and corporate failures, including those of major liability insurance companies, are likely. As physicians and the producers of vaccines can attest, malpractice and product-liability suits threaten both innovation and useful institutions. But worst of all, the system with all its high costs provides neither compensation nor fairness to large numbers of victims.

A vigorous national search is on for alternatives, including no-fault environmental compensation programs or caps on liability. This search has been encouraged in part by widespread reaction to pictures of liability lawyers descending on Bhopal in search of clients. As the tragedy of Bhopal slowly wends its way through the U.S. court system, it may be instructive to examine a different way of handling an industrial tragedy—the response of the Mexican government to the natural gas explosion on November 19, 1984, at San Juan Ixhuatepec.

The disaster killed at least 500 persons, injured over 2,500, and displaced 200,000 or more. It was marked by a restoration, reconstruction, and compensation process unmatched by responses to natural and technological disasters anywhere in the world. The response of the Mexican government, as chronicled by my colleague Kirsten Johnson, was remarkably prompt.[17] The delivery of rapid, albeit rough and ready, aid and compensation in the San Juan Ixhuatepec episode provides an example of an alternative to more traditional judicial processes that may be exceedingly fine but are also exceedingly slow.

In the beginning the relief effort was marred by the same misplaced generosity that characterizes many disasters—unwanted clothing and undistributed food. But within three days of the disaster, large quantities of building materials were delivered to the site to be given without charge to all residents with damaged property. This led to an immediate spate of self-help reconstruction and general neighborhood improvement. One section of the explosion site was made into an instant park, ostensibly to commemorate the victims, but also to preserve the area as a buffer to separate residential areas from land suitable for a future industrial site and to replace the scenes of the disaster with greenery and games. A health facility was put in place and a community center will follow, providing a minimal type of community compensation. To provide housing to about 200 displaced families, part of a newly completed housing development was acquired by the government, and before the week was out the first homeless families received permanent housing. Within another week 80 percent had been housed.

On Christmas Day five weeks later, the Mexican national oil corporation (PEMEX) denied liability but accepted responsibility for the disaster and pledged to pay compensation of more than $10,000 per death victim. There were no precedents for such payments, so the payment schedule was adapted from the workman's compensation code for the various classes of death and injury. Less than three months after the disaster, almost all the victims or their heirs were compensated.

Three months saw a community rebuilt, the homeless housed, and the victims compensated. The speed of the settlement came at the price of an authoritarian uniformity—all victims received the same housing. Some were better off than before, others were surely worse off and removed from their former community. Some nonvictims benefited from the free materials. An occupational compensation scheme was adapted for public use that in litigation might have provided higher settlements. And the park was placed on the destroyed living area and not on the industrial site as some residents expected, thereby forcing the permanent removal of the homeless. But for most victims and for the public at large, some justice was done.

A final example of a behavioral fix needed for the next decade lies in the development of third-generation ethics. The first generation of hazard-management ethics was the ethics of nonmaleficence—do no harm to person or nature. These ethics were celebrated on Earth Day 1970, enshrined in regulatory law, and finally institutionalized in corporate codes of ethics indistinguishable from those of the Audubon Society—all in the space of a little more than a decade.

The second generation of ethical issues attempted to weigh harms—to consider both benefits and the value of lost benefits as well as risk. The ethical underpinnings of this approach, an extension of cost-benefit analysis, were dominated by principles of utility.

Third-generation issues concern equity, fairness, and distributive justice. They are concerned not only with the overall balance of benefit and harm but with their distribution to specific groups or individuals, with the fairness of the process as well as the outcome. These issues are prominent in many situations. Trying to avoid the exposure of women of reproductive age to toxic chemicals poses complex questions of sex discrimination, invasion of privacy, and protection of the unborn, as well as of establishing permissible levels of exposure.[18] A pervasive and ethically unjustified double standard in the protection of workers and the public exists almost everywhere. The standards of workers' exposure to toxic materials are 10 to 1,000 times greater than those applied to the general public.[19] The Bhopal disaster illustrates still another double standard in safety performance, that of industrialized and developing countries.[20] Hazardous waste disposal practices concentrate wastes gathered over a large area in someone's backyard and pass on a legacy of care and risk to future generations.[21]

All of these are third-generation problems requiring ethical analysis capable of illuminating policy choices in modern hazard management. Thus, it is particularly troubling that the widely praised National Science Foundation Program on Ethical Values in Science and Technology (EVIST) may be abolished or dismembered even though the work done under its aegis—the development of a competence for ethical analysis and technological choice—promises to combine rigor and compassion.

As we eschew the single fix, be it technological or behavioral, we should also avoid the choice of a single ethic. It is possible to create a process that addresses the different needs of groups at risk, leading not to a perfect resolution of ethical dilemmas, but to a fairer distribution of the risks and benefits of technology. Scientific and technological fixes can also help by reducing the overall risk or by identifying and protecting groups at greater risk.

## VII

Two centuries after the beginning of the industrial-scientific revolution in the design, production, and use of technology, modern societies began the comprehensive management of the technological hazards created in its wake. Whether one dates the beginnings of this effort with the popular outcry of Earth Day 1970, as I have, or from the early warning of Rachel Carson's *Silent Spring* in 1962, or from the classic paper of Chauncey Starr in 1969[22] that started the professional development of comparative hazard management, the movement is less than a quarter of a century old. The real changes in the way society handles technological hazards are less than 15 years old. But so profound has been the shift in attitudes, institutions, and activities that, in retrospect, these changes may well be viewed as no less revolutionary than the technological revolution that preceded them.

Over the next 15 years the changes will be less profound but the problems may be no less important. I foresee four sets of concerns. The first I have discussed extensively—the limitations, strains, and contradictions of the first 15 years of activity and the search for alternatives to ease or resolve them.

Another set of concerns relates to the major changes under way in the design, production, and use of technology. New products will bring new hazards. Old products and processes in new locales will bring new hazard problems. The rapid restructuring of world industrial production will reduce the hazards in places that have learned to cope with them and move hazards to places where the knowledge and resources for control are not available. At the same time the potential for closed-cycle production, inherently clean or safe processes, and robotics will provide new opportunities for hazard reduction.

The new institutions and activities developed over the last 15 years to cope with hazards in our own country have proved inadequate so far to cope with the newer regional- and global-scale problems exemplified by the biogeochemical cycles of carbon, nitrogen, phosphorus, and sulfur—the basic elements of life. Research over the last 15 years has led to quantitative estimates of the degree of human modification of these natural cycles.[23] The annual release of carbon dioxide to the atmosphere from the consumption of fossil fuels equals about 10 percent of that being used by plants for photosynthesis. The formation of nitrogen oxides and nitrate in the course of fuel combustion and fertilizer manufacture equals about half of what the biosphere produces naturally. The amount of sulfur oxides released to the atmosphere, primarily from fossil fuel burning, exceeds the natural flux from decaying organic matter. These seem to be large alterations in natural cycles, but their long-term implications and synergistic interaction are uncertain. Over the next 15 years we will surely learn more about these fundamental processes, but our science is likely to exceed our social and political capacity to act upon such understanding.

Finally, there will be surprises—surprises that in turn will generate new concerns and activities. There will also be other concerns and surprises unrelated to technological hazards: international tensions, social change, and resource needs. As in the past, these will replace technological hazards on center stage, but the work in the wings will continue. The fundamental attitudinal and institutional changes of recent years have acquired a momentum of their own. The effort to compensate the past, make safe the present, and protect the future will continue.

## NOTES

[1] This article was adapted from a paper delivered at a June 3–4, 1985, National Academy of Engineering symposium on "Hazards: Technology and Fairness." A report on that symposium will be published in book form by the National Academy Press. In preparing this article, I have drawn extensively from the collective research and insight of the Clark University Center for Technology, Environment, and Development Hazard Assessment Group and particularly Christoph Hohenemser, Kirsten Johnson, Jeanne X. Kasperson and Roger E. Kasperson, and Mary Melville. In addition, I have had the benefit of thoughtful comments from Jesse Ausubel, Meredith Golden, and Howard Kunreuther.

[2] *State of Environment: An Assessment at Mid-Decade* (Washington, D.C.: The Conservation Foundation, 1984), 9.

[3] William D. Ruckelshaus, "Risk, Science, and Democracy," *Issues in Science and Technology* I (Spring 1985): 20.

[4] The research supported by the Russell Sage Foundation consists of a series of case studies of plant, corporation, or trade association hazard management.

[5] Peat, Marwick, Mitchell and Co., "An Industry Survey of Chemical Company Activities to Reduce Unreasonable Risk," prepared for the Chemical Manufacturers Association, Feb. 11, 1983.

[6] Robert Cameron Mitchell, "Public Opinion and Environmental Politics in the 1970s and 1980s," in *Environmental Policy in the 1980s: Reagan's New Agenda*, ed. Normal J. Vig and Michael E. Kraft (Washington, D.C.: Congressional Quarterly Press, 1984), 51–74.

[7] James Tuller, "Economic Costs and Losses," in *Perilous Progress: Managing the Hazards of Technology*, ed. Robert W. Kates, Christoph Hohenemser, and Jeanne X. Kasperson (Boulder, Colo.: Westview Press, in press), 157–74.

[8] National Research Council, *The Effects on the Atmosphere of a Major Nuclear Exchange* (Washington, D.C.: National Academy Press, 1985), 185–88.

[9] Bruce N. Ames, "Dietary Carcinogens and Anticarcinogens: Oxygen Radicals and Degenerative Diseases," *Science* 221 (April 29, 1983): 1256–64.

[10] Christoph Hohenemser, Robert W. Kates, and Paul Slovic, "The Nature of Technological Hazard," *Science* 220 (April 22, 1983): 378–84.

[11] Charles Perrow, *Normal Accidents: Living With High-Risk Technologies* (New York: Basic Books, 1984).

[12] Crawford S. Holling, "Director's Corner: Surprise!" *Options* (Laxenburg, Austria: International Institute for Applied Systems Analysis, 1983–84).

[13] Gilbert F. White et al, *Changes in Urban Occupance of Flood Plains in the United States* (Chicago, Ill.: University of Chicago, Department of Geography Research Paper No. 57, 1958).

[14] Leon S. Robertson, *Injuries: Causes, Control Strategies and Public Policy* (Lexington, Mass.: D.C. Heath, 1983), 92–94.

[15] Alvin M. Weinberg, "Science and its Limits: The Regulator's Dilemma," *Issues in Science and Technology*, II (Fall 1985).

[16] National Research Council, *Toxicity Testing: Strategies to Determine Needs and Priorities* (Washington, D.C.: National Academy Press, 1984).

[17] Kirsten Johnson, "State and Community During the Aftermath of Mexico City's November 19, 1984 Gas Explosion" (Worcester, Mass.: Clark University Center for Technology, Environment and Development, unpublished paper, June 1985). This report was made possible by a quick-response grant from the University of Colorado Natural Hazards Research and Applications Infor-

mation Center with support from the National Science Foundation. I am solely responsible for the inferences drawn from Kirsten Johnson's findings.

[18] Vilma R. Hunt, *Work and the Health of Women* (Boca Raton, Fla.: CRC Press, 1979).

[19] Patrick Derr, Robert Goble, Roger E. Kasperson, and Robert W. Kates, "Responding to the Double Standard of Worker/Public Protection," *Environment* 25, no. 6 (July/August 1983): 6–11, 35–6.

[20] Thomas M. Gladwin and Ingo Walter, "Bhopal and the Multinational," *Wall Street Journal*, Jan. 1, 1985.

[21] Roger E. Kasperson, ed., *Equity Issues in Radioactive Waste Management* (Cambridge, Mass.: Oegleschlager, Gunn & Hain, 1983).

[22] Chauncey Starr, "Social Benefit Versus Technological Risk," *Science* 165 (Sept. 19, 1969) 1232–38.

[23] Martin W. Holdgate, Mohammed Kassas, and Gilbert F. White, eds., *The World Environment 1972–1982: A Report by the United Nations Environment Programme* (Dublin, Ireland: Tycooly International, 1982): 623.

# PART V

# EMERGING AND RECURRING ISSUES

As we learn more about risks and hazards, we discover that many of them are not so new after all, nor can they be separated from broader based environmental, social, or political problems. Yet some new hazards and risks are discovered, such as ozone depletion and electric magnetic fields (EMFs). Even old hazards like lead are being rediscovered as government programs to remove it from inner city buildings are renewed.

Chronic hazards, those whose impacts take a long time to materialize, are becoming more pervasive in the 1990s than ever before. While there are rapid-onset events such as earthquakes, industrial accidents, and hurricanes, it is chemical contamination, hazardous waste, pollution, biotechnology, drought, climate change, and hunger that beckon for the hazards manager's attention. Moreover, it just seems like we solve one hazard problem and another one pops up in its place. Once we clean up one hazardous waste site, where do we put the hazardous materials? The hazards associated with our hazardous waste policies is one such example.

In their pioneering work, Couch and Kroll-Smith (Reading 23) try to differentiate human responses to sudden-onset, immediate-impact natural events versus the chronic technological disaster. Using a number of examples, including Love Canal and Centralia, they conclude that chronic technological disasters often produce increased conflict and deleterious long-term strain on communities, much more so than their natural disaster counterparts.

In his review of chemicals as hazards, Greenberg (Reading 24) asks two simple questions: (1) How important is chemical pollution as a causal agent in human cancer, and (2) How much scientific knowledge do we really have about the hazardousness of chemicals? Throughout his paper, Greenberg hints that risk assessments have limited utility, especially for the planning profession—hazard managers who often are involved in the siting of hazardous facilities or locally unwanted land uses (LULUs). Because of a shift away from the acute problems of chemicals (flammability, explosivity) to more chronic ones (toxicity), increased pressure is placed on local officials to ameliorate or reduce the impacts of facilities especially before they are sited and also long afterward.

One specific example of chemical hazards is the U.S. military's Chemical Stockpile Disposal Program. After decades of buildup, U.S. chemical weapons have become remnants of old technologies. They are unstable and have outlived their usefulness. The incineration and disposal of these weapons of mass destruction are now a priority for a downsizing military. Despite the relatively benign nature of the likely environmental impacts from incineration (based on the Army's detailed environmental impact assessments), public opposition is overwhelming in many communities near some of the storage sites. Rogers et al. (Reading 25) describe the emergency planning efforts in response to the stockpile disposal program. One striking conclusion they reach is repeated many times in different locations and with different hazards: Emergency response planning as a mitigation tool creates burdens that host communities are ill equipped to meet.

Not all chronic hazards are derived from failed technological systems, nor are their impacts concentrated at the local level. The interaction between the environment, social structure, and individual action also leads to chronic environmental changes such as drought, soil erosion, and deforestation. Liverman (Reading 26) explores the concept to vulnerability and its usefulness in understanding global environmental change. She illustrates her discussion by examining how the Green Revolution (technology and economic resources) and land reform affected the relationship between natural climate variability and agricultural production in Mexico. Liverman found differential effects of drought on different land tenures, illustrating how the interaction between physical and human use systems may actually increase vulnerability. She argues for moving beyond a biophysical interpretation of vulnerability to a more comprehensive approach that integrates the biophysical with the political, economic, and social.

The application of risk management which O'Riordan and Rayner say is really "people management" to global environmental change is a natural extension of the hazards research (Reading 27). This essay illustrates some of the pitfalls in applying risk analysis to a series of global environmental changes: biospheric catastrophes, climate perturbations, undermining basic needs, and long-term degradation from pollutants. The best use of risk analysis according to O'Riordan and Rayner may be in helping to frame the political debates. They are adamant that diversity in institutional responses (drawing an analogy to the key role that biodiversity plays in ecosystems) is essential when dealing with the complexity and high degree of uncertainty that characterizes global environmental change issues.

Finally, Hewitt's article (Reading 28) illustrates another recurring hazard: civil strife and warfare. Threats to the direct destruction of civilians, settlements, and habitats occur not only from so-called natural and technological sources, but also from acts of aggression. One has only to think about the Persian Gulf War— the burning of Kuwaiti oil fields and, oil spills in the Gulf—to understand the local and regional impacts of these human-induced hazards. Social hazards have many of the same characteristics as other hazard events: People are forcibly removed from their settlements, fatalities are great, and places and resources are often destroyed. Recovery can occur if hostilities cease quickly, but they often don't. In this regard, civil strife is analogous to the chronic types of hazards, those that evoke a constant recognition of threat, a sense of terror. In using case studies from World War II involving Eastern Europe and Japan, Hewitt details how the destruction of human settlements and the targeting of civilian populations and nonmilitary areas has become incorporated into strategies of modern warfare, as the contemporary events in Afghanistan, Sri Lanka, Somalia, and Bosnia and Hercegovina illustrate.

## DISCUSSION POINTS

1. Scale is an important element in hazard and disaster studies. Explain how the impacts of hazards may be accelerated as one moves from the local to the regional to the global level.

2. What is the difference between an acute versus a chronic hazard? Can you provide examples of each?

3. How might the notion of equity between generations be applied to hazards management? Provide some specific examples.

4. Can you think of some contemporary examples of social hazards resulting from civil strife? Where might these social hazards be most prevalent in the year 2005?

5. What is vulnerability? How might this concept be used to explain human susceptibility and response to hazardous waste? To global climate change?

## ADDITIONAL READING

de Castro, J. 1975. *The Geopolitics of Hunger*. New York: Monthly Review Press.

Clark, W. C. and R. E. Munn (Eds.), 1986. *Sustainable Development of the Biosphere*. New York: Cambridge University Press.

Glantz, M. H. (Ed.), 1987. *Drought and Hunger in Africa: Denying Famine a Future*. New York: Cambridge University Press.

Greenberg, M. and R. Anderson, 1984. *Hazardous Waste Sites: The Credibility Gap*. New Brunswick, NJ: Center for Urban Policy Research.

Jasanoff, S., 1986. *Risk Management and Political Culture*. New York: Russell Sage Foundation.

Kates, R. W., J. H. Ausubel, and M. Berberian (Eds.), 1985. *Climate Impact Assessment: Studies of the Interaction of Climate and Society*, ICSU/SCOPE Report No. 27, Chichester: John Wiley.

Newman, L. F., W. Crossgrove, R. W. Kates, R. Mathews, and S. Millman (Eds.), 1990. *Hunger in History: Food Shortage, Poverty, and Deprivation*. Oxford: Basil Blackwell.

Wildavsky, A., 1988. *Searching for Safety*. New Brunswick, NJ: Transaction Publishers.

World Commission on Environment and Development. 1987. *Our Common Future*. Oxford: Oxford University Press.

# 23

# The Chronic Technical Disaster: Toward a Social Scientific Perspective

*Stephen R. Couch* ❖ *J. Stephen Kroll-Smith*

Dioxin and radon contamination, toxic chemical leachates, and underground mine fires are disaster agents quite unlike hurricanes, floods, or tornados. A hurricane strikes quickly and disappears, leaving visible evidence of destruction, and is interpreted as natural in origin or as an "act of God." A mine fire under a town or toxic chemicals leaching into basements and backyards are not natural but technical disasters. Human intervention in the environment created the disaster, and technical intervention is required to abate, extinguish, or otherwise dispose of the disaster agent. Furthermore, such disasters tend to be chronic. A tornado "strikes" a community and vanishes, while radon contamination is "discovered" and may take decades to successfully abate or clean up. The chronic technical characteristics of these disasters, it may be argued, affect individuals and communities in ways discernibly different from natural disasters that strike quickly and disappear.

Since 1963, the Disaster Research Center at Ohio State University has been engaged in organizing, conducting, and publishing research on immediate-impact natural disasters and the emergency social systems constructed in response to them (see Quarantelli and Dynes, 1973, 1977). Several periodicals and journals (including *Disaster* and the *International Journal of Mass Emergencies and Disaster*) were established to promote natural disaster research, and numerous articles have appeared in more general social scientific journals. Very little research, however, has focused on chronic technical disasters. While a few scholars have recognized the existence of chronic disasters (cf. Barton, 1969; Erikson, 1976; Bates, 1982), the chronic technical disaster has not been carefully defined or analyzed. Moreover, what analytic and descriptive accounts there are of chronic technical disaster situations have not considered systematically the role of the disaster agent in determining or influencing local, regional, and national response (cf. Brown, 1980; Gibbs, 1982; Levine, 1982).

This paper offers a preliminary conceptualization of the social dimensions of the chronic technical disaster. First, a nominal definition of the chronic technical disaster is developed. Then, utilizing secondary sources and drawing on our research on the mine fire

Reprinted from *Social Science Quarterly* Volume 66 (1985), pp. 564–83. Used by permission of the authors and the University of Texas Press.

in Centralia, Pennsylvania (Kroll-Smith and Garula, forthcoming [1984]) and on recent research on the toxic waste problem at Love Canal, New York, we will suggest ways in which we expect the social impact of chronic technical disasters to differ from that of immediate-impact natural disasters. In the final section of the paper we discuss some ways in which the pre-impact social structures of affected communities are likely to influence community response to chronic technical disasters.

## SOME TENTATIVE DEFINITIONS

A premise central to cultural ecology is that a human settlement is in an exchange relationship with its environment. "There is," in the words of Sahlins (1964), "an interchange between culture and environment, perhaps continuous dialectic interchange, if in adapting the culture transforms its landscape and so must respond anew to changes that it had set in motion" (p. 133). It is reasonable to assume that a breakdown or destructuring of the relationship between a community and its environment will produce more or less severe reverberations in the psychological, social, and cultural life of the community.

This breakdown will be nominally defined as an *ecodisaster*. When an immediate crisis or long-term deterioration in human system—ecosystem relations is perceived by the community, or sectors therein, as jeopardizing the health and safety of residents, the existing pattern of social relationships, and the shared understandings, beliefs, and ideas in terms of which that interaction takes place, an ecodisaster may be said to exist.

The immediate impact natural disaster and the chronic technical disaster represent two qualitatively different types of ecodisasters. The former is defined as:

An event, concentrated in time and space, in which a society, or a relatively self-sufficient subdivision of a society, undergoes severe danger and incurs such losses to its members and physical appurtenances that the social structure is disrupted and the fulfillment of all or some of the essential functions of the society is prevented. (Fritz, 1968:202)

Practically all the studies which issue from this definition of disasters, or ones like it, have focused attention on those agents "which are quick and unexpected" (Quarantelli and Dynes, 1977:25). By contrast, the following is offered as a nominal definition of the chronic technical disaster:

a slowly developing, extended, humanly produced deterioration in human system—ecosystem relations, in which an entire community or sectors therein perceive and/or incur danger to health and safety and the disruption of ongoing patterns of social and cultural relations.

Essentially, we propose that the crucial differences between the two types of disasters are due in large measure to differences in time and in human/technological involvement. The chronic technical disaster develops slowly and persists for a relatively long time. In addition, while the effects of natural disasters are often influenced by human factors (e.g., population density, agricultural practices, building design), chronic technical disasters are *caused by* human-technological intervention in the environment, and further technical human intervention is required to *contain or abate* the disaster agent itself. There are, of course, many additional dimensions along which disaster agents can be compared and contrasted (e.g., scope of impact, intensity). Certainly, differences along these dimensions will influence the response of the affected community. However, *both* types of disasters discussed here have these additional dimensions as variables. For purposes of our discussion, then, we hold them constant, and look only at the dimensions of time and human technological involvement.

Based on our observations of Centralia, Pennsylvania, a close reading of Levine's (1982) important work on Love Canal, New York, and a study of Love Canal by the Federal Emergency Management Agency (Fowlkes and Miller, 1983), we are prepared to argue that the unique pattern of psychological, social, and cultural disruption accompanying the chronic technical disaster is, by and large, understandable in light of the incessant, gradual quality of the disruption agent and the degree and nature of human influence producing it and/or required to abate it.

## THE DISASTER AGENT AND COMMUNITY RESPONSE

A community's response to a disaster is influenced by the nature of the disaster agent itself. The following discussion attempts to delineate some of the ways in which differences between the immediate impact natural disaster and the chronic technical disaster, are likely to influence the response of affected communities.

### Collective Definitions and Response

If a disaster agent strikes quickly, and within moments or hours disappears, inflicting momentary, dramatic, and visible devastation within a community, then a clear and unambiguous distinction will exist between a previous period of relative stability and the resulting unstructured situation. With little collective doubt as to what occurred and how it should be interpreted, a "therapeutic community" can be expected to emerge wherein citizens and their organizations expand their ordinary roles within the community to meet the immediate needs of the injured, homeless, and grief stricken (cf. Turner, 1967; Barton 1969: chap. 5; Forrest, 1973).

However, if a disaster agent emerges gradually, sporadically advancing, more or less by increments, inflicting minor damage in one place, threatening greater devastation in another place, then the community will lack a clear-cut and unambiguous reference point from which to collectively determine its identity as a community in crisis. Disagreements can be expected to emerge over both the correct interpretation of the situation and the proper role of citizens and community organizations in responding to the problem. At Love Canal, for example, the movement from ignorance of the problem to a realization of the danger was slow and uneven (Levine, 1982:193). According to Fowlkes and Miller (1983), "Only the most minimal consensus was achieved among officials, experts and the resident population (regarding the presence and potential danger of the buried chemicals)" (p. 135). In Centralia, after the condemnation and destruction of over thirty homes by the government due to the mine fire, a significant number of residents, both inside and outside of the impact zone, still believed that the fire was not burning under the borough. The slow emergence of, and struggles between, different community groups with different perspectives on the disasters was evident in both situations (Levine, 1982:203). Levine (1982:179) found that since the Love Canal situation was neither cataclysmic nor dramatic there was no outpouring of aid from private community organizations. While Centralia itself has few private organizations which would help, aid from organizations in nearby communities has not been significant, and an often-expressed view is that Centralians are making too much of their problem.

### Emergency Social System

Once an immediate natural disaster is over, community and extracommunity efforts can be uniformly directed toward reestablishing routine and order in and around the disaster site. The emergency social system which emerges in response to the disaster will be relatively short-lived (Barton, 1969:103).

However, because of the extended duration of chronic technical disasters, those community and extracommunity efforts that can be organized in spite of the ambiguous situation will be channeled toward the disaster agent itself and the health and safety of residents. Reestablishing order and routine must wait until the disaster agent is brought under control. The emergency social system which emerges in response to the chronic technical disaster can be expected to be one of extended duration. The ambiguity of individual and organizational roles exists for some time, creating many long-term problems (Gibbs, 1983). At Love Canal, uneasy relationships between public and private physicians caused problems in health service delivery, increasing the sense of mistrust among residents in making decisions on vital issues (Levine, 1982:27, 69). Many came to fear that a cover-up was underway (Levine, 1982:27, 155; Fowlkes and Miller, 1983:64–65). In Centralia one-third of the adult population believed that a conspiracy existed to keep the fire burning, and the vast majority thought the government was not doing a good job in responding to the situation.[1]

## Conflict and Accountability

The immediate-impact natural disaster is a natural, not a technical, disaster. If the disaster is seen as being natural in origin, then it is less likely that the community will divide over the question of blame. While there may be human involvement in the sense of, for example, building homes on a geological fault, neither humans nor their technology causes earthquakes or tornados, nor are humans expected to be able to stop or control the disaster agent. There may be conflict over blame for lack of preventative measures or inadequate provision of services after the fact, but not over the cause or lack of control of the disaster agent itself (Bucher, 1957:467–75).

By contrast, in its origin and/or continuation, the typical chronic technical disaster is, at least in part, due to some measure of human error. If a disaster agent is interpreted by both the affected population and outside agencies as caused at least in part by, among

other things, human ignorance, apathy, and greed, then it is more likely that community conflict will emerge over the question of assigning accountability or blame in an effort to gain perspective on an otherwise ambiguous situation. It is also more likely that extralocal government and assistance organizations will disagree over who is to assume financial, technical, and material responsibility for meeting the life needs of the victim population and its surrounding environment. At Love Canal, Hooker Chemical Corporation consistently denied responsibility for the problem, while the state and federal governments sued Hooker to attempt to establish accountability (Levine, 1982:19, 25, 138, 191). In addition, both at Love Canal and Centralia, there was constant bickering between government agencies over which branch (if any) should pay to aid residents (Levine, 1982:24, 206).

## "Expert" Intervention

In cases of immediate-impact natural disasters, extralocal government and assistance organizations are prepared to offer immediate financial and material resources to the survivors. Special bureaucratic mechanisms, through legislation or organizational policies, exist to provide and administer aid to victims (Levine, 1982:62, Commonwealth of Pennsylvania, 1978).

But with chronic technical disasters the situation is without adequate precedent and is much more ambiguous. It is more likely that technical uncertainty and conflicting data regarding human and environmental risk evaluation will permit a wide range of "scientific" interpretations, further frustrating the community's efforts to achieve some control over the situation. At one point in Centralia there were three conflicting engineering reports, which differed over the extent, location, and direction of movement of the fire, as well as over how it could be abated or extinguished. The Pennsylvania Department of Environmental Resources, the agency responsible for the residential monitoring of toxic gases, changed the standards for "acceptable risk" on three separate occasions. In both Centralia and Love Canal the effects on public health and safety were matters of serious debate and differing conclusions among experts and community residents alike (Levine 1982:20, 124–26).

## Information Flow

A community disaster of any type can be expected to interrupt the flow of accurate and reliable information at just the point at which accurate and reliable information is most needed. If a disaster agent is of relatively short duration and natural origin, then it is likely that the interruption of reliable and accurate news will be relatively short-term, giving the community access, with only a brief delay, to information critical in forming an adequate collective response to the crisis (cf. Waxman, 1973:751–58). Also, while technical knowledge may be important concerning prediction of the disaster and response to it after the fact, it will not be expected to be able to control the disaster agent itself. In addition, technical knowledge will play a minor role in the ongoing interpretation of the event, reducing the chance for confusion at the local level due to disagreements and conflicting reports among experts (see Nelkin, 1979).

On the other hand, if a disaster agent is technical in nature, with disagreements arising at the extralocal level over the questions of responsibility and the correct technical means to address or abate the disaster, then the accurate reporting of technical knowledge will be critical in the ongoing interpretation of the event, placing role demands on the local news agencies that may be beyond the capacities of those agencies to manage effectively. In the cases under consideration, the news media were critical in keeping the problems before the public (Levine, 1982:189, 191, 193). However, at least in Centralia, the accuracy of the

reporting was a major point of contention. One-third of the population believed that the media made too much of their problem, and over forty percent thought the information they received about the fire was inaccurate. In one instance, a group of local citizens visited the editor of a nearby newspaper to attempt to have a reporter removed from the story because of inaccurate reporting which they argued was resulting in the furtherance of personal anxiety and collective discord.

## Psychological Reactions

A disaster of whatever type threatens the psychological stability of the affected population. If a disaster is natural in origin, striking quickly, forcefully, and suddenly disappearing, then survivors can be expected to adapt a coping style which is consistent with the quality and quantity of the disaster agent. Acute stress, delusion, and hysteria may follow any disaster. Consistent with the temporary presence of the disaster agent, however, acute stress, delusion, and hysteria, when present, can be expected to persist for only a relatively short period of time (Drabek and Stephenson, 1971).

However, if a disaster is human-technical in origin and advances slowly and erratically, then the coping style of the victim population can be expected to reflect the chronic, persistent nature of the disaster agent (Lang and Lang, 1964; Erikson, 1976:255). Anxiousness and delusion, or the readiness to hold on to perceptions contradicted by available evidence and common sense, may become embedded in the character structure of some or all of the victim population. Levine (1982:87, 184–85) found that anxiety, fear, and feelings of helplessness were prevalent among Love Canal residents, and that even "home" lost its deep-seated meaning as a refuge in a hostile world (see also Fowlkes and Miller, 1983; Gibbs, 1983). Similarly, two-thirds of adult Centralians have some fear that their safety is threatened by the fire. A mental health satellite facility was placed in the community and, according to the director, is frequently used by a growing number of residents. Delusion and denial are also present, with some believing that no fire exists while others sleep with packed suitcases under their beds prepared to flee the town at a moment's notice.

## PRE-IMPACT COMMUNITY STRUCTURE

A number of disaster studies (see Quarantelli and Dynes, 1977:23–49) suggest that not only is the disaster agent itself of significance in organizing community response; of considerable import as well is the structure of the community prior to the onset of the agent. This variable is particularly important in the chronic technical disaster, which places the community in an *extended*, if not indefinite, period of collective tension and stress. The strengths and limitations of the pre-impact community structure will have ample opportunity to dramatize themselves in the context of an extended period of collective crisis or tension.

Discussion thus far has centered on the likely influence of types of disaster agents on a community. It should be emphasized, however, that a collective response is a product of a community's *interaction with* the disaster agent. If we can delineate some common social characteristics of communities most likely to be confronted with a certain type of disaster, we can move from an abstract consideration of the disaster agent to an examination of specific response patterns to be expected from a certain type of community.

Of course, if all disasters strike at random, this task would be impossible, for we could not determine any distinctive characteristics of communities which are likely to be affected. However, this is not the case. We argue that chronic technical disasters are much more likely to strike predominantly lower-class communities than other types of settlements.

Most all disasters impact hardest on the lower classes (Bates, 1982:8; 1983). For example, lower-class housing is less likely to withstand a hurricane; lower-class residents are less likely to have adequate insurance protection, or to possess adequate financial resources so as to be able to rebuild or escape from the scene. At the same time, it appears that disasters vary concerning the likelihood that they will strike a certain segment of the population at all. Some disasters possess little "class bias," striking all classes randomly, while others are more likely to impact the lower classes, and to greater or lesser degrees.

At one extreme are tornados, and hurricanes, which are as apt to strike upper- and middle-class areas as they are lower-class settlements. Floods can be categorized as having a moderate class bias; the well-to-do are less likely to build homes in a flood plain, but they are by no means immune. At the other extreme, we find chronic technical disasters such as mine fires and toxic waste dumps, which are much more likely to occur in lower-class areas—in communities more likely to have a history of dependence on and exploitation by corporations and governments with centers of power located far away from the community itself (Lewis, 1970; Walls, 1976, Nyden, 1979). A recent study by the General Accounting Office found that of four hazardous waste landfills currently threatening communities in one state "at least 26 percent of the population in all four communities have income below the poverty level" (GAO, 1983:1). In addition, these communities are often dying even before the disaster occurs, thereby possessing certain structural and cultural patterns which are exacerbated by the disaster itself (Gallaher and Padfield, 1980). Ironically, the economic and political weaknesses which allowed the chronic technical disaster to begin compound the problems faced by these communities in responding to the disaster. We hypothesize that these include the following:

1. *Governments at the extralocal level are more likely to be paternalistic, rather than cooperative.* The

long history of dependence of chronic technical disaster communities on outside corporations and governments leads to the institutionalization of paternalistic structures and attitudes on both sides. This makes it more likely that decisions concerning the disaster situation will emanate from the top down, rather than from the community itself. Levine (1982:26, 27, 35, 84, 123) found numerous cases of citizens not being consulted or taken seriously by the government concerning decisions affecting their lives. Similarly, Centralians were also excluded from the decision-making process, having little input and virtually no formal power.

2. *The historically dependent nature of these working-class communities makes it less likely that effective local leadership will develop*—i.e., leadership that can move the centers of power to act in the interests of the community, and/or unite and mobilize the members of the community toward effective action in their own interests. The leadership which does emerge is likely to be far removed from the center of power, and lacking in significant support from the community itself. The in-fighting occurring between citizens' groups in Love Canal and Centralia was documented above. In addition, the official local leadership was quite ineffective, necessitating the development of informal citizens' groups possessing inexperienced emergent leadership, which itself had trouble obtaining legitimacy from the communities and government agencies (Levine, 1982:36; Gibbs, 1983).

3. *Members of these communities are likely to hold ideologies which impede concerted collective action in their own behalf.* In the ideology of the working class, for example, patriotism and a belief in the soundness of the system is combined with distrust of and cynicism about large corporations and big government, a feeling of powerlessness in the face of seemingly overwhelming obstacles, and frontier individualism, all of which work against effective collective action (Padfield, 1980). In addition, a strong sense of roots and loyalty to the community conflict with strong values attached to biological family and individualism, creating cross-pressures which are emotionally disturbing and which increase the likelihood that while expressing community loyalty residents will be unlikely to sacrifice their own self-interests for the good of the community. Rancorous personal conflict between individuals and families with different interests in relation to the disaster are likely to result. In Centralia, such personal conflict has included shunning, vituperative personal exchanges at public meetings, telephone threats, slashing tires of community leaders, and even the firebombing of a business owned by one community activist.

4. *Chronic technical disasters are more likely to occur in communities with highly adapted cultures.* A majority of Love Canal wage earners worked at the nearby chemical plants. The cadence of day-to-day life was directly linked to the production of chemicals. Centralia today bears the legacy of a single-extraction-industry town. The community's inception, growth, and decline is a mirror image of the fate of anthracite coal on the world energy market (cf. Rose, 1981). As a coal town, Centralia constituted a specialized culture, well adapted to its narrow economic base.

A highly adapted culture is biased; "its design has been refined in a special direction, its environment narrowly specified, how it shall operate definitively stated. The more adapted a culture, the less therefore it is adaptable" (Sahlins, 1964:138).

Cultural specialization, in other words, subtracts from the potential for innovative response. This rigidity is perhaps nowhere better expressed than in the Roman Catholic church's response to the crisis. Eighty-three percent of Centralians attend the Roman Catholic church. The Catholic church, however, has remained silent on the fire issue for the two-decade duration of the fire. In the words of the parish priest, "this is not a spiritual issue, but a political one; the church has no business here." By and large, this attitude is supported by the local congregation, who are content to have the church limit itself to its traditional specialization, the cyclic administration of the sacraments. In short, Centralia's high degree of adaptedness has insured a contradiction between its received religious culture and its crisis predicament.

5. *While few local communities have the economic resources to successfully combat disasters of any magnitude, communities in which chronic technical disasters are most likely to occur are in a particularly poor monetary position.* Their economic status is unlikely to allow them the financial resources to mount a concerted effort to receive outside economic aid to alleviate their problems. In addition, the economic results of such a disaster are apt to be especially insidious. For example, if relocation of part of the community takes place due to the disaster, the already precarious tax base of the community may be strained to the breaking point. This problem has been a point of constant concern in

Centralia, where a significant majority of the residents wish to save the community, but where the only certain way of extinguishing the fire would force relocation of a substantial number of borough taxpayers.

Overall, then, the economic, cultural, and structural conditions of communities most likely to experience chronic technical disasters make them those least able to respond effectively. Powerless to combat the disaster or significantly influence adjustments to it, such communities are at the mercy of others who have not only the community's but their own interests with which to be concerned.

## CONCLUSION

This paper has begun the task of conceptualizing the chronic technical disaster and its impact on human settlements. A nominal definition of the chronic technical disaster was suggested, followed by an examination of some factors which show the disaster agent to be qualitatively different in many critical ways from that of immediate-impact natural disasters. It was suggested that because the nature of the disaster agent and the nature of the pre-impact community are distinctive in the case of the chronic technical disaster, the disaster's impact on the host community is distinctive as well.

Study is needed, both to test and refine the ideas contained in this paper and to add life to what may seem now as a rather static model. A chronic technical disaster is processual, occurring over time (in some cases, many years), and is a product of constant dialectical interaction between a disaster agent, a community, government agencies, and legislators. Only by studying actual cases in depth can the process of chronic technical disasters be understood.

Given our nascent state of awareness concerning the chronic technical disaster, and the paucity of research on the topic, the conclusions reached in this paper must be tentative. We offer evidence from Centralia and Love Canal as illustrative, not as proof. Ob-

viously, there are differences between mine fires, toxic waste sites, and other chronic technical disasters. And, of course, each individual case is distinctive. However, we believe that there are distinctive patterns involved with these disaster agents which influence their social effects and the human responses to them. Studies of additional cases are needed to confirm or refute this argument and to set out in more detail how variations within these patterns affect human responses to them.

While we see study of the chronic technical disaster as being of theoretical interest, we also believe it can have practical importance. As a nation traditionally insensitive to the long-term environmental consequences of short-term profit seeking, more and more human settlements will find themselves facing the ravages of ecological disasters. By breaking down the interdependence between human settlements and their environments, these disasters create several social, economic, cultural, and psychological problems for the people involved. Moreover, if our initial observations and reasoning are correct, the communities most likely to be hit by chronic technical disasters are those least likely to be in a position to respond effectively to them. It is hoped that this essay will encourage scholars to look closely at environmental disasters both as fertile areas for theoretical speculation and as immediate areas of practical concern.

*Acknowledgments* This research was partially funded by a Research Initiation Grant from the Pennsylvania State University, as well as grants from the University's Faculty Scholarship Support Fund and the College of Liberal Arts Research Fund and a grant from the Pennsylvania National Bank and Trust Company. We wish to thank Frederick L. Bates, Roland Pellegrin, and three anonymous reviewers for their comments on an earlier draft of this paper and Michael A. Kryjak and Richard Sachse for their research assistance. Editor's note: Reviewers were Russell Dynes, Larry Lyon, and Lee Sigelman.

## NOTE

[1] Statistics reported on Centralia are taken from a survey of all adult residents of the borough. Conducted by the authors in August 1982, a total of 368 questionnaires were completed, representing a response rate of 55 percent. Other information reported was gathered through participant observation and intensive interviewing.

## REFERENCES

Barton, Alan. 1969. *Communities in Disaster: A Sociological Analysis of Collective Stress Situations* (Garden City, N.Y.: Doubleday).

Bates, Frederick L. 1982. "Disasters, Social Change and Development," in Frederick L. Bates, ed., *Recovery, Change and Development: A Longitudinal Study of the 1976 Guatemalan Earthquake*, vol. 1 (Athens: University of Georgia Press): pp. 1–36.

———. 1983. Personal correspondence with the authors.

Brown, Michael. 1980. *Laying Waste: The Poisoning of America by Toxic Chemicals* (New York: Pantheon).

Bucher, Ruth. 1957. "Blame and Hostility in Disaster," *American Journal of Sociology*, 62 (March): 467–75.

Commonwealth of Pennsylvania. 1978. "Emergency Management Services Code," *Pennsylvania Consolidated Statutes*, part V, title 35, as enacted by the Act of 22 November 1978, PL 1332, no. 323, "Disaster."

Drabek, Thomas E., and J. S. Stephenson. 1971. "When Disaster Strikes," *Journal of Applied Social Psychology*, 1 (April): 187–203.

Erikson, Kai. 1976. *Everything in Its Path: Destruction of Community in the Buffalo Creek Flood* (New York: Simon and Schuster).

Forrest, Thomas R. 1973. "Needs and Group Emergence," *American Behavioral Scientist*, 16 (January): 413–24.

Fowlkes, Martha B., and Patricia Y. Miller. 1983. *Love Canal: The Social Construction of Disaster*. Report prepared for the Federal Emergency Management Agency, RR-1.

Fritz, C. E. 1968. "Disaster," in D. Sills, ed., *International Encyclopedia of Social Sciences*, vol. 4 (New York: Macmillan): pp. 202–7.

Gallaher, Art, Jr., and Hartland Padfield, eds. 1980. *The Dying Community* (Albuquerque: University of New Mexico Press).

General Accounting Office. 1983. *Siting of Hazardous Waste Landfills and Their Correlation with Racial and Economic Status of Surrounding Communities* (Gaithersburg, Md.: Document Handling and Information Services).

Gibbs, Lois. 1982. *Love Canal: My Story* (New York: Grove Press).

———. 1983. "Community Response to an Emergency Situation: Psychological Destruction and the Love Canal," *American Journal of Community Psychology*, 11 (April): 116–25.

Kroll-Smith, J. Stephen, and Samuel Garula. Forthcoming. 1984. "The Real Disaster Is Above Ground: Community Disorganization and Grassroots Organization in Centralia," *Small Town*.

Lang, Kurt, and Gladys Lang. 1964. "Collective Responses to the Threat of Disaster," in D. Grosser, et al., eds., *The Threat of Impending Disaster* (Cambridge: MIT Press): pp. 48–65.

Levine, Adeline. 1982. *Love Canal: Science, Politics and People* (Lexington, Mass: Lexington Press).

Lewis, Helen M. 1970. "Fatalism or the Coal Industry? Contrasting Views of Appalachian Problems," *Mountain Life and Work*, 46 (December): 232–47.

Nelkin, Dorothy. 1979. "Science, Technology and Political Conflict: Analyzing the Issues," in Dorothy Nelkin, ed., *Controversy: Politics of Technical Decisions* (Beverly Hills: Sage): pp. 9–22.

Nyden, Paul J. 1979. "An Internal Colony: Labor Conflict and Capitalism in Appalachian Coal," *Insurgent Sociologist*, 8 (Winter): 33–43.

Padfield, Harland. 1980. "The Expendable Rural Community and the Denial of Powerlessness," in Art Gallaher, Jr., and Harland Padfield, eds., *The Dying Community* (Albuquerque: University of New Mexico Press): pp. 159–85.

Quarantelli, E. L., and R. R. Dynes. 1973. "Editors' Introduction," *American Behavioral Scientist*, 16 (January): 305–12.

———. 1977. "Response to Social Crisis and Disaster," in Ralph Turner and James F. Short, eds., *Annual Review of Sociology*, vol. 3 (Palo Alto, Calif.: Annual Reviews): pp. 23–49.

Rose, Dan. 1981. *Energy Transition and the Local Community* (Philadelphia: University of Pennsylvania Press).

Sahlins, Marshal. 1964. "Culture and Environment: The Study of Cultural Ecology," in Sol Tax, ed., *Horizons of Anthropology* (Chicago: Aldine): pp. 136–47.

Turner, Ralph H. 1967. "Types of Solidarity in the Reforming of Human Groups," *Pacific Sociological Review*, 19 (Fall): 60–68.

Walls, David S. 1976. "Central Appalachia: A Peripheral Region within an Advanced Capitalist Society," *Journal of Sociology and Social Welfare*, 4 (November): 232–47.

Waxman, J. J. 1973. "Local Broadcast Gatekeeping During Natural Disaster," *Journalism Quarterly*, 50 (Winter): 751–58.

# 24

# HEALTH EFFECTS
# OF ENVIRONMENTAL CHEMICALS

*Michael R. Greenberg*

Some chemical substances can cause effects ranging from nausea, loss of appetite, vertigo, and headaches to immediate death, genetic damage, birth defects, and cancer. They can also cause fear that may or may not have any real basis. This is because the public's fears of health effects from chemicals are based on perceptions that distort the reality of the risk (Lowrance 1976). We greatly fear things that are not known with certainty—especially if they can lead to such delayed, irreversible, and dreaded effects as cancer. Also, risks that are faced in a non-work setting and are borne involuntarily are not well tolerated. The public's fear of chemicals has highlighted the need for many planners to learn how to evaluate and control land uses involving the transportation, use, storage, and disposal of chemicals. This review article focuses on the two scientific questions that both underlie our perceptions of the health impacts of chemicals and have strongly influenced government policy toward chemicals. They are: (1) How important is chemical pollution as a cause of

cancer? (2) How much do we know about the hazardousness of chemicals?

The debates about these two questions are so voluminous that is it not feasible to provide the full flavor of the clash of views. A consensus of the literature is presented, arguments for and against it are offered, and explanations for different viewpoints are suggested. The major intersection of chemical hazards and planning practice is briefly analyzed.

## HOW IMPORTANT IS CHEMICAL POLLUTION AS A CAUSE OF CANCER?

The public's fear of cancer borders on, and sometimes crosses into, the terror zone. In a study of the public's perception versus reality, nine of the ten most overestimated causes of death were found to be dramatic and easily sensationalized causes such as accidents, tornadoes, and homicide (Slovic, Fischhoff, and Lichtenstein 1979). The other overestimated cause is a chronic disease: cancer. The point is that cancer is the only major chronic disease that is sensationalized like sudden and dramatic causes of death. In 1984 the National Cancer Institute reported that 46 percent of

Reprinted by permission from *Journal of Planning Literature*, Volume 1 (1), pp. 1–13. Copyright © 1986 by the Ohio State University Press. All rights reserved.

the public believes that "there is not much a person can do to prevent cancer" (*Nation's Health* 1984). It is distressing to learn that almost half of the population does not recognize the effects of cancer prevention programs.

The literature about cancer and chemicals is a tale of misunderstandings and advocacy about the environmental contribution to cancer. John Higginson (1980), the first director of the International Agency for Research on Cancer, reported that "the concept of the environment has been used since the time of Hippocrates to cover all factors which affect man and his health." Yet it was Higginson's research on environmental factors in cancer that contributed to the important and persistent misunderstanding about the role of environment in cancer.

The misunderstanding began with a World Health Organization (1964) report on cancer prevention stating that extrinsic or environmental factors were directly or indirectly responsible for more than three-fourths of cancer incidences and that the majority of cancers are potentially preventable. The first link to chemicals came some five years later, when Boyland (1969) categorized the causes of cancer as biological, chemical, and physical, and argued that 90 percent of cancer was chemical in origin. Five years later, citing these and a few other papers, Samuel Epstein (1974) stated that "there is now growing recognition that the majority of human cancers are due to chemical carcinogens in the environment and that they are hence ultimately preventable." Many other scientists and writers have reiterated this thesis, some in brief scientific papers (Wynder and Gori 1977, Schneiderman 1978), others in long scientific treatises (Fraumeni 1975, Hiatt, Watson, and Winsten 1977, Doll and Peto 1981), and still others in books written for the general public (Agran 1977, Highland, Fine, and Boyle 1979). Few argue against the environment as the major factor contributing to cancer, but they do argue about the definition of environment and chemicals.

There is a good deal at stake in what on the surface appears to be quibbling about semantics. John Higginson summarized the argument and the stakes when he was asked the following question about the role of environment in causing cancer by Thomas Maugh II (1979), a reporter for *Science*: "So, in effect your conclusions have been misinterpreted all along?" Higginson's reply was in part:

They have been misinterpreted. A lot of confusion has arisen in later days because most people have not gone back to the early literature, but have used the word environment to mean chemicals. . . . There's one other thing I should say that has led to the association of the term environment with chemical carcinogens. The ecological movement, I suspect, found the extreme view convenient because of the fear of cancer. If they could possibly make people believe that cancer was going to result from pollution, this would enable them to facilitate the cleanup of water, of the air, or whatever it was.

The issue, in short, is advocacy, not semantics. The consensus position on the definition of environment is that it includes all exogenous factors such as tobacco products, alcohol, nutrition, work, air and water pollution, sexual habits, sunlight, and any other stresses. Looked at another way, environment includes everything that is not due to a genetic or a prior health condition of the individual (Last 1983).

There is also a problem about the definition of *chemical*. A chemical is often thought to mean a synthetic, organic, industrial compound. It also means chemical substances in cigarette smoke, foods, drugs—some of which are manmade and others of which occur naturally.

This bland consensus hides a literature characterized by vitriolic prose. For example, on the one hand, Wilhelm Hueper (1966), former head of the Environmental Cancer Section of the National Cancer Institute, arguing for occupation and pollution as a cause of cancer, stated:

It should be obvious that any wide acceptance

of such scientifically unsound and socially irresponsible claims concerning the principal role of cigarette smoking in the causation of cancers, especially respiratory cancers, would paralyze . . . a legitimate and urgently needed pursuit into the various environmental factors including such cancers, particularly the many industry-related pollutants of the urban air. . . .

On the other hand, two eminent British epidemiologists, Richard Doll and Richard Peto (1981) responded in the following way to a study sponsored by the U.S. government which claimed that between 20 and 38 percent of cancers were related to occupational exposure:

It seems likely that whoever wrote the OSHA paper (it has a list of "contributors" but no listed authors) did so for political rather than scientific purposes. . . . Although its conclusions continue to be widely cited the crucial parts of the arguments for these conclusions have, perhaps advisedly, never been published in a scientific journal nor in any of the regular series of government publications. The OSHA paper should not be regarded as a serious contribution to scientific thought and should not be cited or used as if it were.

The strong opinions expressed by Hueper and Doll and Peto are not unusual for this literature. In general, the literature on chemical pollution and cancer is marked by an initial finding, affirmations of that finding, and then denials and alternative explanations. An illustration is drawn from the air pollution and cancer literature. Stocks and Campbell (1955) found urban cancer death rates in England and Wales in the mid-1950s to be twice as high as rural rates. They attributed half of the high rate in the urban areas to smoking and 40 percent to air pollution. Similar urban-rural differences were reported in Iowa in 1958 (Hagstrom, Sprague, and Landue 1967), in a sample of white, middle-aged American men (Hammond and Horn 1958), in New York State (Levin 1960), and with data about migrants from Britain to South Africa (Dean 1961) and New Zealand (Eastcott 1961). Furthermore, carcinogens were found in city air during this period (Sawicki and Westphal 1961).

The urban air pollution factor was challenged. Wynder and Hammond (1962) argued that urban-rural differences in lung cancer were better explained by urban-rural differences in smoking, occupational exposures, and accuracy of reporting of death. Their position was supported by Doll and Hill (1964) and, as indicated by the consensus in Table 1, by most other researchers.

The scientific and verbal battle continues. Today, it focuses on our responsibility to prevent cancer through changing our habits versus government responsibility to regulate chemicals in tobacco, food, drugs, and other places. This writer strongly recommends the exchanges in several issues of the journal *Nature*. For instance, in his massive book, *The Politics of Cancer*, Epstein (1979) argues for strong government regulation of carcinogens in consumer products and the environment. Richard Peto (1980) criticized Epstein's way of judging and presenting evidence and argued that lifestyle changes are crucial to lowering cancer rates. Epstein and Swartz (1981) attacked Peto's lifestyle arguments. Cairns (1981) joined the argument with an explanation of cancer causation that is considerably more complex than that offered by Epstein. Most recently Ames has exchanged sharply worded comments with Epstein and Swartz on the role of diet in cancer (Epstein, Swartz, and Ames 1984). This writer suspects that the vitriolic exchanges will continue in an effort to sway public opinion.

In summary, the water we drink, food we eat, land we live on, and air we breathe are contaminated to some extent with products of our urban-industrial society, but not every chemical causes cancer. Many chemicals are therapeutic, and others are critical for survival. A key goal of science is to determine, and then educate the public about, which substances are harmful and which are beneficial in what forms and doses for what periods.

**TABLE 1**   The Causes of Cancer: A Consensus

| Type of Cause | Percent of Cancers | |
| --- | --- | --- |
| | Consensus | Typical Range |
| *Environment* | | |
| Diet | 35 | 10–70 |
| Tobacco smoking | 30 | 25–40 |
| Other personal: alcohol, sexual behavior, and other | 10 | 4–12 |
| Occupation | 4 | 2–10 |
| Sunshine | 3 | 2–4 |
| Medicines, medical practices, food additives, ionizing radiation | 3 | <1–6 |
| Environmental Pollution | 2 | <1–5 |
| *Nonenvironment* | 13 | — |

*Source*: Office of Technology Assessment 1981, Higginson and Muir 1979, Doll and Peto 1981, Wynder and Gori 1977.

## HOW MUCH DO WE KNOW ABOUT THE HAZARDOUSNESS OF CHEMICALS?

There is a major credibility gap in the management of chemicals in the United States. It adds to the great fear of cancer. The gap begins with the astounding number of chemicals in general use compared to the few that have been tested. The National Research Council (1984) recently published a three-year study of a random sample of 675 out of the fifty-four thousand chemicals to which people are exposed in commerce, pesticides, foods, drugs, and cosmetics. The expert panel concluded that only 18 percent of the drugs and substances in drugs, 10 percent of the pesticides and inert chemicals in pesticides, 5 percent of the food additives, and virtually none of the chemicals in commerce have been tested sufficiently to allow an adequate assessment of their hazard. The panel further noted that we are losing ground because five hundred to two thousand new chemicals enter the market every year. No one, not even the chemical industry, seems to dispute these findings.

### Chemical Testing Methods

Added to the legacies of tens of thousands of untested, inadequately tested, and forthcoming chemicals is the problem of how chemicals should be tested. Tests for flammability and explosiveness are not a problem. Tests for cancer, mutation, and reproductive effects are. The tests can be categorized as human, in vivo (animal bioassay), and in vitro (test tube). They are summarized in Table 2. Briefly, case-control studies are the most useful, but as Table 2 suggests and the case study will illustrate, conclusive results are rare because of serious data constraints. Also, studies of lower organisms can be screening devices to signal the need for animal research. However, the results should not be directly extrapolated to humans because humans are so different from bacteria and other simpler organisms. Chemical testing of animals (animal bioassay) is the main method of testing for human health effects. However, this method lacks credibility in the eyes of many people because animals and humans differ in metabolism and in other ways that affect response to chemicals. Furthermore, animal studies are made with much larger doses than humans would receive, and these doses are enhanced by the use of solvents that humans would rarely encounter. Readers interested in detailed treatments and critiques of these methods and other methods not summarized here should review the following works: Office of Science and Technology Policy (1979), Office of Technology Assessment (1981), Lowrance (1981), Hayes (1982), National Re-

**TABLE 2** _Major Methods Used to Assess the Risk of Chemicals_

| Type | Description | Length of Wait for Results | Advantages | Disadvantages |
|---|---|---|---|---|
| Retrospective (case control) | Identification of cases with the disease and controls who are similar but do not have the disease; different attributes of diseased and non-diseased are compared | Often decades before disease is manifested | Human data are the most credible; works well with small populations and strong agents | Long latency of disease; costly; exposures usually not known; differences among hosts impractical for thousands of agents |
| Bioassay of animals | Application of material in question to animals in a laboratory environment | 6 months to 3 years | Animal model more credible than lower organism models | Expensive, $250,000 to $750,000; credibility less than human studies because of genetic differences and large and enhanced doses |
| Lower organism, in vitro | Application of material to bacteria and other lower organisms in laboratory | Few days | Inexpensive, $200 to $2,000; used for screening chemicals | Little credibility if extrapolating directly to humans |

search Council (1983), and National Research Council (1984).

Whether the data come from human studies or from bioassays, the most controversial phase of the process is extrapolating dose and response relationships to people. This is because almost all of the data about doses and responses are based on responses to high doses. The low dose effects are extrapolated from high dose data.

One opinion of extrapolation is represented by Jacobs (1980):

"Everyone knows" that extrapolation is statistically unsound and that curves apply only to regions validated by data. In environmental issues extrapolation is routinely used, since no data exist in some important areas. The fallacy becomes obvious in looking at the case of trace elements in the human body. Copper and zinc, for example, are absolutely necessary to life. In larger doses they are dangerous poisons. No extrapolation could predict this reversal at low dosage. We need data, not bad guesses.

A contrary position is represented by the U.S. Environmental Protection Agency (EPA), which uses a linear, nonthreshold model to estimate the effects of different doses. This method is strongly opposed by industry and many scientists because it results in no safe doses. Every molecule can cause cell damage. The model assumes no threshold below which the body can defend itself against chemical assault. Schneiderman (1981) accurately summarizes the uncertainty about modeling of dose-response relationships: "The dose-response relationships that many of us considered 'conservative,' the linear, no threshold concepts are being challenged on both sides as both overstating and understating the risks."

Summarizing, human exposure, animal bioassay, and lower organism tests have been used to determine the hazardousness of chemicals. The evidence against many substances has taken decades to accumulate, and many more years have passed before the pattern of their use has been modified. This painfully slow process of testing substances stands in sharp contrast to the demands for quick and decisive action sought by the public and the federal legislation of the late 1970s.

## Lung Cancer and Diesel Auto Emissions

Dioxin, asbestos, formaldehyde, benzene, DDT, aldrin, malathion, ethylene dibromide, and many other chemical substances have made front page headlines. Diesel fuel extracts have not; but they offer an illustration of the difficulties of chemical risk assessment that should be of interest to transportation and environmental planners.

Light-duty diesel automobiles and trucks are expected to comprise between 10 and 50 percent of the motor vehicles in the United States in the year 2000 (Springer 1982). The increase is expected because diesel-powered motor vehicles are cheaper to operate than gasoline-powered vehicles. However, the economic benefit could result in increased illness. In the late 1970s, animal tests showed that organic particulate constituents of diesel exhaust caused mutations (Huisingh et al. 1978). Other studies have shown mutagenic (capable of causing changes in genetic material) and carcinogenic effects of extracts of diesel emissions (Hoffmann et al. 1965, Wynder and Hoffmann 1962). These findings are especially distressing because diesel engines emit thirty to one hundred times more particulates than gasoline engines. Most of these are fine particulates, many of which are inhaled and retained in the body.

The U.S. EPA asked the National Academy of Sciences (NAS) to study the potential health effects of diesel exhaust. After reviewing the few studies of human populations exposed to diesel exhaust in the workplace, Harris (1983) focused on the lung cancer incidence of diesel bus garage workers, bus drivers, and bus conductors for the London Transport Authority (LTA) (Raffle 1957, Waller

1979). A human exposure study of the effect of diesel fumes should have a population of exposed and unexposed people. In addition, confounding factors like smoking habits should be known. The researcher should know the duration and magnitude of the exposure of the study population to diesel fumes. A pathologist should verify the mortality from the disease(s) in question. Lastly, the population at risk should be followed for a period of time of two or four decades in order to allow for the latency period of cancer (Lilienfeld and Lilienfeld 1980).

Only the LTA study even approaches these data goals. The population at risk were male workers age forty-five to sixty-four, whose lung cancer rates were compared to lung cancer rates of other company workers and to men living in the Greater London area.

After twenty-five years, the at risk population suffered only 79 percent of the expected number of lung cancer deaths (Harris 1983). No effect? Maybe—but this study had some important shortcomings that limit its utility. One was that there was no follow up on employees after they left the LTA. This is important because some researchers believe that cancer is a multistage process (Doll and Peto 1978) that increases rapidly in probability with age, which means that the LTA workers might not show the effects until years after they left the LTA.

Second, little information was collected about confounding factors such as smoking habits and ethnicity. Third, estimates of concentration of diesel exhaust in the LTA working spaces were not based on precise and replicated measurements (Commins, Waller, and Lawther 1957, Waller 1979). Any of these limitations is sufficient to discredit these data in the eyes of many people.

The second approach used in vivo and in vitro bioassays to compare the carcinogenic potential of diesel extracts and known human carcinogens (coke oven, roofing tar, and cigarette smoke emissions) in the same animals (Albert et al. 1983, Harris 1983). The results were then projected to humans. The best judgment was that lifetime exposure (seventy years) to diesel particulates would annually produce between eighty and fifteen hundred excess lung cancers. In comparison, there were over one hundred thousand deaths from lung cancers in the United States in 1980. In short, according to this risk assessment, the diesel engine will cause a health problem, one that could range from small to moderate, and one that is much smaller than other automobile-related problems, such as the widely cited estimate that twelve thousand Americans are killed each year in auto accidents because they do not use seat belts.

The study of the diesel engine illustrates great uncertainty more than it provides a credible estimate of health effects. There are major uncertainties in the diesel emission studies on both the health and the public policy and economic sides. On the health side, the human data base was clearly inadequate. The best mathematics could not compensate for the absence of follow-up and data about confounding factors. The number of animal studies has been limited. Third, there is always uncertainty about applying to humans ratios derived from animals.

The estimated health effects also depend upon the replacement of the gasoline-powered fleet by a diesel fleet. Diesel-powered vehicles may not increase as rapidly as expected because technological changes may make the purchase of gasoline-powered vehicles more attractive or the purchase of diesel-powered vehicles less attractive (Cuddihy and McClellan 1983). Politics could play an important role. If the public or government decide that diesel-fueled vehicles are dangerous, the public could refuse to buy them. Alternatively, the government could regulate them, which, in turn, would increase prices and reduce purchases.

Whatever conclusions can be drawn from this research—and frankly other than noting that diesel exhaust seems to be similar to other combustion products as a potential contributor to cancer this writer could draw none—

must be tempered by the fact that there are other health and safety considerations that were not included in the NAS study. Schwing, Evans, and Schreck (1983) of General Motors argue that diesel-fueled cars are safer than gasoline-fueled cars because their added weight will reduce mortality from accidents. On the other hand, diesel exhausts contain $NO_2$, $SO_2$, and aldehydes, which are irritant gases that will affect the 15 to 20 percent of Americans suffering from emphysema, asthma, bronchitis, and chronic sinusitis. Moreover, a recent epidemiological study of truck drivers showed an excess of bladder cancer (Silverman et al. 1983). Perhaps the lung is not the primary organ affected by diesel emissions.

By now it would not be unreasonable for the reader to think that these risk assessments have little, if any, value. Our efforts to determine the effects of existing chemicals (the same is true of biological and physical agents) in two to four decades may at best yield clues that will help us draft public policy and influence public attitudes. Mistakes are bound to be made. For example, thalidomide was marketed after it showed no problems on bioassay tests, and this writer's reading of the literature suggests that cyclamates, which were taken out of circulation, are probably less dangerous than saccharin. Nevertheless, risk assessments are better than plunging ahead blindly and ending up with chemical, biological, or other technological hazards that could have been prevented.

## THE INTERFACE OF HAZARDOUS CHEMICALS AND PLANNING PRACTICE

Planners must work with fire departments, environmental commissions, health officers, and private organizations to design legal mechanisms that will lead dangerous facilities to the safest areas and require them to use manufacturing, transportation, and residual management methods that limit the chances of chronic emissions and accidents, and to preparations for accident response. This will require technical and administrative skills.

## Technical

Planners must not hesitate to ask questions, even if they are told that a proposed facility or addition to an existing facility will use state-of-the-art technology and emit a negligible discharge. In most cases, the best technology will protect public health and the environment, but most cases are not all. Moreover, many proposals have a weak link. Sometimes it is processing, sometimes transportation, and other times disposal. Frequently it is residual management and plant closure. In short, state-of-the-art does not necessarily mean safe. Negligible means different things to different people. Under the worst circumstances a small discharge of a common chemical can contaminate the drinking water of a small city (Montague 1984). Ask these five questions of the developer: (1) Is the substance flammable, volatile, poisonous, carcinogenic, mutagenic, or teratogenic (i.e., does it have reproductive effects)? (2) Can it become airborne? (3) Is it soluble in water or in fatty tissues? (4) Will it not degrade the environment within a few days? (5) Will it concentrate in any place in the food chain at levels in excess of the surrounding environment?

If the answer to the first question is yes, there could be a problem. If the answer to the first question and to any of the other questions is yes, then a detailed analysis is essential.

Technical information about emissions during normal operations and during accidents can be provided in a variety of forms. The most useful is in the form of occupational or environmental standards derived from human exposure and bioassay studies. Most of the standards and criteria are for workers. A rule of thumb is that ambient levels should be 1/10 and 1/100 of a "worker" standard. This is because workers are supposedly healthier than non-workers and are assumed to spend only the working hours exposed to the chemical in question. The general public includes many young, elderly, and others with impaired immunological systems, and is exposed to other chemicals through eating, smoking, breathing, and recreation. Often, even 1/100 of a worker standard is an order of magnitude greater than background concentrations. The 1/10 and 1/100 ratios, thus, are not wholly protective. As a rule of thumb, if a new facility will lead to a violation of a federal or a state standard or criteria, or increase the background concentration by ten times (or if one is very conservative, two times), one should be very concerned.

The performance record of the developer should be probed. Data from existing plants showing that the developer has a track record of operating facilities with similar processes and by-product technologies are important. An unblemished or slightly tarnished track record does not insure problem-free operation. However, it inspires more confidence than an inexperienced developer proposing a new technology or a technology that is to be upgraded from a pilot plant to a full operating facility.

We have seen in the recent performance of some nuclear power stations that administration may be a weak link. Thus, the planner must carefully scrutinize the personnel who will be responsible for the legal and economic decisions about the proposal. What permits have they secured? Do they know what permits must be secured? What is the company's record of building, operating, and closing facilities? These questions are important because the planner does not want an abandoned facility loaded with hazardous chemicals as a legacy.

The industry's efforts at site planning may be revealing. Does it have adequate information about the ownership of the property, the accessibility of the site, size of the plot, type of soils, the terrain and geological conditions? Have local community attitudes been surveyed? Have they even been considered? If there are community concerns, how does the industry plan to deal with them? Would it be willing to negotiate directly with citizens?

Although planners should not be expected to be familiar with toxicology literature, it should be consulted when a proposal appears. Sax's (1979) *Dangerous Properties of Industrial Materials*, a massive, thousand page reference work on chemical toxicity, lists chemicals alphabetically along with the following information: characteristics; hazards; counter measures; storage, handling, and transportation methods; and regulations. Sittig's (1981) *Handbook of Toxic and Hazardous Chemicals* provides information for about six hundred chemicals, including standards. Lewis and Tatken's (1978) *Registry of Toxic Effects of Chemical Substances* is a fourteen hundred page volume which lists chemicals, their synonyms, and toxic properties. Data on background levels of chemicals in the environment are available in Bowen's (1979) *Environmental Chemistry of the Elements*.

Information about chemicals changes rapidly. One way of keeping up is by reading the following journals: *Science, Environmental Science and Technology, Science News*, and *Chemical and Engineering News*. A more direct way is to use the Chemical Substance Information Network (CSIN, pronounced *see-sin*). Established by the federal government, it is a computerized data base of information about chemical hazards.[1] Also, because literature may be inconclusive, the planner may need the advice of experts who are not affiliated with government. Then advice is often helpful.[2]

## Uncertainty and Facility Siting: Administrative Processes

A new literature is being created on the problems of siting facilities commonly known as noxious, heavy industry, controversial, and LULUs (locally unwanted land uses) (Popper 1983). These facilities have strong opponents and proponents.

Facilities that use, transport, store, and especially those that dispose of, hazardous chemicals are certainly LULUs in the public eye. The association of the acronym LULU is now firmly entrenched in the literature with two other unfavorable acronyms that also describe facilities handling hazardous chemicals: NIMBY (not in my backyard) and LP/HC (low-probability, high-consequence) (Harthill 1984, Armour 1984, Morell and Magorian 1982, Popper 1983).

Proponents of siting chemical and chemical by-product facilities attack the NIMBY syndrome by charging that, along with excessive government regulation of production, it has hurt the American economy as a whole and badly hurt some states. After a four-year study of over one hundred siting cases, Duerksen (1983) reported that this myth is true only for a few industries that he characterizes as declining or slow-growth industries. Despite this finding, planners will continue to feel the pressure from opponents who fear environmental catastrophes and proponents who argue for the local economic benefits of jobs from the chemical and chemical by-product industries.

There are long-standing and new approaches for dealing with the siting of controversial facilities. Seley (1983) summarizes nine: (1) interagency coordination of all facility siting so that a few areas are not overburdened; (2) government exercise of eminent domain or other powers to take land; (3) letting the courts decide; (4) government rules and regulations to control siting; (5) muddling through with incremental decisions when necessary; (6) finding areas willing to accommodate the facilities or with insufficient power to oppose them; (7) frequently changing the siting process to keep everyone off balance; (8) conflict resolution, including mediation and arbitration; and (9) compensation and incentives.

Many of these nine are familiar to planners. Much of the recent and the most interesting literature focuses on approaches 4, 8, and 9. The U.S. EPA has regulations that will lead new facilities that store, transport, reuse, and dispose of hazardous chemicals to be far less dangerous than their existing counterparts. EPA and state regulation of existing and abandoned sites is less effective in the eyes of government, semi-official, and academic writers (Office of Technology Assessment 1983, National Materials Advisory Board 1983, Greenberg and Anderson 1984). The literature about government regulation of the chemical industry, especially its hazardous waste management components, will continue to be argumentative for the foreseeable future. This is because of the great uncertainty in our knowledge about chemicals described in the first two parts of this article and the legacy of mismanagement of chemical wastes.

Nonadversarial-based conflict resolution and compensation are two old approaches from labor and management relations that are being redefined for siting of LULUs. In the 1960s and 1970s this literature, as applied to planning, focused on very mathematical dispute resolution mechanisms. Raiffa (1982), well-known expert on mathematics and conflict resolution, argues that we do not need new mathematical methods. He stresses the use of simple methods, such as questions, to structure negotiations.

Straus (1984), former president of the American Arbitration Association, argues that mediation should be aimed at resolving technical parts of decision making, such as determining where a plume of contaminated chemicals will spread if there is an accident at a site. Straus reasons that if agreements can be reached on technical issues facing parties in conflict, then the parties to a dispute will have to face each other and try to resolve their underlying differences in values that are hidden behind technical arguments. Straus contends: "In my self-interest, it is more important first to understand how the system works, even if it means that I must collaborate with perceived enemies to reach this understanding, than to win a victory based on my present, and perhaps inadequate, understanding of the system." We should expect this type of innovative thinking to occur as experts in negotiations suggest ways of resolving local facility siting conflicts.

The new compensation and incentives literature is interesting and important to planners (Duerksen 1983, Seley 1983, Morell and Magorian 1982, O'Hare, Bacow, and Sanderson 1983, and McMahon et al. 1982). It should not come as a surprise that hazardous chemicals are a major or sole concern of much of this literature.

The work of Morell and Magorian (1982) and O'Hare, Bacow, and Sanderson (1983) is exemplary. Arguing against state preemption, Morell and Magorian call for a siting process involving state and local governments. The local government can accept or reject a proposal. If it accepts the proposal, it can ask for a variety of incentives; but it cannot if it rejects the proposal. The state may affirm the decision of the local government, may affirm it with modifications, or may override the local decision. The incentive for the local government to approve a proposal is the right to ask for incentives as part of its approval. The authors also include provision for participation by residents nearby to proposed sites.

O'Hare, Bacow, and Sanderson (1983) make local residents a major feature of their proposals for compensation. They call for an explicit negotiating process for compensating neighbors. This process was included in the State of Massachusetts Hazardous Waste Facility Siting Act (appended to their book). Another part of their work that planners will find unusually interesting are their lists of characteristics of disputes that will make disputes easy to resolve. This writer believes that compensation and incentives and negotiated settlements are likely to replace preemption, muddling through, and finding areas with powerless people as major administrative mechanisms for dealing with the siting of hazardous chemical LULUs.

Overall, a decade ago concern with dangerous chemicals focused on acute problems caused by flammable, explosive, and odorous substances. With increasing public recognition of potential chronic and acute health effects, including cancer, pressure has been created for detailed preassessment of proposals and continuous surveillance of hazardous facilities after they are located. This writer hopes that planners will want to assume an important role in these processes at the local and state levels.

*Acknowledgments* The author would like to thank Frank Popper, Department of Urban Studies, Rutgers University for his helpful comments.

NOTES

[1] One can request information of CSIN by chemical name by calling, Bolt, Beranek, and Newman, Inc., at 703-524-4870.

[2] Two sources of chemical experts are the Scientists Institute for Public Information in New York City (212-661-9110) and the Environmental Policy Center in Washington, D.C. (202-547-5330).

REFERENCES

Agran, L. 1977. *The cancer connection*. Boston: Houghton Mifflin.

Albert, R., J. Lewtas, S. Nesnow, T. Thorslund, and E. Anderson. 1983. Comparative potency method for cancer risk assessment: Application to diesel particulate emissions. *Risk Analysis* 3: 101–17.

Armour, A., ed. 1984. *The not-in-my-backyard syndrome*. Downsview, Ontario: York University.

Bowen, H. J. 1979. *Environmental chemistry of the elements*. New York: Academic Press.

Boyland, E. 1969. The correlation of experimental carcinogenesis and cancer in man. *Progress in Experimental Tumor Research* 11: 222–34.

Cairns, J. 1981. The origin of human cancers: Review article. *Nature* 289: 353–57.

Commins, B., R. Waller, and P. Lawther. 1957. Air pollution in diesel bus garages. *British Journal of Industrial Medicine* 14: 232–39.

Cuddihy, R., and R. McClellan. 1983. Evaluating lung

cancer risks from exposure to diesel engine exhaust. *Risk Analysis* 3: 119–24.

Dean, G. 1961. Lung cancer among white South Africans. *British Medical Journal* 2: 852, 1599.

Doll, R., and A. Hill. 1964. Mortality in relation to smoking: Ten years' observation of British doctors. *British Medical Journal* 1: 1460–67.

Doll, R., and R. Peto. 1978. Cigarette smoking and bronchial carcinoma: Dose and time relationships among regulator smokers and life-long nonsmokers. *Journal of Epidemiology and Community Health* 32: 303–13.

———. 1981. *The causes of cancer*. New York: Oxford University Press.

Duerksen, C. 1983. *Environmental regulation of industrial plant siting*. Washington, D.C.: The Conservation Foundation.

Eastcott, D. 1961. Other airborne factors in cancer. In *The air we breathe*, S. Farber and R. Wilson, eds. Springfield, Il.: Thomas.

Epstein, S. 1974. Environmental determinants of human cancer. *Cancer Research* 34: 2425–35.

———. 1979. *The politics of cancer*. New York: Doubleday.

Epstein, S., and J. Swartz. 1981. Fallacies of lifestyle cancer theories: Review article. *Nature* 289: 127–30.

Epstein, S., J. Swartz, and B. Ames. 1984. Letters: Cancer and diet. *Science* 224: 658–70, 757–60.

Fraumeni, J., Jr., ed. 1975. *Persons at high risk of cancer*. New York: Academic Press.

Greenberg, M., and R. Anderson. 1984. *Hazardous waste sites: The credibility gap*. New Brunswick, N.J.: Center for Urban Policy Research.

Hagstrom, R., H. Sprague, and E. Landue. 1967. The Nashville air pollution study: VII. Mortality from cancer in relation to air pollution. *Archives of Environmental Health* 15: 237–48.

Hammond, E., and D. Horn. 1958. Smoking and death rates. *Journal of the American Medical Association* 166: 1294–1308.

Harris, J. 1983. Diesel emissions and lung cancer. *Risk Analysis* 3: 83–100.

Harthill, M., ed. 1984. *Hazardous waste management: In whose backyard?* Boulder, Co.: Westview Press.

Hayes, A., ed. 1982. *Methods in toxicology*. New York: Raven Press.

Hiatt, H., J. Watson, and J. Winsten, eds. 1977. *Origins of human cancer*, three volumes. Cold Spring Harbor, N.Y.: The Laboratory.

Higginson, J., and C. Muir. 1979. Environmental carcinogens: Misconceptions and limitations to cancer control. *Journal of the National Cancer Institute* 63: 1291–98.

Higginson, J. 1980. Editorial: The environment. *The Science of the Total Environment* 15: 1.

Highland, J., M. Fine, and R. Boyle. 1979. *Malignant neglect*. New York: Alfred A. Knopf.

Hoffman, D., E. Theisz, and E. Wynder. 1965. Studies on the carcinogenicity of gasoline exhausts. *Journal of the Air Pollution Control Association* 15: 162–65.

Hueper, W. 1966. *Occupational and environmental cancers of the respiratory system*. New York: Springer-Verlag.

Huisingh, J., R. Bradow, R. Jungers et al. 1978. Application of bioassay to the characterization of diesel particulate emissions. In *Application of Short-term Bioassays in the Fractionation and Analysis of Complex Environmental Mixtures*, by M. Waters, et al. New York: Plenum Press.

Jacobs, L. 1980. Environmental analysis. *Science* 207: 1414.

Last, J., ed. 1983. *Dictionary of epidemiology*. New York: Oxford University Press.

Levin, M. 1960. Cancer incidence in urban and rural areas of New York State. *Journal of the National Cancer Institute* 24: 1243–57.

Lewis, R., and R. Tatken, eds. 1978. *Registry of toxic effects of chemical substances*, 1978 edition. Washington, D.C.: U.S. GPO.

Lilienfeld, A., and D. Lilienfeld. 1980. *Foundations of epidemiology*, second edition. New York: Oxford University Press.

Lowrance, W. 1976. *Of acceptable risk*. Palo Alto, Ca.: Wm. Kaufmann.

———, ed. 1981. *Assessment of health effects at chemical disposal sites*. Los Altos, Ca.: Wm. Kaufmann.

Maugh, T., II. 1979. Cancer and environment: Higginson speaks out. *Science* 205: 1363–66.

McMahon, R., C. Ernst, R. Miyares, and C. Haymore, 1982. *Using compensation and incentives when siting hazardous waste management facilities—Handbook*. Washington, D.C.: U.S. EPA, SW-942.

Montague, P. 1984. Evaluating chemical and radioactive hazards. Paper presented for the Council for the Advancement of Science Writing, Oak Park, Il.

Morell, D., and C. Magorian. 1982. *Siting hazardous waste facilities*. Cambridge, Ma.: Ballinger Press.

National Materials Advisory Board, National Research Council. 1983. *Management of hazardous industrial wastes: Research and development needs*. Washington, D.C.: National Academy Press.

National Research Council. 1984. *Toxicity testing: Strategies to determine needs and priorities*. Washington, D.C.: National Academy Press.

National Research Council. Committee on Institutional Means for Assessment of Risks to Public Health. 1983. *Risk assessment in the federal government: managing the process*. Washington, D.C.: National Academy Press.

*Nation's Health*. 1984. April, p. 8.

Office of Science and Technology Policy. 1979. *Identification, characterization, and control of potential human carcinogens: A framework for federal decision-making*. Washington, D.C.: Office of Science and Technology Policy.

Office of Technology Assessment. 1981. *Technologies for determining cancer risks from the environment*. Washington, D.C.: Superintendent of Documents.

Office of Technology Assessment. 1983. *Technologies and management strategies for hazardous waste control*. Washington, D.C.: Superintendent of Documents.

O'Hare, M., L. Bacow, and D. Sanderson, 1983. *Facility siting and public opposition*. New York: Van Nostrand Reinhold Company.

Peto, R. 1980. Distorting the epidemiology of cancer: The need for a more balanced overview. *Nature* 284: 297–300.

Popper, F. 1983. LP/HC and LULUs. The political uses of risk analysis in land-use planning. *Risk Analysis* 3: 255–63.

Raffle, P. 1957. The health of the worker. *British Journal of Industrial Medicine* 14: 73–80.

Raiffa, H. 1982. *The art and science of negotiation*. Cambridge, Ma.: Harvard University Press.

Sawicki, E., and D. Westphal. 1961. *Symposium on the analysis of carcinogenic air pollutants*, three volumes. Springfield, Va.: NTIS.

Sax, N. I. 1979. *Dangerous properties of industrial materials*, fifth edition. New York: Van Nostrand Reinhold.

Schneiderman, M. 1978. Eighty percent of cancer is related to the environment. *Laryngoscope* 88: 559–74.

———. 1981. Extrapolation from incomplete data to total or lifetime risks at low doses. *Environmental Health Perspectives* 42: 33–38.

Schwing, R., L. Evans, and R. Schreck. 1983. Uncertainties in diesel engine health effects. *Risk Analysis* 3: 129–32.

Seley, J. 1983. *The politics of public-facility planning*. Lexington, Ma.: Heath-Lexington Books, Inc.

Silverman, D., R. Hoover, S. Albert, and K. Graff. 1983. Occupation and cancer of the lower urinary track in Detroit. *Journal of the National Cancer Institute* 70: 237–45.

Sittig, M., ed. 1981. *Handbook of toxic and hazardous chemicals*. Park Ridge, N.J.: Noyes Data Corporation.

Slovic, P., B. Fischhoff, and S. Lichtenstein. 1979. Rating the risks. *Environment* 1: 14–39.

Springer, K. 1982. Diesel emissions, a worldwide concern. In *Toxicological Effects of Emissions from Diesel Engines*, J. Lewtas, ed. New York: Elsevier Publishing Co.

Stocks, P., and J. Campbell. 1955. Lung cancer death rates among non-smokers and pipe and cigarette smokers. *British Medical Journal* 2: 923–28.

Straus, D. 1984 forthcoming. Collaborating to win. In *Protecting Public Health and the Environment*, M. Greenberg, ed. New York: Guilford Press.

Waller, R. 1979. Trends in lung cancer in London in relation to exposure to diesel fumes. Paper presented at EPA Conference on the Health Effects of Diesel Engine Emissions, Cincinnati, Ohio, December.

World Health Organization. 1964. *Prevention of cancer*. Technical Report Series, 276. Geneva: World Health Organization.

Wynder, E., and D. Hoffmann. 1962. A study of air pollution carcinogenesis. III. Carcinogenic activity of gasoline engine exhaust condensate. *Cancer* 15: 103–8.

Wynder, E., and E. Hammond. 1962. A study of air pollution carcinogenesis. II. Analysis of epidemiological evidence. *Cancer* 15: 79–92.

Wynder, E., and G. Gori. 1977. Contribution of the environment to cancer incidence: an epidemiologic exercise. *Journal of the National Cancer Institute* 58: 825–32.

# EMERGENCY PLANNING
# FOR CHEMICAL AGENT RELEASES

*George O. Rogers ❖ John H. Sorensen*
*John F. Long, Jr. ❖ Denzel Fisher*

## INTRODUCTION

Emergency planning for the Chemical Stockpile Disposal Program (CSDP) is the central mitigative measure of the Final Programmatic Environmental Impact Statement (FPEIS) (U.S. DA, 1988a). The Draft Programmatic Environmental Impact Statement (DPEIS) characterized the stockpile and its hazardous potential, concluding that the concept of "no risk" was inappropriate (U.S. DA, 1986). The average maximum fatalities estimated under conservative most likely meteorological conditions exceeded 500 deaths and ranged from a low of 1 to a high of over 1400 within 20 km of the facilities.

Response plans at the Army installations predate the DPEIS, but emphasize on-site response to relatively small releases with limited or no off-site consequences. The DPEIS (Appendix L) also raised a series of issues about both on- and off-post emergency preparedness for accidental chemical agent releases, including:

Reprinted from *The Environmental Professional*, Volume 11 (4) (1989), pp. 396–408. Used with permission of the National Association of Environmental Professionals.

- How long will it take to recognize the nature of the accident?
- Who will be notified of an emergency?
- What means of communication will be used?
- What emergency operations center will be established?
- Who will decide when or whether to warn the public?
- How will the warnings be communicated?
- What protective actions will be considered?
- How will the decisions be made concerning what protective actions will be considered?
- Who is in charge of emergency response?

The DPEIS clearly indicated that accidental releases could reach off-site locations and underscored the lack of existing emergency preparedness to mitigate such events.

The complexity of the disposal program, the potential hazards associated with storage and disposal, and the adequacy of emergency response all contribute to the uncertainty about what is needed. The public was asking about the specifics of an emergency response program; yet even on such key issues as warning, the uncertainty was evident. Some people and emergency managers were concerned about the specifics of a warning system that they felt was clearly needed, but others felt no warning system was needed.

## EMERGENCY RESPONSE CAPABILITIES AND NEPA

Emergency response capability became a central part of the National Environmental Policy Act (NEPA) documentation and decision-making process for the CSDP when the principal impacts identified in the FPEIS were those associated with accidents, the consequences of accidents were estimated to be catastrophic, and the probability of accidents was considered credible (i.e. $\geq 10^{-8}$). Although the impacts associated with accidents are not a necessary condition for using emergency response capability to mitigate potential impacts, the credible potential for catastrophic off-site consequences that would overwhelm existing response resources is sufficient to lead to emergency planning as a mitigative measure. The FPEIS commits the Army to implementing a comprehensive emergency response program at each of the stockpile locations regardless of the disposal alternative selected. Carnes (this issue) discusses these alternatives in detail.

Disposal alternatives involving transportation of chemical agents pose additional emergency response requirements for potential accidents occurring in transit, which require unique considerations beyond the fixed-site program requirements for the stockpile locations. Emergency response concepts for each transportation alternative were considered unique and were independently presented in the *Emergency Response Concept Plan* (ERCP) (U.S. DA, 1987). The implications of disposal alternatives on program activities and emergency planning are summarized in Figure 1. The FPEIS concludes that if the selected disposal alternative involves off-site transport of chemical agent, it is impossible, due to logistical and financial

factors, to develop emergency response capabilities along the corridors that are commensurate with those at fixed-site locations (U.S. DA, 1988a, pp. 4–166).

Emergency response programs may not mitigate the effects of chemical agent releases in transit and because of the large geographical areas potentially affected by such accidents, the emergency response programs cannot be as effective as those for a fixed location. The Under Secretary of the Army James R. Ambrose concluded that even though the risk of catastrophic accidents is relatively low for all programmatic alternatives, on-site disposal is the safest alternative; it was therefore selected as the environmentally preferred alternative. The statistical usefulness of the FPEIS analysis is limited somewhat by the fact that the accident events considered have low probability of occurrence, but have potentially large adverse consequences. The risks of these events are never entirely absent from the installation population and surrounding area despite the alternative chosen, whereas they could be avoided completely along the transportation corridors if the on-site alternative was selected (Ambrose, 1988).

Although the Record of Decision to dispose of chemical agents was consistent with the environmentally preferred alternative, "the hazards and risks analyses presented in the FPEIS were a contributing but not determining factor" in the decision (Ambrose, 1988, p. 4). Ambrose found the logistical complexity of security from terrorism, safety, and emergency response more important aspects of the decision: emergency response to protect people along transportation corridors is far more difficult and less effective than for the people near fixed locations.

Because impacts of potential accidents can be reduced through emergency preparedness and the affected communities generally have inadequate preparedness capabilities, communities will need to upgrade their preparedness capabilities for response to accidental

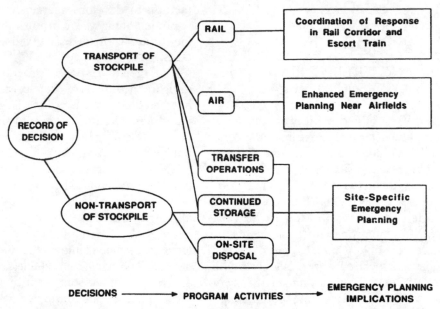

**FIGURE 1** Implications of Disposal Alternatives for Emergency Planning

releases and will be affected by the expenditures required for these upgrades. Moreover, the Army committed to fulfill its responsibility in the event of an accident/incident that exceeds the capabilities of local and state governments to respond. The Army also committed to immediate improvements in (1) local off-site planning and response capabilities and (2) coordination, communication, and decisionmaking by the installation and off-site officials. These improvements will include:

- developing an improved technical basis for planning;
- developing a streamlined decision and communication command structure for off-site notification in an emergency;
- developing improved operating procedures, monitoring, and accident assessment and decision support systems;
- developing improved exercise design criteria for evaluating emergency preparedness;
- conducting emergency exercises involving off-post participation and having an independent review and evaluation of exercises to determine needed improvements; and
- establishing an oversight review board to coordinate planning among the sites and to ensure

that guidelines specified in the ERCP are being implemented on schedule.

The Army is also pursuing other means to mitigate the costs of emergency preparedness enhancements to local governments. First, the Army is providing assistance to improve existing emergency response plans, coordination, communication, and decisionmaking. Second, the Army is developing site-specific emergency response concepts and implementation plans conforming with the programmatic ERCP, including exploring alternative ways of funding the emergency response planning efforts. The Army informed Congress on March 15, 1988, that the implementation of emergency response concepts for each site will cost an estimated $100 million (U.S. DA, 1988b).

## THE EMERGENCY RESPONSE CONCEPT PLAN

The ERCP (U.S. DA ED, 1987) was a programmatic analysis of emergency preparedness implications of the CSDP. It was not intended to provide a detailed, site-specific analysis of emergency preparedness measures or to represent definitive criteria for the development of an emergency response program. Rather, the ERCP identified emergency preparedness issues common to the CSDP, discussed alternative strategies by which they are addressed, and provided emergency response concepts that can be implemented to protect the public. The ERCP established the concepts for the subsequent development of emergency response programs. It addresses accidental chemical agent releases during all phases of the CSDP and for all disposal alternatives.

The emergency response concepts for the on-site and transportation alternatives are based on established principles of emergency management. Sound models for these programs exist in the Radiological Emergency Preparedness Program currently implemented for fixed nuclear facilities by the Federal Emergency Management Agency (FEMA) and the Nuclear Regulatory Commission, in the Chemical Emergency Preparedness Program of the U.S. Environmental Protection Agency, and in the Hazardous Materials Emergency Planning Guidance of the National Response Team. The emergency response concepts developed vary among stockpile disposal alternatives in that each alternative poses a unique set of problems and circumstances that must be considered in fashioning appropriate response programs. Planning for accidents at fixed, defined sites can be accomplished in greater detail and with much greater assurance of an effective response capability than planning for accidents during transportation. However, the fixed-site programs do not provide complete assurance that loss of life can be prevented.

The emergency response concepts presented in the ERCP include a general description of how such programs could be implemented. An important process of cooperative interaction involving local, state, and federal agencies and organizations must be accomplished for these emergency response concepts to be effectively implemented. For fixed-site

emergency response planning, the relationship between the U.S. Army command at each stockpile site and the surrounding community organizations is central to development of programs for the protection of the surrounding population. Further guidance regarding site-specific emergency planning will be developed and reported in the site-specific NEPA documentation. Even though the transportation alternatives were eventually eliminated, the associated emergency response planning involved a broad range of local, state, and federal agencies. The complexity and logistics of planning for accidents that could occur across a potentially expansive area affecting literally thousands of jurisdictions, agencies, and populations were specifically mentioned by Under Secretary Ambrose in the Record of Decision as a significant factor in selecting the on-site disposal alternative.

## EMERGENCY PLANNING ZONE CONCEPTS

Emergency planning zone (EPZ) concepts were identified to support the development of emergency response concepts for fixed-site and transportation alternatives. The EPZ is made up of three subzones depicted in Figure 2: the immediate response zone (IRZ), the protective action zone (PAZ), and the precautionary zone (PZ). EPZs are developed in consideration of the risk analysis, available response time, distance, and protective action options. The EPZs reflect the differing emergency response requirements associated with the potential rapid onset of an accidental release of agent and the amount of time that may be available for warning and response. They were developed in recognition of the importance of comprehensive emergency response planning and support systems for rapidly occurring events and the critical nature of such programs in areas nearest the release point (Table 1).

The EPZs are intended to guide the development of emergency response concepts

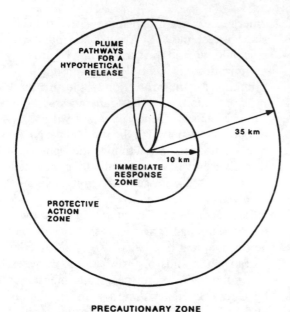

PLUME
PATHWAYS
FOR A
HYPOTHETICAL
RELEASE

35 km

10 km

IMMEDIATE
RESPONSE
ZONE

PROTECTIVE
ACTION
ZONE

PRECAUTIONARY ZONE

**FIGURE 2** *Summary of Emergency Planning Zone Concepts for Chemical Agent Release*

and should not be applied mechanically or inflexibly to specific sites or alternatives or to a specific accident scenario. The development of actual EPZs takes into account the unique political, social, geographical, and stockpile characteristics at each site. Conceptually, the criteria for establishing the EPZs are applied consistently across the program; however, specific configurations and associated distances may vary from site to site. The subzones discussed in the FPEIS were based on the types of accidents identified at each site and the amount of time available to pursue appropriate protective actions. The EPZs currently being developed are based on the hazards posed by a given stockpile and meteorological, topographical, and demographic conditions at each site.

## PRECAUTIONARY ZONE

The PZ is based on the usual maximum no-deaths distance of approximately 35 km for worst-case accidents involving nerve agent re-

leases (U.S. Army, 1988a, Sec. 4.2). Worst-case accidents having lethal effects beyond 35 km provide sufficient warning and response times to preclude the need for detailed local emergency response planning.

The worst-case meteorological condition is a wind speed of 1 m/s (Miller and Kornegay, this issue). The time required for such a worst-case release to travel 35 km from the stockpile location is nearly 10 h, a sufficient time to implement protective actions without prior comprehensive and detailed local planning efforts. Under conservative most likely (CML) meteorological conditions (wind speed of 3 m/s), the time required for a chemical agent release to reach 35 km is slightly more than 3 h; however, under these conditions, substantially greater mixing takes place. The maximum no-deaths distance for most accidents would be substantially less than 35 km under CML meteorological conditions. Hence, precautionary actions outside 35 km are planned via wide-area information dissemination mechanisms for all disposal alternatives.

Given substantial warning and response times for areas beyond 35 km, precautionary measures can be planned and implemented at a state or regional level, although coordination of local emergency managers may prove useful. The PZ is the outermost EPZ and extends theoretically to a distance where no adverse impacts to humans would be experienced in the case of a maximum potential release under any conditions. This distance may vary substantially, based on the circumstances of an accident occurrence and would be determined on an accident-specific basis.

Protective action considerations in the PZ are limited to precautionary measures and actions to mitigate the potential for food-chain contamination as a result of an agent release. Precautionary sheltering in the PZ would reduce the potential for exposure to non-lethal concentrations of chemical agent. A precautionary evacuation could also be implemented in the PZ. The means for implementing the agricultural protection and other precaution-

**TABLE 1** Summary of Emergency Planning Concepts (With Emphasis by Zone)

| Emergency Planning Concept Description | Emergency Planning Zone | | |
|---|---|---|---|
| | Immediate Response Zone | Protective Action Zone | Precautionary Zone |
| Character of Zone | Prompt response critical; limited response time (<60 min); heaviest concentrations; public protective measures possible. | Primary response evacuation; response time adequate (1–3 h); potentially lethal concentrations; responder protection required. | Sheltering and precautionary evacuation for non-lethal exposures; abundant (3–10 h); agricultural response possible. |
| Recommended Implementation Area | <10 km | 10–35 km | >35 km |
| Emergency Response Management | | | |
| Coordination | + + +* | + + | + |
| Command and Control | + + + | + + | + |
| Communications | + + + | + | |
| Accident Assessment | + + + | + + | + |
| Protective Action | + + | + | |
| Decisionmaking | + + + | + + | + |
| Protective Actions and Responses | | | |
| Protective Action Options | + + + | + + | + |
| Public Alert and Notification | + + + | + + | + |
| Access and Traffic Control | + + | + + | |
| Special Populations | + + + | + + | |
| Emergency Worker Protection | + + + | + + | |
| Resource and Information Management | | | |
| Emergency Medical Services | + | + + | + + + |
| Transportation | + + | + + + | |
| Community Resources | + + + | + + + | + + |
| Public Information | + + + | + + + | + + + |
| Evacuee Support | | + | + + + |

*A general indicator of the importance each emergency management function has in each zone. Triples (+ + +) indicate the function is critically important in the zone and is characterized by intense program emphasis. Doubles (+ +) indicate more general importance and will be characterized by significant program emphasis. Singles (+) indicate general importance, and these functions will receive adequate attention but with limited program focus.

ary activities can be based principally on broad-area dissemination of emergency public information at the time of an accidental release of agent.

## PROTECTIVE ACTION ZONE

The PAZ defines an area in which the primary emergency response is evacuation although other options are considered. Hence, the available emergency response times and the hazard distances associated with them are sufficiently large to allow most people to respond to an emergency effectively through evacuation. Operationally, a 35-km PAZ is an appropriate basis for emergency response planning. The comprehensive range of programs and support mechanisms described in the ERCP is applied within the PAZ.

Evacuation, the principal emergency response, must be considered carefully to ensure effective implementation and enhance protection of public health and safety. Evacuation is likely to be the most effective emergency response in the PAZ if time is sufficient to permit orderly egress. However, evacuation, like other protective actions, requires warning. Because time is limited in the PAZ, effective warning systems are required to both alert people to the potential for harm and notify them of the most appropriate actions to take. Available time for protective action varies with agent type, accident, and meteorological conditions at the time.

The capacity to implement timely emergency response is critical in the PAZ. Some areas in the innermost areas of the PAZ will require detailed planning for emergency response; other areas near the PZ will require less resource concentration. However, the degree of needed emergency response capabilities depends on specific agent type; geographical configurations such as transportation routes, terrain affecting dispersion, choke points such as bridges, and available egress; and social characteristics such as the location and type of schools, hospitals, and other institutions.

## IMMEDIATE RESPONSE ZONE

Those areas nearest to the stockpile locations require special consideration because the warning and response times available may be very limited within those areas. An IRZ is defined for the development of emergency response concepts that make immediate response possible in areas nearest the site. Because of the potentially limited warning and response time available in the event of an accidental release of chemical agent, the IRZ extends to a distance of approximately 10 km from the storage/disposal site. This area usually encompasses the no-deaths distances associated with non-external events and has less than 1-h response time for most accidents under CML meteorological conditions. This area is most likely to be affected by an accidental release of chemical agent. These impacts occur within the shortest period of time and result from the heaviest concentrations of agent.

The full range of available protective action options and response mechanisms is considered appropriate for the IRZ. The principal protective actions (sheltering and evacuation) will be considered carefully, along with supplemental protective action options that can significantly enhance the protection of public health and safety. Sheltering may be the most effective principal protective action for the IRZ because of the potentially short period of time before the effects of agent exposure may be expected. In-place protection is particularly important in areas within the IRZ nearest to the release point because the time to complete an evacuation may not be available for people within downwind areas of the IRZ. The suitability of sheltering depends on a number of other factors, including the type(s) and concentration(s) of agent(s); expedient or pre-emergency measures taken to enhance the various capacities of buildings to inhibit agent infiltration; the availability of individual protective devices for the general public; the accuracy with which the particular area, time, and duration of impact can be projected; and the ability to alert and communicate instructions to the public quickly and effectively.

The capability to implement the most appropriate protective action(s) very rapidly is critical within the IRZ. A thorough analysis of the IRZ specific to each storage/stockpile location should be conducted, and a precise methodology for determining the appropriate protective action(s) under various accident scenarios will be established to ensure that time to make decisions is minimized at the time of a chemical agent release. This analysis should identify certain areas within the IRZ that would implement sheltering for most accident scenarios, evacuation being available as a precautionary measure before a release occurs. Sub-zone areas may be defined to accommodate the selective implementation of different protective actions within portions of the IRZ. Given a reasonably effective capability to project the area of impact and to predict levels of impact at the time of a release, it may be appropriate to implement sheltering in areas close to the release point within the plume and evacuation in areas not immediately affected.

## PROTECTIVE ACTIONS AND RESPONSES

The emergency response program should establish the systems, resources, and response capabilities needed to implement appropriate actions for protection of the civilian population. Capabilities should be established for three principal methods of public protection: evacuation, sheltering, and individual protection. Evacuation may be implemented when sufficient time is available to move people without exposing them to lethal concentrations of chemical agent. Sheltering may be implemented when sufficient time for evacuation is not available; however, because sheltering alone does not provide adequate protection from lethal concentrations of chemical agent, additional measures that provide a high degree of protection will be considered for populations in close proximity to the stockpile sites. Measures should be taken to improve the sheltering protection of certain buildings near each site that house sensitive populations or that may be difficult to effectively evacuate, such as schools, hospitals, and nursing homes. Protective equipment is another means of protection available for populations in close proximity to the storage and disposal locations. Protective equipment can include civilian face masks and mouthpiece respirators, hooded jackets for young children, and protective infant carriers. Individual protective measures have the added benefit of being useful in com-

bination with other protective actions, such as evacuation or sheltering.

A number of support systems and capabilities should be put in place to ensure that the means for implementing various protective actions are available (Sorensen, 1988). Most significant is the installation of a public alert and notification system to alert civilian populations to an accident and notify them of the actions they should take to protect themselves. Although none of the specific alert and notification systems has been defined, systems should be established using a combination of alerting devices to provide the most effective capability for each site. Outdoor warning sirens and indoor warning devices, including tone-alert radios and automated telephone alerting systems, should be installed and supported by the Emergency Broadcast System, cable television override systems, and other public information mechanisms to provide good overall coverage in both indoor and outdoor alerting and notification. At greater distances, one or more of the alerting systems or mechanisms can be used to provide sufficient warning time for the public to take appropriate protective action(s).

Evacuation of the civilian population requires comprehensive planning and a number of support mechanisms. Detailed traffic and access control plans will be developed to facilitate the implementation of evacuation. Emergency plans should be developed for all special facilities such as schools, hospitals, and nursing homes. These plans should include the transportation resources required to evacuate the facilities identified, and agreements should be obtained for their use in an emergency. So that special assistance can be provided, a survey can facilitate the identification of individuals within the general population that have special needs, such as the handicapped, hearing and sight impaired, and medically disabled people, and those without a means of personal transportation. Protective equipment may also be offered to all civilian emergency workers who have a role in emergency response to ensure that they are protected from exposure to chemical agent while performing their duties. This equipment includes full protective clothing and respiratory protection, as well as auto-injectors for self-administration of agent antidotes if necessary (Chester, 1988).

## ASSESSING SOCIOECONOMIC IMPACT

A survey of existing emergency preparedness was conducted in February and March 1987. The resulting assessments are summarized in Table 2. On-site emergency preparedness was generally found to be adequate, and preparations for the most likely events were usually organized reasonably well. However, the coordination with off-site authorities and the preparations for supplying off-site authorities with the information they would need to make decisions and to respond effectively were generally found to be lacking. Although capital investments in equipment and facilities may be needed less at the installations than in the communities, facility and equipment improvements are urgently needed; and while there is considerable variation among sites, capital upgrades are badly needed in three areas: emergency operations centers (EOC), communications equipment, and warning systems.

The organizational element of emergency preparedness was found to be seriously deficient. The Army organization on-site typically has a well-organized chain of command and authority, but it generally lacked community awareness. On-site officials were not fully aware of the kinds of information and assistance the surrounding communities would need to make decisions, to warn the public, and to respond effectively to emergencies. Some installation personnel seemed to nurture the sentiment that accidents affecting off-site populations were improbable. To some extent this sentiment was shared by off-site officials; at one site, the local official indicated in a FEMA hazard assessment survey that no fixed-site chemical hazard existed in the area. From the community's perspective, however, there were additional problems: Recognizing the hazard, what kind of emergency response capability could be established with the limited existing resources? The perceived futility of detailed planning in the face of extremely limited warning and response capacity discourages the development of a sound planning and preparedness program with existing resources.

The problem of estimating the costs of upgrading the emergency response capabilities at each site is multidimensional. For some elements of effective emergency response, the cost for individual elements is reasonably well known. For example, a mouthpiece respirator costs about $13, if it is determined that 100,000 people require this kind of protection, the total cost would be $1.3 million. The problem with estimates of this kind is the criteria used to determine who needs this kind of protection. Even though these determinations are made, some people will prefer other kinds of protection. To ensure that some protective action is taken, even though a more desirable alternative exists, officials will yield to these preferences.

Another genre of estimation problems involves less-definitive cost elements, such as the cost associated with an EOC. The costs of an EOC can be arbitrarily estimated, but these estimates may not adequately take into account the degree of renovation of existing facilities and local variations in the cost of land, construction materials, and labor. When these estimates are then multiplied by the number of jurisdictions affected to obtain a program estimate, the original estimation problems are further compounded because this kind of estimation fails to consider the potential for shared facilities or political constraints.

**TABLE 2** Summary of Emergency Preparedness at CSDP Sites* (as of February/March 1987)

| | ANAD | APG | LBAD | NAAP | PBA | PUDA | TEAD | UMDA |
|---|---|---|---|---|---|---|---|---|
| **Plans and Procedures** | I**: Adequate C***1: Defines operating concepts C2: None | I: Well organized C1: General; update required C2: General | I: Designed to protect I mainly C: Draft not coordinated | I: No continuity, quick access, and none C: needs C: None identified | Detailed plans; no decision criteria; limited coordination | I: Adequate C: Not clearly defined or organized | I: No continuity, and quick access C: Not TEAD/CSDP specific | I: Adequate; but contradictions C: Not complete/coordinated with I |
| **Facilities** | I: Equipped EOC; weak link with C C1: Adequate EOC C2: Very modern EOC | I: EOC well equipped and dedicated link with C C1: EOC C2: EOC exceptional | I: EOC excellent C mobile EOC C: EOC weak dual-use and small | I: EOC equipped; No dedicated link with C: no EOC identified | I: EOC excellent C: Command post concept | I & C EOCs adequate; no dedicated I-C communications | I: EOC well equipped; C: linked via I-command to C C:Duel-use EOC | I & C EOCs well equipped; dedicated I-C link; no alternate EOC |
| **Organization** | I: Clear C1: Not well defined C2: Personnel identified | I: Clear C1 & C2: Legal authority defined | I: Clear with alternates C: Complete but no alternates | I: Identified structure C: EMS, police, and fire not coordinated | I & C: Well defined and staffed | I: Not clearly defined C: Exceptionally well organized | I: Not clearly defined C: Not defined; I & C not coordinated | I: Clearly defined C: General, not specific authorities |
| **Accident Assessment** | Agent presence; monitoring teams verify real-time air dispersion model | No procedures specified; plume models and met available | No real-time assessment, plume models, or agent monitoring | Nonreal-time lethal/nonlethal dose monitors | Lethal dose detection MCE criteria used | Real-time monitoring not linked to plume models | Lethal-dose real-time monitoring plume models | Real-time hazard monitoring; no plume prediction |
| **Decisionmaking** | I: Authority identified C1 & C2: Based on I recommendations | I: Authority and procedures identified; C1 & C2: Unclear | I: Authority but no procedures—"intuitive" C: Reluctant | I: Authority but no procedures C: Unclear | I: No criteria C: Will accept recommendations | I: No clear criteria C: Limited to accept/reject | I: Authority identified; no criteria C: No procedures | I: Authority identified C: Limited to accept/reject |
| **Emergency Classification** | I: 2 Level; code-names similar C1: May share w/I C2: None | I: 4 level linked with I response C1 & C2: None | I: General 3 level; limited guidance C: None | I: Two 3-level; no guidance to C C: None | Accident, incident minor leak, occurrence | I: 4 level/one off-site C: 4 level not shared with I | I: 3 level; not guide C: Response C: None | I: 4 level/one off-site C: Potential and actual releases |
| **Accident Notification** | Conduct assigned contacts listed with C1, but not C2 | Authority and conduct assigned dedicated link; C1 notifies C2 | No procedures or responsible dedicated link | Authority, but not conduct assigned; no dedicated link to C | Addressed, but not well defined | No clear procedures; contradictory authority; 911 | Authority and conduct assigned but separated; no alternates | Procedures and responsibilities defined; radio, 911 backups |

**TABLE 2** (Continued)

| | ANAD | APG | LBAD | NAAP | PBA | PUDA | TEAD | UMDA |
|---|---|---|---|---|---|---|---|---|
| **Public Alert/Notification** | | | | | | | | |
| | Local EBS yields partial coverage and tone-alerts to institutions | Limited sirens, route alerting, and EBS; some protocols exist | EBS, weather alert radios, route alerting, limited sirens | EBS, media contacts listed; protocols calm, not alert public | Route alerting | EBS, route alerting, tone alerts to schools and hospitals | EBS, route alerting protocols exist | Limited sirens, EBS and route alerting; no procedures |
| **Protective Actions Implementation** | | | | | | | | |
| | Traffic access and control, evacuation and mass-care resources identified by C1 | C1: Evacuation routes assigned C1 & C2: Have no procedures | No traffic access and control; evacuation resources not well identified | No procedures or planning identified | Ad hoc; no designated mass care or access control | Exceptionally well developed | Traffic access and control, evacuation and mass-care resources identified | Responsibility assigned, but no procedures or specifics |
| **Health/Medical** | | | | | | | | |
| | I&C coordinated; I will supply and train; no decontamination | I&C coordinated to handle non-agent injuries; no decontamination | No training or supplies; no coordination, no training | I & C limited coordination and supplies provided | I & C coordinated training and supplies adequate | I & C coordinated some supplies and decontamination provided | I: Being built; personnel available C: Not addressed | I & C coordinated training and supplies C: No decon |
| **Special Populations** | | | | | | | | |
| | None identified; no hospitals/ nursing homes known; schools identified | None identified; day care/ schools procedures exist; but no transport | Institutions and disabled identified; no procedures | Not systematically identified; some known; institutions unknown | Some identified; no special plans | Some identified; exceptions are: day care and nursing homes | None identified; no institutions in close proximity | None identified |
| **Public Information** | | | | | | | | |
| | I: Preplanned new releases; media contacts C1 & C2 not joint | I: Responsibility C1 & C2: Rumor control; not joint with I | Joint release; center designated joint operations unclear | No procedures; none except hearings | Joint activity agreements; no location or procedures | I: Preplanned news releases; media linked C: Not joint with I | I: Procedures not identified; no location C: None | I: Preplanned news releases; media contacts C: Not joint with I |
| **Training, Drills, and Exercises** | | | | | | | | |
| | I: Regular C1: Periodic C2: Annual no joint | I: 4 years; C1: REP C2: EOC every 5 years C1 & C2 not CSDP; no joint | I: 6 years C: Limited joint annual | I: Regular C: Unknown no joint | I & C annual joint; I response focus | I & C adequate; no joint | I: Regular C: None; no joint | I: Frequently; table-top exercise in 1980 with C |

*ANAD—Anniston Army Depot; APG—Aberdeen Proving Ground; LBAD—Lexington-Blue Grass Army Depot; NAAP—Newport Army Ammunition Plant; PBA—Pine Bluff Arsenal; PUDA—Pueblo Depot Activity; TEAD—Tooele Army Depot; MDA—Umatilla Depot activity.

**I refers to installation or Army emergency preparedness.

***C refers to preparedness characteristics of nearby communities; C with numbers are used to designate one community from another.

Another set of problems associated with cost estimates involves the interpretation of the estimates. Table 3 presents the Army's cost estimates for each installation and compares them with the number of people likely to benefit most from emergency planning at each site (i.e., within 35 km). On the average, emergency planning upgrades are estimated to cost about $50 per person; however, emergency planning upgrades at one site are estimated to require less than $20 per person, whereas costs at others are about $500 per person. It remains unclear at this time whether the levels of protection obtained over the range of costs are equivalent, although this is certainly the goal. Although the costs at the remaining sites fluctuate less dramatically, the fact remains that upgrading the emergency planning at some sites will require more investment than others. When the costs associated with some aspects of establishing/upgrading emergency response capabilities are estimated, cost per person will vary to some degree, but substantial variations highlight the potential for perceived inequality. In addition, once site estimates are released, local officials may expect that money to be allocated directly to them. One local official indicated to reporters that his county would be getting the total amount allocated for that site, even though other jurisdictions in the area have emergency response needs and their program costs have to be deducted from the total amount.

## EMERGENCY PLANNING AS MITIGATION

Emergency management mitigates the risks of accidents to the nearly 1.9 million people living within 35 km of the facilities, but the degree of mitigation is difficult to assess. A quantitative estimation of the benefits associated with the emergency response program is not possible because of the numerous individual circumstances that affect human behavior in emergencies. Although there is some certainty that emergency preparedness programs will improve survival chances and thereby mitigate the effects of potential accidents, the improvements are not quantifia-

**TABLE 3**  *Estimated Costs of Emergency Response Upgrades by Location*

| Locations | Estimated Costs* ($10^6$ $) | Population within 35 km** ($10^3$) | Cost per Person ($) |
|---|---|---|---|
| ANAD*** | 9.98 | 192.4 | 51.87 |
| APG | 22.15 | 1175.1 | 18.85 |
| LBAD | 18.29 | 128.2 | 142.67 |
| NAAP | 8.98 | 101.8 | 88.21 |
| PBA | 8.97 | 113.0 | 79.38 |
| PUDA | 6.72 | 116.2 | 57.83 |
| TEAD | 10.62 | 20.4 | 520.59 |
| UMDA | 14.29 | 28.9 | 494.46 |
| Total | 100.00 | 1876.0 | 53.30 |

*Constant FY 1988 dollars as adopted from U.S. DA, 1988b.
**Adopted from the U.S. DA, 1988a.

| | |
|---|---|
| ***ANAD—Anniston Army Depot | PBA—Pine Bluff Arsenal |
| APG—Aberdeen Proving Ground | PUDA—Pueblo Depot Activity |
| LBAD—Lexington-Blue Grass Army Depot | TEAD—Tooele Army Depot |
| NAAP—Newport Army Ammunition Plant | UMDA—Umatilla Depot Activity |

ble in terms of potential lives saved or protected.

Hence, a qualitative approach to assessing the benefits of enhancing emergency preparedness is summarized in Table 4. Accidents were characterized in terms of agent, mode of release, and wind speed. Agent was classified as being VX, GB, or HD; mode of release is characterized as instantaneous, semicontinuous, or a spill; and wind speeds were hypothesized as slow (1 m/s), moderate (3 m/s) or fast (6 m/s). For all kinds of chemical agent under all modes of release, the benefits of emergency response programs were found to be the greatest under relatively slow wind speeds. Similarly, benefits were assessed as greater under moderate wind speeds than under fast wind speeds. This general assessment reflects the nature of emergency response: more time allows a greater response, and one of the key elements of enhanced response with additional time is available warning time. Rogers and Sorensen (1988) examined the impact of limited warning time on receipt of warning and found that even with combinations of advanced warning systems under rapidly moving events, people near the source would likely be exposed to the hazard before receiving the warning.

## PROGRAM DIRECTIONS

The Under Secretary of the Army James R. Ambrose was briefed on the status of emergency response planning in August 1987. At that briefing, he committed the Army to immediate upgrades not requiring capital expenditures to enhance emergency planning at all eight sites. As a result, the Army is providing assistance to communities for enhanced emergency preparedness in the vicinity of the eight locations. In addition, site-specific concept studies are underway for each site. These concept studies are expected to be released in the fall of 1989 and will examine the generic

recommendations contained in the ERCP in light of site-specific data and information.

Technical support studies on such critical topics as hazard assessment, public information, emergency worker training, and protective action are being undertaken. The Army also has signed a comprehensive Memorandum of Understanding with FEMA to "take the lead in working with state and local governments in developing off-site emergency plans" and provide "liaison, coordination and oversight between the Army and federal, state and local governments in developing off-site emergency preparedness plans" (Becton and Shannon, 1988, p. 4). FEMA also has agreed to take the lead in developing planning standards and evaluation criteria, providing training, developing and conducting exercises, and providing public information. The Army has agreed to provide technical assistance and to take the lead in developing exercise design and criteria, conducting site-specific hazard analyses for emergency preparedness plans, and reviewing FEMA assessments of preparedness. FEMA and the Army have agreed to pursue jointly the determination of emergency preparedness funding rquirements by fiscal year to facilitate the Army's request for necessary funding. In addition, a FEMA/Army Joint Steering Committee has been extablished to review the status of the joint emergency preparedness programs; to discuss, consult on, and resolve major policy issues; and to provide direction in meeting the overall goals of the program.

The purpose of the memorandum was to draw on FEMA's expertise and experience with state and local governments in developing and implementing emergency preparedness programs and to integrate that capability with the Army's considerable technical expertise with these extremely dangerous chemical agents. The Memorandum of Understanding exemplifies the need for cooperation and understanding between all parties to ensure a successful program. The need for an integrated approach to emer-

**TABLE 4** Qualitative Benefits of Fixed-Site Emergency Response Programs for Nerve Agent Accidents

| Accident Scenario | Wind Speed: Slow (1 m/s) | Wind Speed: Moderate (3 m/s) | Wind Speed: Fast (6 m/s) |
|---|---|---|---|
| Moderate instantaneous release (100 kg VX) | Fatalities possible to 33 km; potential for multiple fatalities is high within IRZ; lower in PAZ. High reduction in fatalities at all distances at greater distances | Fatalities possible to 10 km; potential for multiple fatalities is high within 2 km; lower to 10 km. Low reduction in fatalities to 2 km; higher at greater distances | Fatalities possible to 15 km; potential for multiple fatalities is high within 4 km; lower to 15 km. Low reduction in fatalities to 5 km; higher at greater distances |
| Small instantaneous release (10 kg VX) | Fatalities possible to 7 km; potential for multiple fatalities is high within 2 km; lower to 7 km. High reduction in fatalities at all distances | Fatalities possible to 4 km; potential for multiple fatalities is moderate within 1 km; low to 4 km. Low reduction in fatalities to 2 km | Fatalities possible to 6 km; potential for multiple fatalities is moderate within 2 km; lower to 6 km. Low reduction in fatalities at 5 km |
| Large semicontinuous release (1000 kg VX) | Fatalities possible to 100 km; potential for multiple fatalities is very high within IRZ; lower in PAZ and PZ. High reduction in fatalities at all distances | Fatalities possible to 25 km; potential for multiple fatalities is high within IRZ; lower in PAZ. Low reduction in fatalities to 2 km; high at greater distances | Fatalities possible to 15 km; potential for multiple fatalities is high within IRZ, lower in PAZ. Low reduction in fatalities to 5 km; high at greater distances |
| Moderate semicontinuous release (100 kg VX) | Fatalities possible to 45 km; potential for multiple fatalities is very high within IRZ; lower in PAZ, and very low in PZ. High reduction in fatalities at all distances | Fatalities possible to 8 km; potential for multiple fatalities is moderate within 2 km; lower to 8 km. Moderate reduction in fatalities at 2 km; greater to 8 km | Fatalities possible to 7 km; potential for multiple fatalities is high within 2 km; lower to 7 km. Low reduction in fatalities at 2 km; moderate reduction to 5 km |
| Large spill (900 kg GB) | Fatalities possible to 12 km; potential for multiple fatalities is moderate within 3 km; lower to 12 km. High reduction in fatalities at all distances | Fatalities possible to 8 km; potential for multiple fatalities is moderate within 2 km; lower to 10 km. High reduction at distances | Fatalities possible to 8 km; potential for multiple fatalities is moderate within 2 km; lower to 8 km. Moderate reduction in fatalities to 2 km; greater to 8 km |

**TABLE 4** (Continued)

| Accident Scenario | Wind Speed: Slow (1 m/s) | Wind Speed: Moderate (3 m/s) | Wind Speed: Fast (6 m/s) |
|---|---|---|---|
| Moderate instantaneous-release (900 kg HD) | Fatalities possible to 7 km; potential for multiple fatalities is high within 2 km; lower to 7 km. High reduction in fatalities at all distances | Fatalities possible to 2 km; potential for multiple fatalities is moderate within 5 km; low to 2 km. Low reduction in fatalities to 2 km | Fatalities possible to 1 km; potential for multiple fatalities is low. Potential for precautionary measures is low |
| Small instantaneous release | Fatalities possible to 1 km; potential for multiple fatalities is low. High potential for taking precautionary measures | Fatalities possible to 0.2 km; potential for multiple fatalities is very low. Low potential for precautionary measures | Fatalities possible to 0.15 km; potential for multiple fatalities is very low. Precautionary measures unlikely |
| Large semicontinuous release (4000 kg HD) | Fatalities possible to 100 km; potential for multiple fatalities is very high within IRZ; lower in PAZ and PZ. High reduction in fatalities at all distances | Fatalities possible to 10 km; potential for multiple fatalities is high within 2 km; lower to 10 km. Low reduction in fatalities to 2 km; high at greater distances | Fatalities possible to 7 km; potential for multiple fatalities is high within 2 km; lower to 10 km. Low reduction in fatalities to 5 km; high at greater distances |
| Moderate semicontinuous release (100 kg HD) | Fatalities possible to 8 km; potential for multiple fatalities is very high within 2 km; lower to 8 km. High reduction in fatalities at all distances | Fatalities possible to 2 km; potential for multiple fatalities is moderate within 5 km; lower to 2 km. Moderate reduction in fatalities at 2 km | Fatalities possible to 1.5 km; potential for multiple fatalities is high within 5 km; lower to 1.5 km. Low reduction in fatalities at 1.5 km |
| Large Spill (900 kg HD) | Potential for fatalities is very low. High likelihood of successful precautionary measures | Potential for fatalities is very low. High likelihood of successful precautionary measures | Potential for fatalities is very low. High likelihood of successful precautionary measures |

324

gency management has never been more apparent than in the CSDP.

## REFERENCES

Ambrose, James R. 1988. Record of Decision: Chemical Stockpile Disposal Program, Department of the Army, Under Secretary of the Army. February 23.

Becton, Julius W., and John W. Shannon. 1988. Memorandum of Understanding Between the Federal Emergency Management Agency and the Department of the Army, Subject: The Chemical Stockpile Disposal Program. August 3.

Chester, C. V. 1988. Technical Options for Protecting Civilians from Toxic Vapors and Gases. ORNL/TM-10423. Oak Ridge National Laboratory, Oak Ridge, TN.

Rogers, G. O., and J. H. Sorensen. 1988. Diffusion of Emergency Warnings. *The Environmental Professional* 10(4):281–294.

Sorensen, J. H. 1988. Evaluation of Warning and Protective Action Implementation Times for Chemical Weapons Accidents. ORNL/TM-10437. Oak Ridge National Laboratory, Oak Ridge, TN.

(U.S. DA) U.S. Department of the Army. PEO-PM Cml Demil. 1988a. Chemical Stockpile Disposal Program Final Programmatic Environmental Impact Statement. Aberdeen Proving Ground, MD. January.

(U.S. DA) U.S. Department of the Army. 1988b. PEO-PM Cml Demil. Chemical Stockpile Disposal Program Implementation Plan. Aberdeen Proving Ground, MD. March 15.

(U.S. DA) U.S. Department of the Army. 1987. PEO-PM Cml Demil. Emergency Response Concept Plan. Aberdeen Proving Ground, MD. July.

(U.S. DA) U.S. Department of the Army. 1986. PEO-PM Cml Demil. Chemical Stockpile Disposal Program Draft Programmatic Environmental Impact Statement. Aberdeen Proving Ground, MD, July 1.

(U.S. DA ED) U.S. Department of the Army, Engineer Division. Huntsville, AL. 1987. Draft Emergency Response Concept Plan. Prepared for the Program Executive Office–Program Manager for Chemical Demilitarization, Emergency Response Concept Plan. Aberdeen Proving Ground, MD. June, July.

# VULNERABILITY TO GLOBAL ENVIRONMENTAL CHANGE

*Diana M. Liverman*

In the southern Mexican state of Chiapas, thousands of hectares of tropical forest have been destroyed in the last hundred years. Timber companies, followed by ranchers and coffee producers, converted biologically diverse, soil-protecting forests into barren and desiccated landscapes. Additional forest has been destroyed in the search for petroleum resources in the region, and by an influx of impoverished refugees from Central America and migrants from other parts of Mexico seeking farmland (Gonzalez Pacheco 1983).

In Mexico City, contamination of air and water has already reached fatal or disabling levels for many residents of the metropolis, with increases in respiratory and cardiac diseases, gastro-intestinal problems, cancer, poisonings, and other toxic exposures (Ortiz 1987; Puente and Legoretta 1988). Pollution has become a major political issue and has prejudiced the international business and tourist reputation of the city (*Excelsior* 1983).

In northern Mexico, 1989 has been a year of severe drought and heat stress. As in the US in 1988, people have connected this drought to the predictions of global warming associated with the greenhouse effect and it has been linked to forecasts of the lowest agricultural production in nine years and a corresponding increase in malnutrition and expensive food imports.

These are localized examples of the types of global environmental changes that may face the world in the coming decades. Environmental changes in Mexico are typical of many countries, and have similar social causes and consequences. Who will be most vulnerable to these changes? In forest regions like those of Chiapas, the most direct and severe impacts of deforestation are probably felt by the indigenous peoples of the region, such as the Lacandon and Maya (Lobato 1980). Forest clearance decimates the animal, bird, fish, and plant populations that provide the food and fiber base of traditional hunting and gathering livelihoods (Szekely and Restrepo 1988). Environmental changes also imperil the health and nutrition of the poorer refugees and migrants who are trying to survive on small, infertile plots of deforested land. Larger farmers and ranchers, as well as timber companies,

Reprinted from *Understanding Global Environmental Change: The Contributions of Risk Analysis and Management*, Edited by R.E. Kasperson, K. Dow, D. Golding, and J.X. Kasperson. Worcester, MA: Earth Transformed Program, Clark University, 1990, pp. 27–39, 41–44. Used with permission of the author and Clark University.

are seeing diminishing returns to their unsustainable use of the Chiapan forests. Local and national government are clearly concerned about political unrest stemming from the social and ecological impoverishment of the region (Aguayo 1987). The loss of biological diversity and vegetation cover in tropical forest regions, such as Chiapas, may also jeopardize the food and health security of the world as a whole, and alter global nutrient and hydrological cycles (Myers 1984).

Although both rich and poor in Mexico City are exposed to ambient air and water pollution, the former can afford air conditioning, bottled water, rural retreats, good nutrition, and health care. Certain occupations—such as factory worker or driver—entail greater exposure than others, and pregnant women, children, and the infirm are often more vulnerable to pollution (Puente and Legoretta 1988). In addition to these individual health risks, air and water pollution incur economic costs to individuals, corporations, and governments in terms of prevention, clean-up, labor absenteeism, and material damage (Vizcaino 1975).

In the 1989 drought, newspapers reported that many communities in northern Mexico were without water and children were dying from dehydration during the peak of the heatwave. In important urban-industrial centers, such as Chihuahua and Monterrey, water supplies were rationed. The food supplies and livelihood of many small and cooperative agricultural producers of rainfed crops were in jeopardy, and conflict over irrigation rights was intense.

## WHY A CONCERN WITH VULNERABILITY?

These cases illustrate the variety of individuals and institutions affected by environmental degradation. They could be used to argue that almost everything and everyone is vulnerable to global change. Many of the recent statements on global change do employ *spaceship earth* analogies to emphasize that we are *all in this together* and that nobody will be able to escape the impacts of the greenhouse effect, ozone depletion, or deforestation (Brown 1989; Myers 1988). Although such sentiments carry an appropriate urgency, and promote international cooperation, they tend to evade the issues of differential susceptibility to global changes, the distribution of blame, and the possibility that some may benefit from environmental transformation. At the other extreme, some analyses suggest that the burden of global change will fall almost entirely on the poor people and countries of the world, because political economic structures are such that hardships are always passed on to the oppressed (Wijkman and Timberlake 1984). Although there are good political reasons for promoting either of these perspectives, neither is particularly helpful or profound in designing responses to changes in specific places. Since it seems unlikely that we will marshall the resources to prevent all global changes or to compensate everyone for their impacts, we need more precise estimates of who is vulnerable in order to decide where, when, and how most effectively to focus our responses.

Many people may want to understand and reduce vulnerability to global change for purely humanitarian reasons. But it is important to recognize that some people and institutions will be opposed to detailed assessments of vulnerability which highlight inequality and conflicts within society, because the solutions to these problems will threaten their own interests. For example, some groups have opposed studies of famine vulnerability in Sudano-Sahelian countries because such assessments reveal exploitation, discrimination, and government failure (Franke and Chasin 1980; Somerville 1986). Similarly, studies of the unequal impacts of land reform and agricultural development are suppressed in parts of Central and South America (Burbach and Flynn 1982; Montgomery 1984).

However, several recent discussions of

global environmental change have suggested that concern about the social causes and consequences of environmental degradation beyond one's own country or locality is in the national and individual self-interest (Caldwell 1985; Maguire and Welsh-Brown 1986; Mathews 1989; Myers 1988). Reinforcing the ability of other countries to cope with environmental problems is seen not only as a way to develop strategic allies, but also as a means to insure the stability of import and export trade. Linkages among environmental degradation, economic decline, and the debt crisis have been used to demonstrate the vulnerability of global financial institutions to environmental change (*The Ecologist*, 1987). Self-interest is at stake in protecting global commons such as the atmosphere and oceans and in preventing the destruction of biodiversity and other ecologically and economically important resources. International security concerns also arise from the politically destabilizing influence of environmental impoverishment and activism, and from the possibility of thousands of *environmental refugees* moving across borders (Jacobson 1988).

Such arguments are clearly influencing policies towards deforestation and global warming and attitudes towards regions such as Africa and Central America. But there are risks in powerful nations like the U.S. defining their national security interests in terms of their vulnerability to environmental changes in other countries. Perceived threats to sovereignty, heavy-handed financial incentives and discentives, or requests to halt development (made by those already developed) may eventually work against cooperative efforts to respond to global change.

# DEFINITIONS OF VULNERABILITY

Peter Timmerman (1981) has explored the meaning and use of the concept of vulnerability. He locates the concept at the heart of

research into climate change in citing an objective of the World Meteorological Organization's Climate Program as:

> determining the characteristics of human societies at different levels of development which make them either specially vulnerable or specially resilient to climatic variability and change (p. 3)

Timmerman reviews the models of social collapse and uses of the term vulnerability provided in a variety of intellectual traditions, grounding his analysis in the works of philosophers such as Aristotle and Comte. He posits that "vulnerability is a term of such broad use as to be almost useless for careful description at the present, except as a rhetorical indicator of areas of greatest concern" (p. 17). Nevertheless, he does provide a definition of vulnerability as "the degree to which a system may react adversely to the occurrence of a hazardous event" and links it to resilience, "the measure of a system's capacity to absorb and recover from the occurrence of a hazardous event" (p. 21). He concludes, with a touch of irony, that our real vulnerability may lie in the inadequacy of our models of social systems and concepts.

In the nine years since Timmerman's paper was published, vulnerability has been mentioned and redefined many times in connection with climate impact and global change studies (Hewitt 1984; Kates *et al*. 1985; Wilhite and Easterling 1987). It has been related or equated to concepts such as resilience, marginality, susceptibility, adaptability, fragility, and risk. It would be an exhausting, and probably rather meaningless, task to review all the different ways in which people have used the word vulnerability, or similar concepts, in recent studies. Instead, I will discuss some of the analytical frameworks and methods that have been employed to assess vulnerability to global change, using selected examples and some of my own work. My goal is not to further semantic or theoretical debate but to try to find practical ways of defining who is most vulnerable to global change, and why.

## VULNERABILITY AS A BIOPHYSICAL CONDITION

In many cases, the most vulnerable people are considered to be those living in the most precarious physical environments. This may be by implication, when a study focuses only on the vulnerability or degradation of biophysical conditions and extrapolates directly from these estimates to the impact on the human occupants of a landscape. Alternatively, populations at risk may be explicitly delimited according to physical criteria. Implicit, biophysical determination of vulnerability can be found in many of the discussions of global change, where the loss of species, soils, or crop diversity implies adverse human impacts (Ehrlich and Ehrlich 1981). In many of the proposals for earth-system science and global change research, the vulnerability of the planet's biophysical processes is seen to indicate the conditions of its human inhabitants.

A recent study on environmental change in Latin America uses physical criteria to delimit the *fragile* lands of Latin America, and, by implication, the populations at risk in a changing environment (Browder 1989). Fragility is defined according to topography, climate, soils, and natural vegetation, although Denevan (1989) does point out that fragility depends not only on the physical conditions of a landscape but on the technology and intensity of human use. Fragile lands in Latin America are estimated to cover ⅓ of the continent. Global transformations such as climate warming and vegetation loss are likely to extend this area of fragile lands, although there may be areas where rainfall or soil changes may reduce the likelihood of flooding and erosion, or may improve soil fertility.

Research on natural hazards has also used biophysical conditions to define vulnerability (Mitchell 1989). Populations at risk live in areas of high hazard magnitudes or frequencies—severe rainfall deficits, flood plains, coasts with severe storms, basins that trap air and water pollution. Using this type of criteria, groups most vulnerable to global change would include those living in areas likely to experience sea level rise, increasing storminess, drier conditions, or heavier flooding. Those studies of global warming and sea level rise which delimit populations at risk within topographic contours illustrate this approach (Schneider and Chen 1980; Barth and Titus 1984).

Agricultural production and carrying-capacity studies also use biophysical frameworks for vulnerability assessment. For example, the FAO/UNFPA/IIASA study of carrying capacity estimates the agricultural production potential of lands in the developing world based on climate and soil constraints (FAO 1984). A number of regions and countries are identified as *critical zones* (e.g., the Sahel, Bangladesh, Haiti), in which current population demands for food already exceed agricultural production potential. Although the study does take into account different scenarios for population growth and technology, its underlying analytical framework is based in a biophysical determination of food insecurity (Hekstra and Liverman 1986). The study does not take environmental degradation into account in its scenarios for the year 2000, although proposals have been made to use the same climate and soil data base to assess the impacts of global changes in carrying capacity (Chen and Parry 1987). The methodology for such a study would be clear. Climatic data would be changed in accordance with global warming forecasts (for example) and thereby permit corresponding estimates of changes in agricultural potential, critical zones and populations at risk.

In some senses, Parry, Carter, and Konijn's (1988) impressive study of the impact of climatic variations on agriculture is also grounded in a biophysical definition of vulnerability.

The authors distinguish *direct* and *adjoint* approaches to assessing climate impacts. The direct method traces the impact of a specific change in a physical input variable (such as temperature) on yields or biomass, and then, through a series of steps, to impacts on economy and society. The adjoint method focuses on the sensitivity of the exposure unit to a range of climatic variations, for example, the vulnerability of farmers or households to frost or drought. Although the authors thus acknowledge the role of social marginality in determining climate impacts, the actual analytical frameworks used in the case studies (with the exception of Ecuador and Kenya) emphasize the biophysically driven direct method and associate vulnerability with changes in yield, crop failure, or agroecological potential.

These examples represent, I believe, a dominant analytical framework in studies of the human impact of global change. Parallel analyses of desertification, deforestation, or air pollution report changes in the area or intensity of degradation as the primary measure of human impact. The predominance of this framework guides and justifies a research focus on biophysical monitoring, modeling, and modification in global change studies. It implies that in order to understand and delimit vulnerability, we just need to know how and where the physical environment may change. Physical indicators will then provide adequate insight into the populations at risk.

The framework also guides the selection of responses. If biophysical conditions define vulnerability, then vulnerability can be reduced by modifying these conditions, or by moving people away from biophysically marginal areas. This framework commands the bulk of the research funding and the policy discussion concerning global environmental change, at least in the United States.

The demographic factor is, of course, important in many of the biophysical assessments of vulnerability such as the FAO study, described above (FAO 1984). The more people who occupy a biophysically vulnerable location, such as a floodplain or coast, the greater the total population at risk will be. This is clearly illustrated in the increase in the number of critical zones in the FAO study when population growth is included in the estimates of carrying capacity in the year 2000. Cohen (1986), and several of the papers in the recent special issue of *Scientific American* (1989), also combine estimates of environmental and demographic change to identify vulnerabilities and impacts.

## POLITICAL ECONOMY AND VULNERABILITY

A very different framework provides strong critiques of both physical and demographic determinants of vulnerability. This political economy (or neo-Marxist) approach has become increasingly important in climate impact studies. In this framework, vulnerability is defined by the political, social, and economic conditions of a society. According to Susman et al. (1984) vulnerability "is the degree to which different classes in society are differentially at risk." They use a theory of social marginalization to show how underdevelopment (flows of resources out of a region, land expropriations, exploitative labor conditions, political oppression, and other processes associated with colonialism and capitalism) has made people, especially the poor, more vulnerable to disaster, and has forced them to degrade their environment.

A similar analysis emerges from studies of the 1972 drought in the Sahel and other regions. Garcia (1981; 1982; 1986) and others propose a *structural* approach for diagnosing the impact of climate anomalies. By analyzing the historical evolution of social systems in various regions, the studies sought to demonstrate how certain groups become so disadvantaged and exploited that they are unable to cope with drought, or struggle for the resources to overcome environmental stress.

Other authors who use versions of this framework for understanding vulnerability to environmental change include de Castro (1975), Watts (1983), Hewitt (1984), and Spitz (1977). Most of their work has strong links to the broader critical literature on development and famine. For example, the critique of population growth as a major determinant of vulnerability draws on the work of Mamdani (1973) and Michaelson (1981), who show that the impact of population growth is strongly mediated by political economy and the distribution of access to resources. Critiques of the biophysical explanation of disaster often refer to the analyses of the Sahel which showed the predominant role of social factors in the pattern of famine and suffering during the 1970s (Franke and Chasin 1980; Copans 1975; Lappe and Collins 1979).

Recent analyses of the political economy of famine vulnerability have been inspired by Sen's (1981) theory of food entitlements, in which food security depends on a household or individual rights and access to grow, purchase, or share in food resources. Many of these entitlements are determined by power or position in the social structure. As global change threatens food, water, or fuel resources, an individual with limited entitlements to use or purchase these resources would be most severely affected.

## OTHER ANALYTICAL FRAMEWORKS

The biophysical and political economy perspectives provide two contrasting views of who may be vulnerable to global change. Of course, many scholars have actually blended elements of both frameworks. For example, Parry *et al.* (1985) note that it is necessary to include economic and social vulnerability, as well as the sensitivity of the physical system, when studying the impacts of climate on agricultural systems. An integrating approach is the *political ecology* of Blaikie and Brookfield (1987), who focus on the land manager and the physical, technological, economic, and political conditions that may constrain the use of land or the human response to environmental and social changes. Glantz (1987) emphasizes the importance of bringing together environmental and economic development approaches in the study of hunger in Africa.

The relationship between technology and vulnerability can be analyzed from both the biophysical and political economy perspectives. In the former, technology is generally seen to reduce vulnerability to environmental change because, for example, irrigation, improved seeds, and fertilizer are designed to mitigate against rainfall deficits and infertile soils. Dams and other engineering efforts prevent flooding and storm damage and new and old technologies provide alternative livelihoods for those living in degraded landscapes (Slater and Levin 1981).

However, technology does not always reduce biophysical vulnerability. Irrigation can cause the salinization and waterlogging of productive land. In severe drought years, an overdependence on irrigation can mean greater vulnerability when water storage is eventually depleted (Heathcote 1983). Engineering works and populations protected by dams always carry the potential for catastrophic failure and face a consequent high vulnerability to rare but devastating floods.

In the political-economy framework, technology is frequently tendered as increasing social inequality and vulnerability. The Green Revolution for example has provided the archetypical case of the social impacts of a set of technologies (improved seeds, irrigation, chemicals) that were partly designed to reduce vulnerability to environmental variation. Numerous authors have documented the social marginalization of small farmers and the poor associated with the introduction of Green Revolution technologies (Hewitt de Alcantara 1976; Pearse 1980; Barkin and Suarez 1982; Sanderson 1986).

## SUMMARY AND LIST OF POTENTIAL MEASURES OF VULNERABILITY

In contrasting the biophysical and political economy perspectives I have arbitrarily placed authors into opposing categories in order to illustrate some general tendencies. If we take a more synthetic approach to understanding vulnerability, the literature provides us with many factors to consider in assessing who may be vulnerable to global change. Nevertheless, it is important to emphasize that the most vulnerable people may not be in the most vulnerable places—poor people can live in productive biophysical environments and be vulnerable, and wealthy people can live in fragile physical environments and live relatively well. Yet it is often the combination of the two types of vulnerability that is important where impoverished populations live in ecologically marginal environments. Another analytic approach is to map vulnerability in geographical space (*where* vulnerable people and places are located) and in social space (*who* in a place is vulnerable). Vulnerability should also be applied to a specific scale of analysis or reference unit. Are we talking about vulnerable people, vulnerable regions, or vulnerable nations? Although cascades of vulnerability differ from one scale to another, in many cases the factors determining vulnerability may be different.

What conditions and variables seem to be important in determining vulnerability to global change? A preliminary list (which includes people, regions and nations) might include the following:

### ENVIRONMENTAL CONDITIONS

♦ Climatic changes will be less critical in regions with adequate rainfall and moderate temperatures than in those with low or variable rainfall and high temperatures;

♦ Deforestation and storms may have lesser impacts in regions where soils are resilient to erosion and slopes are gentle;

♦ Countries with a wide range of genetic varieties of crops may be less vulnerable to a loss of biodiversity than others;

♦ People living in environments subject to meteorological extremes (droughts, hurricanes, floods) may be vulnerable if global change increases the magnitude and frequency of such events.

### TECHNOLOGICAL CONDITIONS

♦ Irrigation and reservoirs may buffer agriculture and other sectors against moderate precipitation deficits but may actually increase the catastrophic potential of major droughts on intensive agriculture which relies on limited water storage;

♦ Considerable debate wages about the role that improved seeds and fertilizers may play in vulnerability of producers and countries to climatic variability and other changes. Although investigators such as Hazell (1984) and Michaels (1979) conclude that high-yielding varieties and other technological changes may have increased the variability of cereal yields (and the sensitivity to weather variations), others believe that the Green Revolution has reduced ecological and social vulnerability to environmental change (Byerlee and Harrington 1982);

♦ Indigenous agricultural technologies and *low input* or sustainable agriculture are seen by authors such as Wilken (1987) as appropriate buffers against global change, reducing biophysical vulnerability through soil and moisture conserving practices, and economic vulnerability by reducing costs and dependency on purchased inputs;

♦ Dependence on certain types of energy (e.g., imported oil) can create vulnerability in periods of economic and political instability (Brooks 1986).

### SOCIAL RELATIONS

♦ Most neo-Marxists consider class to be the overriding category which differentiates social vulnerability. The appropriate divisions (proletariat, bourgeoisie, etc.) are open to discussion and are occasionally identified with differences in:

- Income, in which wealth can be used to purchase land, food, or other resources to buffer an individual against environmental change;
- Gender, according to Schroeder (1988) and Ali (1984) is an important indicator of vulnerability to drought. Women, because of lower incomes and a lack of rights to land and other resources, are frequently more affected than men by drought and natural disaster (Rivers 1967). Dankelman and Davidson (1989) also describe how changes in water, soil, and forest resources are likely to affect women more than men in the Third World because women often have traditional responsibilities for fuel and water collection and for farming;
- Race and ethnicity can be important factors in determining vulnerability to global change. Timberlake (1985) describes how apartheid has forced the South African black population onto crowded, unproductive lands and increased their vulnerability to drought and soil degradation. Researchers associated with Cultural Survival have shown many times how the loss of land, power, and other rights by indigenous groups in regions such as Brazil, Central America, Australia, and the United States has increased their vulnerability to deforestation and other environmental changes.

## DEMOGRAPHICS AND HEALTH

- Health and age are monitored as important determinants of famine vulnerability in many studies. Pankhurst (1984) lists the elderly, the disabled, pregnant women or mothers, the chronically ill, and children as particularly vulnerable groups in refugee camps. These groups have particular nutritional needs or disadvantages and may be more sensitive to food shortages, climate extremes, and other physical stresses that may accompany global change, and may not be able to move easily;
- High population densities, population growth rates, and pressure on limited food, land, and water resources can make regions very vulnerable to global change.

## LAND USE AND OWNERSHIP

- Insecurity or restrictive conditions of land tenure may increase vulnerability to global change if land degradation creates competition for land that displaces the insecure tenure holds or means that their land can no longer support them;
- Similarly, the livelihoods of sharecroppers and other fixed or proportional renters may be disproportionately threatened by decreases in land productivity;
- In other cases, sharecropping and other relationships that constrain the independence and flexibility of agricultural producers in terms of the crops they grow may restrict the ability of people to respond to environmental change by shifting, for example, from cash or nonfood crops to subsistence cereal production;
- The landless are particularly vulnerable because they cannot produce their own food.

## ECONOMY AND INSTITUTIONS

- Lack of access to markets can make it difficult for people to buy or sell food and other products and can increase vulnerability to environmental change;
- For producers, prices controlled at artificially low levels by government or private buyers can create very difficult conditions. Consumers, in contrast, are vulnerable to high prices;
- Lack of private and government institutional support systems such as disaster relief, insurance, meteorological services, and agricultural extension can create problems in the face of environmental variability and change;
- Debt at the individual or national scale can exacerbate vulnerability to global change.

## CASE STUDIES

Torry (1984) points out the need to support theoretical discussions about vulnerability with empirical studies of cause and effect. Several case studies of the impacts of climatic variations illustrate the ways in which vulnerability to global change might be concretely measured and analyzed in different regions. Such empirical studies provide ideas about the type of variables that are available in vulnerability studies and can also serve to test theories and hypotheses about the causes and distribution of vulnerability.

Several of the case studies in Parry, Carter and Konijn (1988) discuss vulnerability in specific regions of the world. In the Kenya case study, Downing and colleagues discuss smallholder vulnerability and response to drought in terms of household food economy, ecological degradation, land tenure, agricultural technology, and institutional support (see also Downing *et al.* 1989). In Ecuador, the vulnerability of indigenous farming systems is related to farming practices and labor strategies. In the India case study, Jodha (1988a) and colleagues discuss relationships between drought vulnerability, climate variability, agricultural technology, and government policies. Jodha (1988b) has also discussed the general vulnerability of developing countries to drought in terms of their reliance on rainfed subsistence agriculture and inadequate institutional support and infrastructure.

A comparative study of the Tigris-Euphrates, Sahel, and U.S. Great Plains demonstrated shifting patterns of vulnerability to drought associated with changing technology, social conditions, and government policies (Bowden *et al.* 1981). In the case of the Tigris-Euphrates, increased vulnerability was associated with a lack of technological innovation and by inflexible sociopolitical organization. In the Sahel, this study identified problems associated with population growth and export cropping and suggested that food aid reduced local vulnerability. In a related study, Warrick shows how the social consequences of drought in the Great Plains have changed in this century (also see Worster 1979). He suggests that local vulnerability has been reduced in later years through insurance, economic adjustments, and soil and water conservation. But he believes that although technology and social organization have lessened local vulnerability, the regional and global impacts of drought in the Great Plains may have increased (Warrick 1984a; 1984b).

## APPROACHES TO HAZARD VULNERABILITY IN MEXICO

In my own work (Liverman 1990) I have been trying to analyze vulnerability to drought and other natural hazards in Mexican agriculture, using data available from the agricultural census and meteorological service. Florescano (1969; 1980) has proposed that dramatic differences in vulnerability to natural hazards, particularly drought, have been prominent in both colonial and present-day Mexico. He suggests that the economic and land tenure relations imposed by the Spanish crown created a tremendous vulnerability to drought among the poorer and indigenous campesino populations. The colonial political economy allowed the larger landholders and merchants to manipulate the price of staples in drought years to the disadvantage of poor consumers and small producers. In post-revolutionary Mexico, he suggests, differential vulnerabilities to drought, inherited from the colonial land tenure systems, are very evident. He claims that

> the most disastrous effects of drought, as in earlier times, are concentrated in the rainfed agriculture practiced by the poorest ejidatarios and campesinos, lacking credit, irrigation, fertilizers, and improved seeds. (1980, p. 17)

In this century, a steady and unwavering expansion of Mexican agricultural production has been necessary to meet the demands of a rapidly growing population and agricultural export market (Wellhausen 1976). In the drive to modernize and expand production, the Mexican agricultural system has incorporated some techniques (such as irrigation, improved seed varieties, and chemical inputs) that may reduce hazard losses (Yates 1981; Venezian and Gamble 1969). At the same time, these new techniques have replaced some traditional hazard prevention strategies, such as mixed cropping and microclimate modification (Wilken 1987), and have allowed agriculture to expand into areas of high hazard risk—such as deserts, mountains, coastal regions, and the disease-susceptible humid tropics.

Any discussion of hazards and agriculture in modern Mexico will intersect with these two major debates about agricultural development. The first, which ranges beyond Mexico to many other developing countries, concerns the success or failure of the *Green Revolution*, in which a combination of improved seeds, irrigation, fertilizer, and other technologies was introduced into parts of Mexico and other countries with the backing of international foundations and multinational corporations (Yates 1981; Sanderson 1986). Although the *Green Revolution* has been heavily criticized for creating dependency on foreign imports, for causing environmental degradation, and for exacerbating social inequities (Barkin and Suarez 1982; Hewitt de Alcantara 1973; 1976; Pearse 1980), very little research addresses the role of the *Green Revolution* in reducing or increasing vulnerability to natural hazards. Michaels (1979) and Hazell (1984) have suggested that improved wheat varieties in Mexico and other countries may be more sensitive to drought and climate variability than traditional seeds. Others have suggested that the large regions of single-variety high-yielding crops established through the *Green Revolution* are much more vulnerable to pests and diseases than traditional mixed cropping systems.

A second major debate in Mexican agriculture concerns land reform, particularly the performance of the *ejido* sector. The *ejido* land tenure system was established in postrevolutionary Mexico (McBride 1923; Yates 1981). Land is held communally in usufruct by a group of families and although some plots are farmed collectively, most land is farmed individually. Some authors claim that the *ejidos* are inefficient and unproductive (Whetten 1948; Wellhausen 1976), whereas others suggest that, in terms of input use, they are relatively efficient and produce yields equal to the private sector (Dovring 1970; Mueller 1970; Nguyen 1979). The problems of the *ejido* sector have been explained in terms of their lack of political power, difficulties in obtaining access to credit and inputs, bad management, and poor land resource endowment (Hewitt de Alcantara 1976; Coll-Hurtado 1982). Most of the research that compares land reform sectors in Mexico, however, fo-

cuses on economic, political, and productivity issues and ignores questions of environmental conditions or natural hazard vulnerability on landholdings of different size or tenure.

In my research, I have tried to explain differential vulnerability to drought and natural hazards in terms of physical geography, access to technology, and land tenure. To achieve this objective I have raised a series of questions and hypotheses and evaluated them using data for the states of Sonora and Puebla in 1970 and for the whole of Mexico from 1930 to 1970.

*What is the pattern and severity of drought loss?* This is documented and mapped using census reports of area lost to drought and yields of various crops.

*What is the relationship between reported drought losses and the physical pattern of climate?* The hypothesis that drought impacts are determined by the physical severity of drought is evaluated by comparing, statistically and cartographically, reported drought losses to various indices of physical drought, as calculated from monthly rainfall and temperature data.

*What variables can be found to document vulnerability to drought and thus to explain patterns of reported drought losses?* The agricultural census provides a series of technological, economic, and land tenure variables that can be used as measures of drought vulnerability. Mexican government maps also provide information on soils, topography and agricultural potential. The following specific hypotheses are examined:

♦ The hypothesis that irrigation reduces drought losses is evaluated by comparing irrigated acreage with reported drought losses and yields.

♦ The hypothesis that improved seeds or chemical fertilizer reduce drought losses is evaluated by comparing expenditure on seeds and fertilizer with reported drought losses and yields.

♦ The hypothesis that certain crops are more vulnerable to drought than others is evaluated by comparing the relative area in wheat or corn to reported drought losses.

♦ The hypothesis that *ejidos* are more vulnerable than private farms is tested by comparing the relative area in *ejido* tenure with reported drought losses and yields.

*Are there significant differences in reported drought losses between farms of different size and tenure?* Since the census reports *municipio* drought losses and other variables in three land tenure categories—large private, small private, and *ejidos*—it is possible to test for significant differences in drought losses and to try to explain them in terms of differences in irrigation and other variables.

In the case study of drought in Sonora, and Puebla, the physical pattern of drought, expressed in terms of rainfall deficits and evapotranspiration, did not correlate strongly with the pattern of drought loss reported in the census. Both states included *municipios* with high drought losses in areas of low rainfall and some with low losses where there were severe rainfall deficits. I also found that, in general, lower drought losses seemed to be associated with the use of irrigation, high yielding varieties, and fertilizer. However, there were *municipios* with high reported drought losses where irrigation did not seem to buffer the agricultural system against rainfall deficits. The *ejido* sector in both states was, on average, reporting twice the drought losses of the large private landholdings.

In the national study of natural hazard losses in Mexican agriculture, the data indicated that overall natural hazard losses have increased since 1940 in Mexico and 1970 losses were almost double those of 1940 (Figure 4.1). The meteorological record shows no indication of an increasing severity of weather events in the census years. This increase may support the hypothesis that hazard losses have been increasing irrespective of weather severity, because of increases in population vulnerability to natural disasters. For example, the Swedish Red Cross (Wijkman and Timberlake 1984) reports that the average number of people affected worldwide by reported disaster events grew significantly from the decade of the 1960s

to the 1980s. They and others ascribe such increasing vulnerability and losses to increases in the number of people living in hazard-prone areas and to increasing poverty.

Clearly some relationship links the national pattern of hazard losses and physical geography (Figure 4.2). High drought losses tend to occur in the arid northern region, where precipitation is low and highly variable. High flood losses tend to occur in states such as Tabasco and Veracruz along the Atlantic gulf coast, or in the southern regions of the Pacific coast such as Sinaloa, Colima, Nayarit, or Oaxaca. These are all regions that can receive heavy cyclonic summer rainfall and with major rivers that often overflow their banks. Thus any global changes that bring drier conditions in northern Mexico, or increased storminess along coasts, would increase hazard losses in much of Mexico.

How do natural hazard losses relate to other characteristics of the agricultural system? Combining the data for Mexican states over the four census periods for which irrigation and drought loss information are available (1940–1970), a correlation coefficient of $-0.34$ (significant at 0.05 level) is obtained. This relationship is much weaker than that obtained in the Sonora and Puebla studies, partly reflecting the lower use of irrigation as a whole in Mexico. States with high levels of irrigation, such as Sonora and Sinaloa, do have much lower drought losses than many of the states with smaller proportions of irrigated land. Coll-Hurtado (1982) suggests that low drought losses in Baja California are due to the high proportion of irrigated land in this region. However, Aguascalientes and San Luis Potosi, with about 20% of their land irrigated, report drought losses of 70% and 50% respectively. Clearly, irrigation may not always buffer Mexican states against drought.

As noted earlier, some people suggest that the use of improved seeds may be associated with changes in natural hazard vulnerability. In this study, the correlation between expenditure on improved seeds and drought losses

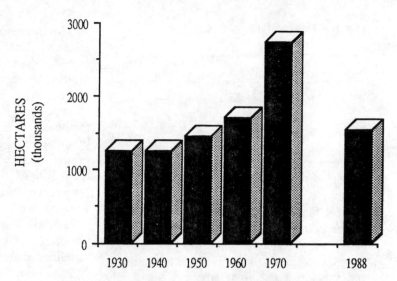

**FIGURE 4.1** Hazard Losses in Mexico

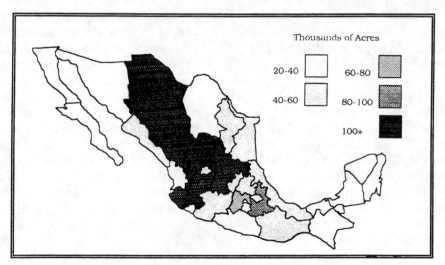

**FIGURE 4.2** Mexico Summer 1970 Drought Hazard Losses (total acres lost)

in 1970 is negative but weak (−0.18), with higher expenditures on improved seeds associated with lower drought losses.

Table 4.1 shows the average losses, at the state level, for the two main land tenure sectors in Mexico. In every census year the average losses in the private sector are less than those in the *ejido* sector. In 1950, losses are almost double on *ejido* lands. Differences between private and *ejido* losses are signifi-

cant for specific hazards, with *ejidos* generally reporting higher drought, flood, and frost losses. Pest losses seem, however, to be the same, possibly because of monoculture and modernization in the private sector.

This seems to confirm the hypothesis of Florescano about the relative vulnerability of the *ejido* sector to hazards and supports the findings of the study of drought losses at the *municipio* level in Sonora and Puebla in 1970

**TABLE 4.1** Hazard Losses and Land Tenure in Mexico*

| Year | Private | Ejido |
|------|---------|-------|
| 1930 | 14.9% | 23.2% |
| 1940 | 12.0% | 16.8% |
| 1950 | 10.2% | 20.2% |
| 1960 | 11.8% | 17.7% |
| 1970 | 26.7% | 28.0% |

*Percent of areas sown but not harvested.

where I found similarly disproportionate losses on the *ejidos*. Possible explanations include more marginal land (mountainous, hazard prone, etc.) given to *ejidos* in the land reform and the lower use of technology, especially irrigation, on ejidos.

## FAMINE EARLY WARNING AND VULNERABILITY

A second example of concrete attempts to assess vulnerability is the use of early warning systems in famine response. Such systems are generally designed to gather on a set of indicators information than can be used to anticipate the onset of famine, to target the people at risk, and to estimate food aid or other relief requirements. Early examples of such systems are the Indian Famine Codes developed to assist colonial administrators anticipate the onset of food shortages (Brennan 1984). The goal of current systems is to reduce the delays in responding to famines such as those in Ethiopia and the Sudan. Some of the early warning systems—such as the FAO's *Global Information and Early Warning System* (GIEWS)—just focus on the physical event in monitoring rainfall deficits and vegetation conditions. In the United States, the Agency for International Development (AID) has established the *Famine Early Warning System* (FEWS) to collect and synthesize existing physical and social information in order to identify regions and populations at risk of famine.

The indicators currently used in FEWS to assess vulnerability to famine include rainfall, crop conditions, remotely sensed vegetation indices, the nutritional status of populations, and the prices of food in markets. All of the current early warning systems pay lip-service to the social explanations of famine, backed by documents citing the importance of identifying vulnerable groups, and downplaying the role of drought. Yet biophysical and crop production indicators predominate in many warning systems because they are easier to collect, are less politically sensitive, and take advantage of existing weather networks and satellite capabilities.

The FEWS system is designed to incorporate an understanding of vulnerability to famine. In the development of FEWS, it was accepted that a *natural hazard* model of famine was inadequate because it did not take account of the prevalent thinking on the social causes of famine. Thus, FEWS shifted to a strong emphasis on the social model of famine, employing the concept of vulnerability, and attempting to measure a wide range of social indicators. Although the social model now used in FEWS, is linked to the structural and political economic analyses of famine vulnerability, the social indicators do not reflect the richness and complexity of the research literature. The emphasis is on measures of individual health and nutrition, and, more recently, on market prices. The information is difficult to obtain, is localized, and is often unreliable in most Sahelian countries. This is partly explained by the FEWS mission to use existing information rather than to collect primary data.

The primary use of the agroclimatological information in a famine warning system is to provide early indications of the magnitude and location of crop production declines and of stressed pastures which might trigger livestock losses and human migrations. The goal of the agroclimatic indicators is to assist in making production estimates as soon as possible, and to use them to focus in on critical regions and vulnerable populations. In most countries the government agricultural statistics service will produce a production estimate just before harvest and a final crop assessment several months later. In certain countries the political sensitivity of production estimates results in considerable delays in the publication of reports and provokes debates about the credibility of the information. In many countries the focus of data collection is on cereals, even in areas where roots or vegetables are important.

The last five years have seen a major advance in agroclimatological monitoring in Africa in the form of satellite remote sensing. The great advantage of satellite imagery is that it provides comprehensive spatial coverage, filling in the gap between meteorological stations and giving general indications of environment and vegetation conditions in advance of ground observation.

Increasingly, the literature on famine vulnerability is emphasizing the value of social indicators. Authors such as Fleuret (1986), McCorkle (1987), and Cutler (1984) suggest the use of market prices and changes in population behavior as warnings of famine conditions. Eldridge et al. (1986) report on the use of many such measures in the Sudan and conclude that rainfall, cereal and livestock prices, and market throughput seem to provide warning of the onset of food crises.

Operational famine early warning systems are always working with compromises between using the best possible measures of vulnerability and using those which are easily, cheaply, and rapidly available. Although we may hope that our vulnerability assessments for global change will be less constrained, it is important to acknowledge the difficulties in acquiring many of the most useful social indicators.

## CONCLUSION

These examples illustrate some of the ways in which we might design vulnerability assessments for global change. Although the ex-amples primarily refer to drought and hunger, many of the key vulnerability measures hold true for other global change issues as well. For example, vulnerability to the accumulation of toxins may depend on geographical location and health status. Vulnerability to soil erosion will depend on farm practices, technology, and land tenure. However, vulnerability assessments of natural ecosystems and human health to other global change issues (e.g., ozone, loss of biodiversity) may depend more on ecological and medical factors discussed in the risk assessment literature than those discussed in this paper.

Such vulnerability assessments must be a key component of global change research and policy. Focusing on vulnerability reminds us of the social context of environmental change. It illustrates why we should disaggregate our analyses of global change by region and social group because it shows us that some people and countries will suffer more, and more quickly, than others from global change (and, of course, that some will benefit). Vulnerability analysis also expands the range of choices in responding to global change. For example, it demonstrates that we can reduce impacts of global change not just by slowing the rate of climate change or ozone depletion but also by reducing the vulnerability of populations to these changes. It may also permit more efficient and economic responses by highlighting regions and peoples who are most vulnerable, or whose vulnerability can be most easily reduced.

Many questions about vulnerability analysis are unresolved. There are few empirical studies of vulnerability to environmental change outside the famine and hazards literature. There is no consensus on the appropriate definitions and indicators. We need to know how to deal with multiple vulnerabilities to multiple threats. We need to decide whether studying the existing pattern and degree of vulnerability to drought and other conditions is an accurate guide to future vulnerability or whether patterns of vulnerability are changing with rapid social, technological, and demographic transformations. And, above all, we need to communicate about vulnerability to other scientists, to policy makers, and to the public. Too many people still believe that the impacts of disasters and climate change are determined solely by the physical characteristics of events—that drought impacts are most severe where there is least rain—or want to believe that famine and soil erosion are uncontrollable acts of nature rather than socially created crises. We can know very little about the social impacts of global change unless we work to understand, document, and communicate the nature of vulnerability.

## REFERENCES

Ali M. 1984. Women in famine, pp. 113–34 in Currey and Hugo 1984.

Aguayo S. 1987. *Chiapas: Las Amenazas a la seguridad nacional.* Mexico: Centro Latinoamericano de Estudios Estrategeticas.

Barkin D., and B. Suarez. 1982. *El fin de autosuficiencia alimentaria.* Mexico: Editorial Neuva Imagen.

Barth M.C., and J.G. Titus, eds. 1984. *Greenhouse Effect and Sea Level Rise.* New York: Van Nostrand Reinhold.

Blaikie, P.M., and H.C. Brookfield. 1987. *Land Degradation and Society.* London: Methuen.

Bowden, M., *et al.* 1981. The effect of climatic fluctuations on human populations; Two hypotheses. In T.M. Wigley, M.J. Ingram and G. Farmer, eds. 1981. *Climate and History: Studies in Past Climates and their Impact on Man.* Cambridge: Cambridge University Press, pp. 479–513.

Brennan, L. 1984. The development of the Indian Famine Code, pp. 91–112 in Currey and Hugo 1984.

Brooks, H. 1986. pp. 325–48 in W.C. Clark and R.E. Munn, eds. *The Sustainable Use of the Biosphere.* Cambridge: Cambridge University Press.

Browder, J.O., ed. 1989. *Fragile Lands of Latin America: Strategies for Sustainable Development.* Boulder: Westview Press.

Brown, L.R., ed. 1977. *Redefining national security.* Worldwatch Paper 14. Washington: Worldwatch Institute.

Brown, L.R., ed. 1989. *The State of the World.* New York: W.W. Norton.

Burbach, R., and P. Flynn. 1980. *Agribusiness in the Americas.* New York: Monthly Review Press.

Byerlee, D., and L. Harrington. 1982. *New Wheat Varieties and the Small Farmer.* CIMMYT Economics Working Paper. Mexico: Centro Internacional de Mejoriamento de Maiz y Trigo.

Caldwell, L.K. 1985. *US Interests and the Global Environment,* Iowa: Stanley Foundation.

Canados Cruz, L., T. Salcedo Solis, R.E. Bravo, and G. Knapp. 1988. The vulnerability of indigenous farming systems to climatic variations. In Parry, M.L., T.R. Carter, and N.T. Konijn, eds. 1988. *The Impact of Climatic Variations on Agriculture.* Dordrecht: Kluwer, Volume 2, pp. 413–28.

Castro, J. de. 1975. *The Geopolitics of Hunger.* New York: Monthly Review Press.

Chen, R.S., and M.L. Parry. 1987. *Climate Impacts and Public Policy.* Laxenburg, Austria: International Institute for Applied Systems Analysis.

Clark, W.C. 1987. National security and the environment. *Environment* 29(5):1,44–45.

Cohen, S.J. 1986. Impact of $CO_2$-induced climatic change on water resources in the Great Lakes basin. *Climatic Change* 8:135–53.

Coll-Hurtado, A. 1982. *Es Mexico un pais agricola?* Mexico: Siglo XXI.

Copans, J., ed. 1975. *Sécheresses et famines du Sahel.* 2 vols. Paris: Maspero.

Currey B., and G. Hugo., eds. 1984. *Famine as a Geographical Phenomenon.* Boston: Reidel.

Cutler, P. 1984. Famine forecasting. *Disasters* 8(1):48–56.

Dankelman, I., and J. Davidson. 1989. *Women and the Environment in the Third World.* London: Earthscan/IIED.

Denevan, W.H. 1989. The fragile lands of Latin America, pp. 3–25 in Browder, J.O., ed. *Fragile Lands of Latin America: Strategies for Sustainable Development.* Boulder: Westview Press.

Dovring, F. 1970. Land reform and productivity in Mexico. *Land Economics* 46(3):264–74.

Downing, T. 1988. Smallholder vulnerability and response to drought in Central and Eastern Kenya. In Parry *et al.* op. cit. 1988, vol. 2.

Downing, T.E., K.W. Gitu, and C.M. Kamau, eds. 1989. *Coping with Drought in Kenya.* Boulder: Lynne Rienner.

Ecologist, The. 1987. Special issue on deforestation and debt. *The Ecologist* 17(4/5)

Ehrlich, P.R., and Ehrlich A.H. 1981. *Extinction: The Causes and Consequences of the Disappearance of Species.* New York: Random House.

Eldridge, E., C. Salter, and D. Rydjeski. 1986. Towards an early warning system in the Sudan. *Disasters* 10(3):189–96.

Excelsior. 1983. (Series on environmental issues in Mexico City). *Excelsior* 18–23 January 1983.

FAO (Food and Agriculture Organization). 1984. *Potential Population Supporting Capacities of Lands in the Developing World.* Rome: FAO.

Fleuret, A. 1986. Indigenous responses to drought in Sub-Saharan Africa. *Disasters* 10(3):224–29.

Florescano, E. 1969. *Precious del maiz y crisis agricola en Mexico*. Mexico: Colegio de Mexico.

Florescano, E. 1980. Una historia olivadada: la sequia en Mexico. *Nexos* 32 9–18.

Franke, R., and Chasin, B.H. 1980. *Seeds of Famine; Ecological Destruction and the Development Dilemma in the Western Sahel*. Montclair, New Jersey: Allenheld Osmun.

Garcia, R.V. 1981, 1982, 1986. *Drought and Man: The 1972 Case History*. 3 vols. New York: Pergamon Press.

Glantz, M.H., ed. 1987. *Drought and Hunger in Africa*. Cambridge: Cambridge University Press.

Gonzalez Pacheco, C. 1983. *Capital extranjero en la selva de Chiapas 1863–1982*. Mexico: Institut de Investigaciones Economicas.

Harrison, B.A., ed. 1988. *Famine*. New York: Oxford University Press.

Hazell, P.B.R. 1984. Sources of increased instability in Indian and U.S. cereal production. *American Journal of Agricultural Economics* 66:302–11.

Hazell, P.B.R., ed. 1986. *Summary Proceedings of a Workshop on Cereal Yield Variability*. Washington DC: International Food Policy Research Institute.

Heathcote, R.L. 1983. *The Arid Lands: Their Use and Abuse*. New York: Longmans.

Hekstra, G.P., and Liverman, D.M. 1986. Global food futures and desertification. *Climatic Change* 9:59–66.

Hewitt, K., ed. 1984. *Interpretations of Calamity*. Boston: Allen and Unwin.

Hewitt de Alcantara, C. 1973. The Green Revolution as history: The Mexican experience. *Development and Change* 5:25–44.

Hewitt de Alcantara, C. 1976. *Modernizing Mexican Agriculture*. Geneva: UNRISD.

Jacobson, J. 1988. *Environmental refugees*. Worldwatch Paper 86. Washington DC: Worldwatch Institute.

Jodha, N.S. 1988a. Introduction to Part V. In Parry, M.L., T.R. Carter, and N.T. Konijn, eds. 1988. *The Impact of Climatic Variations on Agriculture*. Dordrecht: Kluwer, vol. 2, pp. 503–21.

Jodha, N.S. 1988b. Perspectives from the developing world. In N.J. Rosenberg, W.E. Easterling III, P.R. Crosson, J. Darmstadter, eds. *Greenhouse Warming: Abatement and Adaptation*, pp. 147–58. Washington DC: Resources for the Future.

Kates, R.W., J.H. Ausubel, and M. Berberian, eds. 1985. *Climate Impact Assessment*. SCOPE 27. New York: Wiley.

Lappe, F.M., and J. Collins. 1979. *Food First: Beyond the Myth of Scarcity*. New York: Ballantine.

Liverman, D.M. 1988. Agroclimatology and famine early warning in Africa. Technical Annex to U.S.A.I.D. Project Paper on the Famine Early Warning System (FEWS). Unpublished.

Liverman, D.M. 1990. Drought in Mexico: Climate, agriculture, technology and land tenure in Sonora and Puebla. *Annals of the Association of American Geographers* 80(1):49–72.

Lobato, R. 1980. Social stratification and destruction of the Lacandona Forest in Chiapas. *Ciencia Forestal* 5:21–38.

Maguire, A., and J.W. Brown. 1986. *Bordering on Trouble*. Bethesda, MD: Adler and Adler.

Mamdani, M. 1973. *The Myth of Population Control*. New York: Monthly Review Press.

Mathews, J.T. 1989. Redefining security. *Foreign Affairs* 68(2):162–77.

McBride, G.M. 1923. *The Land Systems of Mexico*. New York: Conde Nast.

McCorkle, C.M. 1987. Food grain disposals as early warning famine signals: A case from Burkina Faso. *Disasters* 11(4):273–81.

Michaels, P.J. 1979. The response of the Green Revolution to climatic variability. *Climatic Change* 5:255–79.

Michaelson, K.L., ed. 1981. *And the Poor Get Children*. New York: Monthly Review Press.

Mitchell, J.K. 1989. Risk Assessment of Environmental Change. Working Paper 13, Environment and Policy Institute, East West Center, Honolulu: East West Center.

Montgomery, J.D., ed. 1984. *International Dimensions of Land Reform*. Boulder: Westview Press.

Mueller, M.W. 1970. Changing patterns of agricultural output and productivity in the private and land reform sectors in Mexico 1940–1960. *Economic Development and Cultural Change* 18(2):262–66.

Myers, N. 1984. *The Primary Source: Tropical Forests and Our Future*. New York: W.W. Norton.

Myers, N. 1988. Environment and security. *Foreign Policy* 74:23–41.

Nguyen, D.T. 1979. The effects of land reform on agricultural production, employment and income distribution: a statistical study of Mexican States 1959–1969. *The Economic Journal* 89:624–35.

O'Keefe, P., K. Westgate, and B. Wisner. 1976. Taking the naturalness out of natural disaster. *Nature* 260:15 April.

Ortiz, M.F. 1987. *Tierra profanada: Historia ambiental de Mexico*. Mexico: INAH/SEDUE.

Pankhurst, A. 1984. Vulnerable Groups. *Disasters* 8(3):206–13.

Parry, M.L., T.R. Carter, and N.T. Konijn. 1985. Climate change: How vulnerable is agriculture. *Environment* 27:4–5, 43.

Parry, M.L., T.R. Carter, and N.T. Konijn, eds. 1988. *The Impact of Climatic Variations on Agriculture*. 2 volumes. Dordrecht: Kluwer.

Pearse, A. 1980. *Seeds of Plenty, Seeds of Change*. Oxford: Clarendon Press.

Puente, S., and J. Legorreta, eds. 1988. *Medio ambiente y calidad de vida*. Mexico: Plaza y Valdes.

Rivers, J.P.W. 1967. Women and children last: An

essay on sex discrimination in disasters. *Disasters* 6(4):256–67.

Sanderson, S. 1986. *The Transformation of Mexican Agriculture: International Structure and the Politics of Rural Change*. Princeton: Princeton University Press.

Schneider, S.H., and R.S. Chen. 1980. Carbon dioxide warming and coastline flooding: Physical frameworks and climatic impacts. *Annual Review of Energy* 5:107–40.

Schroeder, R. 1988. *Gender Vulnerability to Drought in Northern Nigeria*. Natural Hazards Working Paper. Boulder: Natural Hazards Research and Applications Information Center.

*Scientific American*. 1989. Managing planet earth: Special Issue. *Scientific American* 261(3)(September).

Sen, A. 1981. *Poverty and Famines: An Essay on Entitlement and Deprivation*. New York: Oxford University Press.

Slater, L.E., and S.K. Levin. 1981. *Climate's Impact on Food Supplies: Strategies for Climate-Defensive Food Production*. Boulder: Westview.

Somerville, C.M. 1986. *Drought and Aid in the Sahel*. Boulder: Westview Press.

Spitz, P. 1977. *Silent Violence: Famine and Inequality*. Geneva: UNRISD.

Susman, P., P. O'Keefe, and B. Wisner. 1984. Global disasters: A radical interpretation, pp. 264–83 in Hewitt 1984.

Szekely, M., and I. Restrepo. 1988. *Frontera agricola y colonizacion*. Mexico: Centro de Ecodesarrollo.

Timberlake, L. 1985. *Africa in Crisis*. Earthscan: Washington.

Timmerman, P. 1981. *Vulnerability, Resilience and the Collapse of Society*. Environmental Monograph No

1. Institute for Environmental Studies, University of Toronto.

Torry, W. 1984. Social science research on famine. *Human Ecology* 12(3):227–45.

Venezian, E.L., and W.K. Gamble. 1969. *The Agricultural Development of Mexico: Its Structure and Growth Since 1950*. New York: Praeger.

Vizcaino, M.F. 1975. *La contaminacion en Mexico*. Mexico, D.F.: Fondo de Cultura Economica.

Warrick, R.A. 1984a. Drought in the Great Plains: Shifting social consequences? in Hewitt 1984.

Warrick, R.A. 1984b. The possible impact on wheat production of a recurrence of the 1930s drought in the US Great Plains. *Climatic Change* 6:5–26.

Watts, M. 1983. *Silent Violence: Food, Famine and the Peasantry in Northern Nigeria*. Berkeley: University of California Press.

Wellhausen, E. 1976. The agriculture of Mexico. *Scientific American* 235(3):128–50.

Whetten, N.L. 1948. *Rural Mexico*. Chicago: University of Chicago Press.

Wijkman, A., and L. Timberlake. 1984. *Natural Disasters: Acts of God or Acts of Man*. London: Earthscan.

Wilhite, D.A., and W.E. Easterling, eds. 1987. *Planning for Drought*. Boulder: Westview Press.

Wilken, G.C. 1987. *The Good Farmers: Traditional Agricultural Resource Management in Mexico and Central America*. Berkeley: University of California Press.

Worster, D. 1979. *Dust Bowl: The Southern Plains in the 1930s*. New York, Oxford University Press.

Yates, P.L. 1981. *Mexico's Agricultural Dilemma*. Tucson: University of Arizona Press.

# 27

# CHASING A SPECTER: RISK MANAGEMENT FOR GLOBAL ENVIRONMENTAL CHANGE

*Timothy O'Riordan* ❖ *Steve Rayner*

Global environmental change is both a concept and a process that changes in meaning with scientific discovery, public concern, and political responsiveness. It is the relationship between the problems as perceived and the various institutions that help shape and adapt to such problems that defines global environmental change. There is a kind of race between scientific detective work and political adjustment to lessen the likely impacts that predictive science is trying to verify. That race cannot be *won* because the contestants are one and the same. Risk analysis, because of its capacity to recognize this relationship in many spheres of problem identification, can contribute to the political debate, mostly by proposing institutional redesign of the relationship among scientific research, public entry, and experimental readjustments to consensus formation and international action.

Reprinted from *Understanding Global Environmental Change: The Contributions of Risk Analysis and Management*, Edited by R. E. Kasperson, K. Dow, D. Golding, and J. X. Kasperson. Worcester, MA: Earth Transformed Program, Clark University, 1990, pp. 45–62. Used with permission of the authors and Clark University.

## IDENTIFYING THE CONTESTANTS

We begin with the assertion that humanity as a whole does not really know what the phenomenon *global environmental change* is. As it draws toward an apparent consensus, the character and meaning of global environmental change modulate into a new specter. This is a definitional problem of even greater dimensions than the well known frustrations of pinning down concepts such as *risk* and *sustainable development*. All of these notions have two features that render definition elusive:

◆ The state of scientific knowledge is growing impressively but is still inadequate: at the boundaries of knowledge huge chasms of *chaos* (i.e., unpredictable uncertainty) loom, but they can only be appreciated at the edge of the chasm. The struggle to the edge of understanding does not produce knowledge; it reveals the despair of future darkness.

◆ Scientific analysis moves along pathways that distort the character of the phenomenon being studied. This is because the very acts of modeling and prediction (the essence of risk analysis) strip away the essential interconnections that defy any known methods of forecasting.

The *biosphere* in Vernadskian terms is not modelable.[1] This is why global environmental change will always remain a specter.

By introducing the metaphor *specter*, we imply a shadowy and elusive phenomenon whose indistinctness poses challenges to management response. It is the very indeterminacy of the concept, yet its very real meaning in the hearts of millions of people, that keep it high on the contemporary political agenda. If global environmental change could be *packaged*, it would lose its qualities for risk assessment.

What we are witnessing is a fascinating race between scientific indecision and political prevarication. On the one hand, there is a faith in science to promote, in sufficient time, enough formal understanding and degrees of forecastable accuracy to allow difficult and agonizing political and economic decisions to be taken with public (electoral) compliance. On the other hand, the inclination the world over, regardless of ideological persuasion, is to define problems in pragmatic terms isolating them from the wholes of history, culture, and power relationships in which they are embedded, so that some identifiable action can be taken without disturbing too many electoral harmonies.

To illustrate this point, let us look at the response to two aspects of global change, namely CFC elimination and $CO_2$ reduction. It is relatively easy to isolate CFCs as a prime cause of stratospheric ozone depletion. The politician in industrialized societies can focus on the private sector to come up with a technical-fix solution and make the consumer pay. It is much more problematic to fund poorer countries so that they can afford to invest in the much higher costs of alternative chemical formulations before they have even enjoyed the manifest benefits and the cheaper period of utilizing CFC-emitting technology. Hence, the stalling by the wealthy nations over a fund for CFC alternatives to compensate the Chinese and the Indians for not producing low-cost refrigerants for much-sought-after cooling apparatus. On the $CO_2$ front one has to magnify this dilemma to concerted political action, on a truly global scale, to reduce future energy demand to current levels by behavioral changes involving higher personal expenditures and a greater personal inconvenience. Risk management becomes a truly participatory requirement involving sustained and sympathetic commitment on the part of billions of consumers. It is little wonder that politicians are investing heavily in $CO_2$-climate warming modeling and associated panels and conferences. They seek scientific legitimacy before committing themselves to policies that will transform the economy and public behavior.

So the global risk management race lies between the *risk identifiers* and the management *solvers*. They are envisioned as running side by side as separate participants. In fact, they are one and the same contestant. The *race* is within us, not between artificially yet conveniently labelled antagonists.

Another example illustrates this point. More than two-thirds of tropical moist forests are nowadays being felled or burnt by landless families, half of whom have no history of forest management or agriculture in such areas.[2] The other half are ecological refugees forced out of desiccated former forestlands by the very drying process that their earlier clearance has helped to create. This climate impact research is being coordinated by Salati and his colleagues.[3] To establish a forest protection fund by *debt-for-nature* swaps or through some kind of leaseback arrangement using the secondary banking market of discounted debt repayments will merely displace the problem elsewhere and make it worse.[4] The supply of the landless will not be reduced by forest protection funds or by tree planting to compensate for new fossil fuel burning power stations.

Arguably, tropical forest removal is one of the greatest contributors to global environmental change. Yet, the problem is still primarily addressed as a management issue, namely how to protect the best of the remaining stands of *pleistocene rainforest*, those with the maximum of diversity of species, or how to create *sustainable tropical forests* in a simplified ecological structure. This is certainly the view of the British government. Its former overseas Development Minister, Chris Patten, was widely credited with an environmental diplomatic coup when he signed a deal with his Brazilian counterpart along these lines, emphasizing the role of ecological research linked to ethno-ecological history in the improved understanding of how to create and manage a semi-natural (secondary) tropical forest on a sustainable basis. Mr. Patten was rewarded with the Environmental portfolio in the Thatcher Cabinet for his justifiable achievements. But the solutions he proffered go no way toward dealing with the tropical forest if the relationships that give rise to tropical-forest depletion outlined in Figure 5.1 are to be believed.

The real management solutions to the tropical forests lie in changing the structure, incentives, and purposes of international finance, in arms trading arrangements, in corporate investment in developing countries, and in the corrupt power structures of many Third World governments whose leaders steal global wealth for deposit in biospherically meaningless havens.[5] All of these forces, coupled with much international aid, actually reinforce the destruction and landlessness that give rise to tropical rainforest predation and colonization on a scale that is beyond policing or education. Yet, one will never find even the most enlightened environment minister talking such language.

## THE CHALLENGE OF RISK MANAGEMENT

The concept of *risk* has evolved over the past twenty years in three important respects.

*From Externalization to Internalization.* To begin with, risk was judged to be an identifiable physical hazard, whether man-made or natural, that was external to the human condition and for which some form of externalized *blame* could be attributed. This was implicit in much of the early work on natural hazards research where *coping* strategies of adaptive response were sought in the face of an externalized physical event that impinged on societies and institutions. It was also implicit in the early work on risk perception which emphasized the physical concreteness of danger (guns in violent hands, electromagnetic radiation from power lines, reformulated organisms from genetic engineering) to which people in different economic and political contexts might be expected to respond.

*Vulnerability and the Abuse of Power.* This was followed by the work of political theorists and anthropologists who sought to encapsulate risk in the political culture of history, social relationships, and political justice. They saw the distribution of risk as disproportionately dependent on whether people were able to control the environmental conditions of their existence or whether they were not in such a fortunate position. Thus, drought adaptation strategies of multiple cropping in a variety of geographical areas coupled to interfamilial or intertribal reciprocal arrangements to provide a culturally sustaining *safety net* would only work in the context of one set of power relationships.[6] Shift those relationships to dependency on external aid technology or arms trading, couple it to the opening up of cash payments, taxation, and the fencing of *common* land to endure both administrative management of resources and resource use, and these stabilizing and sustaining relationships will be shattered beyond cultural repair. The drought became a hazard, not because it was meteorologically any different, but because its relationship to society and economy drastically altered. Vulnerability came to be imagined as a human relationship with the surrounding world, not a physical relationship. Hewitt expands this point with case examples.[7]

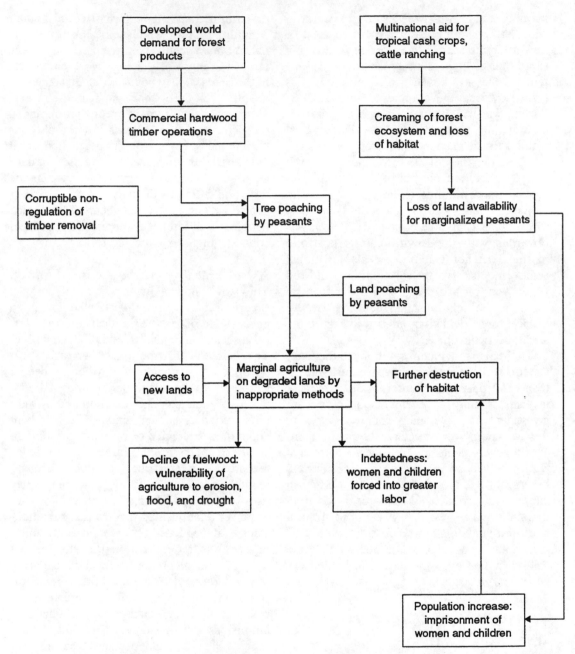

**FIGURE 5.1** The Tropical Forest Depletion Scene

***Risk Communication and Empowerment.***
This third stage of risk management has been
dominated by how best to communicate danger to the suspecting or unsuspecting so that
a *best* management mix can be achieved. Be-

ginning with signal theory and fairly unsophisticated communication idioms, recent work
has attended the lessons of phase one and two
of the risk management experience by looking
at issues to do with power, social relation-

ships, and innovative mechanisms of providing people with a sense of control over the risks they believe they face. The work of Roger Kasperson and his associates,[8] together with various commentaries, have begun a constructive debate on the changing role of risk communication.[9] This means, in some instances, providing independent risk mediating arrangements (risk ombudsmen, community risk assessment officers, community risk liaison panels and the like) that give a degree of delegated power to affected communities and allow them to present their view of appropriate safety measures in ways which they believe are culturally appropriate (i.e., suitable, given their social mix and political relationships).

Changing the bias of power changes the way in which actors think and act. The tropical forest will be saved only if local people are allowed to believe and to conduct their affairs as though they have a stake in its future and the future of that forest is important as much for their grand-children as it is to them. Without resource entitlement, by which we mean ready access to the basic necessities of survival (fertile soil, potable water, sustainably generated energy at reasonable cost, shelter, health, and education), they are not enabled to think or to act beyond the present. Understandably, they are not in a position to have a stake in their future enjoyment of the natural assets of soil, water, and forest upon which the livelihood of their children depend. By the same token, communities without risk-protection entitlement, that is, a degree of effective control over safety and post-accident compensation, are not enabled to respond to the threat or the reality of radioactive waste disposal or a toxic waste repository in their neighborhood. Modern risk management is about thinking right and acting right, and that means reorganizing the basis of power and institutional responsiveness to allow a collective view on a solution to emerge through enlightenment and legitimate discussion and political action.

## GLOBAL CHANGE AS ENVIRONMENTAL RISK: TOWARD A HOLISTIC INTERPRETATION

We examine the conditions that are generally held to constitute global environmental change and then look at a possible typology of global change for which different risk management strategies would be appropriate.

◆ Global environmental change is caused primarily by human activity, superimposed on any underlying organic evolution of biogeochemical processes.

◆ The effects of these changes are expressed, in a variety of ways, globally. This means that they have direct, or all-encompassing, implications: or the cost of trying to alter the changes is so great as to affect the viability of the present or future global economy.

◆ The rate of change is so rapid as to be identifiable within a feasible political lifetime. This means within 50 years when decision makers responsible for today's choices will be sufficiently remembered (or in some cases, even still alive) as to be held culpable. There is no such excuse as *not in my term of office*. For a politician, the removal of such a convenient excuse is a major consideration.

◆ The scale of change is so pervasive and so persistent as to be visualizable as being potentially irreversible—or at least so expensive to remedy as to be economically crippling.

These principles do not take us very far even in the realm of climate perturbation, which arguably forms the core of global environmental change. So we need to add two further dimensions. One relates to this *wholeness* of earth's physical and social processes—referred to as *biospheric* processes if you are a Vernadskyite, or as *gaian* processes if you are a follower of James Lovelock.[10] The other implies to the *levels* at which change can be seen to occur, where the notion of *level* has both an environmental and a geophysical meaning.

It is arguable that humanity will eventually come to accept a biospheric wholeness to the workings of this planet. This is a matter of science and instinctive intuition. To believe in the biosphere, we must couple science with faith. These two dimensions of human understanding of the cosmos have always been part of a totality, but for two hundred years many conventional scientists have sought to separate the two. For global environmental change to be understood adequately, science and faith will have to be reconnected.

Figure 5.2 attempts to put together the links that could allow visualization of the earth as something more than a sum of biogeochemical processes. It also seeks to provide for risk management a perspective that takes us beyond even stage three of the evolutionary pathways in risk management postulated earlier.

*Homonology* is from the Greek words for interdependent systems of trophic self-reliance but intertrophic dependence. A microbe is part of a myriad of fellow microbes. Each microbe acts to carry out a task: that task may be to promote collective well-being, for example, acting as antibodies against foreign microbial invasions. Yet the collectivity of microbes may be acting to provide energy or materials (carbon or sulphur, for example) for other trophic kinds of organic or inorganic life. Thus, the sulphur flux of the ocean surface is microbially determined, biogenic in

character, yet responsible for cycling an inorganic mineral from regions of surplus to regions of deficit. Thus, homonology is the totality of relationships that make the biosphere tick.

The subatomic equivalent to the biosphere is the *superforce*—the unity of forces acting at the level where energy and mass are interchangeable to produce a homonological pattern of energy and mass relationships. We really do not yet know how far this *micro-biosphere* works; nor, indeed do we know how it relates to the *macro-biosphere*. But faith and instinct tell some of us that they are part of a totality of unimaginable simplicity and complexity.[11]

At the margins of the macro- and micro-biospheres is *chaos*, the relatively recent recognition that unpredictability is for real and that minor changes in initial states can lead to unanticipatable outcomes.[12] The beauty of the emerging chaos theory lies in its combination of a paradox of wholescale unpredictability with a belief (faith) that there is some kind of order in randomness. Patterns can be found in the uncertainties, yet for the most part these patterns defy explanation on conventional statistical or mathematical grounds. Chaos, therefore, tells us that meddling with biospheres cannot be measured with

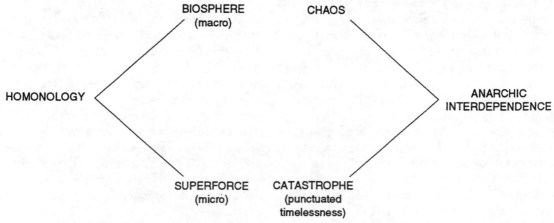

**FIGURE 5.2** A Holistic Representation of Risk Management and Global Environmental Change

any surety, and that every run of every model, should any input variable be altered even at the third decimal place, will produce an unexpected outcome.

Chaos is difficult enough to contain in physical systems. Yet it works just as effectively in human systems, probably even more so. If scientists cannot be sure of turns of events with infinitesimal changes in input variables, imagine how risk managers must feel about tampering with human response where the vagaries of political inputs are the equivalent of an order of magnitude. Who would have predicted even two years ago a linkage in Soviet foreign policy between ecological sustainability and limitations on international arms spending and trading? Who would have thought, even last year, that the Group of Seven would devote half a communique to environmental sustainability and the control of global environmental change? Admittedly, all this is still at the level of rhetoric, but political attitudes *are* changing rapidly and unpredictably. Electoral dissatisfaction at the slow rate of political response is beginning to show up in voting patterns of these democracies where proportional representation is the norm.

Completing the discussion of Figure 5.2 is the introduction of the concept of *punctuated timelessness*. By this we mean the recognition that the biosphere is for all interests and purposes timeless. But perturbations do take place, and *Gaianists* talk of modulated Gaian phases (pre-oxygen; post human) where catastrophic flips alter the phase states but the principles of eonic homeostasis still hold good. This suggests that states maintain fluctuating dynamic equilibrium over eons, altering catastrophically on occasion, but retaining the basic objective of evolving life on earth.

Punctuated timelessness allows us to consider the positive virtues of catastrophic danger—the threat of cataclysmic change so great as to remove most, but crucially not all, species. The key to catastrophe is that the biospheric linkages between the organic and the inorganic are never broken: no matter how much of the organic is lost, the scope for life restoration via the metamorphosis of the inorganic to new organic forms remains secure. Catastrophes can, therefore, be postulated as a cleansing process, rather like a liver diet where only fruit is consumed for five days to clear out unwanted residues in an overworked organ.

There are plenty of examples of extreme and dramatic environmental perturbation cleansing and rendering healthy *diseased* ecological states. For example, fire is essential to maintain heathland, moorland, grassland, and certain species of conifer, notably the Douglas Fir. Wholesale removal of vegetation, in a regime adapted to irregular extreme conditions, purifies this seed bed and prepares the ground for successful recolonization. Similarly, many rivers do their real erosive and sediment-removal work in the 0.1 percent of the time they are in catastrophic flood, leaving 99.9 percent of the time to reorder the *mess* the flood created.

So, maybe, this is the case with global environmental change. Maybe this is the *crisis* humanity really needs. Maybe humanity will see it for real, a cumulative set of processes that could irreparably alter the habitability of the globe. Maybe this is the challenge that the modern breed of risk managers is really awaiting: A foreseeable peril of such magnitude on such a relatively short time fuse and where issues of social justice and resource entitlement finally have to be faced and dealt with if the crisis is to be avoided. Maybe, just maybe, global environmental change for all its indefinability is the ultimate risk that will make risk management work.

If that is the case, then Figure 5.2 closes on the right hand side where it opened on the left. This may lie in seeking management solutions that address questions of redistribution of power, or control over one's environmental futures, via a combination of anarchic self-reliance at the personal and communal level (along the lines initiated by the Chipko movement in the Himalayas[13] which connects individual action from a localized, commitment to a meaningful response to global change). Here is where Rayner's notion of *invisible networking* is relevant.[14] Actions for global change may have to come about outside formal governmental initiatives and laws. People will begin to do things *instinctively* because they know, or feel, it is right for the biosphere. Hence, our use of the phrase *anarchic interrelatedness*, where anarchy means beyond governmental action. Globally speaking, responsive action has to be felt, not forced; the joy lies in sensing, not proving, that others, possibly billions of others, are feeling the same commitment and acting according to culture and environment in an appropriate way. This gives a sense of joy and liberation to globally sustainable behavior. We suggest a turnaround of the well-known phrase *think globally, act locally* to *act globally, think locally*.

Coupled with this refreshing networking would be the introduction of wholesale new international arrangements for corroborating across natural boundaries in order to safeguard the biosphere of which humanity is a part, but only a part. So we can imagine carbon levies, resource-depletion taxes, cross-border Superfund arrangements to pay to clean up yesterday's mess, and international liability arrangements that endure environmental good performance bonds on all users of the global commons, whether *renewable* or *non-renewable*. These ideas have recently been well developed by the World Resources Institute. One challenge is to make codes of environmental good practice internationally enforceable with international means of proving liability and collecting compensation even when present generations do not demonstrably lose but future generations could. Another challenge is to persuade the drafters of such codes, as well as the practitioners, that the indeterminacies of prediction must be couched in terms outlined in Figure 5.2 and subsequently understood.

# A POSSIBLE TYPOLOGY OF RISK MANAGEMENT AND GLOBAL CHANGE

All the discussion so far may be acceptable as an abstract intellectual exercise, but it lacks realistic analysis and illustrative case studies. To become more specific, we seek to identify a typology of global environmental change to which various risk management strategies might apply. All of this must be set in the framework provided in Figure 5.2.

We propose four *levels* of global environmental change, where the concept of *level* suggests scale of alteration and character of societal response.

*Biospheric Flip.* Here we are talking cumulative and aggregative onslaught either already at work or imaginable within our present generation of perturbations sufficient to cause a human-induced catastrophic Gaian *flip* or change of phase. As yet no one is talking precisely of this eventuality, but it is the highest level at which global environmental change could be postulated. Just as people living next to a dreaded facility believe that it will malfunction, so a growing number of people, at least in electoral terms those who actually support green parties worldwide, believe that the biosphere could malfunction as a result of human-induced change. They feel that outcome instinctively and through cultural and social normals of day-to-day communication. They are skeptical of the delivery of expertise, though they recognize that without scientific grounding concerted action is unlikely. This link between scientific positivism and culturally determined *extended facts* is reasonably well-known in the risk management literature.[15] It is also a relationship that has been well canvassed over the years. It is an important contribution that the risk management community can make to global change studies.

*Climate Perturbation.* For a number of

analysts, climate change is global environmental change. Anyone reading the useful survey of global change published by *Scientific American*[16] will readily see that climate perturbation is a subset but only a part of global environmental change. Nevertheless, climate change is of sufficient specificity to be regarded as a risk arena in its own right. This is partly because so much atmospheric chemistry and global kinetic energy is locked up in climate patterns, and equally because climate oscillations have actual and potential economic repercussions, (both positive and negative) of major dimensions.

*Undermining the Provision of Basic Needs.* We have defined basic needs in terms of adequate soil fertility, water, and energy to provide food and shelter for survival together with enough surplus to allow for time to be educated, to act communally rather than individually, and to behave generationally rather than for today. Basic needs go beyond mere survival: they apply to sustainable survival. Hence, a manageable surplus for reciprocal use and for personal and social development must be incorporated. Global change occurs if such basic needs are not met for such a large proportion of the population that the cumulative outcome is degradation of ecosystems whose health is essential for the health of the globe. Here the change is not definable on an *ad hominem* basis or even possibly at the level of the community. Risk management needs to address the problem in terms of aggregations of actions undertaken by individual actors who cannot see, or who are not allowed to see, the total or intergenerational picture.

*Micropollutants with Chaotic Long-term Environmental Consequences.* Possibly the least understood feature of global environmental change is the geobioaccumulative properties of persistent and toxic micropollutants that emanate from dispersed (and un-

controllable) sources via concentrations at or below the level of detection. These can then unpredictably (chaotically) accumulate or magnify, in food chains or air/ocean currents or in rainfall and soil-forming processes in such a way as to become catastrophically hazardous. Many volatile organic compounds fall into this category, as do trace quantities of heavy metals and the polychlorinated compounds. No one can predict their long-term consequences because there is neither the monitoring apparatus, nor the ecotoxicological models to provide even the crudest estimates. In any case, the pattern of their accumulation may be truly chaotic: in which case their peril is unimaginably unpredictable.

One can begin to see, therefore, that global environmental change is not only elusive. It acts at various levels where biogeochemical processes and human agency play crucially differing roles. Yet the overarching difficulties, stated at the outset, of wholesale unpredictability, of action and reaction acting interchangeably but treated as analytically separable, and of unwillingness or incapability to contemplate the fundamental interconnections when identifiable and politically manageable segmental cause-and-effect relationships can be politically more comfortably treated—all these remain to plague risk managers at all four levels.

## DECISION MAKING CULTURES AND GLOBAL RISK MANAGEMENT

Elsewhere, Rayner,[17] following Ravetz and Funtowicz, has suggested a typology of three kinds of science that can be applied to different levels of decision *stakes* or scale of implications for society and economy if decisions turn out to be wrong, and different degrees of uncertainty in scientific knowledge. Figure 5.3 depicts this typology. At low levels of decision significance and uncertainty, con-

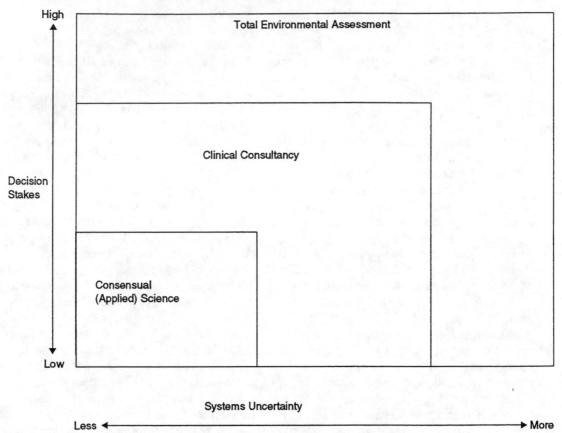

High

Total Environmental Assessment

Clinical Consultancy

Decision
Stakes

Consensual
(Applied) Science

Low

Systems Uncertainty

Less ◄──────────────────────────────► More

**FIGURE 5.3** *Three Kinds of Science (Source: J. Ravetz [1986])*[18]

sensus comes relatively easily, and much scientific work is of the *follow-on* kind of applied and programmatic investment. Examples include drinking-water treatment and noise-reduction technology. At an intermediate level of decision significance and systems uncertainty, consensus is more difficult to achieve. Much has to be done through clinical or experimental research where discovery by carefully monitored trial and error becomes the mode of proceeding. The highest levels, where uncertainty is great and the consequences of misunderstanding or failure are serious, require the introduction of different aspects of science. These include ethical positions, consensus formation by public involvement and organized public support programs and, above all, a new sense of the concept of *trust* in de-

cision procedures. The summary of these positions appears as Table 5.1.

The issue of public *trust* is of crucial significance. It relates closely to the concept of anarchic interdependence outlined in Figure 5.2. Trust is felt through a combination of experience and expectation. The world of familiarity, upon which all cultures depend in order to cope with the myriad everyday relationships, relies upon a spontaneous and inherent trust for its adaptive functioning. That trust operates through the person, based on a combination of knowledge and value judgement. It also applies through the cultural and political norms and institutions that frame the relationships of a person and a culture with everything else, including nature. These management systems function because people

**TABLE 5.1** Strategies for Risk-Related Problem Solving

| Level | Example | Procedures Applied | Criterion for Decision-Making | Problem | Purpose of Communication |
|---|---|---|---|---|---|
| Technical problems | Safety engineering | Rules | Evidence | Confidence in information | Information transmission |
| Clinical problems | Medicine Site selection | Skills | Experience | Credibility of experts | Mixed |
| World view problems | Societal choice of energy systems | Ethical standards | Argument | Trust in decision process | Creating shared meaning |

expect them to function properly. People do not need to know what others are doing: they simply have to feel confident that others are acting with the same sense of trust that the *systems* will work in such a way as to improve social well-being. That *system* could be a government, or a nongovernmental organization, or a network of presumed similar behaviors.

Luhmann has suggested that trust provides a recourse for linking an innate sense of present certainty to an equally innate concern for future uncertainty.[19] The bridging connection is a faith in a shared belief in a *right* way of behaving that fuses the science of verification and positivistic modeling to the instructive or felt *science* of experience and expectation. This is why modern risk management has potentially so much to offer. It has a track record of connecting the two elements of scientific endeavor to redesigned institutions responsible for forecasting and coping with uncertain outcomes, in which successful achievement relies as much on the trust of personal commitment to a universal ideal of biospheric health as it does to a willingness to accept consensually imposed new norms and regulations. This in turn means new ways to reconstruct institutions that develop and rely upon trust in a shared sense of a life world, and in the patterns of government necessary to achieve it.

Table 5.2 summarizes three decision making cultures that, in part, may serve to assist this crucial process of institutional adaptation.

The first is a *market culture* that relies on personal level of trust and cooperation and proceeds from the expectation that everyone else will be thinking and acting to promote shared ideals. The table suggests how such a culture would operate with regard to criteria that are relevant to risk management and global environmental change, notably those to do with procedures and consensus.

The second cultural norm is *hierarchical*, based on a recognition of systems trust. Here the systems of government, of regulation, of education, and of economic investment, to name but a few, can operate on society as a whole only if everyone affected accepts and is willing to live by the hegemony of that systemic action. This in turn means adequate measure of system accountability and responsiveness, based on agreed rules, participatory and educational procedures, and shared meanings.

The third level of culture lies between the two. It is *collectivist* in the sense that it rests on common action, but it is also *egalitarian* in the sense that deployment of justice, fairness, and, where necessary, redistribution allows those otherwise unable to gain access to basic needs to do so in order to become true global citizens. Here the rules rest on participation, equity principles of sharing, and agreed codes of redistribution, and cooperation lies in a fusion of both personal and system variants of trust.

We are not quite out of the woods yet. These different interpretations of decision

**TABLE 5.2**  Summary of Three Decision-Making Cultures

| Decision-Making Culture | Markets | Hierarchies | Egalitarian Collectives |
|---|---|---|---|
| Sovereignty | Consumer | Institution | Collective |
| Applied procedures | Skills | Rules | Ethical standards |
| Criteria | Experience | Evidence | Argument |
| Consent | Revealed | Hypothetical | Consensus |
| Liability | Loss spreading | Redistributive taxation | Strict fault |
| Trust | Successful individuals | Procedures | Participation |
| Time depth | Short | Long | Compressed |
| Future generations | Self-sufficient | Resilient | Fragile |

making and of trust can be joined into new institutions for bestowing authority on individual and collective action. Here we acknowledge again the work of Funtowicz and Ravetz.[20] They explore three types of uncertainty: the *technical* (based on disputes over observation and interpretation of conclusions), the *methodological* (based on the principles of discovery), and the *epistemological* (applied to the very basis of knowing and organizing knowledge). Table 5.3 suggests how each of these three variants of uncertainty can be applied to the manifestation of authority and the style of problem solving.

At the technical level, authority comes from evidence and instruction, and problems are solved in a reductionist manner. At the methodological level, authority derives from experience, and problem solving is experimental and pragmatic, with scope for monitoring and reassessment. At the epistemological level, where indeterminacy reigns, faith dictates authority and problem solving operates biospherically or holistically. Trust of both a hierarchical and a collectivist kind is especially important here.

We are now in position to combine the four levels of global environmental change, based crudely on the system states outlined in Figure 5.3, with the three forms of decision culture applied to the emerging concept of trust summarized in Table 5.2 and Table 5.3, but very much dependent on the themes raised in Figure 5.2. For ease of clarification, we present our results in the form of two tables. Table 5.4 is the *clean* version, and Table 5.5 the more *complex*.

Table 5.4 suggests that for each of the types or levels of global change, important differences in orientation, in regulation, in type of communication, in economic restructuring, and above all, in political power redistribution, will apply. The most problematic differences are at the top and bottom. Because few, if any, politicians in power see global environmental change in Vernadskian or Lovelockian terms, the idea of addressing the *problematique* as a *whole* risk issue has not received serious attention. Such an approach would probably require genuinely unconventional styles of risk management where faith would come before science, where a Gaianistic ethos placing humans within the totality of life-protecting processes would prevail, where meanings transcending time and present-day human understanding of the *value* of the globe would become the discourse of risk communication, and where major efforts at redistribution of wealth, opportunity, and control over resource use would prevail. Nothing on the political horizon suggests that any of this is under serious consideration. But wise risk managers should develop paradigms that allow for it because both chaos theory and biospheric analytical perspectives allow for its manifestation in a foreseeable future.

The derivation of a risk management strategy of biospheric homeostasis would be a formidable task. All of these risk management approaches would have to be redeveloped or at least extended. We would need to devise more flexible mechanisms for calculating probabilities and outcomes across multiple physical, biological, and social dimensions. Given an absence of ready devices for allocating blame, we would have to work on the basis

**TABLE 5.3** Conditions for Authority and Styles of Problem Solving

| Type of Uncertainty | Conditions for Decision Making | Prerequisite for Authority | Source of Authority | Problem Solving Style |
|---|---|---|---|---|
| Technical | Risk | Information | Instruction | Reductionist |
| Methodological | Uncertainty | Respect | Experience | Pragmatic |
| Epistemological | Indeterminacy | Faith | Revelation | Holist |

**TABLE 5.4** Global Environmental Change and Risk Management Strategies (simple)

| Level of Change | Criteria for Decision Making | Application and Approaches | Type of Discourse | Economic Requirements | Political Requirements |
|---|---|---|---|---|---|
| Biospheric | Faith first, science second | Gaianistic ethos | Creating shared meaning | Common pools of cross-national funding | Global institutions |
| Climatic | Faith and science equal | Argument over evidence | Creating shared action | Specific taxes and management funds | Multinational institutions |
| Basic Needs Plus | Science first, faith second | Evidence and experiment | Executing shared action | Aid and action funds | Bilateral institutions working collectively |
| Micropollutants | Metamorphosis of faith and science | Intuition and foresight | Transcending meaning with advance action | Regulation at source irrespective of cost | Science and technology under political regulatory control |

**TABLE 5.5** *Global Environmental Change and Risk Management Strategies* (complex)

| Level of Change | Criteria for Decision Making | Application and Approaches | Type of Discourse | Economic Requirements | Political Requirements |
|---|---|---|---|---|---|
| Biospheric | Faith first science second | Gaianistic ethos | Creating shared meaning | Common pools of cross-national funding | Global institutions |
| (Worldview) | Collectivist holist | Collectivist holist | Collectivist holist | Collectivist pragmatic | Hierarchical holist |
| Climatic | Faith and science equal | Argument over evidence | Creating shared action | Specific taxes and management funds | Multinational institutions |
| (Clinical) | Market pragmatic | Market pragmatic | Collectivist pragmatic | Hierarchical pragmatic | Hierarchical holist |
| Basic Needs Plus | Science first faith second | Evidence and experiment | Executing shared action | Aid and action funds | Bilateral institutions working collectively |
| (Technical) | Reductionist hierarchical | Reductionist hierarchical | Pragmatic collectivist | Pragmatic hierarchical | Collectivist pragmatic |
| Micro Pollutants | Metamorphosis of faith and science | Intuition and foresight | Transcending meaning with advance action | Regulation at source irrespective of cost | Science and technology under political regulatory control |
| (Technical/Clinical) | Market holist | Market holist | Collectivist pragmatic | Reductionist pragmatic | Collectivist pragmatic |

357

of adaptive management within a context of transferring control over resource futures to more self-sustaining communities, yet within viable international institutions.

This is possibly why the climate-change components of global environmental change offers so much promise. It is a specific and identifiable subset of problems and outcomes to which more conventional risk management approaches can be applied. Yet, in every case, innovative, even radical, variations of risk-management strategies are needed. The main challenge would be to devise a communications arrangement that would create shared meaning over cause and outcome, to which the application of new approaches to carbon taxation and subsidies for low impact energy technologies would operate through multilateral institutions such as the World Bank or some new climate fund of the kind advocated by Gro Brundtland.[21]

The *basic needs plus* level is the core of sustainable development. We are nowhere near achieving that goal but it is at least the most manageable in terms of technical possibility and economic realism. What is still missing is that level of discourse, which is essentially biospheric, that enables politicians to realize that deprivation of basic needs plus constitutes a loss to global survival, not just to individual or community survival. This is why the two levels above are at least relevant for providing essential context for this most urgent, and arguably most solvable, aspect of the global problematique. Part of the key is the link between nonsustainability (or basic needs minus) and military instability on the one hand, population growth on the other. Try as they might, analysts cannot prove a connection among these three. Here is where scientific principles of evidence gathering and responsible production are required because it is only upon evidence and experiment that bilateral agencies and voluntary organizations will be able to show that redirection of resources from military spending to basic needs plus (i.e., including education, ethnic and gender rights, and contraceptive advice) will result in net gains for sustainable development and global stability.

Finally, we come to the awkward issue of multiple micropollutants and the possible implications for biospheric stability. There is no easy answer; not even faith is sufficient. Here we have to metamorphosize faith and science in a new variant of precautionary science that operates on the basis of intuition and foresight and that accepts that regulation must take place at the point of product design, not the point of emission. This means a shift away from toxic waste *creation*, not merely the achievement of nontoxic waste *discharges*. It may also come to mean that the cost of regulation, in the form of wholly reorganizing industry in certain sectors towards nontoxic waste technologies and wholly isolated toxic substance streams, could be exorbitantly high, beyond any convention cost-benefit trade-off. In turn, this might well mean that final application of science and technology operate within political control, where the new politics would encompass the community as a whole along the lines of Figure 5.2.

Currently, risk management of micropollutants aimed at reformulating industrial processes and waste streams is at a most embryonic stage. Maybe this special form of management cannot cross the political barriers that require scientific evidence, before uncomfortable decisions are taken, where experience must always precede foresight. We hope not, and our guess is that it will not. So society will need to develop a new risk management of foresight specially tuned to an arena of science and political action, the fusion of which is inescapable if major global change is to be averted before its discovery proves that we have acted too late.

Table 5.5 attempts to put all our arguments together. It seeks to show how risk management strategies should adapt to different compositions of institutional structure and procedure to match the changing nature of verifiable science and public trust. The extreme complexities of interaction among all the variables presented compound the complications.

Table 5.5 attempts to show how a richness of management strategy can help to marry the most promising of the risk-management legacy with the emerging challenge of global problem-solving. The message is surely that by concentrating on the generation of trust through careful experimentation with different structures and procedures, humanity at least stands a chance of adapting to global change in a purposeful, and (we hope) successful, manner.

## NOTES

[1] See Jacques Grinewald Sketch for a history of the idea of the biosphere, in Peter Bunyard and Edward Goldsmith (eds.) *Gaia: The Thesis, the Mechanisms and the Implications*. Wadebridge Ecological Centre, Worthyvale Manor, Wadebridge, Cornwall, 1988, pp. 1–34.

[2] International Union for the Conservation of Nature. *Tropical Forest Action Plan* IUCN, Gland, Switzerland 1987.

[3] E. Salati, D.E. Oliveira, N.O.R. de Schubart, F.C. Novacs, M.J. Dourojeanni, and J.C. Umana. Changes in Amazon over the past 300 years. In B.T. Turner, R.W. Kates, and W.M. Meyer (eds.), *The Earth as Transformed by Human Action*. Cambridge University Press, Cambridge, 1990.

[4] See World Resources Institutes *Natural Endowments: Financing Resource Conservation for Development*. WRI Washington, DC 1989.

[5] For a provocative analysis of this issue, see Susanna Hecht and Alexander Cockburn. *The Fate of the Forests*. Verso Press, London 1989.

[6] This topic is well aired by Michael Watts, On the poverty of theory: Natural hazards research in context, in Kenneth Hewitt (ed.) *Interpretations of Calamity*. Allen and Unwin, London, 1983, pp. 231–62.

[7] Hewitt, *op. cit.*, Note 6.

[8] Roger Kasperson *et al.*, The social amplification of risk: A conceptual framework. *Risk Analysis* 9(2), 1988 177–88.

[9] See Helmut Jungermann, Roger E. Kasperson, and Peter M. Weidemann, (eds.). *Risk Communication*. Kernforschungsanlage (KFA) Julich, West Germany, 1989.

[10] For discussion of the Gaia hypothesis, see James Lovelock *The New Ages of Gaia*. Oxford University Press, Oxford, 1988, and Elizabet Sabtouris, *Gaia: The Human Journey from Chaos to Cosmos*. Pocket Books, New York, 1989.

[11] See Paul Davies, *The Cosmic Blueprint: New Discoveries in Nature's Creative Ability to Order the Universe*. Touchstone Press, New York 1988.

[12] James Gleick, *Chaos: Making a New Science*. Penguin Books, London 1989.

[13] Vandana Shiva and Juyanta Bandyopadhyay, Chipko: Rekindling India's forest culture. *The Ecologist* 17(1), 1989, 26–35.

[14] Steve Rayner, Risk communication in search of a global climate management strategy. In H. Jungermann *et al. op. cit.*, Note 9, pp. 169–77.

[15] Silvio Funtowicz and Jerome Ravetz, The communication of quality of hazard information. In H.B.F. Gow and H. Otway (eds.). *Communicating with the Public about Major Accident Hazards*. Elsevier Science Publishing, Barking, UK, 1990.

[16] *Scientific American*. Managing planet earth. Special Issue 261(3); 1989.

[17] Steve Rayner, Muddling through metaphors to maturity. *Risk Analysis* 8(2), 1988, 201–4.

[18] Jerome Ravetz, Usable knowledge, usable ignorance: Incomplete science with policy applications. In William C. Clark and Ralph W. Munn (eds.) *Sustainable Development of the Biosphere*, Cambridge University Press, Cambridge, 1986, pp. 415–32.

[19] Niklas Luhmann, *Trust and Power*. John Wiley, Chichester, UK, 1979.

[20] Silvio Funtowicz and Jerome Ravetz, The arithmetic of scientific uncertainty, Joint Research Centre, Ispra, Italy.

[21] Gro Harlem Brundtland, Global change and our common future. *Environment* 31(5), 1989, 16–25.

# 28

# THE SOCIAL SPACE OF TERROR: TOWARDS A CIVIL INTERPRETATION OF TOTAL WAR

*K. Hewitt*

## WAR AND CIVILIANS IN THE TWENTIETH CENTURY

In the major conflicts of our time, civil society has been as fully engaged as the military. The greater part of modern mass armies is drawn from the civil population. When soldiers die, or are injured, there is a permanent loss to civil life. The fear of it is a major burden for those left behind. Conscription also profoundly alters the composition of civil society and its problems in wartime. Essentially, it leaves families, institutions, services, and economic activities to carry on without the majority of their able-bodied menfolk.

Industrialized warfare leads to ever more complete mobilizing of the resources and the workforce of nations. In relation to war or its threats, totalitarian methods of government are adopted everywhere, which bring complete control over civil life, and an introduction of military methods or of war-fighting attitudes in all institutions. Moreover, it allows war leaders to impose ever more of the risks and the stresses of war upon civilians.

Reprinted from *Environment and Planning D: Society and Space*, Volume 5 (1987), pp. 445–74. Used with permission of the author and Pion Ltd.

For such reasons as these, "total war" is widely assumed to be the usual, perhaps inevitable form of conflict in this century (Ludendorff, 1935; Earle, 1943; Liddell-Hart, 1967; Aron, 1955; Paret, 1986). During the Second World War, leaderships spoke of total war meaning the involvement of every last man, woman, and child: a war over the entire fate of peoples and continents. Nowadays, it is difficult to find a war, "hot" or "cold," in which the same language is not in use.

A singular development associated with all of this is the escalating threat of direct destruction for civilians, their settlements, and their habitat. In many recent conflicts, from the Spanish Civil War to Laos, East Timor, and Afghanistan, the majority of those "at the sharp end" of war have been civilians rather than soldiers. In part, that follows from the increased destructive power and the mass production of weapons. In part, it reflects the enormous expansion in the geographical scope of motorized warfare. Artillery, tanks, naval guns, rockets, flamethrowers, and air strikes can do colossal damage to land and life. So can the now formidable fire power of the infantryman—classically, the one "at the sharp end." All of these have done great harm to civil society, as a side effect of the clash of armed forces.

That side effect must be distinguished from deliberate assaults upon civil populations and nonmilitary areas. These have grown to unprecedented levels, whether in aerial bombardment of cities and villages, in scorched earth policies, or in environmental warfare. When unarmed populations and undefended areas are subject to armed violence one usually refers to this as *terror*, whether carried out by small, dissident groups or by state forces (Chomsky and Herman, 1979; Walzer, 1977).

For the West, at least, three wars stand out as revealing progressively deeper layers of threat and terror to civil society. And few events have caused a greater spread of fear and grief among civil populations. The First World War is distinguished by the annihilation of men; millions of them, in mass armies sent to do battle with industrially produced weapons (Howard, 1986). The men were nearly all civilian volunteers and conscripts. For the military this was "war of attrition." For civil life it can, perhaps, be summed up in the phrase "the lost generation" (Wolff, 1959; Fussell, 1975). It was not terror as I will use the term, but it did reveal the extraordinary capacity of modern states to mobilize, discipline, and destroy manpower in great numbers.

In the Second World War military casualties were much larger, but its special message lay elsewhere, in the mass destruction and uprooting of civilian populations themselves. Almost as many unarmed men and women, young and old, were killed or maimed by armed assault, as were soldiers. Violent death took about 16 million in this way (Urlanis, 1971). More than 12 million others were either deliberately killed, in "security measures" by forces of occupation, or died from the privations of war (Elliot, 1972). For similar reasons, tens of millions fled, or were forced from their long-time homes, many into permanent exile (Vernant, 1953; Proudfoot, 1957).

The military sense of these events is given in such phrases as "strategies of annihilation," "pacification programmes," or simply "total

war" (Weigley, 1973). For civil life it meant holocausts, genocide, and what I have termed "place annihilation" (Hewitt, 1983a). It was warfare that strove towards, if it did not always achieve, an end of the settled historic places that have been the heart of civil life, and an extermination of entire civil communities.

The USA's Vietnam war, the Second Indo-China war, reinvented, as it were, each of the forms of assault on civilians noted above, and elaborated them. There are many parallels between the experiences and literature of its veterans, and the aftermath of "The Great War" (Remarque, 1929; Lifton, 1973). Relatively, even greater destruction of settlements, civilian uprootings, and civilian casualties occurred than in the Second World War. The war's special achievements, however, lay in a systematic assault upon the land and habitat, backed by the scientific and industrial resources of the United States. This was a huge, deliberate, military strategy. It sought to destroy the living cover of the earth, the fertility of agricultural areas, and to exploit natural forces as agents of war (Westing, 1976; 1984). The means—biocides, napalm, heavy earth-moving equipment, aerial spraying, and bombardment—were unique products of recent research and industrial production. In military terms, the actions were a response to the supposed requirements of "counterinsurgency warfare" (Blaufarb, 1977; Shy and Collier, 1986). It was strategy intended to deny the enemy cover, resources, and friends in the countryside. From a civil perspective, it became the annihilation of *living space* or, more comprehensively, "ecocide"—an assault upon the very biological bases of survival (Lewallen, 1971; Bunge, 1973).

The unique significance of thermonuclear weapons is, of course, the ability to achieve, *and inability to avoid*, all these levels of annihilation simultaneously! To the extent that there is relative risk, it is the reverse of the old sense of war, with habitat, settlements, and civil populations more vulnerable than military systems. That too has its apparent rationale in modern strategy. For civil life it creates the permanent, overarching threat of "omnicide" (Somerville, 1985).

In this part, where I am exploring what a civil perspective on war involves, I will examine two aspects of the civilian predicament in recent years. Two examples from the Second World War provide the empirical basis. The choice is not entirely arbitrary. Unlike the aspects of the nuclear threat, Vietnam, or First World War raised above, the events and issues are rarely considered in the readily available literature. Yet, the aspects singled out seem peculiarly relevant to a human geographer or anyone concerned with "society and space" in war. They are, the enforced uprooting of long-settled populations, and the annihilation of urban places. Both processes, often interrelated, seem peculiarly to threaten the survival of civil life.[1]

To date, the Second World War was undoubtedly the occasion of the largest and the most widespread upheavals and devastations of civil life. And it included the populations, cities, and heartlands of industrial powers. However necessary the fight, and however satisfactory the Allied victory, in civilized and civilian terms, the war was an immense disaster. If there is a precedent, a "proving ground" for what civil societies will face in another great war, it was here, and not only at Hiroshima.

If the war, as a whole, saw unprecedented destruction and atrocity for civil life, the extreme projection of this came near the end, especially in 1945. In part, that reflected the state of exhaustion in many of the peoples and the armed forces involved. Mainly, it reflected the fact that the two materially decisive factors of the war, war production and mobilized manpower, only reached their peak late in 1944. This was so even for Germany and Japan [see United States Strategic Bombing Survey (USSBS), 1945a; 1946]. Additionally, many of the restraints that had kept the war somewhat less than "total," were abandoned in a fight to the finish.

Only then were the iron expectations, and the plans for total war with the newest weapons, realized. Moreover, it was the civil populations of Germany and Japan that felt the fullest impact. With their own armies overextended or decimated, and their leaderships increasingly in disarray, they came under intensifying direct assault by the greatest array of armed might the world had seen in action— the combined air, sea, and land power of the Allies. These civilians, their settlements and habitats, felt the full impact of the latest weapons of mass devastation.

## MASS UPROOTING

Modern warfare, "blitzkrieg" and its relatives, is so indiscriminately destructive and most civilians are so lacking in means of protection, that there is rarely any option but to flee from it. All of the campaigns of the Second World War thrust great masses of civilians before them (Scott, 1968, pages 153–156). Not content with that, war plans and the activities of occupying forces called for mass deportations, concentration, exterminations, and expulsions, or large-scale shipments of people as forced labour. The plight of these folk stands among the greatest calamities of our time. Moreover, a large fraction of those involved went through the same experiences twice or more. In Eastern Europe, for example, great numbers of civilians from all nations experienced at least two flights or deportations, as the fortunes of war shifted for and then against Germany.

Between 1939 and May 1945, more than 40 million Europeans, which included people from 21 nations, were made refugees (Proudfoot, 1956, page 32). That does not include people evacuated from, or bombed out of, their homes in the air war—probably a further 25 million in all. There were the further millions shifted about as forced labour. And there were the first five million German people who left Eastern Europe. They will be the focus of attention here. They represent the first wave of some 15 million Germans eventually expelled from Eastern Europe (Figure 1) (Vernant, 1953; De Zayas, 1977). These were those who fled the battlegrounds and atrocities of the Eastern Front (Thorwald, 1953; Toland, 1966).

## The Great Offensive

Few episodes involved more human misery than the flight of the residents of Eastern Europe who were in the path of the Great Offensive begun by the Soviet armies in June 1944.

And that is particularly true of that of the German nationals and ethnic German communities within and around the borders of Poland.

It was in October 1944 that Soviet troops had made their first thrust into German soil, near Gumbinnen in East Prussia. They had been quickly repulsed, yet reports of what happened in this brief encounter spread terror among the civilians who lived in the path of the Soviet armies. The ferocity of murder and pillage, stories of the burning of homes and the raping of women, convinced German folk they must, at any cost, try not to fall into Soviet hands.

To a great extent, what happened here was a culmination of the uncompromising hatreds and long trail of violence between Fascism and

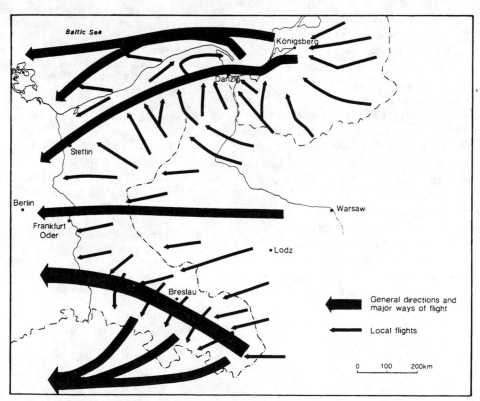

**FIGURE 1** Mass flight of the East German civilians during the advance of the Red Army to the Oder and Neisse, January–April 1945 (source: Schieder, 1953).

Communism, between Slav and German, as well as an already unparalleled story of atrocity since the German invasion of Poland in 1939. Soviet soldiers were encouraged by official propaganda to treat *all* Germans as enemies; to rape, loot, and kill them at will. By no means did all soldiers feel or act in accordance with such values (Schieder, 1953; Terkel, 1984), but enough did, so as to produce a nightmare for civilians in their paths. A key source is a long poem by a Russian soldier involved in the campaign, see Solzhenitsyn's *Prussian Nights* (1977).

After the October episode, East Prussian authorities evacuated some 600,000 civilians. This was itself a tremendous upheaval. To it was added a large unofficial flight of those who had come here as evacuees from the bombing of cities in the Reich itself. After that, however, officialdom discouraged, sometimes violently, further evacuation. The full import of that only emerged when the winter offensive began. Then, even such orders for evacuation as were issued, generally came too late for organized assistance.

On 12 January 1945 the Soviet armies began their drive to the River Oder. Between them and the river there were some 12 million German civilians. About five million would flee westwards as the offensive developed. A greater number were overrun. Hundreds of thousands already in flight were caught to be killed, abused, or deported eastwards to face years of forced labour. Some thousands, who felt they could not make the journey—especially old people but also many young women—committed suicide as the Soviets drew near. Hundreds of thousands died, on the treks westwards, of starvation, exposure, sickness, or in air, land, and sea attacks, or through other misfortune.[2]

### The Flight in Winter

From the "Documents of the Expulsions" (Schieder, 1953) a compelling portrayal of the civilian experience can be gained. The recollections of one woman from East Prussia encompasses the story of so many others that I will base my discussion around it.

The woman, her two small children and aged parents, were called to assemble in the town offices of Sensburg. It was January with heavy snowfall outside. The sounds of the Soviet artillery could be heard in the distance, their soldiers twenty-four hours away.

> It was (she recalled) a sad picture in the big hall of the Landrat's office. Aged people, sick people, lame people and children were waiting around, and desperate mothers tried to pacify their crying babies . . . (Schieder, 1953, page 137).

Official transport never materialized and they began their journey thus:

> When the window panes were shattered and the bullets whistled near, my father who was 74 years old had a fit of shrieking. Making up our minds with lightning speed we prepared the children, we took our ruck sacks and handbags and in spite of all ran to the highway. . . . It was icy cold and there was a continuous snow-storm—panting we dragged the preambulators (sic) through deep snow. . . (page 138).

The woman and her family were later assigned to one of the thousands of treks. She and her children sat in a covered wagon, but her parents were in an open sleigh:

> . . . my father was not able to endure the journeys because of the cold. As early as February 2, we had to leave him . . . at a village inn, which was full of wounded and refugees. He could no longer stand up. My mother remained with him. It was one-and-a-half years before I learned that my father had lived another nine days and then been put by the Russians in a mass grave. . . (page 138).

As with most of the treks, this woman's was cut off to the west in the rapid encirclement by the Soviet Second Army. She and her family were among 1.5 million German civilians trapped in the East Prussian pocket. Their only remaining hope of escape was via the Baltic Sea.

A German army unit then placed her in another trek going northwards, ". . . the leaders of which," she adds, "received us very unwillingly" (page 140). Like others, this trek was continually subject to aerial bombardment, to attrition from sickness and breakdown, and to a nightmarish level of uncertainty:

> The children were getting continually more tired owing to the cold and the small amount of food and did not want to come out of the cart. . . . They became ill with a dysentry-like diarrhea, which was called the 'highway illness.' We all became victims of this disease. . . (page 140).

Finally, they reached a geographical feature of very special meaning for any who experienced or heard of these events—the *Frisches Haff*, a body of water ponded behind a coastal spit some 100 km long. Today's atlases called it the *Vislinskij Zaliv*. To reach the coast, in hope of being evacuated by ship, the treks had to cross the frozen surface of the Frisches Haff. Many wagons and horses went through the ice, often through holes where Soviet aircraft bombed it to prevent movement. The woman recalls looking out from the shelter of her cart,

> . . . over the wide extent of the Haff and the dark grey sky of night spreading over it. Occasionally, the way was indicated by torches. Then one could see the endless rows of the treks, which were proceeding at long intervals in silence and with inconceivable slowness . . . (page 140).

Even when she did reach a staging point for official evacuations, her problems were not over and in some ways became worse:

> The roads inside the camp were indescribably dirty, my children lay ill in straw in the hut. The National Socialist Welfare organization was a complete failure. *Only people who were alone and healthy*, could, with any chance of success, queue up the whole day for bread and watery soup. I could not leave the children so long alone and no longer had any utensils to fetch food in . . . (page 142, my italics).

## Civilian Predicaments

It is appropriate to pause and consider some general features of this woman's experience. As noted, the German authorities, once so determined to control every aspect of civil life, failed nearly everywhere to order evacuations in time or to provide organized assistance. They feared the effect upon "morale." As so often in "civil defense," officialdom viewed the plight of civilians through the eyes of those preoccupied with the military situation and the problems of their own bureaucracy.

Second, and as a result of official mishandling, the majority of the civilians had to flee, in haste, as the last moment. They fled as families and individuals. They lacked the provisions which were adequate for the journey, or anything but the vaguest notion of where they were going.

The journeys, even for persons who got away by train, were under conditions of extreme hazard. They taxed the survival abilities of well-equipped fit adults. But then there were many on foot, travelling in severe cold, over icy ways and deep snow or, later, through the equal miseries of the thaw and the cold rains. Constant hunger, thirst, and illness made a life-threatening situation for all, even without the strafing and bombing they endured and without playing hide-and-go-seek with Soviet troops and Polish partisans.

One must emphasize that the majority of those in flight were women, commonly with children and aged relatives. Young women led many of the treks which were formed out of those forced to flee on foot and in carts. As a woman on a later trek described it, ". . . we were defenseless, for there were practically only women, children and very old men in our column. . ." (Schieder, 1953, page 293).

In the whole episode, the mortality of the elderly was very high: higher even than in the bombing war on German cities where it was disproportionate. To be old when a modern war begins, is a singular calamity—unless, perhaps, you are a president or a general! An unknown number, but certainly several hundred thousand old people died, in all the treks and expulsions from East Europe at that time, out of a total death toll estimated to exceed 2 million (Schieder, 1953; Elliot, 1972; De Zayas, 1977).

This woman's plight, being responsible for children and aged parents, provides evidence of something which has quite general significance for a civilian view of war. War or no war, the everyday requirements of nurture, health, and caring for dependents go on. Governments may make special arrangements to assist those responsible for such needs. Many people do, in fact, recall wartime as one of unusual "togetherness," of unlooked for kindness and caring; as often appears in natural calamities too. Yet, the conditions of war—the draining away of manpower and resources at home, no less than enemy blockade or assault—put the needy in special jeopardy. And when all assistance fails, as it so easily does in a war zone, only those unencumbered with dependents will readily move away and find sustenance. Women with small children, the old, and any who have care of the sick will be at great risk.

Like the woman's husband, 1.5 million soldiers had been conscripted from these eastern regions into the German army. Then, in June 1944, Hitler had ordered a mass induction of all German menfolk up to the age of sixty-five, as labour to build fortifications. In October, these men, and any others who could be found, were pressed into the so-called *Volkssturm*. Apart from the doubtful military value of either the fortifications or these untrained forces, their absence greatly increased the helplessness of, and harm to, their families who were forced to flee across the war-devastated winter lands.

After more journeying, more sickness, and more air attacks, this woman and her children were among those who reached the sea. They were placed in an evacuation ship and brought safely to Denmark. Thus, we know her story.

## The Baltic "Dunkirk"

The seaborne evacuation of these people from the Baltic ports, between January and May 1945, was the most astonishing, and in this case, most organized evacuation of the war. It was an evacuation under the command of the German naval authorities but largely against the will of the National Socialist leadership. It moved twice as many soldiers and wounded as the Dunkirk evacuation of June 1940 (Table 1). But the number of civilians moved was *four times* as large. In all, nearly two million persons were transported to the Reich and to other Baltic states.

The ports, and ships tied-up in them, were constantly bombarded from the air, and often from land and sea too. The sea lanes, ice-infested and shrouded with fog, were alive with Soviet submarines and other warships. Communications were an appalling mess. Evacuation needs inevitably took low priority compared with the needs of the military calamity that was developing to the south, and in western Europe. There was virtually no German air cover, and very little assistance from their own warships. Vessels which were capable of carrying the numbers involved were mostly old liners, troop, and hospital ships. If they were sunk, there was little chance of rescue, certainly not within the survival time of people in waters with temperatures near freezing point. In this and in other ways, the shore and maritime conditions of the Baltic rescue suggest the sort of stress and losses that an organized evacuation could face during a nuclear exchange.

Again, one should remember that most of those who were drowned were civilians, and mainly women, children, and the elderly. Brustat-Naval (1970) gives a blow-by-blow account of the people, the ships, and the conditions. He describes how the trekkers arrived at the seaports already in a wretched state; the bombed-out chaos of the ports; the desperate wait for a ship by hungry, thirsty,

**TABLE 1** German Naval Evacuations from the Eastern Baltic Sea, January–May 1945 (source: Brustat-Naval, 1970).

| Port | Fleeing Civilians | Wounded | Soldiers | Totals |
|---|---|---|---|---|
| Libau | 500 | 19 717 | 31 215 | 51 432 |
| Memel | 10 670 | 7 000 | — | 17 670 |
| Königsberg | 40 319 | 1 290 | 2 040 | 43 649 |
| Pillau | 291 151 | 99 335 | 27 975 | 418 461 |
| Danzig | 119 069 | 45 971 | 4 988 | 170 028 |
| Gotenhafen | 316 333 | 83 460 | 7 024 | 406 817 |
| Hela | 247 134 | 163 363 | 85 313 | 495 810 |
| Elbing | 4 000 | — | — | 4 000 |
| Stolpmünde | 32 760 | — | — | 32 760 |
| Rügenwalde | 5 560 | — | — | 5 560 |
| Kolberg | 116 717 | 1 915 | 9 950 | 128 582 |
| Swinemünde | 68 590 | 13 323 | 51 748 | 133 661 |
| Stettin | 2 050 | 900 | 900 | 3 850 |
| Greifswald | 1 000 | 1 200 | — | 2 200 |
| Warnemünde | 5 055 | 2 092 | 1 748 | 8 895 |
| Stralsund | 7 512 | 4 341 | 3 790 | 15 643 |
| Sassnitz | 7 552 | — | 13 106 | 20 658 |
| Rostock | 7 450 | 150 | 350 | 7 950 |
| Wismar | 4 400 | 7 000 | 1 042 | 6 142 |
| Travemünde | 4 100 | — | — | 4 100 |
| Totals | 1 291 922 | 444 757 | 241 189 | 1 977 868 |

Total ships involved: 790

| Total losses: | ships | tonnage[a] | mortality[b] |
|---|---|---|---|
| | 123 | 464 340 | 19 152 |

[a]Measured in long tons.
[b]Includes over 7000 prisoners of war (POWs).
Note: Dunkirk operation, May–June 1940; 338226 soldiers rescued.
Totals are slightly different from those in the source.

sick folk, packed together where they hoped a bomb would not reach them; the agony of waiting when a ship was there, hoping to load, and leave before being hit; the miseries of those packed together in the holds of ships that might be holed and sunk at any moment (compare with Toland, 1966).

Nine of the greatest maritime disasters of history occurred, four of them involving between two and four times as many deaths as those in the "Titanic" disaster (Table 2). The sinking of the "Goya," torpedoed on 16 April, is the single largest loss of civilian life in a maritime disaster.

Yet, against all odds, the great majority of trekkers who reached the coast were brought away to safety (Figure 2). It was an astonishing achievement, snatching a special kind of victory from defeat as at Dunkirk. In many ways, it vindicated the role of organized assistance, despite lack of planning and effective overall control. Strangest of all, unlike Dunkirk or most heroic rescues of the war, this was largely a saving of civilians and the least "useful" of them.

## Uprooting as Strategy

The flow of refugees from Eastern Europe did not cease with the war's end, but became a still greater flood of desperate humanity. A further 10 million German-speaking folk were

**TABLE 2** Major Losses of Life and Shipping in the Baltic Sea Evacuations, January–May 1945 (source: Brustat-Naval, 1970).

| Date | Ship | Location | Cause | Casualties (comments) |
|------|------|----------|-------|----------------------|
| 30 January | Wilhelm Gustloff | near Stolpmünde | torpedo | ~4000[a] (fleeing civilians) |
| 10 February | Gen. von Steuben | 55°09′N: 10°37′E | torpedo | 3000[a] (civilians and wounded) |
| 17 February | Eifel | near Libau | ? | 677 |
| 7 March | Robert Möhrung | Safsnitz Hafen | bomb | 350 |
| 12 March | Andros | near Swinemunde | bomb | 570 (mostly women and children) |
| 25 March | Weser | Neufahrwasser | bomb | 250 |
| 10 April | Posen | near Hela | ? | 300 |
| 10 April | Neumark | Danzig Bay | torpedo | 800 (wounded and some civilians) |
| 11 April | Moltkefels | near Hela | ? | 400 |
| 11 April | Vale | Pillau | bomb | 250 |
| 13 April | Karlsruhe | near Stolpmünde | ? | 900 |
| 16 April | Goya | 55°13′N: 18°20′E | torpedo | 5900[a] (fleeing civilians and wounded |
| 16 April | Cap Guir | near Libau | | 774 |
| 3 May | Cap Arkona | near Neustadt | | ~5000 (mostly POWs)[b] |
| 3 May | Thielbek | near Neustadt | | ~2000 (mostly POWs) |

[a]Compare with 'Titanic' disaster, in which 1513 lives were lost.
[b]Prisoners of War.

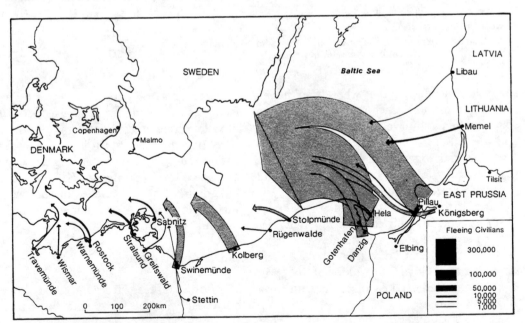

**FIGURE 2** The Baltic "Dunkirk": naval evacuations of German civilians from Baltic Sea ports, January–May 1945 [source: based upon figures in Brustat-Naval (1970)].

displaced after VE Day (Proudfoot, 1957). By then they were openly designated "expellees," systematically forced from their homes and driven westwards in treks hardly less awful, sometimes worse, than those of the war. The entire process involved expulsions from Czechoslovakia, Poland, and other states, as well as prewar German territories. These millions were crowded into the now much contracted area of the two postwar Germanies.

It is important to repeat, however, that before and after the war's end this was an integral part of Allied strategy.

The Germans who fled from Eastern Europe did not discover until later how Allied policy had already made them permanent exiles, to be joined later by twice as many "expellees" (Vernant, 1953; De Zayas, 1977). That was an important part of the hidden agenda, the long range geopolitical and racial strategy of the Great Offensive.

Its broad philosophy was already being discussed when Roosevelt and Churchill met in Quebec, August 1943. The specific intention to expel German folk and relocate them within a smaller demilitarized, deindustrialized Germany, were matters of accord between the Big Three at Yalta and specifically of Article 13 of the protocols signed at Potsdam (De Zayas, 1977). True, the Western Allies underscored the notion of "orderly and humane transfers," but they could not ensure, and had every reason to know and expect, that Stalin would not shrink from utterly ruthless methods (Conquest, 1970; 1971; Fitzgibbon, 1975; Tolstoy, 1977).

In any case, what is, or could be, "humane" about massive forced uprootings? Under the most favourable conditions, to be obliged by outside forces to leave one's long-time home and homeland is deeply disturbing; being forced out is traumatic, and being expelled is devastating. For long-settled, traditional, urban and rural communities, as most of these were, it was bound to be worse. For families without their menfolk, for children, and for the elderly, it was doubly threatening, apart from the environment of strife, severe weather, war damage, and vengefulness into which they were thrust. It could not have been anything but a prescription for the human catastrophe it actually became, and continued to be for years afterwards in Germany (Gollancz, 1947).

This kind of uprooting is a peculiarly "geographical" calamity, but only Bowman seems to have spoken out strongly against the "principle" of population transfers (Bowman, 1946). Others who did speak out at that time, like Gollancz (1946) and Schweitzer (quoted in De Zayas, 1977, page xix), seem also to have been voices crying in the wilderness. For mass transfers, or "forced draft" "pacification" programs, as they were called in Vietnam (Huntington, 1968), became commonplace methods of the armed forces which served virtually every shade of political complexion in subsequent wars. Meanwhile, if any sort of survival of the civil population were possible in a future world war, evacuations too horrible to contemplate, and permanent uprootings from today's major centres of populations, are deemed inevitable. That is to say, they are an integral part of everybody's war plans. The 'DPs' (displaced persons), saddest figures in the devastated European landscape of the postwar years, speak to the fate of any who survive the next great war.

## PLACE ANNIHILATION

Through systematic policies of urban attack, the Allies razed about 290 square miles in total (750 square kilometres), of city areas within the prewar boundaries of Germany and in the Japanese Home Islands (Hewitt, 1983a). As many as 1.5 million civilians were killed in these raids alone, and more than two million were severely injured. To the millions uprooted by official and unofficial evacuations, were added over 16 millions made homeless by bomb destruction. No other national areas suffered so extensive and irrevocable loss of urban places.

Although great numbers of towns and cities have been destroyed by bombardment as part of the general clash of military forces, here I am concerned with bombing specifically directed at populous cities which were more or less remote from the battle zones. Moreover, I am dealing with a systematic policy of "planned destruction" of urban areas, prepared over more than a decade, and perfected in hundreds of raids. It represents a sharp, if not a wholly distinct, development in the use of armed force to destroy peoples and places. It is the lineal ancestor not only of our thermonuclear predicament, but also of the practices adopted in Vietnam, Afghanistan, Lebanon, the Persian Gulf War, and others.

In the Second World War, what is often called "area," "obliteration" or "saturation" bombing involved mass raids, usually by hundreds of heavy bombers arriving in waves. Their "aiming point" was commonly just the heart of a city's built-up area, and their task to lay down as thick a carpet of bombs as possible. Even then, high-explosive bombs proved unsatisfactory and these attacks increasingly assumed the form of fire raids. The spread of fires from the incendiaries the bombers dropped was the main cause of damage and casualties. This defines the essential character of *area bombing*, even in the final form of the A-bomb raids [Bond, 1946; Stockholm Peace Research Institute (SPRI), 1975]. For the victims, fire more than explosion, burns more than other types of wounds, were the definitive experience. This also placed the raids firmly in the category of *terror bombing* (Ford, 1944; Rumpf, 1963; Veale, 1962).

Dresden and Hiroshima are commonly perceived as the worst, most damaging, and "final" of these raids, and one will not deny their unique place in the annals of atrocity. However, the extreme result to date—in overall devastation and casualties, in the area of a city burnt out in a single raid, and in numbers of civilians who lost their homes—occurred elsewhere, but also in 1945.[3]

## The "Big Fire" Raid

On the night of 9 March, a force of B-29 bombers was sent against Japan from newly won air-bases in the western Pacific Islands. The B-29 or "Superfortress" was then the newest, most powerful, and most expensive weapons system in the air. Its range of operation was over 4800 km, with a payload, in this instance, averaging about 5.5 tonnes of bombs. It bore an impressive array of guns, and novel electrical, hydraulic, and electronic equipment, to defend and guide itself through the skies of war. More important still, the aircraft was now available in many hundreds of copies.

In the raid of 9 March, 334 planes were involved. They began taking off from airstrips in the Marianas at 17.35 h. The lead aircraft were over the target area at about midnight. If by no means the largest bomber raid of the war, it was a formidable array of force.

However, this force was not being applied in a battle against another military force. It might have encountered resistance from aerial defenses, but it was not being sent against them, or directly against the military bases, or the war industries of the enemy. Its mission was to destroy a city (USSBS, 1947b; Daniels, 1975).

The bombers' "aiming point" was the congested Asakusa district of Japan's capital city, Tokyo. Here, along the banks of the Sumida river just north of Ginza, was one of the densest concentrations of humanity in the world. Air force intelligence estimated an average of 40 000 persons per square kilometre, which rose, in places, to more than 55 000 (Craven and Cate, 1953). The ratio of roofed-over area to total area, or "built-upness ratio," was also exceptional. At about 50%, it was several times that for the inner area of most Western cities. Equally significant, the area was seen to contain countless flimsy close-packed structures, of which their composition made them highly susceptible to fire. Fire was to be the main means of destruction.

The bombers were carrying incendiaries only; about 2200 tonnes in all. The marker bombers, which formed the first wave, set a scatter of fires with M47 napalm bombs. The main force followed with 240 kg clusters of delayed-fuse M69 napalm bombs. A few days later, an article in the *New York Times* would reassure its readers that everything possible was being done to bring Japan to its knees, by describing the action of these bombs:

> . . . one of the principle instruments of destruction in the fire attacks has been the M-69 incendiary . . . (or) jellied oil bomb . . . containing gel-gas, a resin-type jelly (or) . . . a composition mixed with gasoline . . . cheesecloth impregnated with flaming jelly is spewed out in all directions over a radius of twenty-five yards shortly after the bomb strikes. The material burns fiercely at a heat of about 3,000 degrees F for eight to ten minutes (*New York Times*, 1945).

The bomber crews' orders were to lay down a carpet of not less than 11 tonnes of bombs per square kilometre in the target area, or about 5000 M69s (Craven and Cate, 1953). The aircraft came in at a daringly low altitude of 1500–3000 m. Visibility was good and aerial defenses proved to be negligible. The result was an unprecedented concentration of incendiaries. This, the flammable nature of the neighbourhoods attacked, and a rising wind, produced a vast conflagration.

In a report to the National Fire Protection Association of the United States, Major F. J. Sanborn, a member of the Strategic Bombing Survey, observed that:

> In a conflagration, the pillar of the mass fire, once it had been established, slanted appreciably to leeward and the hot burning gases contributed much to the ignition of combustible materials on the ground. The chief characteristic of the conflagration was the presence of a fire front, an extended wall of fire moving to leeward preceded by a turbulent mass of pre-heated vapours. The progress and destructive features of the conflagration were, therefore, much greater than those of the fire storm. . . . The conflagration of the 9th March in Tokyo

was the most notable example of this type of mass fire. . . . An extended fire swept over 16 miles (25.7 km) in six hours. Pilots reported the air was so violent that B-29s turned over completely at 6,000 ft (1,820 m) and the heat was so intense they had to put on oxygen masks. The destruction was complete; not a single building escaped damage in the area affected. The fire had spread largely in the direction of the natural wind (Bond, 1946, page 181).

By mid-morning, on 10 March, the fires had done most of their work. The main fire had burnt itself out. In all, an area of forty-one square kilometres was laid waste. Under the smoke lay tens of thousands of dead civilians. The exact number remains in doubt. The USSBS figure of 83,600 seems minimal (1947c). Later studies suggest over 100,000, and some have argued that 200,000 is nearer the mark. Severely injured civilians numbered about a quarter of a million. Even so, given the density of population, the congestion, lack of organized help, and rapid spread of the fire, these figures may seem miraculously low.

Some 270,000 buildings were destroyed. Most were civilian housing, but nearly two-thirds of the city's commercial establishments and one-fifth of the industrial were also consumed in the fire (Figure 3).

What, however, were the conditions on the ground that lay behind such bare statistics. Tens of thousands of women and children were trying to flee through the burning streets; fighting their way over narrow bridges. The mind can hardly comprehend one, let alone the vast array of terror experienced. One can only guess at the multitudes of those who struggled, failed, and died, in attempts to rescue their own from burning homes, or at the state of mind of those who crowded, as a last hope, into the few open spaces in these districts (Caidan, 1960).

The behaviour recorded can, all too readily, be called "panic." It surely felt like the end of the world to those involved, and should not be treated like the classic scenes from, say, London's "Blitzes." The scale was vastly

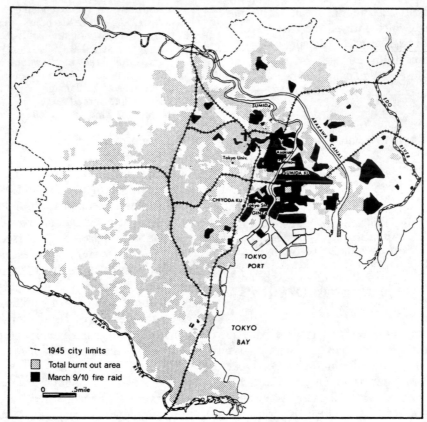

**FIGURE 3**  Tokyo areas burnt out in fire raids, 1945 (based on USSBS, 1947b).

greater than anything Londoners had to cope with. This was one of the earliest raids experienced by the residents of Tokyo. There had not been that crucial period of adjustment experienced by civilians in British or German cities subject to recurrent bombing (see Schmideberg, 1942; Harrisson, 1976). There was almost no assistance from trained personnel. Great numbers of families, lost children, mothers carrying babies, injured, aged, and infirm persons alike, had to struggle alone, amid vast crowds of frantic people. There were no teams of valiant rescue workers to help free them; rarely any police or soldiers to guide those hurrying to places of safety, or to control the appalling congestion at the many bridges. Some three hundred fire engines were on the scene early, but they were completely over-whelmed, as were most of the first-aid posts (Bond, 1946). Many rescue units did not, or could not, move in from districts outside the stricken areas.

Meanwhile, during the full-scale fire raid, civilians on the ground had other problems to contend with. Problems of escape or rescue were greatly magnified by continued bombing. In this instance, it went on for about three hours. The air was repeatedly filled with a rain of fire which was created by the spilling canisters of jellied gasoline. New centres of fire were being started all the time. These were standard tactics in urban area raids, and intended to frustrate any efforts at rescue and at fire fighting.

People running for refuge were trapped by the bombings ahead and around them and were encircled with flames and black smoke. They looked for protection to the canals and rivers, but in some districts the shallow canals were boiling from the heat which seemed to be compressed by the wind, and the canals were full of people. In some places one swarm of humanity after another crowded into the water and by the time a third or fourth wave of frantic people had jumped, the first wave lay on the bottom . . . (USSBS, 1947a, page 70).

Kazutoshi Hando, then a schoolboy in grade three of the middle school, found himself alone:

My family's house was burned to cinders, and I escaped only by jumping in a nearby river, where I stayed the whole night long. When dawn broke, I saw that the banks of the river were piled high with charred bodies, and many who had taken refuge in the water were dead as well (Pacific War Research Society, 1972, page 13).

An eye-witness—from a distance during the attack, but later in the burnt-out areas—was R. Guillian, a Frenchman who remained in Japan during the war (Guillian, 1947). He tells how hundreds of folk converged in desperation upon the grounds of the great Senso-ji or Asakusa Kannon Temple. A Buddhist foundation of the seventh century, dedicated to the "Goddess of Mercy," it was a major landmark in the city and a place of pilgrimage. People who ran there in the 1923 fire, were said to have been saved. On this night, the temple and all who sought refuge in its grounds perished in the flames (Caidan, 1960).

For the "profane space" of the Yoshiwara Yukwaku, the "nightless city" of the Gieshas and courtesans, matters proved equally final. When the raid began they closed the metal fire doors, but the doors were not proof against the form of incendiarism they were facing. The "ladies of the night," whose story is no doubt more one of exploitation and enforced bondage than of pleasure (see Hane, 1982), perished with their clients. And it may well be that from among the latter came the largest

compliment of the military casualties suffered in the whole raid!

In fact, many who sought protection in air-raid shelters were worse off than those in the open, as few shelters were adequate to withstand these fires. Rather, they turned into death traps were thousands died of asphyxiation, carbon monoxide poisoning, or heat stroke; the commonest killers in incendiary raids everywhere (USSBS, 1945b; 1947c).

The scenes from that night are, to a great extent, like those from the better known events at Hiroshima, where fire was also the largest cause of death and destruction (Osada, 1959). The Tokyo fire was, however, on a much larger scale, although there were not the additional horrors of radiation.

An important point to note is that, throughout the war, mass raids against cities were rarely very successful when confronted with determined anti-aircraft and aerial defenses. Even where many bombs were dropped, the disruption of bombing patterns, and harm to the attacking force, would prevent critical concentrations of incendiaries. All the really devastating raids, from Lübeck and Hamburg to Dresden and Nagasaki, showed the defenses unprepared or in disarray. In the present case, even as the Japanese military were flinging suicide planes at Allied warships, they provided no credible defenses for their capital city. The Tokyo raid showed, months before the two A-bombs, that Japan's cities and their civil population were virtually defenseless against the threat they faced. In the words of the USSBS team:

The overall picture of civilian defense in Japan, was not a happy one. It is hard to conceive of a nation's undertaking a major war and paying so little heed to the protection of its vital industries, to the continuance of its essential economic life *and to the safety of its people*. . . (1947a, page 140, my italics).

The raid of 9 March was just a beginning. By June, nearly one third of Japan's urbanized area was in ruins. In a dozen or so raids, upon the five major cities—Tokyo, Yokohama, Osaka, Nagoya, and Kobe—some 259 square kilometres of built-up area had been laid waste by fire (Table 3). By the end of July, a further 166 square kilometres and fifty-seven lesser cities had been similarly destroyed (Figure 4). Some ninety-four other cities reported civilian fatalities in excess of a hundred persons, and damages that exceeded the total for Britain during the war (USSBS, 1947a). And only then did the A-bombs do their work; a further sixteen square kilometres of devastation, mainly caused by fire, and some 180,000 civilians dead (Committee for the Compilation of Materials on Damage Caused by the Atomic Bombs at Hiroshima and Nagasaki, 1981).

Thus, the greater part of the urban area of an industrialized nation was eliminated, and in a relatively small number of blows over a short timespan. The devastation encompassed the central parts, and often much more, of all but two of the important towns of Japan. More than eight million civilians had their homes destroyed in the raids (Table 4). Still greater numbers were forced to flee into an already overcrowded and impoverished countryside. The whole episode is remarkable in the extent of unrestrained assault upon human settlements and noncombatant resident populations. Equally, perhaps, it is remarkable for the lack of awareness shown by, or qualms of conscience in, the Allied nations then or since (Veale, 1962).

And if there is a precedent for our thermonuclear prospects, it is this whole episode of "overkill" and annihilation that provides it, rather than just the A-bomb raids.

## Place Annihilation as Strategy

The devastation in Tokyo and elsewhere was not without precedent; indeed it was something air forces had been struggling to achieve for years. It was "planned destruction," a form of strategic bombing nurtured by decades of preparations and, ultimately, by enormous investment in the means to carry it out. The tactics and exact means to use in the raids had emerged from years of wartime practice in the saturation bombing of cities. The almanac of great urban fires already embraced dozens of other cities, before the inclusion of those that began in Japan in March 1945 (Hewitt, 1982a).

It was the aeronautics pioneer W. F. Lanchester who, as early as 1914 (published 1916), had predicted what one sees occur in the Second World War. In the use of air power against cities, he stated that:

> The critical point, and the point to be aimed at as an act of war, is that at which the fire-extinguishing appliances of the community are beaten and overcome. Up to this point, the damage done may be taken as roughly proportional to the means and cost of its accomplishment: beyond that point the damage is disproportionately great: the city may be destroyed *in toto* (Lanchester, 1916, page 121).

The firestorm at Hamburg was, perhaps, the first to demonstrate fully the accuracy of this prediction (Middlebrook, 1980). From that, and other raids, bombing strategists knew that the degree of damage from incendiaries was as dependent upon the density and type of built-up area attacked, as upon the concentration of bombs. The degree of crowding together of structures was important; so too was the type of building materials used, notably the amount of wood and other biomass materials. The presence of many wallpapered rooms cluttered with fabrics and furniture, of domestic fuel stores and home fires, of many numbers of outlets for electricity and gas, all aided the development of fires (Bond, 1946). In a word, it was found that dense *residential* neighbourhoods of the inner city were ideal places to start a mass fire that would fulfill Lanchester's prediction.

If the larger threats a particular civil society faces in air war are especially the weapons and actions of an enemy, the doings of home governments bear directly upon the harm their own citizens may suffer. Failure of powerful

**TABLE 3** Civilian Impacts of Major Fire Raids, Japan 1945 (sources: USSBS, 1947a; 1947b; 1947c; Craven and Cate, 1953).

| Date | City | Comments |
|---|---|---|
| 25 February | Tokyo | 27,970 buildings destroyed.<br>2.5 km² burnt out including Kanda University |
| 10 March | Tokyo | 267,171 buildings destroyed.<br>38.2 km² burnt out.<br>130,000–200,000 civilian deaths.<br>Over 50,000 severe injuries.<br>Over 1 million made homeless.<br>66% of commercial area destroyed.<br>Numerous hospitals, clinics, schools, temples, etc, destroyed. |
| 14 March | Osaka | 135,000 houses destroyed.<br>20.7 km² burnt out.<br>~4000 deaths and 8500 injured. |
| 17 March | Kobe | 65,051 houses destroyed.<br>~2700 deaths and 11,300 injured.<br>7.5 km² burnt out.<br>242,000 persons made homeless. |
| 15 April (twice) | Tokyo–Kawasaki | 238,732 buildings destroyed.<br>24.8 km² burnt out and 4.0 km² of Yokohama.<br>841 deaths. |
| 15 and 17 May | Nagoya | 113,460 buildings destroyed.<br>32 km² burnt out.<br>3866 deaths.<br>472,701 rendered homeless.<br>(Note: includes damages from earlier but much smaller attacks.) |
| 23 and 25 May | Tokyo | 221,160 buildings destroyed.<br>14 km² and 43.5 km² burnt out, respectively.<br>83,000–100,000 deaths (according to Brodie, 1973, page 51)[a].<br>(Note: in 25 May raid largest area burnt out in any air raid.) |
| 29 May | Yokohama | 89,073 buildings destroyed.<br>23 km² burnt out.<br>~4500 deaths. |
| 1 June | Osaka | 136,107 houses destroyed.<br>8 km² burnt out.<br>3960 persons dead or missing.<br>218,682 persons rendered homeless. |
| 5 June | Kobe | 51,399 buildings destroyed.<br>17.27 km² burnt out. |
| 7 June | Osaka | 55,333 buildings destroyed.<br>6 km² burnt out. |
| 15 June | Osaka–Amagasaki | 4.9 km² of Amagasaki burnt out. |

[a]This may be an error; he may have mistaken this for the 10 March raid.

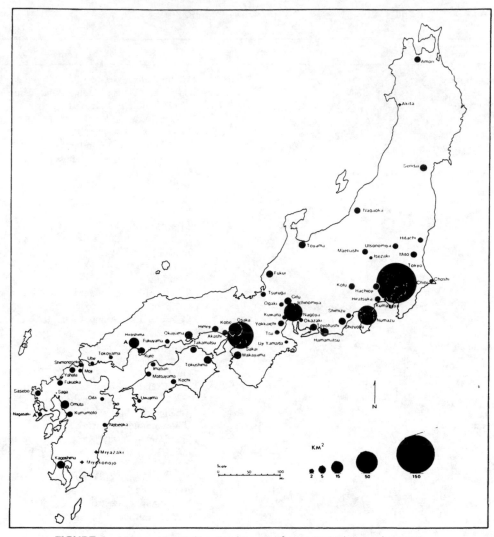

**FIGURE 4** Urban devastation: the impact of area bombing on Japanese cities, by area.

states to outlaw and remove means of terror bombing are part of that, but it also has two other aspects. First, there are direct civil-defence measures which, as noted, were quite inadequate in Japan (Hewitt, 1985). Second, there is the responsibility of governments that do develop and use weapons of terror, in that they provide an enemy with reasons to assail their home areas with the same. This raises Richardson's arguments that ". . . in a round-about way the bombing airplanes are a danger to the nation that owns them" (1960, page 229).

The Japanese Air Force had pioneered long range strategic bombing, essentially in the form of terror, against cities and civil populations in China. Early in the Sino-Japanese War, their aircraft began raids that were, at the time, unparalleled in distances flown and in destructiveness. They assailed the central congested areas of cities, and razed large areas of Canton, Nanjing, Hangzhou (Hangkow),

**TABLE 4** Some Aspects of Civil Damages of Area Bombing Raids on Japanese Cities (sources: as given in table).

| | |
|---|---|
| *Bombs dropped* (tons); USSBS (1947b) | |
|    all of Japan | 160 500 |
|    urban area | 104 000 |
|    all Tokyo attacks | 16 500 (12500 incendiaries) |
| *Civilian casualties* | |
|    deaths: all raids | 0.9–1.3 million (Hewitt, 1983a) |
|          fire raids | 0.65–0.8 million (Hewitt, 1983a) |
|          A-bombs | 180 000 (immediate) + 160 000 (to 1950)[a] |
|    severely injured | >1 500 000 (Hewitt, 1983) |
|    fire burns as cause of death | 56–84% (five cities) (USSBS, 1947c) |
|    total casualties | ~8 million (Kosaka, 1972) |
| *Built-up area destruction* (62 cities): | |
|    complete destruction: | |
|       total | 425 km² (USSBS, 1946) |
|       proportion | ~50% (USSBS, 1947b; Craven and Cate, 1953) |
|    buildings destroyed | ~2.2 million (USSBS, 1946) |
|       Greater Tokyo | 0.86 million (USSBS, 1947b) |
|       Hiroshima | 70 000[a] |
|       Nagasaki | 19 587[a] |
|       by structure evacuation | 614 000[a] |
| *Specific damages for civil life* | |
|    housing units destroyed | 2.5 million (Havens, 1978) |
|    persons made homeless (raids) | 8.3 million |
|    structure evacuation | 3.5 million (Havens, 1978) |
|    persons evacuated (to October 1944) | 2.1 million (USSBS, 1947a) |
|               (to August 1945) | 8.3 million (USSBS, 1947a) |
|           Tokyo (to August 1945) | 4.1 million (57%) (USSBS, 1947a) |
|    hospitals destroyed (fire raids and A-bombs) | 969 (USSBS, 1947c, page 10) |
|    hospital beds lost (fire raids) | 51 935 (USSBS, 1947c, page 10) |
|    pharmaceutical factories destroyed | 200 (32%) (USSBS, 1947c, page 10) |
|    food stores destroyed | 221 891 tonnes (5%) (USSBS, 1947b) |
|    service professionals killed in Hiroshima[a] (%) | |
|       physicians | 90 |
|       pharmacists | 80 |
|       dentists | 86 |
|       nurses | 93 |

[a]*Source:* Committee for Compilation of Materials on Damage Caused by the Atomic Bombs at Hiroshima and Nagasaki, 1981.

and Chongqing, mostly with fires started by incendiaries (Linsay, 1975). Nonmilitary areas and civilians were the main focus of destruction (Hsu and Ming-Kai, 1971).

Although the U.S. Air Force generally avoided incendiary and area bombing raids in Europe, they had long been preparing for their possible use against Japan. From 1943, trials were carried out at Eglin air base in Florida, and the Dugway Testing Ground, Utah, in which mock-ups of ". . . model urban areas

typical of Japanese construction" were burned (Bond, 1946). They studied the great Tokyo fire, that which followed the 1923 Kanto Plain earthquake, as a "model" for incendiary attacks against the city (Bond, 1946). Napalm was used in an area bombing raid of singular devastation, against Hankow in China on 18 December 1944 (Craven and Cate, 1953; SPRI, 1975; Hewitt, 1982a). Although held by the Japanese army, most casualties were Chinese civilians. The French seaside town of Royan was also wiped out in a napalm raid when American bombers were experimenting with its use in Europe (Zinn, 1970). Although part of an attack on the remaining German forces, the casualties were again (French) civilians. Finally, the raid of 9 March on Tokyo, was not only a "highly successful" action, but also the decisive *experiment* that confirmed 21st Bomber Command's use of fire raids as its preferred role—its "final solution" to Japan's resistance.

In many respects, the unique significance of the fire raids on Japan, like Dresden or Royan, lies *only* in their impact upon civilian space and urban places. Other raids, much earlier in the war, had involved many more bombers and far greater bomb loads. Some continued over several days and were, for the aircrews involved, "battles" in every sense, in which they came up against powerful aerial defences. Apart from the appalling statistics of damage, most of the fire raids were militarily uninteresting—a "piece of cake" as the phrase goes.

Although I fail to see how a civilized approach to these events can ignore the morality they presuppose, that is not my focus here. Rather, it is the iron implications for the survival of civil society and its places. Indeed, the experience of civilians on the ground, and the fate of their places, is what defines the most basic meaning of this style of war. It was, regardless of other military benefits expected or hoped for, not just a strategy of annihilation, but it was specifically one of (civil) *place annihilation*. This is seen in just whom and what the raids harmed most, and most often. Here is a sort of negative listing of the ingredients of civil "ecology" and its living space (Hewitt, 1983a). The consequences of the assault of Japanese cities were mainly:

1. Large concentrated fatalities among resident civilian populations.
2. A predominance of women, children, the aged, and infirm among the casualties (that is, "non-combatants").
3. "De-housing," or destruction of *homes* of ci-

vilians, as a common objective, and the predominant form of physical destruction (Ikle, 1958).
4. Indiscriminate destruction involving schools, shops, banks, libraries, hospitals, theatres, temples, and landmarks, and including buildings of great civic and artistic significance.

The implications of this incendiarism for civilization are further expressed in Ienaga's remark that: "No one could ever count the books, documents, paintings and other treasures that went up in the flames . . ." (1978, page 2). It was indeed a process of "de-civilizing" an entire people, and that was not lost upon advocates of the policy. US Air Force General Chennault spoke of ". . . burning the guts . . ." out of Japan (Craven and Cate, 1953, page 144). US Secretary of War Stimson spoke of ". . . bring(ing) the heart of Japan under the guns and bombs of the Army Air Forces . . ." (Craven and Cate, 1953, page 144; *New York Times*, 1944, 16 June). The architect of the fire raid strategy, and commander of the forces that carried it out, General Curtis LeMay, is notorious for describing the objective as bombing Japan ". . . back into the Dark Ages" (Kantor, 1965, page 565). Later, as the head of Strategic Air Command, he would advocate bombing North Vietnam ". . . back into the Stone Age . . ." and a similar approach to the Soviet Union and China (see Stone, 1967, pages 92–104). On the evidence of what his bombers had done to Japan's cities, this is not just a figure of speech (compare with Branfman, 1972).

## THE SOCIAL SPACE OF TERROR

It is often inferred that "total war" blurs if it does not eliminate the distinction between civilians and soldiers (McReavy, 1941). It is usually assumed that weapons of terror and mass devastation, being indiscriminate in ethical and targeting senses, produce "wall-to-wall" destruction. They strike rich and poor, shabby and salubrious neighbourhoods, alike.

These notions must erode the sense that war can be restrained by civilized rules, and they tend to make social, geographical, and historical understanding redundant. It is my sense of the materials examined above, that these notions can be challenged. As is found with natural disasters, notions of "wall-to-wall" destruction are media overdramatizations or technocratic fictions (see Hewitt, 1982b; 1983b). And for a social scientist it is actually imperative to ask just *who* dies and *whose* places are destroyed by violence.

It has already been shown that area bombing assailed the central, most congested, areas of cities. Their destruction led to some of the most populous parts of cities being emptied and abandoned. Where their flight took urbanites beyond the untouched suburbs or where they were evacuated to the countryside, it led, in effect, to a deurbanization of society. This, in itself, is a huge subject. For one finds that twentieth-century violence everywhere, has the effect of either herding vast numbers of rural folk into cities, or forcing urban folk *out of* cities. It either accelerates, or slows and reverses urbanization (Thrift and Forbes, 1986). Allied bombing "deurbanized" Germany, but the great majority of German folk who were expelled from Eastern Europe were from rural and farming backgrounds, and most eventually ended up in urban areas of the postwar Germanies.

It is important to add that even saturation bombing and firesetting were rarely very effective against dispersed urban areas or suburbs. Mass fires could not be generated in the less dense neighbourhoods. Critical and newer industries, also often in the suburbs or dispersed to the countryside, and heavily defended, could rarely be hit at all, nor reached by mass fires set elsewhere. Few strategists will deny that the ability to knockout key war industries, energy sources, military bases, airfields, and communications nodes is a far more effective and desirable use of strategic air power. Unfortunately, it is well documented that the bombers were rarely able to hit or destroy these; at least, not without unacceptable casualties (Hastings, 1979). And so, frustration and losses among the much-championed and expensive bomber fleets was a large factor in turning them against targets that they could hit and could seem to cause a great impression upon. These were the city centres. Only then did "barn-door" targetting and carpet bombing provide impressive statistics of destruction.

One also finds that the bombings and uprootings were not indiscriminate with respect to demographic, social, and economic conditions among the victims. In the case of the bombing of cities, it is often assumed that the industrial "war workers" were the only or the main targets. If that were true it would itself involve a specific social space. In fact, it was the remaining city-centre civil populations *in general* that were assailed. Persons whose work lay in local government and services, law enforcement and nonwar businesses or enterprises, were also exposed. Commonly, they were more at risk than the industrial workers because they were more heavily concentrated in city-centre areas. Workers in key war industries were better protected and fared better than most during the air raids (USSBS, 1945a; 1945b; 1946). Again, it is often assumed women who were not evacuated successfully to the countryside were in war work; "Rosie the Rivetter," etc. Important for women as this development was, at least in Britain and North America, its relative role has been exaggerated. In Germany and Japan, the overwhelming numbers of women, especially in the cities, were "homemakers" (Stephenson, 1981; Havens, 1978). Those with work outside the home—itself often of a part-time or a voluntary type, over and above domestic duties—were in services (education, health care), in small businesses (bakeries, tobacconists), and in civic, religious, and cultural occupations.

Meanwhile, conscription of men from the cities was rarely balanced by evacuation of women, so that the latter formed the larger fraction of the urban populations which were subject to bombing. A disproportionate concentration of air raid deaths and injuries among women is well established for most German cities (Rumpf, 1963; Hewitt, 1983a).

Women were almost certainly the larger fraction of casualties in the fire raids on Japan, although data have not yet come my way to substantiate this claim for most cities. However, Hiroshima, object of minute examination, fully shows this to be the case. There, death was proportionately much greater in "spouses" (wives) than in "household heads" (husbands), and among daughters compared with sons (Committee for the Compilation of Materials on Damage Caused by the Atomic Bombs at Hiroshima and Nagasaki, 1981).

It may be noted that near the time of the raids, as in East Prussia just before the flights and expulsions, the Japanese government ordered a final wave of conscription that took many remaining men away, who had been previously considered too old or disabled for military service. Japan had its *volkssturm* too! The military also stripped civil society of most of its physicians and other service professionals, not to speak of medicines and of other facilities that would mean the difference between life and death for thousands when the fire raids happened (Hewitt, 1985; 1986).

Although organized evacuation in Japan did remove a large fraction of very young children, and officials encouraged the elderly to evacuate, great numbers of older children and old people remained behind, especially those in lower-income groups. In October 1944, just as the main period of bombings began, about 14.5% of people in the major cities had been evacuated (USSBS, 1947a). By August 1945, 57.6% had been evacuated, about 35% in organized or voluntary movements, and the remainder in precipitate flights during or after the raids (USSBS, 1947a). That is to say, millions of "ordinary" housewives, children, youths, and the elderly, were in each major city for many or for all of the raids. As was

already well-known in Britain and Germany, and is dimly realized by those planning mass evacuations for nuclear war, there are very severe constraints on where, and how many people can be supported outside the cities of an urban-industrial state—and this is if the crisis continues only for weeks, let alone years. Studies of demography, rationing, and other services in wartime, support these observations about the gender and age of city populations. They are rarely put together with the raid data (but see USSBS, 1945a; 1945b; 1947a; 1947c; Titmuss, 1950; Havens, 1978).

A further aspect is the income (or class) distribution of victims in congested city-centre areas (Hewitt, 1983a). In German and British cities, urban targetting led to disproportionate casualties among low income or poor families. To the extent that "the German industrial worker" or his British counterpart were affected, it was usually the unskilled or semiskilled. Mainly, however, it was the families of servicemen in the lowest ranks who were being struck at, and, of course, this happened while the men were away, again in disproportionate numbers, at the battle fronts. American advocates of air power do not seem to have followed their British counterparts who advocated attacks on "working-class neighbourhoods" (Webster and Frankland, 1961, pages 331–332; Snow, 1961, pages 47–48). Nevertheless, by assailing the densest central areas of Japanese cities, their actions had a similar effect. The better-off urbanites lived mostly in more open and surburban areas, or were easily evacuated to comfortable properties in the countryside.

A recurring observation of even better-off respondents, to the thousands of interrogations and questionnaires of the USSBS, stresses this aspect. They speak of "slum bombings," or of the much greater plight of poorer neighbourhoods (Ibuse, 1969, page 71). And they suggest that had the bombings affected more expensive neighbourhoods, much more fuss would have been made, and this would have required more *effective* government action.[4] This was paralleled in Britain, especially in early phases of the bombing (Harrisson, 1976).[5]

In the enforced flights from Eastern Europe, farm families tended to fare better than urbanites. Their menfolk often remained at home, because of special consideration for food producers, and they often had draft animals for the journeys. Still, the majority of trekkers were women, children, and the elderly.

The expulsions raise another dimension of the social space of terror: the ethnic or racial. Total war may blur the distinction between citizenship of a state and linguistic or racial diversity within it, even if there is difficulty in the absorption, or in the fair treatment, of the minorities within its boundaries. In war, the tendency is not to differentiate in the case of the enemy state—to the dismay of, say, the people of Hamburg who could not understand why Britain burnt their city when they were "Saxons" not Prussians (Middlebrook, 1980)! But conquerors or forces of occupation will exploit group differences or pursue racialist policies. In the East European expulsions, like the Hitlerite treatment of Jews, Gypsies, and Poles, one can note enemies identified by race and language. Whether that is even rational, let alone "civilized," is doubtful, but it is one of a host of examples in which peacetime prejudice and wartime hatreds are used to justify assailing people on such grounds. When armed force is applied, one encounters genocide as the term is usually understood.

## The "Definitive" Civilians

What the evidence seems to show is that the wartime civil populations, urban and rural, were skewed towards rather than away from those typically regarded as noncombatants. This applies not only to their being unarmed, and of an age or sex or health status that excluded them from fighting, but also to their being in occupations which were marginal, or irrelevant, to the conduct of war, and which gave them little or no say in that conduct.

The civilian majorities I have encountered as victims of bombings and uprootings, were predominantly female, the young, and the elderly. They were most occupied in domestic and service roles. Most of them came from the least wealthy, and hence least influential, classes in their respective societies. To be sure, their menfolk had provided the bulk of the armed forces, and it could happen, when enemy soldiers invaded and occupied their land, that wives, children, and grandparents would obey their governments' call to take up arms. The history of seige, although usually emphasizing martial prowess, actually shows that civilians have always willingly defended their actual homes. Perhaps Allied soldiers would have faced Japanese matrons armed with spears had the A-bombs not brought a final capitulation (Havens, 1978)! But that must be distinguished from the kinds of warfare outlined earlier, and is, in any case, a desperate resort that no civilized power should require, even of its enemy.

However, the victims of terror I have examined were mostly of a truly civilian status. Because of age, gender, occupation, and social position, no less than being unable to fight back, they can be fairly termed *"definitive" civilians*, both in ethical and in functional interpretations of war. In wartime they were largely engaged in "peaceful" activities, whether domestic, neighbourhood, civic, or economic. Their lives maintained, at home, in schools, stores, offices, churches, museums, and places of entertainment, whatever vestiges of "peacetime" ways of life could survive during war.

However, it was also these majorities of "ordinary" civilians who bore the brunt of the bombings and the uprootings. *Their* places were destroyed without restraint. Their home areas became "killing grounds" where vast numbers were massacred and maimed. They were shifted around or driven out with little or no regard for what that meant to them. It is my sense of all the wars that have happened since 1945, that a similar disproportionate impact of destruction and terror has been visited upon such civilian majorities. They have included Laotian, Afghan, and Ethiopian peasants, North Korean and Lebanese townfolk.

In one's attempts to provide a broad conceptualization of what all this means, it may be useful to suggest certain similarities of this to a more familiar debate. The civilian majorities, which I have identified here, go through a parallel process to that of many "traditional" folk caught in the so-called "development trap." The second is the underside of economic development. This is where it exploits, destabilizes, and marginalizes the established lives and livelihoods of millions outside the industrial world today. The parallels with civil populations in wartime are of the following kinds.

There is a shift to wage labour (war work) in a traditional (peacetime) society, to commercial and industrial (war production) economies, and to strong central (war planning) government—the matters that concern the bulk of the development (war) literature. These tend also to shift a huge burden of inescapable domestic, traditional, and cultural ("home front") work onto the female, the young, the elderly, and the poorly educated ("ordinary civilian") parts of the population. And much of the last's often enormously arduous and aggravated work, is in the "hidden" or "shadow" (nonwar, "black market") economies.

Those who most readily adapt to commercial work/enterprise (war work, air raid duties, etc) receive the more favourable attentions of government. And it is against this background that the more devastating changes in indigenous (civilian) societies take place. They include "drift to the cities," resettlement, land expropriation (evacuations, "concentration"), destructive collapse of traditional land uses and environment ("scorched earth," "resource denial"), and the famines

and high mortality (bombings). And *the main victims* have already been largely written off or written out of the Five-Year Plans, Foreign Loan conditions, etc (war councils, war allocations), or redefined in official euphemisms as targets of another sort—"underdeveloped" areas ("the war-making potential of the enemy").

At this point, I can only propose these analogies as a basis for discussion; a way to mobilize concepts from areas in which there is more vigorous and conceptually grounded debate. For it is my sense of the literature that these "definitive civilians," in this phase of their lives, are most often written out of war studies in general, and strategic bombing or other uses of terror in particular—as, until recently, their parallels have been written out of the literature on economic development.

It seems fair to conclude that the main victims of terror, encountered or caused by male-dominated, industrial, totalitarian states at war, are those whose gender, age, class, and occupation have rendered them invisible. They certainly were so in the military states of Japan and Germany late in the Second World War. No less than their menfolk in the "other ranks" of the military, these civilians were expected to obey, if necessary to die, but not to be heard in, or to have any influence upon, the affairs of the state at war.

Meanwhile, the "great events"; the fascination which we scholars, no less than the popular studies, have with leaderships, weapons, battles, spying and such, dominates the literature. Not only does this distract attention from the plight of national civilian majorities, but also it has tended to make them, and their roles and needs, appear pathetic if not banal. Their problems appear as unfortunate side-effects, if not boring irrelevancies in the clash and decisions of "great forces." And that placement of the problem is hardly noticed even when the forces stooped to a policy of terror towards these civilians. Nowhere is the sense of this captured with more poignancy than in Ibuse's *Black Rain* (Ibuse,

1969; compare with Liman, 1986; Hewitt, 1986).

Quite apart from the assumptions and ethics involved here, my investigations show that the meaning of terror in war is not solely a question of the weapons and goals of the attacking forces. If the *attacks* were indiscriminate, the *vulnerability* of different places and segments of society was not. Indeed, my own discussion probably overstresses the "impacts" in terms of the attacks, in that it presents only the losses, damages, and flight of civilians. A full portrait of these events involves also an extraordinary history of adjustment to, courage in, and assertion of will by, civilians struggling to survive and protect their own. And if they often failed, or finally had to flee, then this diminishes their achievements no more than when soldiers die or hold steady in battles they eventually lose.

## Concluding Remarks:
## A Defenceless Space?

Destruction of places, driven by fear and hatred, runs through the whole history of wars, from ancient Troy or Carthage, to Warsaw and Hiroshima in our own century. The miseries, uprootings, and death of civilians in besieged cities, especially after defeat, stand among the most terrible indictments of the powerful and victorious. In that sense, there is, despite the progress in weapons of devastation, a continuity in the experience of civilians from Euripides' *Trojan Women* or *The Lamentations* of Jeremiah, to the cries of widowed women and orphaned children in Beirut, Belfast, the villages of Afghanistan, and those of El Salvador today. A vital difference, however, concerns the defense of cities and civil populations. Often in the past, it has been the first priority of civilized societies.

In his Preface to a translation of Weber's *The City* (Weber, 1958), Martindale asserts that, ". . . it is of decisive importance, which units of social life are able to maintain themselves by armed force . . ." (page 60). It is

a principle held widely in the history of political thought. He stresses the vulnerability of contemporary cities to destruction in war. He emphasizes the gulf that separates them from the military situation enjoyed, or at least sought after, by most cities and civilizations in the past. And I suggest that at the heart of this is, on the one hand, an inability or unwillingness of twentieth-century military powers to protect civil society and its settlements, whereas, on the other, they place enormous investment and faith in the instruments that destroy them. Here is another paradox of urban-industrial societies and their warmaking. From Aristotle to Machiavelli, von Clausewitz, Engels, and Mackinder, politicomilitary thinkers in the West seem to share one central premise; that the methods of war developed by a state should be appropriate not only to its material means but also to its political form. Many developments in twentieth-century warfare, certainly among the ostensibly civil democratic states, whether of the left or right, seem profoundly at odds with their ideologies and material lifestyles.

Total wars involve and threaten everyone. However, I believe it is not difficult to show and document how the "home fronts"—the voice and needs of civilians in particular—have been consistently slighted in comparison with offensive forces (Hogg, 1978; Snyder, 1984). This applies to the provision of resources and ingenuity, and the serious concern of governments. The home defences of major powers were allowed to languish, long before it could be shown they were a lost cause.

The events considered were not just out-of-control warring when leaderships were overstretched and morally exhausted. In a certain sense they were unusually successful consequences of military strategy. That is because they had been goals, however misguided, of long-range planning and investment. Moreover, the events, although extreme projections of harm to civil life in wartime, were fully symptomatic of the growing threats of modern warfare. Examples go back at least to the American Civil War—often called the "first industrialized war"—and to the shelling and starvation of Paris in the Franco-Prussian War. Colonial military actions of Britain, France, Germany, Italy, Belgium, and other powers were commonly against whole peoples and settlements (Gottman, 1943; Porch, 1986), as was Stalin's strategy in the Ukraine and elsewhere (Elliot, 1972; Conquest, 1986). They included experiment and deployment of such weapons as the machine gun (Ellis, 1975), and aerial bombing of settlements (Divine, 1966), both precedents for what happened in the world wars.

The First World War, fought on all sides with an ethos of the offensive (Howard, 1986), was an overwhelming testament to the superiority of defensive warfare. Instead of capitalizing on that, military men and belligerent politicians saw it as a fatal impasse. They eagerly sought ways to restore the power of the offensive, the "war of movement," and the concept of decisive victory. The result was "blitzkreig," an even longer and more destructive world war, and the events I have examined. Again, although slighted in nearly every way, developments in civil, aerial, and other defensive measures in the Second World War could be highly effective. In the Battle of Britain, the Battle of the Atlantic, or at Iwo Jima, and in a host of other better defended military and urban centres, defensive measures defeated or demanded extraordinary sacrifices of offensive forces. The achievements of counter-city strategic bombing and other attempts to devastate and terrorize civil populations in Europe were themselves extremely costly in men and material, often for small and indecisive results (USSBS, 1945a; Webster and Frankland, 1961).

Germany's aerial defenses were remarkably effective until the Luftwaffe was finally overwhelmed, largely because of "attrition" of men and machines on the Eastern Front. In going against them, in order to bomb German cities, 55,573 aircrew of RAF Bomber Command were killed—a loss every bit as tragic as the "lost generation" of junior officers and NCOs killed on the Western Front in the First World War.

Whatever fraction of the defeat of the Axis powers can be attributed to counter-city bombing, one at least knows that the war was ultimately decided by the clash of the armed forces. Only the final crushing defeat of the Axis air, land, and sea forces by overwhelming Allied military strength brought the war to a conclusion. The expulsions, the fire raids, and V-weapons were in warfare, if not political, terms—"side-shows."[6] The debate continues as to whether the A-bombs made a significant difference in ending the Pacific War (Miles, 1985). They certainly look like "overkill," too much too late, when far more devastation in much more vital cities had already occurred. Moreover, because the attacks were on cities occupied largely by civilians, to suggest they saved Allied soldiers' lives, as appealing as that may be, appears to condone a "crime of war." One may not sacrifice noncombatants in hope of saving military lives *and* remain within "the laws of civilized warfare."

Valid or not, it may seem naive to think these reflections have relevance any longer. The prophets of doom are having more of a field day now, than in the 1930s. There is total pessimism about, and lack of real action to enable, the defense of civil society. And perhaps that is valid for all-out nuclear war. It is not true of the many wars since the Second World War, or of the civil implications of today's arms trade. It is not true of many other, not unrelated emergencies (Chernobyl?). And the civilian complexities that "omnicidal" war plans obliterate, in thought as well as in deed, *are* required evidence for a revitalized sense of the need for a durable peace. They could help reinstate the terms of a (lost) sense of the "laws," or civilized morality, in warfare—once considered the West's greatest contribution (Veale, 1962; Laarman, 1984). In an age when most war plans, conventional as well as nuclear, threaten mass devastation to settlements and their populations, and when these are virtually defenceless against such assaults, perhaps it would be valuable to know, in a detailed and concrete way, just what civilians stand to lose because of that.

*Acknowledgments.* A large part of the article involves materials presented to the Phillip Uren Memorial Lecture 1985, at Carleton University, Ottawa. Thanks to the Social Science and Humanities Research Council (Canada), and the Office of Research, Wilfrid Laurier University, for funding parts of the work. Thanks to Mr. B. Gabbei and Ms. D. Senese for research assistance, and to P. Schaus and P. Carnochan for map preparation.

## NOTES

[1] I assume that by "civil society" most people will take me to mean "non-combatants" in the context of a war. Rather than employ an extended discussion of that, I will proceed empirically, and use the wartime evidence to reconstruct what else it implies. I realize, however, that this begs some important questions. Obviously, the point about "total war" is that it blurs, and for some eliminates, the civil–military distinction. Meanwhile, in other contexts civil society means very different things to different people or disciplines. Much of the literature on material and cultural history treats "civil" and "civic" as virtually the same. It identifies them in urban-based communities. It emphasizes "citizenship" as against slavery or unenfranchised persons. It emphasizes a form of community, in which public life serves to defend the personal security of citizens and their property through equality before the law. That connects with war through the duties of citizens in "defense" taxes, conscription, and so forth. Here, civil society is also often identified with the rise of modern commerce and capitalism (Pirenne, 1956; Black, 1984). Marx seems to have adopted Hegel's view that "civil" is synonymous with "bourgeois" or "burgher" society. However, some of the socialist literature stresses a contrast between "civil society" and "the economy," as well as "the state" (Urry, 1981; Keane, 1984). Then, there is the whole area of "civility" as a way and an ethic of living, with or without a sense of its material and political roots. That is linked to the philosophical and artistic preoccupations with "civilization" à la Sir Kenneth Clark. All of these issues will emerge as relevant to the predicament of unarmed populations which are subject to military assault. In a sense, my investigation here turns upon how it defines just *who* and *whose* "space" one is referring to as "home populations" during wartime. One will begin to realize who are the "definitive civilians," what they are usually doing in wartime, how far it bears on "total war," and what they suffer and lose under armed attack.

[2] Sources on these events are few and not readily available. Most of the literature in English, that I cite, deals with the theme tangentially, in terms of legal–political or "refugee" problems (De Zayas, 1977; Vernant, 1953). The only exception in the popular literature involves sections of Toland's *The Last 100 Days* (1966). Selections from the immense documentation of expellees made in Germany were translated into English (Schieder, 1953). I am aware of no reference to this in the geographical literature, even when it concerns Germany, or in most other fields.

There is no difficulty finding descriptions of the military campaigns which took place here, and maps swarming with arrows to show the movement of troops (Michel, 1975; Salmaggi and Pallavisini, 1979). And it was an extraordinary offensive. In many ways it outdid Hitler's virtually unopposed thrust in the other direction. Soviet military casualties were enormous too. Nevertheless, it was arguably an even more "decisive" affair for the civil populations of these regions.

[3] My argument for devoting space to the description of events here, is again the lack of readily available material in English. Ten years after Daniels (1975) made the same point, I believe it is still valid. Certainly, the geographical and urban literatures miss these events even when discussing Japan. There is, of course, a sizeable "raid" literature, in both the official histories and other military sources. The Reports of the US Strategic Bombing Survey have long been available for those who are interested, but are rarely, if ever, cited in the literature one uses (Daniels, 1981). The vast quantity of civil information and analysis in their archive (Modern Military Archives; National Archives, Washington, DC) remain largely untapped and uncited after several years in the public domain. That is disturbing when the surveys were, without doubt, used to assess and *improve* strategic bombing of settlements and civilians. I cite some of the materials here, having worked on these archives in recent years. Apart from Hiroshima and Nagasaki, it may be noted that most Japanese cities, including Tokyo, have published detailed studies by local scholars that have never been translated.

[4] Interrogation 419, sections 2, 2d—documents 1 and 2, Morale Division, USSBS—these are unpublished documents in an archive.

[5] Unpublished Mass Observation Archive, report 8/J, page 3.

[6] One can never raise this subject without the question of postwar reconstruction, the Japanese and German "economic miracles," being raised. On the one hand, it is amazing and ironic how, within a decade or two, where the destruction was most complete, the record, on the ground, was least. Or rather, it lay in prodigious reconstruction that dwarfed the war memorials. On the other hand, to treat that, as many seem to, as the essential message and even as a vindication of the bombing (as if often heard about London's South Bank Site), seems to me an extreme measure of the brutalizing of modern thought by the technocratic planners. At least, to see it

as such is an equally valid ethical and "aesthetic" judgement. Otherwise, one might start trading off "lost generations" against postwar "baby booms." One might find in the state of Israel or in efforts to outlaw overt anti-semitism, compensation for the death camps! Without trying to trade atrocities, I think the postwar reconstruction of firebombed cities merely emphasizes also the futility and criminality of the untimely death for so many millions and the irrecoverable loss to civilization. Meanwhile, the recoveries themselves have taken place in concert with the development of ever greater weaponries which are able to destroy all cities and all civil majorities (Hewitt, 1983a).

## REFERENCES

Aron, R. 1955 *The Century of Total War* (Beacon Press, Boston, MA)

Black, A. 1984 *Guilds and Civil Society in European Political Thought from the Twelfth Century to the Present* (Cornell University Press, Ithaca, NY)

Blaufarb, D. S. 1977 *The Counterinsurgency Era: US Doctrine and Performance* (Free Press, New York)

Bond, H. (Ed.). 1946 *Fire and Air War* (National Fire Protection Association, Boston, MA)

Bowman, I. 1946 "The strategy of territorial decisions" *Foreign Affairs* 24 177–94

Branfman, R. 1972 *Voices from the Plain of Jars: Life under an Air War* (Harper and Row, New York)

Brodie, B. 1973 *War and Politics* (Macmillan, New York)

Brustat-Naval, F. 1970 *Unternehmen Rettung: Letztes Schiff nach Westen* (Koehlers, Herford)

Bunge, W. 1973 "The geography of human survival" *Annals of the Association of American Geographers* 63 275–95

Caidan, M. 1960 *A Torch to the Enemy* (Ballantine Books, New York)

Chomsky, N., Herman, E. S. 1979 *The Political Economy of Human Rights: Volume I. The Washington Connection and Third World Facism* (Black Rose Books, Montreal)

Committee for the Compilation of Materials on Damage Caused by the Atomic Bombs at Hiroshima and Nagasaki, 1981 *Hiroshima and Nagasaki: The Physical, Medical and Social Effects of the Atomic Bombings* (Basic Books, New York)

Conquest, R. 1970 *The Nation Killers: The Soviet Deportation of Minorities* (Macmillan, London)

Conquest, R. 1971 *The Great Terror: Stalin's Purge of the Thirties* revised edition (Penguin Books, Harmondsworth, Middx)

Conquest, R. 1986 *The Harvest of Sorrow: Soviet Collectivization and the Terror-famine* (Oxford University Press, Oxford)

Craven, W. F., Cate, J. L. (Eds), 1953 *The Army Air Forces in World War II. Volume 5: The Pacific, Matterhorn to Nagasaki* (University of Chicago Press, Chicago, IL)

Daniels, G. 1975, "The great Tokyo air raid, 9–10 March, 1945," in *Modern Japan: Aspects of History, Literature and Society* Ed. W. G. Beasley (University of California Press, Berkeley, CA) pp 113–31

Daniels, G. (Ed.), 1981 *A Guide to the Reports of the United States Strategic Bombing Survey* (Royal Historical Society, London)

De Zayas, A. M. 1977 *Nemesis at Potsdam: The Anglo-Americans and the Expulsion of the Germans; Background; Executions, Consequences* (Routledge and Kegan Paul, Andover, Hants)

Divine, D. 1966 *The Broken Wing: A Study in the British Exercise of Air Power* (Hutchinson, London)

Earle, E. M. (Ed.), 1943 *Makers of Modern Strategy* (Princeton University Press, Princeton, NJ)

Elliot, G. 1972 *Twentieth Century Book of the Dead* (Penguin Books, Harmondsworth, Middx)

Ellis, J. 1975 *Social History of the Machine Gun* (Pantheon Books, New York)

Fitzgibbon, I. 1975 *Katyn: Triumph of Evil* (Anna Livia, Dublin)

Ford, J. C. 1944, "The morality of obliteration bombing" *Theological Studies* 5 261–309

Fussell, P. 1975 *The Great War and Modern Memory* (Oxford University Press, Oxford)

Gollancz, V. 1946 *Our Threatened Values* (Victor Gollancz, London)

Gollancz, V. 1947 *In Darkest Germany* (Henry Regnery, Chicago, IL)

Gottman, J. 1943, "Bugeaud, Gallieni, Lyautey: the development of French colonial warfare," in *Makers of Modern Strategy: Military Thought from Machiavelli to Hitler* Ed. E. H. Meade (Princeton University Press, Princeton, NJ) pp 234–59

Guillian, R. 1947 *Le Peuple Japonais et la Guerre* (Juillard, Paris)

Hane, M. 1982 *Peasants, Rebels and Outcasts. The Underside of Modern Japan* (Pantheon Books, New York)

Harrisson, T. 1976 *Living Through the Blitz* (William Collins, London)

Hastings, M. 1979 *Bomber Command* (Pan Books, London)

Havens, T. R. H. 1978 *Valley of Darkness: The Japanese People and World War Two* (W W Norton, New York)

Hewitt, K. 1982a, "Air war and the destruction of urban places" DP-8041, Office of Research, Wilfrid Laurier University, Waterloo, Ontario, Canada N2L 3C5

Hewitt, K. 1982b, "Settlement and change in basal zone ecotones: an interpretation of the geography of earthquake risk," in *Social and Economic Aspects of Earthquake* Eds. B. G. Jones, M. Tomazevic; Proceedings of the Third International Conference, Institute of Earthquake Engineering and Engineering Seismology, Skopje, Yugoslavia, pp 15–42

Hewitt, K. 1983a, "Place annihilation: area bombing and the fate of cities" *Annals of the Association of American Geographers* 73 257–84

Hewitt, K. 1983b, "The idea of disaster in a technocratic

age," in *Interpretations of Calamity* Ed. K. Hewitt (Allen and Unwin, Jemel Hempstead, Herts) pp 3–32

Hewitt, K. 1985, "Strategic air power against the Japanese civilian, 1945; the threat and its preconditions" RP-8574, Wilfrid Laurier University, Waterloo, Ontario, Canada N2L 3C5

Hewitt, K. 1986, "The meaning and loss of place in war: Japanese civilians in World War Two," in *FUDO: An Introduction* Eds. A. V. Liman, F. S. Thompson (University of Waterloo Press, Waterloo, Ontario) pp 44–58

Hogg, I. V. 1978 *Anti-aircraft: A History of Air Defence* (Macdonalds and Jane's, London)

Howard, M. 1986, "Men against fire: the doctrine of the offensive in 1916," in *Makers of Modern Strategy: From Machiavelli to the Nuclear Age* Ed. P. Paret (Princeton University Press, Princeton, NJ) pp 510–26

Hsu, L. H., Ming-Kai, C. 1971 *History of the Sino-Japanese War 1937–1945* (Chung Wu Co, Taipei)

Huntington, S. P. 1968, "The bases of accommodation" *Foreign Affairs* **46** 642–56

Ibuse, M. 1969 *Black Rain* (Kodansha International, New York)

Ienaga, S. 1978 *The Pacific War: World War II and the Japanese, 1931–1945* (Pantheon Books, New York)

Ikle, F. C. 1958 *The Social Impact of Bomb Destruction* (William Kimber, London)

Kantor, M. 1965 *Mission with LeMay: My Story* (Garden City, New York)

Keane, J. 1984 *Public Life and Late Capitalism. Towards a Socialist Theory of Democracy* (Cambridge University Press, Cambridge)

Kosaka, M. 1971 *100 Million Japanese: The Post-war Experience* (Kodansha International, Tokyo)

Laarman, E. J. 1984 *Nuclear Pacifism: 'Just War' Thinking Today. American University Studies* (Peter D. Lang, New York)

Lanchester, F. W. 1916 *Aircraft in Warfare* (Constable, London)

Lewallen, J. 1971 *Ecology of Devastation: Indochina* (Penguin Books, Harmondsworth, Middx)

Liddell-Hart, B. H. 1967 *The Revolution in Warfare* (Yale University Press, New Haven, CT)

Lifton, R. J. 1973 *Home from the War: Vietnam Veterans, neither Victims nor Executioners* (Simon and Schuster, New York)

Liman, A. V. 1986, "Furusato: place in modern Japanese literature," in *FUDO: An Introduction* Eds. A. V. Liman, F. S. Thompson (University of Waterloo Press, Waterloo, Ontario) pp 81–97

Linsay, M. 1975 *The Unknown War: North China 1937–1945* (Begstrom and Boyle, London)

Ludendorff, E. 1935 *Der Totale Krieg* (Ludendorffs, Munich)

McReavy, L. L. 1941, "Reprisals: a second opinion" *Clergy Review* **20** 131–38

Michel, H. 1975 *The Second World War* (Praeger, New York)

Middlebrook, M. 1980 *The Battle of Hamburg: Allied Bomber Forces Against a German City in 1943* (Allen Lane, London)

Miles, R. E. 1985, "Hiroshima: the strange myth of half a million American lives saved" *International Security* **10** 121–40

*New York Times*, 1944, "Our superforts bomb cities in Japan," 16 June, reprinted in *Japan: The Great Contemporary Issues* Ed. E. O. Reischauer (Arno Press, New York) pp 127–28

*New York Times*, 1945, "Fire bombing Japan," 21 March, reprinted in *Japan: The Great Contemporary Issues* Ed. E. O. Reischauer (Arno Press, New York) p 130

Osada, A. 1959 *Children of the A-Bomb* (Uchida Rukakuho, Tokyo)

Pacific War Research Society, 1972 *Japan's Longest Day* (Kodansha International, Tokyo)

Paret, P. (Ed.), 1986 *Makers of Modern Strategy: From Machiavelli to the Nuclear Age* (Princeton University Press, Princeton, NJ)

Pirenne, H. 1956 *Medieval Cities: Their Origins and Revival of Trade* (Doubleday, New York)

Porch, D. 1986, "Bugeaud, Gallieni, Lyautey: the development of French colonial warfare," in *Makers of Modern Strategy: From Machiavelli to the Nuclear Age* Ed. P. Paret (Princeton University Press, Princeton, NJ) pp 376–407

Proudfoot, M. T. 1956 *European Refugees: A Study of Forced Population Movement* (Faber and Faber, London)

Remarque, E. M. 1929 *Im Westen nichts Neues* (Propyläen, Berlin)

Richardson, L. 1960 *Arms and Insecurity: A Mathematical Study of the Origins of War* (Boxwood Press, Pacific Grove, CA)

Rumpf, H. 1963 *The Bombing of Germany* (Holt, Rinehart and Winston, New York)

Salmaggi, C., Pallavisini, A. 1979 *2194 Days of War: An Illustrated Chronology of the Second World War* (Mayflower Books, New York)

Schieder, T. (Ed.), 1953 *The Expulsion of the German Population from the Territories East of the Oder–Neisse Line; A Selection and Translations of "Dokumentation der Vertreibung der Deutschen aus Ost–Mitteleuropa"* Federal Ministry for Expellees, Refugees and War Victims, Bonn, FRG

Schmideberg, M. 1942, "Some observations on individual reactions to air raids" *International Journal of Psychoanalysis* **23** 146–76

Scott, F. D. (Ed.), 1968 *World Migration in Modern Time* (Prentice-Hall, Englewood Cliffs, NJ)

Shy, J., Collier, T. W. 1986, "Revolutionary war," in *Makers of Modern Strategy: From Machiavelli to the Nuclear Age* Ed. P. Paret (Princeton University Press, Princeton, NJ) pp 815–62

Snow, C. P. 1961 *Science and Government* (Oxford University Press, Oxford)

Snyder, J. 1984, "Civil-military relations and the cult of the offensive, 1914 and 1984" *International Security* **9** 108–46

Solzhenitsyn, A. 1977 *Prussian Nights: A Poem* (Farrar, Straus and Giroux, New York)

Somerville, J. 1985, "Nuclear 'war' is omnicide," in *Nuclear War: Philosophical Perspectives* Eds. W. A. Fox, L. Groarke (Peter D. Lang, New York) pp 3–10

SPRI, 1975 *Incendiary Weapons* Stockholm Peace Research Institute (MIT Press, Cambridge, MA)

Stephenson, J. 1981 *The Nazi Organization of Women* (Croom Helm, Beckenham, Kent)

Stone, I. F. 1967 *In Time of Torment* (Random House, New York)

Terkel, S. 1984 *The Good War: An Oral History of World War Two* (Pantheon Brooks, New York)

Thorwald, J. 1953 *Flight in the Winter: Russian Conquers, January to May 1945* (Pantheon Books, New York)

Thrift, N., Forbes, D. 1986 *The Price of War: Urbanization in Vietnam 1954–1985* (Allen and Unwin, Hemel Hempstead, Herts)

Titmuss, R. M. 1950 *Problems of Social Policy* History of the Second World War, United Kingdom Civil Series, 4 volumes, Ed. W. K. Hancock (HMSO, London)

Toland, J. 1966 *The Last 100 Days* (Random House, New York)

Tolstoy, N. 1977 *Victims of Yalta* (Hodder and Stoughton, Sevenoaks, Kent)

Urlanis, B. 1971 *Wars and Population* (Progress Publishers, Moscow)

Urry, J. 1981 *The Anatomy of Capitalist Societies: The Economy, Civil Society and the State* (Macmillan, New York)

USSBS, 1945a, "Overall report (European war)" United States Strategic Bombing Survey (US Government Printing Office, Washington DC)

USSBS, 1945b, "The effect of bombing on health and medical care in Germany" United States Strategic Bombing Survey (US Government Printing Office, Washington DC)

USSBS, 1946, "Summary report (Pacific war)" Office of the Chairman, United States Strategic Bombing Survey (US Government Printing Office, Washington DC)

USSBS, 1947a, "Final report covering air-raid protection and allied subjects in Japan" Civil Defense Division, United States Strategic Bombing Survey (US Government Printing Office, Washington DC)

USSBS, 1947b, "Effects of air attack of urban complex Tokyo–Kawasaki–Yokohama" Urban Areas Division, United States Strategic Bombing Survey (US Government Printing Office, Washington DC)

USSBS, 1947c, "The effects of bombing on the health and medical services in Japan" Medical Division, United States Strategic Bombing Survey (US Government Printing Office, Washington DC)

Veale, F. J. P. 1962 *Advance to Barbarism: The Development of Total Warfare from Sarajevo to Hiroshimo* (The Mitre Press, London)

Vernant, J. 1953 *The Refugee in the Post-war World* (Allen and Unwin, Hemel Hempstead, Herts)

Walzer, M. 1977 *Just and Unjust Wars: A Moral Judgement with Historical Illustrations* (Basic Books, New York)

Weber, M. 1958 *The City* (Free Press, New York)

Webster, C., Frankland, N. 1961 *The Strategic Air Offensive against Germany, 1939–1945* History of the Second World War, United Kingdom Military Series, 4 volumes, Ed. W. K. Hancock (HMSO, London)

Weigley, R. F. 1973 *The American Way of War: A History of United States Military Strategy and Policy* (Macmillan, New York)

Westing, A. H. 1976 *Ecological Consequences of the Second Indo-China War* (Almqvist and Wiksell, Stockholm)

Westing, A. H. 1984 *Herbicides in War: The Long-term Ecological and Human Consequences* (Taylor and Francis, London)

Wolff, L. 1959 *In Flanders Fields: The 1917 Campaign* (Penguin Books, Harmondsworth, Middx)

Zinn, H. 1970 *The Politics of History* (Beacon Press, Boston, MA)

 PART VI

# EPILOGUE

Throughout this book, I attempted to integrate the research traditions in natural, technological, and social hazards, disaster studies, and risk assessment. In doing so, I am struck by the great commonalities of purpose and approaches that increasingly blur the disciplinary boundaries. As the risks and hazards become more complex, so too must our approaches to understanding them.

Julian Wolpert wrote an exemplary piece (Reading 29) more than ten years ago that weaves together many of the themes in this anthology. It also highlights the multiplicity of approaches in understanding hazards management and illustrates one person's attempt to come to grips with "The Dignity of Risk."

# THE DIGNITY OF RISK

*Julian Wolpert*

## INTRODUCTION

In the past few years I have returned to the examination of the problem of risk which was of great interest to me many years ago. At that time I was very concerned with the role of irrational and arational decision-making behaviour for which there was ample evidence in the landscape and in the movement of people. In the intervening years I became more interested in the threatened incursion of public "bads" which were unwanted and inequitable and which were sprouting all over the landscape. This preoccupation with the issues of threat and conflict then led me to study groups labelled deviant, including mentally handicapped populations, and the implications of their re-entry into urban communities.

My current research merges much of this previous study of conflict and location behaviour into a concern with competency and dangerousness as reflected by ideas developed to apply to the mentally handicapped but, nevertheless, clearly evident in the behaviour of

Reprinted from *Transactions, Institute of British Geographers*, Volume 5 (4) (1980), pp. 391–401. Used with permission of the Institute of British Geographers.

normal people. These concepts are being formulated through the development and analysis of comparative case studies of man-produced hazardous events. The range of the cases include nuclear accidents, urban bankruptcy and riots, incidents of individual violence and block-busting through gentrification, to name only a few. These are cases of high to low risk and high impacts which must be analysed in comparison with the phenomena of high risk and low impact which have become quite commonplace. The rare events, however, are additive in their occurrence or threat of occurrence and, therefore, constitute an important source of disruption to our residential life. The rare events disturb our sense of competency and sense of security, stability and permanence and, thereby, threaten a very basic and elemental need and source of satisfaction.

The cases demonstrate that our lack of competency in forestalling a significant range of technological and institutional hazards introduces public choices between alternatives, all of which have potentially dangerous outcomes. The major objective of the research, therefore, is to demonstrate that for a number of types of hazardous events, especially nuclear power-plant disasters and urban displacement, insufficient precautions have been taken, given the "state of the art" in analysing risky outcomes. On the other hand, I hope that the demonstration will show that other types of risky outcomes, for example self-harm or harm to others by mentally handicapped people, are treated by excessive over-precaution.

## TECHNOLOGICAL AND INSTITUTIONAL HAZARDS

When the urban resident considers the sources of hazard in his existence, he is likely to feel that the institutional and technical ones seem more ominous than the natural disaster, about which so much has been written in geography. More threatening are the daily doses of air pollution, toxic waste deposits, the threat of arson, the bankruptcy of one's employer or even one's municipality, displacement from one's home and a host of other dangers which clearly measure up to floods, earthquakes and hailstones as conveyors of threat and as contributors to ulcers and strokes. These institutional and technological hazards are man-induced but, nevertheless, claim the geographer's attention. The technological hazards arise from gaps in the enlarged capacity which technology and institutional structures make possible. The significance of these hazards is not in question. More important are their means of prediction, control and management so that impacts can be curtailed or reduced.

### Rare Events

Complex as are the natural hazards, the complexity of the social and technological hazards is probably a good deal more severe. Or it may be that social scientists simply prefer to define their problems so that they are unsolvable. Whichever is the case, I shall be presenting several examples, drawn from current research projects and cases which concern situations hazardous enough to cause temporary or permanent displacement of people.

The hazards which I shall be discussing involve acute episodes, that is, high-impact rare events.[1] These episodes differ from the hazards of high probability and low impact, such as smog, water pollution, noise and petty crime whose frequency permits some degree of predictability as to time of occurrence, duration and magnitude. These everyday hazards are indeed worthy of study and have, in fact, recently received a good deal of attention in the geographical literature as stimulants for disruptive or dysfunctional population movement, which is an important indicator of structural problems. The rare event which is social in origin, however, has only recently been treated with the respect it deserves.

The rare event which is catastrophic in magnitude has either never occurred historically or occurs with such low probability that its next occurrence cannot be predicted. Rare events do not lend themselves generally, as do highly probable hazards, to rigorous estimation, assessment, management and control. Instead, the rare event is treated with what one hopes is over-precaution.

Technology has managed not only to augment in both absolute and relative ways the exposure to risk of our average citizens but may have multiplied the opportunities for measuring and assessing the added risk. So we should have a lot more that we are aware of, about which to worry. The equity implications of risk assessment and management are also not well understood. The outcomes of higher risk and low impact are generally more differentially selective in their distribution of impacts. Catastrophes should be more equalizing. The environmental impact statements and the cost benefit tableaux currently being devised still give little attention to risk assessment and to the quality of the "fail-safe" buffers under extreme conditions.

The interesting phenomenon in the recent research concerned with risk impact is the shift in emphasis from the individuals to the collectivity, the population subgroup, the people at one place, the institution, those affected by a new legislative programme or new technology. Not only is there a greater awareness of communality of risk incidence, but we speak more frequently about the specialized risks associated with being elderly, handicapped, poor or employees of the Chrysler Corporation. Only a decade or so ago the risk literature dealt extensively with the virtues and dignities associated with a modicum of risk, as a stimulus for creativity in solving problems. Exposure to some risk was considered not a privilege but a civil right and was used as an argument against over-protectiveness. To be denied the opportunity to make choices which carried risk was assumed to be anti-therapeutic in raising children and in attending to the needs of the handicapped and incarcerated.

The cases demonstrate how limiting were our traditional programme policies and project evaluation methods such as cost benefit, cost effectiveness and goals achievement. These methods, as they have been used traditionally, give most prominence to certainty or to the high-risk outcomes of known magnitude. Not enough attention in the comprehensive enumeration of the myriad of possible outcomes was devoted to identification of the potential for the catastrophic event, the outcome which is so overarching as to warrant dominating the policy environment. At a stage in which a major role for public policy consists of risk containment and management, the prevention of worst outcomes or catastrophes has become of highest priority.

My comparative analyses of the case studies are at an early stage. The analyses focus on the sources of risk as well as the degree of competency or knowledgeability which exists to mitigate the potential impacts. The cases demonstrate a very general tendency, the need to be highly dependent upon experts, where competency, even by experts, is difficult to attain. The cases also demonstrate the tendencies and incentives for experts to minimize the degree of peril or dangerousness.

Some preliminary comments must be made on the concept of competency, both with respect to the assessment of risks and the mitigation of impacts. Equally important is the meaning of hazard or dangerousness as related to people, places, technology and institutions.

## Population Displacement

The gulf between the limits of our competency and the dangerousness which is inherent in a complex society is demonstrated well in the cases of population displacement we have been observing. Few events seem as threatening as the need to be evacuated or displaced permanently from one's home, but this danger has become more commonplace. What is a greater challenge to our competency than our inability to enjoy some measure of stability or permanence?

Some forms of evacuation are required by the incidence of rare events which are in the category of natural hazard—the typhoons, floods and earthquakes. Others result from the technological and management creations which comprise the man-made hazards. The organizational hazards arise from the potential breakdown of the elaborate system of institutional buffers which have been created to cushion exposure to man-made or man-induced catastrophes. The technological hazards arise from the failure of our machines and production implements. The issue of displacement is singled out for analysis here but is only one of a set of worst outcomes that could be used to illustrate the themes of competency and dangerousness.

Despite the significant number of Americans evacuated from Florida hurricanes, California forest fires and Pennsylvania floods in the past few years, a dramatic shift has occurred in thinking about technologically linked evacuation. This type of evacuation has occurred in the vicinity of the Three Mile Island accident, the Love Canal, the Toronto suburbs, cases we have been analysing in our study. The release of radiation and iodine through core meltdown, the seepage of toxic waste into underground water systems, the release of toxic chemicals into the atmosphere because of ship, train and truck accidents have all been occasions for population evacuation, temporary in most cases but permanent in the case of the Love Canal.

## The Nuclear Hazard

The Three Mile Island incident is perhaps the most ominous because of the sheer magnitude of the potential catastrophe. Following the accident I looked back to the original environmental impact statement to find the risk-benefit statement, but to no avail. "Core meltdown is as likely an event as a meteorite striking Times Square on Easter Sunday, 1980 at 12:01 p.m."

The contrast is very significant between the original environmental impact statement filed prior to the construction and licensing of the nuclear power facility at Three Mile Island and the report of the Kemeny Commission which has just completed its study of the accident. One skips the hundreds of pages of costly impact analysis of high-risk assessment of impacts to fish, wildlife and wages for construction labour, to seek discussion of a catastrophic accident and its likelihood of occurrence; or discussion of evacuation plans. Risk management for such unlikely events was thought to be too alarming or threatening or too risky, politically, to be included. However, the Three Mile Island incident has altered our thinking about the trade-off between high-risk fossil fuel carcinogenics as opposed to the low risks of nuclear accidents. In an age in which we prefer not to think about "worst outcomes" we, as a society, show a strong preference for accepting and tolerating the more risky but less awesome outcomes. We would, however, be better equipped to "contain" the catastrophe once we admit to its finite probability of occurrence and develop management strategies which pay greater attention to the cost or effort of prevention

than to the simpler technology of moderating or compensating for the more subtle environmental impacts.

## Fault-Tree and Event-Tree Analysis

One of the major distinctions between man-made rare events which can lead to catastrophe and the natural hazards is the notion that the event is possibly preventable. The natural hazard, the earthquake or typhoon is not preventable. One can act so as to mitigate impacts but not forestall the occurrence of the natural phenomena. Of course, a continuum may be said to exist between hazards which are totally man-made and those which are exclusively natural, presuming that man's activities are not "natural." The man-made or institutional hazards are typically subject to the development of a logic structure which can be used to analyse the preventability of the rare event. The Rasmussen report on nuclear accidents, for example, used the methods of the fault-tree/event-tree analysis.[2] The fault-tree method allows one to proceed back in time from possible catastrophic accidents to examine components of the sequence which have given probabilities of failure. The event-tree method allows the observer to proceed forward in time from potential component failures to their accident implications. These methods can make the study of such accidents somewhat more systematic. They establish a classification of some potential accident sequence and permit identification of procedures for estimating risks associated with these sequences. Of course, a data base must exist on the risk of failure of critical component elements. The methods cannot be totally comprehensive. The essential element is the attempt at completeness and the ability to provide assurance that only minor contributors to accidents have been left out.

Risk-benefit analysis based upon methods such as fault-trees and event-trees have proceeded a long way in the last several years for examining hazards from nuclear plants, food additives and pharmaceuticals. The largely uncharted area is in the prediction of institutional, legislative and judicial impacts. Not that private market effects are all that well understood, but even benevolent direct public sector programmes and regulation of private sector activities are subject to policy slippage and failures which have enormous and selective consequences. If one assumes that programme failures are not entirely traceable to deliberate sabotage of policy objectives, then much of the residual cause can be linked to inadvertent outcomes which are potentially preventable. The public sector programmes, for example, which are now reinforcing private gentrification of our cities can be subject to risk-benefit impact analysis. Fault-trees can be analysed, as we are now doing, to examine impacts before displacement needs to take place. Methods are not well formulated for analysing institutional failures but much can be accomplished in developing data bases for assessing the risks of policy failure.

Fault-tree studies of reactor accidents have established, for example, that between one-third and two-thirds of hypothesized accidents are caused by human error. What is human error—frailty, vulnerability, incompetency in the assigned task? Human error was largely to blame for the Three Mile Island incident. The assessment of risk from human error is regarded by experts as more difficult than that resulting from other causes. When one superimposes the maps of seismic risk and population density over the maps showing the location of nuclear reactor sites, along with their management and operations safety rating which varies considerably, the aggregate distribution of catastrophe potential is probably high but still to an unknown degree.[3] When one adds to this distribution the potential danger from sites now under construction and those planned to begin construction soon, that composite picture of the distribution of hazard is complex indeed but, nevertheless, susceptible to a rational assessment.

In the Three Mile Island accident almost 30 per cent of the reactor core was damaged, although the core did not melt. Most of the potential risk occurs as a result of cracks, ruptures or explosions of the containment from a complete core melt. A containment breach might release radioactive debris over large distances, depending on wind velocity, direction and rain. These distances might be considerably beyond the ten-mile limit for emergency evacuation now being discussed. Most nuclear plants had no approved plan for evacuation but now mandatory plans will be required for all sites.

## Hazard Management

The important issue to which reference was made earlier is the difference between the high-probability low-level impact and the low-probability catastrophe. With hindsight it is possible to return to each of the rare events we have been investigating and find clues which are portents of the catastrophe. The documentation on Three Mile Island shows frequent examples of operator and management error prior to the accident. The argument presented here is that the indicators of the crisis can invariably be noticed at early stages when outcomes are not so precarious. The problem, of course, is filtering through all of the highly probable minor hazards to determine that string of actions or events which will culminate in the catastrophe.

Just five years ago I worked with a team developing a model for evacuating a population threatened by core meltdown at a nuclear power plant.[4] The object of the study was to provide guidelines on the management of land use and transportation routes in the vicinity of such facilities. The simulation model permitted consideration of a number of scenarios of nuclear accident designed to facilitate planning for rare events. Our team was disbanded, however, being unable to find encouragement or support to continue its work until after the Three Mile Island incident. This near-catastrophic event has unearthed a pre-cautionary fervour of crisis planning which is still unabated.

## Hazard Planning

We cannot overestimate the importance of the occurrence of this rare event because of its extraordinary influence upon policy. Contingency and strategic planning for such rare events has become the highest priority role for the public sector, resembling the era of wartime crises. My objective here is to encourage continued study of the more subtle high risk but low impacts of institutions and technology and simultaneously to convey my sense of the opportunity and impetus for more risk assessment of technological as well as institutional hazards. The high-risk low-impact events are cumulative. They may be seeds or clues of future catastrophes or themselves add up to catastrophic effects.

Hundreds of people were evacuated from the vicinity of Three Mile Island and remained away from their homes for periods of up to one week but a much larger number of Americans suddenly became aware of the nuclear plant in their own vicinity and their own vulnerability. The future time bombs, of course, are the toxic waste disposal sites which are scattered all over the country, and which are the subject of other case studies.

## Urban Displacement

A second type of case drawn from the current project concerns the revitalization of urban areas. Some renaissance of inner cities is now taking place on an international scale. The decades of urban decline have left in their wake some considerable "warehousing" of both people and land sites. In those U.S. cities for which we are able to assemble valid data and in which we can detect renewed interest by the private sector developers, we find significant gentrification. While public sector investments continue to flow into the mod-erate-income neighbourhoods to assist in spot rehabilitation of dwelling units and public services, the private sector is gradually converting low-income areas into havens for the new urbanists—the young and affluent professionals. These processes have occurred in the wake of crisis planning concerning city survival.

The rationale for current public and private activity is that resources are sufficient to insure city survival only if the stable middle-income groups can be induced to stay and the affluent can be attracted back. Current urban development and redevelopment is premised upon continued high-probability low-impact hazards for some groups, namely the poor, but reduced risk of the catastrophe of municipal bankruptcy, from which all groups would suffer. The fault-tree analysis of bankruptcy has settled too easily on an inequitable method of prevention. Here, as in the other cases, the low-income and less privileged population is expected to absorb the higher risk and spill-over effects from the lower- and moderate-level impacts. Here, too, crisis planning has become synonymous with public interest. One can argue that the cumulative effect of all this exposure to impacts which are themselves high-risk low-impact adds up to near catastrophe for some population subgroups, especially for those impacts which are not easily measurable nor compensatable.

The social hazards are not less ominous with respect to population displacement. It is difficult indeed for the individual to understand why his displacement is a necessary component of a structural solution. Displacement of low-income residents to make way for urban revitalization has become a common enough event to have an aggregate impact at least as severe as the catastrophic bankruptcy, which we try so hard to prevent.

In current studies of urban displacement, public sector efforts often extend the effects of gentrification which are rooted in private market forces. Low-income people fear revitalization because it must exclude them—

they are the problem, the failure or fault, for which their displacement is the solution. It is hazardous for a low-income household to live in an area of architectural or historic interest or in a zone peripheral to large-scale commercial development. Somehow, low-income housing is always reduced in supply—a phenomenon which defies solution by the combined talents of the private and public sector. Is it because the organizational task is inherently so complex or is it because known solutions are too complex or too deeply structural to implement? Here, too, one finds elaborate and optimistic legislative and programme impact statements but never a risk-benefit analysis of the expected impacts. Having so recently learned about the awesome consequences of displacement due to urban renewal, highway and airport construction, one would have expected improvement in the skills in associating displacement risk with our new urban programme. Geographers as well as other social scientists should be able to analyse each new programme in terms of its distribution of risky impacts and the resulting equity implications, as they should be able to determine the risk benefits for new sitings of nuclear plants and nuclear waste disposal sites.

## Competency and Dangerousness

The displacement of urban residents owing to redevelopment has strong equivalences in the placement of supposedly high-risk groups such as the mentally handicapped and criminal offender populations, the subjects of other case studies. The argument here is that these individuals need to be displaced for the public good. Their displacement to a protective zone constitutes a "merit good" to the rest of society because the negative spillovers can be internalized at a place removed from the mainstream of society. The ghettoization of "dangerous" people is at least partly intended to provide protective security to those who would otherwise harm themselves or others or be harmed by others. They are perceived as incompetent in the skills of social adjustment and dangerous to the rights of others, a similar rationale to that behind the displacing of the low-income, urban population. Places become less hazardous because of the removal of dangerous individuals and groups.

The common themes that run through our analysis of risk assessment are the issues of competency and dangerousness as reflected by people and places in a technological and institutional context. Competency in the human dimension implies adequate or satisfactory performance levels and a rational approach to risky outcomes, at least as defined by the general social science paradigms of economics, sociology and political science. Inadequate performance implies a shortcoming or deficiency on the part of the actor and may imply dangerousness. Our social science models generally only attempt to characterize competent behaviour by individuals and groups rather than the range of distribution from incompetent to optimal. The place dimension is also characterized on the dimension of competency or knowledgeability. Here, too, our models generally assume competent and functional land use and infrastructure after making allowances for risky outcomes. Even though technology and institutional structures are motivationally neutral, they may also be considered competent or knowledgeable in relation to their functions. Again, the most common paradigm is one which assumes that competent people create functional technology and institutions which shape a functional environmental context.

In our models we dwell upon this rational expectation of competency and pay less attention to the range and deviations. Institutions, programmes and policies generally operate according to a model which presumes competency and directedness. The incompetent and the inadequate are outside the participatory mainstream. The issue, of course, is that the notion of competency results from a process of labelling rather than from any inherent differentiation. Competency is evaluated in relation to an ever-increasing and ever-more-demanding technology and institutional structure which, each year, places more of us in the incompetent group. There are fewer and fewer things, for example, that even mildly retarded people can do any longer, though they have not changed. For marginal citizens and marginal areas, extraordinary remedies are targeted not to the structural change that would make mainstream participation possible but to making participation unnecessary.

The linkage of competency and dangerousness has been discussed most widely in the legal literature specifically on the issue of the catastrophic rare event, of self-harm or harm to others by those in custody. Until very recently, only a narrow band of behaviour was reserved for "normalcy." The civil rights movement and the development of a more open and tolerating society has only begun to widen the band which we label normal. Incompetency implies a sufficient low level of functioning to warrant a custodian or custodial care, the adult version of parental care. Dangerousness implies a high likelihood of committing self-harm or serious harm to specific others. The harm is non-restitutable and non-reversible. The determination of dangerousness, as with incompetency, warrants guardianship, but in this case through incarceration, another form of deprivation of liberty. Incompetency may not be uniform over all functions and roles and neither will dangerousness be constant nor consistent in all social and environmental contexts. In the case of dangerousness, however, the probability of hazard must be both high and not a rare event. Those most likely to be deprived of civil rights are those at the extremes—incompetent in important life functions and likely repeaters of violent acts. The rest of us who are incredibly inept at only certain functions and only occasionally given to committing violent self-harm or violence to others usually go scot-free as normal persons in an open society, as do the managers of nuclear plants.

The legal literature is also quite clear on the role of technology in these determinations.[5] Decisions cannot be made by clinical experts or tests but must be made by a jury of peers and designated judges with only lay expertise, although the expert can be quite influential.

There is little science or technology in the sensitive task of evaluating the probability that a given mentally handicapped person or an offender is likely to be violent to himself or his fellow men. If the expert were asked to show proof of his predictive skills, objective data could not be offered. The most he can say is that he has had considerable experience in dealing with disturbed people who commit dangerous acts, that he has been designated by society to diagnose and possibly treat such individuals and that his skill in treating dangerous behaviour in those diagnosed as mentally ill has generally been appreciated.[6] Psychiatric evaluation of dangerous behaviour, therefore, remains a matter of clinical judgement. This means the examiners must rely upon past experience, anecdotal material, a good many theoretical models and often plain intuition. While clinical judgements must be respected they are hardly a scientific basis for indeterminate commitment.

The diagnosis of dangerousness on the individual level has a high probability of error. Observations by those designated as experts cannot occur in the same situational contexts in which dangerous acts usually occur. Furthermore, the experts know little about predispositions to commit violent acts or the factors which trigger the commission of such acts on the individual level. Neither is much known about recidivism of violence.

Despite this humility the psychiatrist and the clinical psychologist exert a great deal of influence upon a determination of dangerousness and usually in the direction of overprecaution. An equivalent process takes place in the determination of incompetency. A previous history of the use of physical force to resolve conflicts is commonly found in those who ultimately commit dangerous crimes, and it is especially true when such behaviour has occurred with some frequency. Most criminals do not commit dangerous acts. Criminality, however, implies a willingness to solve a problem of adaptation by breaking rules, and further implies an enhanced likelihood to behave in a dangerous manner when confronting a situation in which attacking others will temporarily solve a problem. The clinician is guided by warning signals and makes an effort to define those situations or environmental variables which increase the probability that a given person will harm himself or specific others. The evaluator should make statements which suggest procedures to minimize dangerous potentialities.

This discussion of competency and dangerousness is familiar to those who know the geographical literature on natural hazards and their management and control. This is a literature which has taken seriously the issue of people's competency in controlling, preventing or mitigating the dangerousness of places through appropriate use of technology. This literature has also not limited its analysis to a narrow band of rational behaviour but has explored the irrational, the incompetent and the dangerous responses by man, as well.

The competency to assess dangerousness does not proceed in a straight line but has plateaux, reversals and some positive spurts. The unevenness can be traced to innovations in measurements and instrumentation which improve the knowledge dimension. At the same time the objective hazard or peril may be abating or augmenting. The dynamics of these processes can be illustrated from our example of the mentally handicapped where improved measurement makes it possible to provide a typology according to level of functioning in a number of important life skills. This typology replaces the now obsolete three-fold classification: mentally deficient; trainable; and educable. The substitution of the functioning scales provides a better opportunity for assessing an individual's ability to cope with his external world and, at the same time, the scale provides some indicators of the ad-

ditional skills which must be acquired to permit greater independence and self-sufficiency.

Incompetency and dangerousness can be combined in the same person who may be incompetent in some areas of functioning but, nevertheless, effectively dangerous to himself or others. Incompetency by the operator or manager of a nuclear power plant or by gatekeeper public officials, of course, can affect many more people. It is difficult to say that predictability is now better in the assessment of dangerousness of the individual but recent empirical studies are beginning to give us better information on what is not known and the dangers of relying upon experts.

## The Community As Institution

The issue of dangerousness arising from urban perils is a perplexing one. Urban residents are now much more acutely aware of the degree of overprotection they once experienced, now that deinstitutionalization of the mentally handicapped and the ex-offender has been occurring. Fewer people can be classified as dangerous now that advocacy for civil rights has been extended to these groups. Conflicting goals are maintained in delicate balance: to provide society with protection from violent acts and, simultaneously, to avoid improper incarceration.

For the mentally handicapped and criminal offender population, incarceration was intended to serve a dual purpose, i.e., to rehabilitate and to protect society. A parallel risk assessment differentiates the minor as opposed to the more major outcomes arising from a more liberal incarceration policy. A reduced level of institutionalization has occurred for juveniles, white-collar criminals, the mildly retarded and the less seriously disturbed mentally ill population. This trend has added significantly to the high risk, low level of impact on communities and society in general. The traditional institutions are devoting themselves more exclusively to catastrophe prevention, i.e., separating those individuals from society who are suspect of being a danger to self or others, albeit with a low probability of committing such acts. The standard for incarceration requires "clear and convincing" evidence of likelihood as well as severity of a future dangerous act. A high risk of murder or suicide clearly warrants incarceration but a lower risk of committing such acts usually results in incarceration as well, because of the over-compensating caution assumed by the courts. Attention to these priorities means that families and communities experience more troublesome exposure to deviant behaviour which does not threaten life but are still relatively over-insulated from the rare catastrophic event. The issue reminds us that in an environment increasingly prone to risk, pressures exist to give our greater priorities to the high-impact occurrence, even if the probability is low, rather than to attempt insulation from the minor impacts as well.

Technology is also linked to competency and dangerousness in quite a different manner. Again, a good deal of insight is provided by the literature on the mentally handicapped. Edgerton, for example, has provided us with a concept, "the cloak of competence," which relates as much to normal populations as to the mentally handicapped.[7] The cloak of competence is the denial of the label "incompetent or retarded" and the attempt to "pass" as "normal." Edgerton's longitudinal study of mentally handicapped people discharged from institutions pointed to the investment of a large proportion of time and energy in denying the authenticity of the label. The two basic criteria for evaluating the normality of social adjustment were social competency and independence, as reflected in such general considerations as economic security, social participation, enjoyment of leisure activities and feelings of stigma. Edgerton's conclusion was that the cloak of competence was best maintained initially when the handicapped person was able to locate and hold benefactors. But at a later stage the dependency could be successfully reduced. The other major finding

was the discrepancy between the professionals' and the subjects' assessment of success. The subjects were able to define themselves as normal and competent despite a lack of vocational and material success. Recreation, hobbies, leisure, good times, friends and family contributed more to personal satisfaction than the experts' criteria of normalization.

Normal people have a cloak of competency to cover their inabilities and their failures to perform tasks which the experts have programmed for us as normal. We are not universally competent and do some things better than others and we fail frequently. In this decade we fail mostly in tasks, or in our understanding of functions, which are technical and organizational. We are deficient in relation to the knowledge necessary to adjust to an incredibly complex world which reminds us constantly of the dangerousness of events now taking place and of impending events. Our benefactors have been the technical and management people who provide us with clues, in simple terms, as to why things do not work as intended but who also act as custodians on our behalf.

Dangerousness is a probablistic concept which measures our knowledgeability or competency to predict the occurrence of infrequent events whose impacts are quite well known or at least thought to be well known. The pursuit of interests such as recreation, hobbies, leisure, good times, friends and family may not be perilous but the attempt to manage technology and complex social organizations, successfully, carries hazards which are indeed threatening.

## CONCLUSION

It has been possible only to describe, very briefly, the specific studies which have been undertaken to explore the problems of competency and dangerousness, hazards to people and places and catastrophes resulting from both technological and social causes. The preliminary findings of these studies, however, point to a number of tentative observations. First, in relation to the complexity of our environment, we are equivalent to the mentally retarded with our own cloaks of competence which help us to deny our true condition. Secondly, our dependency upon benefactors, experts and managers to buffer risk carries mixed blessings because of their vulnerability. Thirdly, we are hazardous to one another, whether as low-income people, incompetent people or simply through our own roles and labels. Fourthly, hazards are likely to be interdependent—intervention to reduce one source of hazard is likely to increase the likelihood of other hazards. Fifthly, the pursuit of recreation, hobbies, leisure, good times,

friends and family has not yet become very hazardous.

## NOTES

[1] Kunreuther, H. (1978) (ed.), *Disaster insurance protection* (New York)

[2] U.S. Nuclear Regulatory Commission (1975) *Reactor safety study*, WASH-1400, NTIS

[3] Rubinstein, E. (1979) "Nuclear power in the U.S.," *I.E.E.E. Spectrum*, 16, 11, 49–57

[4] Wolpert, J. (1977) "Evacuation from the nuclear accident," in Odland, J. and Taaffe, R. N. (eds.), *Geographical horizons* (Dubuque, Ia)

[5] Brooks, A. D. (1974) *Law psychiatry and the mental health system* (Boston) pp. 585–9

[6] Halleck, S. L. (1971) *Psychiatry and the dilemmas of crime* (Berkeley, Calif.) pp. 234–44, 313–14

[7] Edgerton, R. B. and Bercovici, S. M. (1976) "The cloak of competence: years later," *Am. J. Ment. Defic.* 30, 485–97

# INDEX

# F

# G

human use system, 83–89, 90
hunting, 67

# I

# J

# K

# L

# M

military strategy:
   place annihilation, 374–78
   forced relocation of civilians, 362–65, 367–69
Mississauga train derailment, 227–28
Mississippi River, 70
mitigation (*see* adjustments)
   adoption of, 184, 188–91
   earthquake, 203
   emergency planning, 319
   loss reduction, 181–82
   models, 127–30
   strategies, 129
models, 82–83, 89, 107–11, 116–18
Monsanto, 275
Mothers Against Drunk Driving, 147
Mothers Against Pesticides, 147
motor vehicle travel, 66

# N

National Academy of Science, 22–23, 302
National Aeronautics and Space Administration (NASA), 246, 250
National Environmental Policy Act (NEPA), 311, 313
National Flood Insurance Program, 181, 183, 191
National Opinion Research Center (NORC), 20–21, 22, 27
National Red Cross, 26
National Research Council (U.S.), 280, 300
National Resources Planning Board, 6
National Science Foundation, 13, 26
natural disasters, 94–96 (*see also* disasters)
   evacuation decision making, 208–21
   historical, 42–43
   risk/benefit, 68
natural events systems, 86, 90–91
natural hazards, 70–73, 91 (*see also* hazards; disasters; natural disasters; flood hazard)
   choice of adjustments, 78–89
   definition, 86
   field studies, 13–15
   losses, 336–38
   models, 78–93, 107–111
   paradigm, 75
   research, 4–16
   responses to, 14–15
Natural Hazards Research Applications and Information Center, 70
Nebraska, 13
negligence, 41
Nevada test site, 121
New Hampshire, 43
New Jersey, 21, 22, 142
New York, 222, 299 (*see also* Love Canal; Shoreham)

New Zealand, 299
Nicaragua, 94
Niger, 94
NIMBY, 114, 305
nuclear accidents (*see* emergency response; evacuation)
   behavioral responses, 222–32
   evacuation behavior, 153
   hazards, 393
   probability of, 259–60
nuclear power plant safety, 65–66
Nuclear Regulatory Commission, 222, 246, 259, 313

# P

PEMEX, 281
Pennsylvania, 174, 392 (*see also* Three Mile Island)
perception, 9–10, 81–82 (*see also* risk perception)
Persian Gulf War, 286, 370
personality theory, 167
place annihilation, 361, 369–78
planning, 96, 212, 303–7 (*see also* mitigation; emergency planning)
Poland, 369
political cultures, 167
political ecology (economy), 75, 330–31
political theory, 167
pollution, 43–45
population displacement, 328, 362–69, 392–93
preferences, 157–59
Price Anderson Act, 265, 266
privity, 41
probabilistic risk analysis (PRA), 34–37, 259–60, 262
probability, 34–37, 258–60 (*see also* quantitative risk assessment)
psychometric paradigm, 141, 151, 157
public:
   attitudes towards hazards, 276–78
   policy, 11–13
   response to risk information, 195–205
   risk acceptance, 163–64

# Q

quantitative risk assessments (QRAs), 2, 58–63, 233

# R

railroad travel, 67
Rapid City flood, 181
rare events, 267, 390, 391–92

# T

Tanzania, 91
technocentrism, 138
technology:
  change, 277–78
  disaster, 285, 288–95
  fixes, 265, 279–82
  and gender, 138–40
  hazards, 138, 288–96, 391–400
  risk, 55–68, 137–47
Tennessee, 8
Tennessee Valley Authority (TVA), 6, 10
terror (*see* social hazards)
Texas, 22
Three Mile Island, 114, 120, 153, 161, 222–30, 245–46, 252, 258–59, 262, 265, 277, 392–95
Tigris-Euphrates, 33, 334
Times Beach, 121
torts, 41, 266
total war concept, 360–61, 378–81
Toxic Substances Control Act, 280
trivial risk, 270
tropical forest degradation, 344–46
Troy, 383
Tylenol poisonings, 277

# U

underdevelopment, 96
United Kingdom, 13, 36, 40, 43, 97–107, 299, 379, 381, 384
United States, 5, 46, 276, 330, 333
urban areas (*see* population displacement)
  devastation, 376–77
  hazards of war, 395–99
  Japan, 370–74
  planned destruction of, 370
urban resettlement (*see* population displacement)
USSR, 13, 364
Utah, 377

# V

Valley of the Thousand Drums, 121
VE Day, 369
Venezuela, 94
Vietnam, 67, 362, 369, 370
  war, 60, 361
vinyl chloride, 239

ISBN 0-13-753856-1

90000

9 780137 538560